Advances in Science, Technology & Innovation

IEREK Interdisciplinary Series for Sustainable Development

Advances in Science, Technology & Innovation (ASTI) is a series of peer-reviewed books based on important emerging research that redefines the current disciplinary boundaries in science, technology and innovation (STI) in order to develop integrated concepts for sustainable development. It not only discusses the progress made towards securing more resources, allocating smarter solutions, and rebalancing the relationship between nature and people, but also provides in-depth insights from comprehensive research that addresses the **17 sustainable development goals (SDGs)** as set out by the UN for 2030.

The series draws on the best research papers from various IEREK and other international conferences to promote the creation and development of viable solutions for a **sustainable future and a positive societal** transformation with the help of integrated and innovative science-based approaches. Including interdisciplinary contributions, it presents innovative approaches and highlights how they can best support both economic and sustainable development, through better use of data, more effective institutions, and global, local and individual action, for the welfare of all societies.

The series particularly features conceptual and empirical contributions from various interrelated fields of science, technology and innovation, with an emphasis on digital transformation, that focus on providing practical solutions to **ensure food, water and energy security to achieve the SDGs.** It also presents new case studies offering concrete examples of how to resolve sustainable urbanization and environmental issues in different regions of the world.

The series is intended for professionals in research and teaching, consultancies and industry, and government and international organizations. Published in collaboration with IEREK, the Springer ASTI series will acquaint readers with essential new studies in STI for sustainable development.

ASTI series has now been accepted for Scopus (September 2020). All content published in this series will start appearing on the Scopus site in early 2021.

More information about this series at https://link.springer.com/bookseries/15883

Federica Rosso • Claudia Fabiani •
Haşim Altan • Mourad Amer
Editors

Advances in Architecture, Engineering and Technology

Springer

Editors
Federica Rosso
Department of Civil, Construction
Sapienza University of Rome
Rome, Italy

Claudia Fabiani
Department of Engineering
Università degli Studi di Perugia
Perugia, Italy

Haşim Altan
Department of Architecture, Faculty of Design
Arkin University of Creative Arts and Design
(ARUCAD)
Kyrenia, Cyprus

Mourad Amer
International Experts for Research
Enrichment and Knowledge Exchange
Alexandria, Egypt

ISSN 2522-8714 ISSN 2522-8722 (electronic)
Advances in Science, Technology & Innovation
IEREK Interdisciplinary Series for Sustainable Development
ISBN 978-3-030-86912-0 ISBN 978-3-030-86913-7 (eBook)
https://doi.org/10.1007/978-3-030-86913-7

This Springer imprint is published by the registered company Springer Nature Switzerland AG
The registered company address is: Gewerbestrasse 11, 6330 Cham, Switzerland

Scientific Committee

Preface

The continuous population growth in nowadays cities, combined with the constantly developing human needs, calls for innovation in urban design and planning. This innovation can be achieved through the implementation of modern paradigms in Architecture, Engineering, and Technology. These three topics were selected for this book due to their interdisciplinary nature and the importance of bringing them together for the sake of a holistic, futuristic, and innovative design. Such design is necessary for the sake of overcoming imminent challenges to modern design such as the depletion of green spaces, increase in carbon footprint, and marginalization of ecosystems in nowadays cities. The book comprises studies in the three aforementioned fields that aim at providing designers and policy makers with the necessary tools for improving urban spaces and modernizing design.

The book is divided into four parts that cover the topics of urban spaces, efficient designs, data analysis, and studies on energy through the perspectives of architecture, engineering, and technology. The studies in this book tackle various challenges and offers solutions and opportunities of improvement for modern design and urban planning. The book opens with a discussion on livability in lost spaces in the Egyptian case and a demonstration of the possible solutions for the inadequacies of tactical urbanism. Furthermore, the book explores the social dimension of design through issues such as the empowerment of women where the first chapter discusses the functionalizing of female empowerment in urban spaces. The first part also discusses the connection between artificial intelligence and urban spaces through a deliberation on the understanding of the new media intervention impact on urban spaces and how it can be utilized in enhancing the urban experience. In "Modern, Environmentally Friendly Electric Transport as a Step toward Improving City Welfare" and "Towards Greener Neighborhoods: A case study for street renovating solutions in Cairo", the authors discuss the questions of good street designs and green mobility in Egyptian urban spaces.

The second part of the book discusses the possibility of fulfilling efficiency in design through the integration of architectural, engineering, and technological concepts. The first chapter discusses the bio-sustainability of modern spaces through the application of a futuristic interior concepts. Such concepts rely on the evolution of biotechnology and its innovative utilization in modern design. Biomimicry is also discussed in "Enhancing the Efficiency of Natural Ventilation Systems by Biomimicry", which discusses the possibility of imitating of the adaptation methods found in flora and fauna for the purpose of enhancing natural ventilation systems. The second chapter expands on the concepts of space design by proposing the use of water as an element of architectural space design while discussing the psychological effect of water on occupants. This part of the book also discusses the cultural aspect of architecture as it overviews the influence of orientalism on both Islamic and western architecture. The book then delves into technological aspects where it discusses, in part three, applied sciences and data analysis.

The third part of the book, entitled applied sciences and data analysis, explores technological applications in the fields of engineering and architecture. The first chapter highlights the importance of the application of technological paradigms in architecture as it discusses the absence of Building Information Modeling (BIM) in Egyptian higher education and proposes the integration of BIM in the architectural engineering curriculum in Egypt. Another technological application in architecture that is discussed in part three is the use of 3D printing in architecture alongside the integration of Building's Energy Simulation Tools (BEST) with Intelligent Decision Support System (IDSS). Furthermore, one of the chapters in this part proposes the use of nano-technology based materials to decrease the cooling loads with special reference to high-income housing in KSA. This part of the book concludes with a discussion on the use of advanced technology in construction to achieve thermal comfort and, consequently, reduce the consumption of energy. All in all, this part focused on how technological applications can further the development of architecture and engineering concepts.

The book concludes with a compilation of studies on energy, its effects, and requirements. The chapters of part four dwell on methods and strategies of reducing energy consumption and energy renewal alongside technological applications in engineering such as friction stir spot welding and corrosion assessment methodologies. The entirety of this book aims at encompassing various fields for the purpose of creating a mutual end that serves modern planning and designs.

The challenges facing humanity in modern societies and cities necessitate innovation and the transcending of disciplinary boundaries toward a holistic design that incorporates different fields, which is attempted in this book. The significance of this book stems from its capacity for offering new paradigms that would enable policy makers to efficiently plan for urban spaces in accordance with the vision proposed by the Sustainable Development Goals.

We would like to thank the authors of the research papers that were selected for addition in this book. We would also like to thank the reviewers who contributed with their knowledge and constructive feedback in hopes of ensuring the manuscript is of the best quality possible. A special thanks goes to the Editors of this book for their foresight in organizing this volume and diligence in doing a professional job in editing it. Finally, we would like to express our appreciation to the IEREK team for supporting the publication of the best research papers submitted to the conference.

Rome, Italy — Federica Rosso
Perugia, Italy — Claudia Fabiani
Kyrenia, Cyprus — Haşim Altan
Alexandria, Egypt — Mourad Amer

Foreword

Advanced technologies, as the world knows, have acted as a main driver for innovation in the architectural and engineering disciplines, providing the industry with a plethora of new tools to take on complex engineering achievements that have become part of everyday life. The current levels of sophistication in engineering research have led to the conversion of scientific discoveries into functional, marketable, and profitable applications, making way for further potential development and a goal of making the world a more prosperous, healthy, and safe place.

From drones, 3D printing and Simulation tools to the Internet of Things, Big Data, Virtual Reality, Artificial Intelligence and more; the Architect and Engineer's toolbox is continuously evolving and, in return, changing the way we create and build. We now can see complex, large, and robust structures constructed with the help of 3D scanning, robots generating multi-functional geometric forms, and simulation tools facilitating the use of fewer and more sustainable materials. Such recent advancements in technology, especially prefabrication tech, and multifaceted programs and software have resulted in an increase in collaboration between engineering and design disciplines, an essential component in tackling current problems and making progress.

Moreover, issues of environmental sustainability in cities call for the deployment of new technology and a mindfulness of the natural environment. In acknowledging that our world and surroundings are ever more artificial, Urbanists are now learning that Clean Environmental Technologies (ET), coupled with Information and Communication Technologies (ICT), must be embraced to contribute to a greener and smarter urban future. Existing studies look into the contribution of such technological manifestations in integrating smart solutions into aspects of sustainable and efficient designs and explore their benefits, while leaving room for further improvement.

Neri Oxman's "age of entanglement", known as the Bermuda Quadrilateral of 2006, states that design does not act by itself but relates to the four "hats" of creativity and innovation—science, engineering, design and art—which represent four different, yet intertwining, systems of being and thinking. This entanglement of disciplines and the emergence of intelligent and applied systems technologies in the AEC industry, along with their impact on social and environmental sustainability, are investigated and assessed in this book. Scientific research in engineering and technology has always played a vital role in the development of civilization and their evolution. As new knowledge and more powerful analytical and experimental methods expand the power of engineering in practice, problems of design and development once considered too complex to be dealt with other than empirically, intuitively, or by trial and error, have become solvable and applicable.

Mourad Amer
Series Editor, Springer
CEO & Founder, IEREK

Contents

Applied Systems and Data Analysis

Effects and Requirements of Energy

Improving Urban Spaces

Tactical Urbanism for Improving Livability in Lost Spaces of Cairo

Basma Elbeah, Abeer Elshater, and Ahmed Toama

Abstract

This paper compares the perception of residents and pass-by users of lost spaces to extract the perceptual factors affecting these spaces. The aim is to demonstrate the need for examining livability for practical applications in the levels of regions, neighborhoods, and small areas. The findings discuss some definitions of the term "livability." The results demonstrate that although there is a common worldwide concern and interest in livable places, there is a lack of an acceptable concept considering those livable spaces and places. Using tactical urbanism initiatives, the authors here proposed a framework that identifies variables coving the meaning of lost spaces from people viewpoint. We also introduce an investigation of livability in the Egyptian case. The concluded remarks demonstrate how urban designers can meet the inadequacies of Tactical Urbanism. The outcomes are structured and developed to test the possibility of creating livable places and spaces in Cairo.

Keywords

DIY Urbanism • Leftover Spaces • City Repair

1 Introduction

One can feel the influence of Haussmann's plans of Paris in the downtown of Cairo. The initial design follows the Garden City Movement. The master plan contains public gardens, parks, and playgrounds, and many recreational places (Elshater, 2018; Elsheshtawy, 2013; Selim, 2016). In the master plan of Cairo Downtown, the rules of building heights were set. However, the suburbs are no longer small and luxurious, as they have expanded to take in part of the growth of the metropolis.

Nowadays, Cairo agglomeration is on the verge of losing its identity and uniqueness. Besides, the suburbs are losing livability, and lost spaces exist. There are growing appeals for developing lost spaces, including the areas located under the elevated paths and highways with available lost spaces beneath most elevated cases. There is also a deficiency of places that encourage people for human communication and interaction. For instance, under-elevated areas consider being an excellent opportunity for local communities if it will be developed considering the livability (Elshater, 2014; Lydon & Garcia, 2014; Lydon, Bartman, Dwoudstra, & Khawarzad, 2010).

Scholarships address the projects in the dead lost spaces to be designed and turned into creative venues for various community facilities and outdoor activities such as libraries, markets, public spaces, playgrounds, art galleries, canteens, seating areas with traditional board games, and so on (Courage, 2013; Shackelford, 2014; Trancik, 1985). Those projects aim to gather points and nodes that assist all types of residents to interact and communicate with each other. The environmental settings are the thing that encouraging locals and passers-by to gather and communicate with each other for cultural and social objectives (Nooraddin, 1998). Those are livable places in which public life can improve and, then, developing the quality of life in Cairo, Egypt. The present work aims to apply the approach of tactical urbanism to reach the result of typically a chain of small-scale interventions that leads to the public realm. The objective is to demonstrate the utility of making and designing lost spaces that provide and present more user-friendly spaces and places for the public also much more livable spaces.

The current research question leads to myriad problems in achieving livability in Cairo. Besides, it is of interest to know which design principles and criteria are suitable and applicable in the case of Cairo. The current work aims to present design criteria to develop the livability in two cases

B. Elbeah (✉) · A. Elshater · A. Toama
Department of Urban Planning and Design, Ain Shams University, Cairo, Egypt

F. Rosso et al. (eds.), *Advances in Architecture, Engineering and Technology*, Advances in Science, Technology & Innovation, https://doi.org/10.1007/978-3-030-86913-7_1

from Cairo by investigating the possibilities of the application of tactical urbanism approach in lost spaces. In this, Cairo can achieve a better livable society and community for citizens to live in, also, solving the problem of the existence of lost spaces in Cairo`s urban fabric to promote the concept of short-term action toward long-term change.

Accordingly, the research is structured into four main steps and objectives. The first is the literature review which explains the concepts of tactical urbanism, livability, and lost spaces. We also present some urban communities which try to apply the tactical urbanism approach and how they integrate the users or the citizens as a participatory part in urban planning and urban design process. Second, it discusses the research methodology, which explains the in-depth detail of the research strategy, method, approach, and methods of data collection. The third part discusses the data collection and to figure out what is the lost spaces. The fourth part analyzes the current sense in the selected local case study and tests the principles and the approach of tactical urbanism. The discussion explores the hurdles faced by them when they applied the approach and how they solved them. The results from analyzing these cases provide some suitable solutions and guidelines to address the research problem in Cairo.

2 Literature Review

2.1 Tactical Urbanism: Theory and Practice

Tactical urbanism is the term that describes a known global phenomenon of formal and informal interventions in the urban fabric in both terms physical and cultural (Lydon & Garcia, 2014). Tactical urbanism is an umbrella term for various and many subgenres of activity like hacktivism, DIY, Pop-Up Urbanism, wiki urbanism, acupuncture, guerilla urbanism, or subversive (Pfeifer & Montreal, The Planner's Guide to Tactical Urbanism, 2013). Tactical urbanism has two sides; it can be grassroots and bottom-up. It aimed at all approaches mentioned above to enhance and improve the living experience through basic and ordered steps and strategies (Elshater, 2014). Those steps are always temporary, low cost, quick to install, informal, spontaneous, driven by people, and the problems of the community itself (Dew-Jones, 2016).

In the twenty-first century, the cities' wide-reaching react to a developing populace, ever-changing economic situations, modern technologies, and, also, climate change. The short-term projects, which are considered as community-based projects, have different scales. These scales start from pop-up parks to open streets actions. They become empowerment and adjustable tool of urban extremist, urban planners, and decision-makers looking for improvements in the cities. These new projects which are often quick in installing, low-cost, and creative projects are the pivot, core, and basis of the tactical urbanism movement. Creating livable plazas overnight and re-designing parking lots as a way of gathering neighborhood`s residents is a new effective way to gain government and decision-makers support. This support can invest in constant projects, which will be forming the urban public spaces in new interesting ways so residents and civic leaders can enjoy it (Elshater, 2014; Lydon & Garcia, 2014).

Tactical urbanism is a term used to describe cheap projects and often temporary projects that try to improve the livability of a small part of the city or make it more enjoyable. The tactics used to improve this small part of the city can be replicable across cities, but in certain instances, it becomes a worldwide phenomenon. Tactical urbanism is a trend/movement come to the field of "City Repair" and "Do It Yourself Urbanism (DIY)." This approach aims to develop the urban living experience through step-by-step action plans for improvements (Courage, 2013).

The tactical urbanism approach is how cities are built or how cities are improving themselves as Lydon, Bartman, Dwoudstra, and Khawarzad (2010) argue, "It is step by step, piece by piece." Tactical urbanism is known as short-term action, long-term change (Lydon, Bartman, Dwoudstra, & Khawarzad, 2010). The goal of this approach is not making or creating a "cool project" which is cleaned up by the city, but to create something even if it is temporary that something will change everything in the place in a working way. Once we figure out that tactic and how it works out the planners think of making this change permanent. Types of tactical urbanism applied in multi-initiatives are as follows:

- **Better block initiative** aims to transform streets using citizen volunteers and low-cost materials. Urban streets and fabric are transformed by mapping and designing places for food booths, pedestrian sidewalks, adding bike lanes, and reducing the streets starting from the Dallas neighborhood in Oak Cliff (Better Block Foundation, 2014).
- **In chair bombing**, building seating for citizens by reusing salvageable materials. The new seatings are put in the places which provide the interaction between the users of the place.
- **Depaving is to** make the city and the neighborhoods more green by demolishing unnecessary pavement to green space by making good use of rainwater (Depave, 2014).
- **Food carts/trucks** provide a small opportunity for ordered street vendors business by creating and providing

food booths to attract people to using urban lost public spaces (Ehrenfeucht, 2017).

- **Open streets** is an initiative that temporarily tries to reserve more safe places in the urban fabric for types of activities such as walking, lines for bicycling, and social interactions. The expected outcome is raising the local economy that also raise resident`s awareness of the impact of transport and cars in the urban spaces. "Open Streets" are considered as an effective term for the area of South American "Ciclovia," which was generated in Bogota (Lydon, Miller, & Hebbert, 2012).
- **Park(ing) day is** an annual event on-street parking where streets are transformed into park-like urban spaces. Rebar Studio launched this lead for the first time in 2005 (BEST, 2011).
- **Pavement to plazas** is an initiative in New York City. It is popular to transform the unusable pavement plazas into livable public space. Closing Times Square to transportation and transforming it into the pedestrian plaza is a good application for a pavement plaza.
- **Pop-up cafes** is a common way to provide sitting areas and eating areas combined with food booths and Pop-up cafes which are installed in parking lots in the cities with a lack of sidewalks and urban public spaces.
- **Pop-up parks** are the way of transforming the city into a green city by planting the underused urban spaces and making them beautiful green gathering places.

2.2 Lost Spaces: Hidden Pockets of Humanity Under the Elevated Roads

Trancik (1985) defines the lost space as an unstructured scenery at the core of high-rise towers or the unused flooded plaza and far of the flow of the activities practiced by pedestrians in cities. Lost spaces are the parking lots that surround the urban root of all types of cities which also separate the connection between the commercial center and residential areas. Lost spaces are also no man`s land, which over the boundaries of freeways nobody cares about maintaining. Lost spaces, also, are the waterfronts which the people neglected, lines of the trains, vacated military lands which are evacuated, and industrial complexes which have moved to the suburbs due to convenient access. Those lands are also unoccupied blight, clearance locations, remnants of the urban renewal days—that were, for a multiplicity of causes, never redeveloped. They are defined as the lands remaining in between districts and loosely collecting sectors of commercial lands and buildings which spring without anyone realizing them. They are deteriorated parks and minor public housing projects that have to be rebuilt because they do not provide their designed aims. In general, lost spaces are urban spaces that are unwelcomed; they should be re-design. They are also called anti-spaces, making no positive donation for both surroundings and users. Lost spaces, also, are ill-defined, without borders that can be measured, and fail to connect elements coherently. Furthermore, those places offer major opportunities to the urban planners for urban renovation and creative infill and for rediscovering the many hidden resources in our cities (Trancik, 1985).

Memarian and Niazkar (2014) describe the lost space as being the left sights without manufacturing, fields, and tall tower lands which are detached from the flow of walking activity, stagnant, and have been left unused. Elshater (2014) adds the other types of unused spaces (e.g., parking lots, areas located at the border of freeways, and highways that are left without attention). In another citation, Elshater (2018) mentions that the lost spaces consist of beaches that are leftover, railway lines, empty military lands, and industrial workshops due to the possibility of easier accessibility. In a similar context, a recent initiative describes the vacuous places which have not been rebuilt or reused and are located on fringe park areas aside from the areas (Anderson, 2018). However, these elevated are sturdy barriers that look like massive scars that tear the overall physical form of the city. Figure 1 shows the spaces under the elevated roadways which have been left without planning, underused, dirty, dark, ugly, without a target, unattractive, deteriorating, and unsafe also frightening. They are lost outdoor spaces that are ignored or denied. They are also the in-between spaces and "non-places" which partition territories in the cities (Gehl, 2011).

The public spaces lost its importance when it lost the human's social interaction as it reflects the social life and interaction of people. Lost space is the outcome of urban space left over by people if it endures, we lose the place. the lost space is all about an unserviceable land in the urban fabric while losing a place does not only mean people's memories being forgotten from space but also the place identity to be lost and the image of the city and citizenship will incline hardly. The perspective of lost spaces is completely different from the passers-by opinion than the resident's cause of the unique aspects of their way of living.

The elements of space, which accomplice the meaning of lost space for citizens, were not identified; so there is a huge gap in picturing lost spaces that bring back the people`s life cycle. Our fabulous elevated infrastructure has undoubtedly contributed to the amplification and growth of the economy in Cairo, "but this elevated infrastructure has a negative impact on different communities, especially low-income and minority neighborhoods physically separated and isolated by elevated highways and rail lines." As Campanella (2015) writes about the elevated paths as they are almost a

A

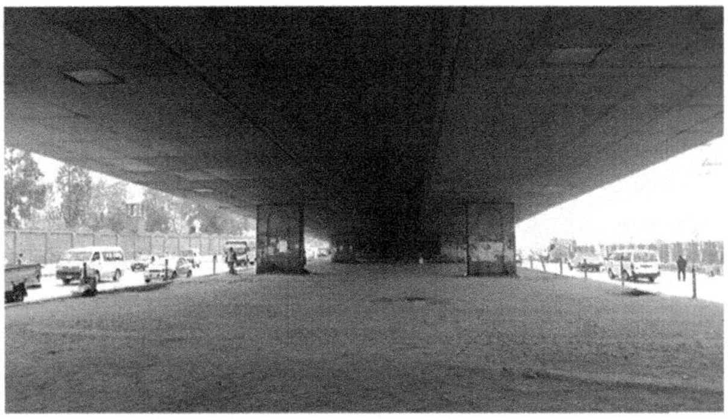

B

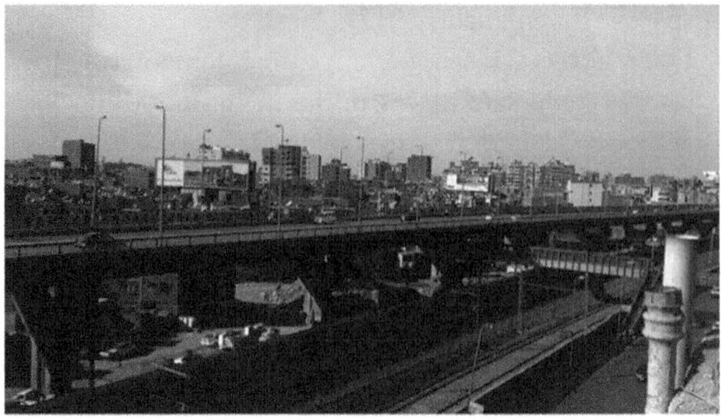

Fig. 1 The different forms of lost spaces under leftover roads. *Source* The authors

deterioration in the city. The elevated and bridges were built to get people in and out of town and offer little but darkness and pigeon excrement in the streets below (Campanella, 2015).

2.3 Livability

A number of authors recognize that creating a livable community is becoming a target around the world (Epley, 2008; Stevens, 2009; Wheeler & Deakin, 2001). The term livability can be defined as the summation of the ingredient which can be added to a community's quality of life. These qualities comprise the built and natural environments, economic prosperity, social stability and equity, educational opportunity, and cultural, entertainment, and recreation possibilities (Abusaada & Elshater, 2019). The livability of places is set by many factors, which are in turn influenced by a variety of tangible and intangible elements—such as good quality design and materials, place identity, accessibility, and so on—concerning the area in question and the surrounding (AAVV, 2017; Appleyard, 1981). Aulia (2016) argued that planning the quality of the residential area where occupants feel comfortable in the present and the future is the goal of the designers. Also, Dempsey (2009) mentions that designing the livable community is needed through some criteria driven by occupant satisfaction housing, neighborhood designing scale, and design indicators. Some schools define the term livability of a domestic environment as one of the measuring standards of a community`s quality of life (Aulia, 2016; Carmona & Magalhães, 2009; Epley, 2008).

Livability takes place at the scale of the everyday physical environment and concentrates on the term "place-making" (Carmona & Magalhães, 2009). Livable communities are where communities are attached to places (Gehl, 2011). In another context, livability is described as the phase to which a person can perform, feel low relief and also enjoy the place where they live. Livability is the phase to which a living environment meets the adaptive storage of a universe. Applied to human society, it indicates the convenience of governmental arrangements with human needs and abilities (Veenhoven, 2014).

The theory of livability explains the happiness observed differences in terms that fit the environment need. Planning for livable communities is diverse due to the scale of planning. Planning for a small scale is completely different than planning livable communities on the regional scale (Aulia, Planning Livable Community with Social Systems Approaches: Medan, Indonesia, 2014). Figure 2 shows the seven characteristics of creating livable communities.

2.4 Under the Elevated

The New York-based Design Trust for Public Spaces organization is the first organization to show interest in the elevated lines with huge number of researches and reports reach to 2002-reports, and has published a heavy new paper titled Under the Elevated that hopes to deal with the similar spaces in creative way under roads and elevate infrastructure across the cities. This report made a partnership with the NYC Department of Transportation, considering the outcome of a two-year study of these types of spaces. They called them "el-spaces."

- ***Division Street (Manhattan)***
  In Chinatown under the Manhattan Bridge at Division Street, this area under the elevated considered as dirty, unsafe, and dark because of vehicle emissions, bridge runoff, and noisy because of the trains above the elevated. In April 2014 the design trust and dot decided to

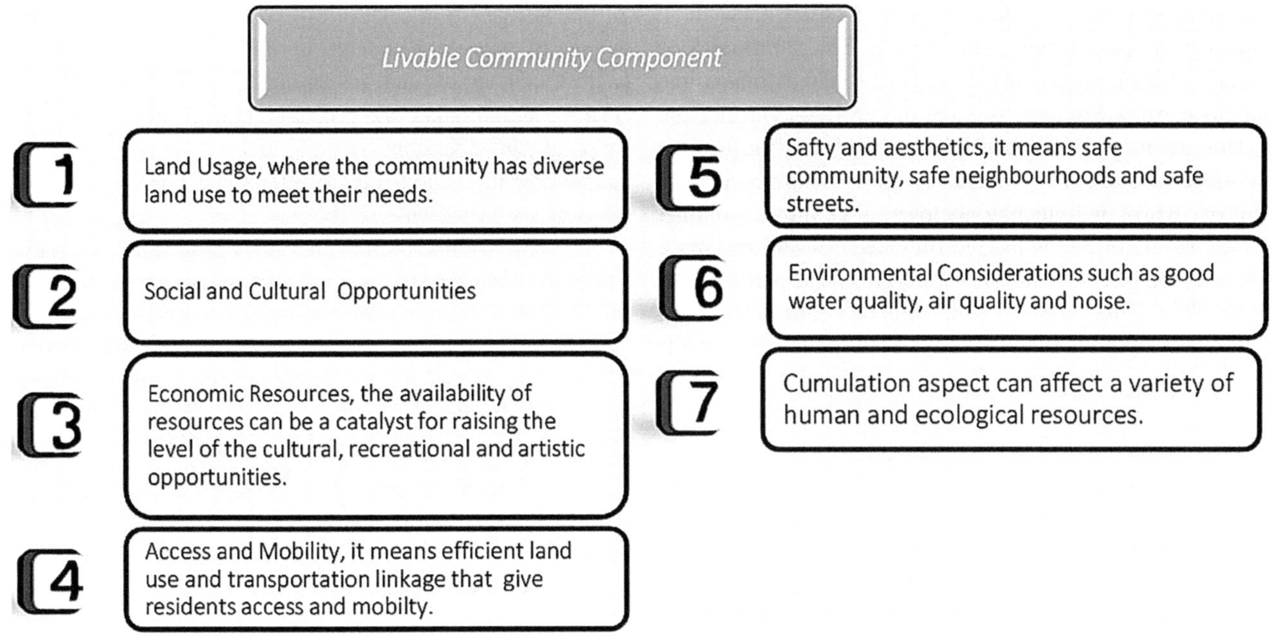

Fig. 2 The seven characteristics of livable communities. *Source* The authors based on (Epley, 2008)

develop this area and make good use of it by creating and installing a pop-up community calendar and designed red led lighting.

The whole idea about installing and developing this area under the elevated is that many people pass by it, but no one stopped at it, as the calendar provided them with the chance, weekly, to post social activities like training classes, exercise programs, jobs, or housing notices, and so on. That kind of project shows that people are going to stop and hang around in a space that does not meet the classical mold of public space as shown in the left side (Fig. 3).

- ***Southern Boulevard (The Bronx)***

  "Boogie Down Booth" is the name given to the project made under the high tracks located Southern Boulevard, in the Bronx in July 2014. The challenges of this project are the noise, from two to five trains that travel along elevated tracks which made this area inhospitable for pedestrians. In order to get rid of the sound, they created a booth that played music from Bronx artists via solar-powered speakers every day and night. It attracts a massive amount of passers-by every week during its pilot run. The surprising feedback of the people was requesting to turn the music up (the right side of Fig. 3).

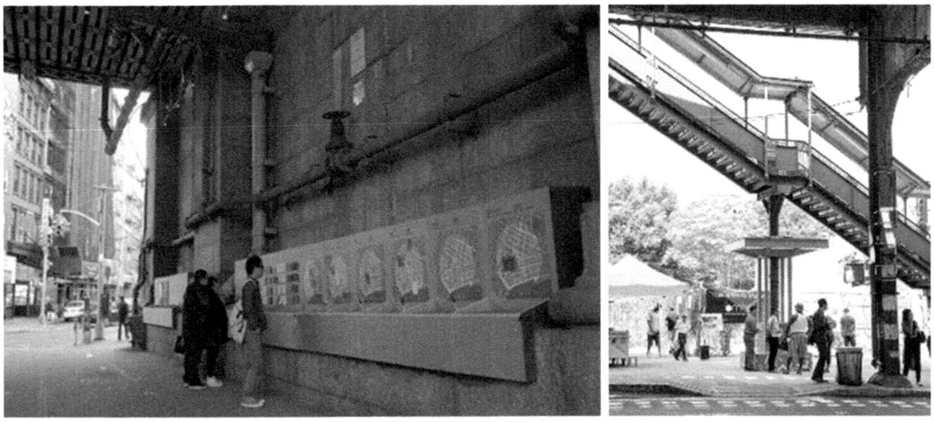

Fig. 3 A community calendar for inhabitants under the Manhattan Bridge, (left). Source: (Neil Donnelly / Courtesy of the Design Trust for Public Space). The "Boogie Down Booth" playing music by local artists, along Southern Boulevard in the Bronx, Source: (William Michael Fredericks / Courtesy of the Design Trust for Public Space) (right)

- ***Queensboro Bridge, Manhattan, and Queens***
  Designers put plans for this area on how it would look like. These designers divide the area of development into two sides: one of them is under the Ed Koch Queensboro Bridge, in Manhattan and Queens; however, most of the bridge located on the Manhattan side containing a tennis court which will also be enclosed a TJ Maxx. Another side of the bridge is located on Queen, where the space under the bridge is full of vacant areas and parking lots. So far, the development concept is to install hubs so the owners of electric cars can charge their cars while enjoying the place. Also, a lounge area for drivers could be located there, and more social activities to do while they are waiting (Fig. 4).

- ***Broadway Junction, Brooklyn***
  At Broadway Junction in Brooklyn, five subway lines create what the under elevated streets are called "tangle of elevated structure." There is a lack of activities going on in the el-space only street vendors. The design trust decided to make good use of this vast area by painting the elevated beams different colors so they could divide this area into sectors with various social activities done on this sector like parks, farmers markets, and street fairs (Fig. 5).

3 Methods

This section explains the research methodology (Fig. 6). The study follows the investigation method in observing, measuring, and determining public concentration zones and reasons behind those and find the surrounding land use data to establish a natural linkage through field survey. With the questionnaire, the study acquires certain data regarding the user's purpose and frequency of visit, typology, and volume of the user group in categories like age limit, income group, mode of traveling, sense of place, and performance. With GIS mapping and behavioral mapping, the study will analyze and simulate the potential leftover spaces. With the help of the literature review and recent urban design guidelines, this study tries to offer some possible interventions at a selected case study.

The primary intention of this study is to find out existing under the elevated spaces at an urban scale for the proposed study area. Observing the naturally occurring public concentration zones and then gather enough data through observation, interview, to sketch out the level of development and changes in *el-space* and circulation for the area. After mapping active land-use and the human behavior information and needs map, this study tries to identify potentially lost spaces and evaluate the problem areas for possible interventions.

3.1 The Case Study

Asher Bus Station was selected for the case study. The site is located in Cairo-Ismailia Desert Road (Ring Rd, Huckstep, Qism El-Nozha, Cairo Governorate) as shown in Fig. 7. Cairo-Ismailia Desert Road is one of the essential eastern entrances to the Cairo City region as it is connecting also new urban communities and cities located on this road such as Salam, Huckstep, Obour City, Shorouk City, New Heliopolis, and 10th of Ramadan City. The length of this main highway till the end of those new urban communities reaches 35 km. This highway also serves a lot of passengers and goods transitions between the regions and governorates in Egypt like Ismailia, Port Said, and North Saini.

The first phase of this multi-lane highway is for a 3.5 km section that starts at El Salam City and runs as far as the El Obour Tunnel, which features a width of 17.1 m. The roadway features are four lanes in either direction with

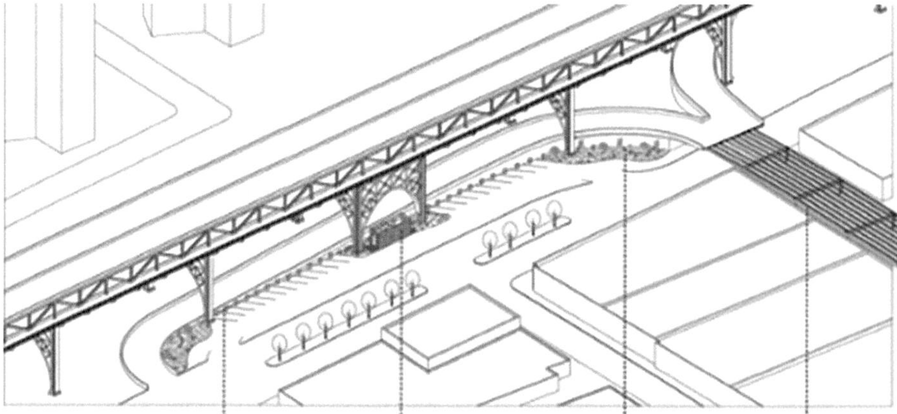

Fig. 4 The design created for the Queensboro Bridge which would install charging stations for cars under the elevated. *Source* Susannah C. Drake/Courtesy of the Design Trust for Public Space

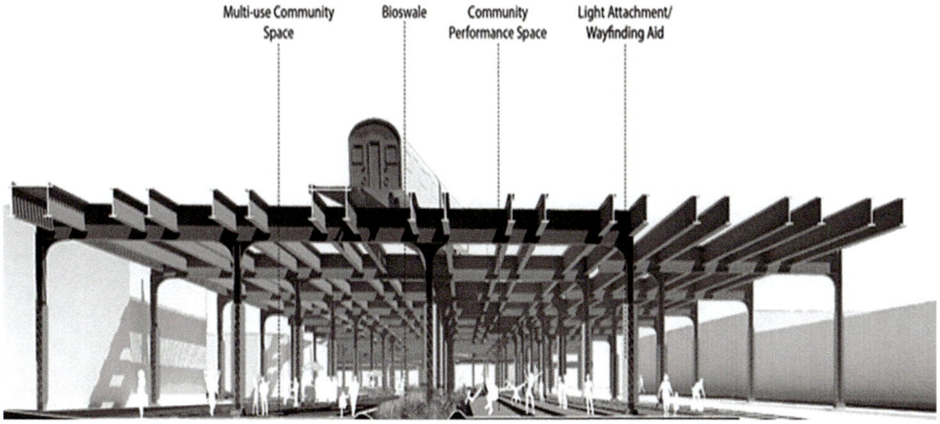

Fig. 5 Imagined idea to reinvent Broadway Junction creating spaces for the community and residents separated by color schemes. *Source* Susannah C.drake/courtesy of the design trust for public space

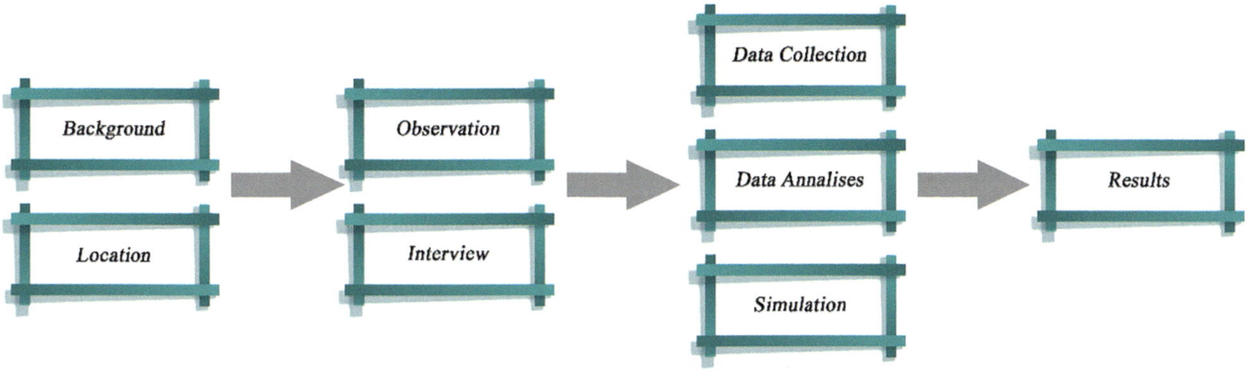

Fig. 6 The method structure. *Source* The authors

paved shoulders, instead of the original old three-lane layout. The second stage runs from the El Obour Tunnel to 10th Ramadan City, a distance of 26.5 km in length and 20.6 m in width. The road has five lanes in either direction with a width of 20.6 m, in addition to the paved shoulders.

There are various activities widespread along the Ismailia Highway Road such as educational activities (institutes, schools, and academies), entertainment activities (e.g., restaurants, amusement parks), and agriculture activities as little small farms. Also, the area has a lot of different human activities daily practiced in el asher bus station like food booths, coffee and tea trucks, and street vendors. Also, there is a huge amount of passers-by without a waiting area for the public transformation (Fig. 8).

3.2 Observations

Researchers of the present work collected data using field observations in the lost spaces under the elevated roads. The observations took place in 2018 in two selected locations shown in Fig. 7. During the year, the daily observation was about 30 min from 8:00 a.m. to 8:30 a.m. The present work focused on the evolution of people's behavior and their different activities is done there. Meanwhile, the reaction of the pass-by pedestrians to this evolution was observed and documented using camera shooting.

The authors also did scheduled meetings with pedestrians and vendors who regularly use this bus station. In order to gather data related to the research problem, the lost spaces under the elevated, and how it affects the people behavior was also investigated. In this, the authors try to find out if it is going to positively affect space livability. The questions of the semi-structured interview are

- For the present location and before moving the Asher Bus Station, do you consider the new location is better for pedestrians' movement and fulfilling users' needs?
- As a pedestrian, how long have you been passing-by the site? Is it daily or randomly?

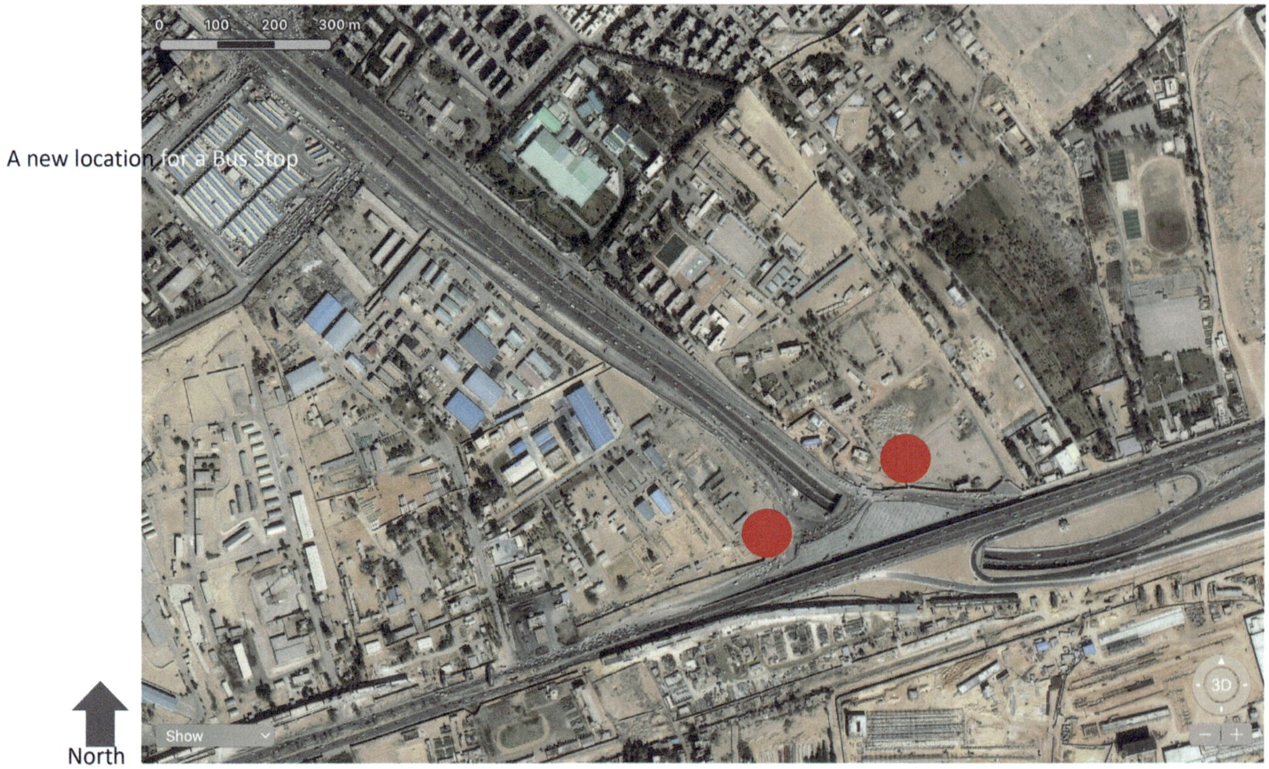

Fig. 7 Asher bus station, where the observation of the first author took place. *Source* The authors from google maps

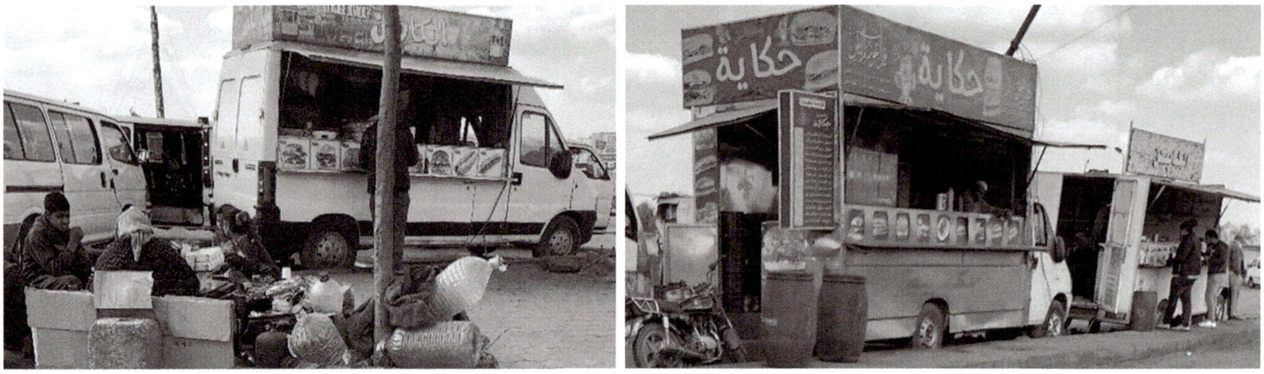

Fig. 8 Example for the different activities in Asher bus station, (left) and example of food booth and trucks which consider as a daily activity in the site. *Source* The authors

- What do you think about the problems at Asher Bus Station?
- Do you have any suggestions for developing the site?

4 Results and Findings

In order to determine whether the authors` hypothesis in the current research about the lack of livability in El-Spaces, data collecting is needed. Further, as the tactical urbanism would help to create more livability and formal spaces, we conducted a series of semi-structured interviews. The interviewers' sample was randomly chosen from some vendors and passers-by inside the case study location in 2018 in EL Asher. With regards to the on-site questionnaire, this was administered to about 30 people, both passers-by, and vendors. Questions are launched mainly to people while waiting for their daily transportation. All the people who are interviewed have answered the questions, even though they have dedicated different time for answering. To the first question "IS the place before moving the Asher bus station is better now is better according to your opinion?" About 70% of people both passers-by and vendors agreed through answers that the Asher Bus Station was better before moving it to El Salam Bus Station. On the other hand, 30% confirm that moving the Asher Bus Station is a good thing. Meanwhile, they see that the Asher Bus station is better now. As regards the second question, "How long have you been a passer-by? Is it daily or randomly?" Answers regard different elements. Almost 65% of the sample are daily passer-by, and 53% of the sample are randomly passers-by. "Unsafe, insecure, the bad vision of livability, noise, and overcrowd" was the answer of one of the passers-by to the third question "What do you think the Asher bus station problem according to your personal opinion?".

The whole sample of the interview (100%) approved and confirmed that the problems of the Asher Bus Station are "unsafe, unorganized, and overcrowded." One of the daily passers-by talked about the lack and absence of waiting areas, and the other one talked about the accessibility for the disabled and providing basic services such as restrooms. Moving to the most important part of the interview which is the fourth and final question "Do you have any suggestions for developing the space?" The answers to this question were remarkable as 75% of the passers-by imagined and suggested to design green spaces, install street furniture, install different booths for vendors, managing cars and pedestrians circulations, divide the area into partitions of different facilities needed their to afford users' need, (e.g., lighting the area), and the existence of gates and police guarding the bus station. Meanwhile, 20% suggested providing the basic public services such as the restrooms and other services also respecting the disabled and older adults people. Twenty percent of the sample, especially girls, indicated making the bus station more ordered, organized, and secured.

The results demonstrated in this paper that the physical, tactical action can be occurred and applied here in the Egyptian context. Here we compare the results of the proposed method with those of the traditional methods. The findings consider the lost space as under the elevated and get the maximum benefit of place and design it for good accessibility. In line with previous studies, the findings give two pillars to create livability. These pillars consist of people, and urban fabric (Fig. 9).

5 Conclusion: Short-Term Action for a Long-Term Change

The authors suggested here a theoretical solution which consists of two main pillars: people (pass-by and vendors), and urban configuration that handles the challenge caused by lost spaces. The research findings include a multidisciplinary framework created to focus on people's needs in the city. Likewise, the suggested initiatives applied in cities of Global North and discussed in literature can make a maxim benefit of the time during "day/holiday/festive/ anniversaries." Further, the configuration of city elements can have a maximum benefit of the urban fabric and elements such as the platforms, road spans, parking lots, etc. These developing projects pay attention to the people whether they passing-by or vendors. Taking into account, considered behavior in further development for the selected case study can create liveability. Getting over the huge impact of the environment on the urban fabric, future development for the site can promote the urban aesthetics across convenient ambiance.

In conclusion, this paper presents a method of applying principles of tactical urbanism to comeover the challenge of lost places by improving livability in these spaces. Qualitatively, the results here used two types of data collection under a particular hypothesis. The findings provide a piece of evidence that the hypothesis can be true in case urban planners taking into consideration essential notes. First, public participation is the most important part of tactical urbanism, but it has two sides as it can be a motivating part in tactical urbanism principles but also some communities disagree with the idea. Second, the application of tactical urbanism should be esteem and consider the cultural context and various human's behavior. Finally, the third note is testing the impact of applying tactical urbanism through a digital model or another efficient way.

The current research follows the analytical approach. This approach concentrates on observations and interviews powered by the authors of the present work. The interview was scheduled with both vendors and bus station pass-by pedestrians in the analytical study. The research has a limitation in the qualitative methods and needs further investigation on spatial analysis of urban configurations. In conclusion, the significance of this paper is to introduce an

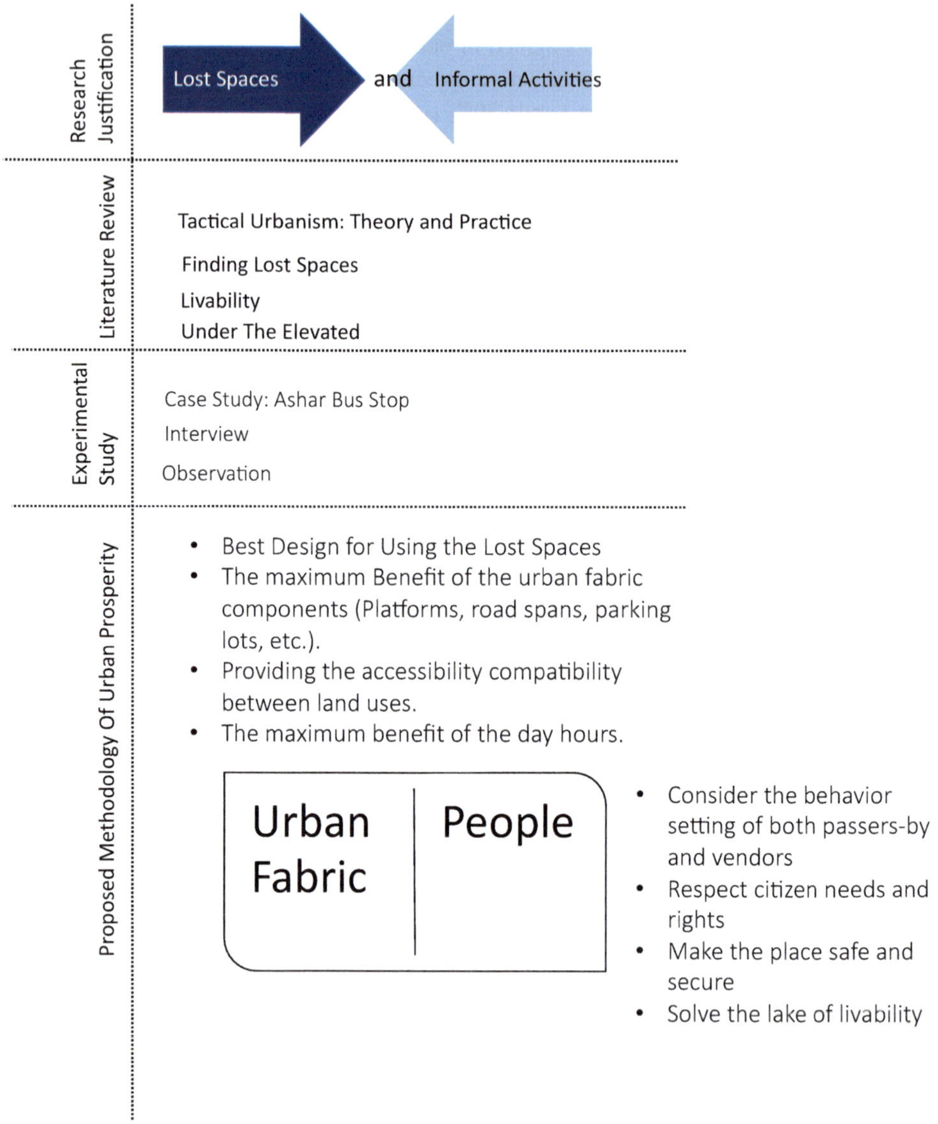

Fig. 9 The proposed pillars within the research structure. *Source* The authors

action plan concluded from the theoretical framework across the problems of lost spaces in Cairo that are recognized by the lack of livability.

References

AAVV. (2017). Health and Urban Design. *Urban Design Group Journal*, 12–39.

Abusaada, H., & Elshater, A. (2019). From chaos to order: Articulating the urban policies for cities of hardship. In: U.G. Benna (Ed.), *Industrial and urban growth policies at the sub-national, national, and global levels.* USA: IGI Global.

Anderson, J. (2018). *Much to do about (vacant) lots*. https://www.nextcity.org/features/view/much-to-do-about-vacant-lots?utm_source=Next+City+Newsletter&utm_campaign=29cf563054-Issue_286_COPY_01&utm_medium=email&utm_term=0_fcee5bf7a0-29cf563054-44084333.

Appleyard, D. (1981). *Livable streets*. University of California Press.

Aulia, D. N. (2014). Planning livable community with social systems approaches: Medan, Indonesia. In: *The 2nd International Conference on Sustainable Innovation (ICoSI).* Yogyakarta: Universitas Muhammadiyah.

Aulia, D. N. (2016). A framework for exploring livable community in residential environment. Case study: Public housing in Medan, Indonesia. In: *AMER International Conference on Quality of Life, AicQoL2016Medan* (pp. 336–343). ScienceDirect.

BEST, H. (2011). http://www.spacing.ca/toronto/2011/09/16/parking-day/.

Better Block Foundation. (2014). http://www.betterblock.org/.

Campanella, J. T. (2015). The spatial ecology of New York elevated. In: *Under the elevated.*

Carmona, M., & Magalhães, C. D. (2009). Local environmental quality: Establishing acceptable standards in England. *Town Planning Review, 80*(4–5), 517–548.

Courage, C. (2013). The global phenomenon of tactical urbanism: Asian indicator of new forms of citizenship. *Engage Journal*, 88–97.

Dempsey, N. (2009). Are good-quality environments socially cohesive? Measuring quality and cohesion in urban neighbourhoods. *Liverpool University Press Online*, 315–345.
Depave. (2014). https://www.depave.org/, https://www.depave.org/learn/why-depave/.
Dew-Jones, T. (2016). *Tactical urbanism: Teachings for portage and main*. http://www.downtownwinnipegbiz.com/wp-content/uploads/2016/03/Tactical-Urbanism.pdf.
Ehrenfeucht, R. (2017). Do food trucks and pedestrians conflict on urban streets? *Journal of Urban Design, 22*(2), 273–290.
Elshater, A. (2018). The philosophy of urban reload. *Asian Journal of Quality of Life, 3*(9), 161–164.
Elshater, A. (2014). Tactical urbanism: A method of community empowerment in Cairo neighborhoods. *SB14*. Sustainable Building (SB14).
Elsheshtawy, Y. (2013). City interrupted: Modernity and architecture in Nasser's post-1952 Cairo. *Planning Perspectives, 28*(3), 347–371.
Epley, D. R. (2008). A method of assembling cross-sectional indicators into a community quality of life. *Social Indicators Research*, 281–296.
Gehl, J. (2011). *Life between buildings: Using public space*. Island Press.
Lydon, M., Bartman, D., Dwoudstra, R., & Khawarzad, A. (2010). *Tactical urbanism: Short-tem action, long term change*. NextGEn.
Lydon, M., Miller, J., & Hebbert, F. (2012). *The open streets guide: Opening streets to people sharing resources transforming communities*. Street Plans.
Lydon, M., & Garcia, A. (2014). *Tactical urbanism*. Island Press.
Memarian, A., & Niazkar, N. (2014). The lost space of architecture in the context of urban lost space. *International Journal of Engineering and Advanced Technology, 3*(5).
Nooraddin, H. (1998). Al-fina', in-between spaces as an urban design concept: Making public and private places along streets in Islamic cities of the Middle East. *Urban Design International, 3*(1–2), 65–77.
Pfeifer, L. (2013). *The planner's guide to tactical urbanism*.
Selim, G. (2016). Between order and modernity: Resurgence Planning in revolutionary Egypt. *Journal of Urban History, 42*(1), 180–200.
Shackelford, K. (2014). *Tactical urbanism: A movement on the rise*. https://www.peoriamagazines.com/ibi/2014/feb/tactical-urbanism.
Stevens, Q. (2009). "Broken" public spaces in theory and in practice. *Town Planning Review, 80*(4–5), 371–392.
Trancik, R. (1985). *Finding lost space*. Van Nostrand Reinhold Company.
Veenhoven, R. (2014). Livability theory. In: A.C. Michalos (Ed.), *Encyclopedia of quality of life and well-being research* (pp. 3640–3645). Springer.
Wheeler, S. M., & Deakin, E. (2001). *Livable communities: Creating safe and livable neighborhoods, towns, and regions in california*. University of California at Berkeley Institute of Urban and Regional Development.

Feminist Non-functional Empowerment in Urban Spaces: An Empirical Study on New-Cairo, Egypt

Indjy M. Shawket

Abstract

Surrounding urban spaces are quite known by how they would affect every user's daily life. For thriving this effect, studies should be generated in bounded issues; one of these are the feminist empowerment in urban spaces, and how would these spaces functionally suit them? In this research, the researcher aims to clarify feminist rights in the built environment and how urban spaces would affect on their daily attitude. The research conducted observations on feminists' contact with urban spaces—density and duration—and the questionnaire survey to records on why they chose or avoid, how much they spent time, and others—based on feminist database—in these spaces located in a new high-class community, New-Cairo, Egypt. The research then works on analysis data to assess the causal relationship between the feminist social aspect, the built environment, urban behavior, and the feminist functional empowerment in urban spaces. This study would provide design guidelines to municipalities' feminist needs in urban spaces to re-conduct functional urban spaces through fulfilling these needs.

Keywords

Feminist • Built-environment • Urban spaces • Functionalized spaces • Street activities

1 Introduction

Links between women's activities and changes in cities cannot be split up. Neither the role of housewife- mother nor those of the children and men who formed her primary concern would have been created in its early twentieth-century form without the separated city. Similarly, urban built-environment is a key factor in which women (and men) organize, work, and change (Andrew & Milroy, 2011). Besides, the world urbanization is one of the most trickily facts, recently, which showed profound changes in every virtual phase of stakeholders' lives scocially. Thus, urbanism as complicated phenomena may require coherence and unity although diversity and conflicts are based on sociological analysis -concerning the ecology, the social organization, and the social psychology of the urban mode of life confirms the fruitfulness (Shawket, 2018a). And as the researcher believes that open spaces in urban centres are not only elements to beautify them, but they bear multiple functional needs and reflect cultural values as well (Yang, et al., 2019), the study will highlight feminist requirements in urban spaces through a literarture review and emperical quantaitative case studies reveal that particular types of women are in need of partícual needs. In different environments, the action or behavior of habitants can be improved or leaded through a functional fulfilling urban space, and this was the reason of chosing New-Cairo, Egypt; although it is a new high class city, its women inhabitants suffer some shortfalls in their community, by which enhancing habitats to the communities make them more livable—through providing and developing services—and rescues the stress on their daily lives. Accordingly, the concolusion of the study will provide design guidelines to municipalities' feminist needs in urban spaces to re-conduct functional urban spaces through fulfilling these needs.

I. M. Shawket (✉)
Modern Academy for Engineering and Technology, Cairo, Egypt

2 Functional Urban Spaces

Urbanization is a process of unconscious or conscious development, based on pragmatic needs and literature backgrounds; besides it is a process of continual socio-spatial transformation not only in isolated "points" or "zones" but also within them. As, for example, streets have nominated different elements, activities, and buildings and each has a

F. Rosso et al. (eds.), *Advances in Architecture, Engineering and Technology*, Advances in Science, Technology & Innovation, https://doi.org/10.1007/978-3-030-86913-7_2

value of its own and an integration with its surroundings. (Lillebye, 1996; Santamaría, 2018). Researchers are working on identifying the functionally different land-uses and examining their association with the informality of social control, where urban-based informal social control is important for preventing unwanted behavior, and solving different communities' problems (Corcoran et al., 2018).

As a setting for interaction among/exposure to people with different backgrounds, values, and opinions, the public realm is an important normal context of a public sphere. However, exposure to diversity of issues and opinions through public spheres is depending on the range of external inputs available from the mass media and as well the everyday interactions embedded within the private, parochial, and public realms (Hampton et al., 2010).

As well as the actual functions of a space may not reflect the intent of the original zoning scheme from designers. But, numerous methods have been proposed—by computerized methods or even social sensing and data collection—to identify the actual urban functional regions (Wei et al., 2019). Melissa Gilbert argued that urban spaces are reproduced through people's everyday activities, where women's daily lives—as an example—being conceptualized entirely through the neighborhood leading to false assumption that neighborhood and community are interchangeable concepts. As well as she demonstrates that the scale of daily life is contingent upon, and mutually constituted through, power relations (Gilbert, 1999). Therefore, a good urban place should include the quality of three ticklish elements: Physical space, Experience, and Activity. The vibrancy of activities—that extended to a vibrant place—distinguishes sufficient or good urban places from others. And for delivering a space they are manifested as four dimensions: "feature," "form," "function," and "image" surrounded by conceived, lived, and perceived forms of space, while considering a globalized context (Yigitcanlar et al., 2018).

Accordingly, previous chart (Fig. 1) will be used in the empirical part when classifying spaces' assessment. The assessment will be divided into two branching: the feature and functional (where the conceptual design is developed and operated) and the second is the form and image (where the user feels the development of space and can percept it).

Indeed, as discussed by (Loukaitou-Sideris & Banerjee, 1998) that urban design would be with privatization and negotiation rather than having a public policy in this era.

Nowadays, in our communities, we can find too many public spaces that has a left over from another function, as traffic, and when the urban land is not owned to a profit-oriented society the space may be subjected to less secureness, increasing privatization, and non-realm. This phenomenon is defined as "SLOAP" means space left over after planning, where functional and organized urban design offers a solution for it. Known that spaces somehow have grown-up organically, through accumulation of different activities added and adaptabilities. Thus, by understanding the ingredients of a successful fulfilling public urban spaces designers should be able to create more good spaces, avoid bad ones, and retrieve existing bad ones. Ultimately, the factor of FEELINGs underpinned all factors affecting the success of space (Shaftoe, 2012). Here we can highlight a question: How do females feel when using functional designed public urban spaces?

3 Feminist Empowerments in Urban Spaces

Women's use of public space has been one of the much recent research subjects nowadays. Existing theories in environmental and behavior studies are inadequate to explain different issues (Day, 2000).

It is more understandable now that women often have more specialized requirements/needs in open spaces than men. Women are found to feel less comfortable using some urban public spaces, especially if they are considered to be uninviting or unsafe. Where the needs for safety, security, and comfort are especially important to be considered when designing urban space for women (Francis, 2003).

As well, the researcher is aware of the complexity and the difficulties that appear when it is chosen to differentiate between women and men into categories. But, this differentiation according to ECEC (early childhood education and child institution) suggests that women seem to be more motivated to work with children than men do in out-door spaces, because they want children to acquire valuable knowledge about their built-environment and the natural world as well. Men, more than women, are reported that being out-doors is important for their own motivation and well-being, and that they would rather work outdoors than indoors. The understanding to the social diversity and the meaning of gender, ethnicity, class, and culture are likely to vary increases the design awareness (Waller et al., 2017). Although many researches have considered women in urban studies, there is a large body of evidence from a number of disciplinary areas which shows that men and women's experience, conception, and use of space have differences (McDowell, 1983).

A spatial division between private and public was the ideal focus for fallacious feminist urbanism. Many efforts have been made to defy that the home is a women's sphere. As the private/public variation and its influence on urban forms of land-use patterns in different cities and towns. Both the nineteenth century's massive industrial urbanization and the

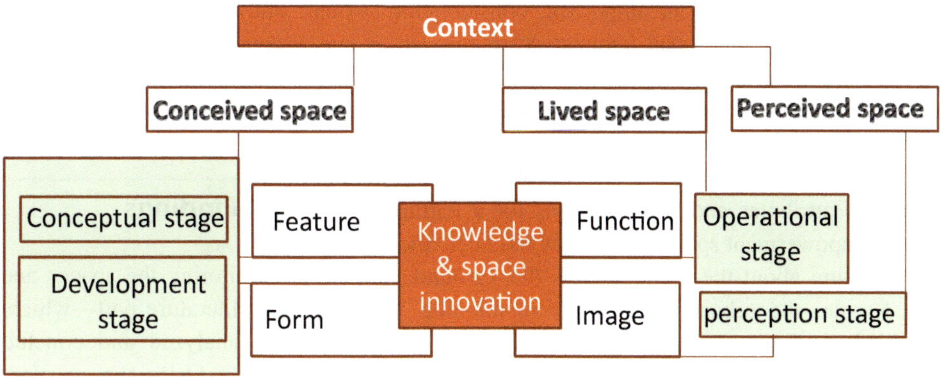

Fig. 1 Delivering a space. *Source* Derived from Yigitcanlar et al. (2018)

rural / urban and international migrations to global cities in the present century have re-written gender relations and as well they proved a source of fascination to feminist urbanism (McDowell, 1993). Cities have been shaped to limit women's role into the family, and the site from which women perform freely, more domestic, and child-care services. Symbolically, it is expressed in the assembly of men with the working in the city/public world, and of women whose family and community are in the suburbs. In other words, male control of, as a result of patriarchal social relations imbedded in marriage and building a family. On the other hand, feminists' responsibility for domestic labor requires their mobility (especially for the transfer of services and goods from and to home) (Peako, 1993). Also, women spend more time in shopping than men where daily purchases are commonly bought by women. Again, women are the majority in using public transport (Hille Koskela, 2000). Accordingly, the empowerment and inequalities are deeply embedded not only in the urban spaces' design, but also in the system of urban settlement planning and various state policies. From this point women's responsibilities for the reproduction of labor power involved in the home and the community. There are a range of psychological and social problems which faces women in defined urban spaces known as "the problem with no name," where women's mutilated use of public urban space is itself seen as a spatial expression of patriarchy given the restrictions on women's behavior resulting from fear of male violence and the state's apparent reluctance to protect them from it and other problems conducted in the urban spaces' design (Peako, 1993). Females in campus communities are feeling safe as these communities are having rules and regulations to actions and behaviors, but do public urban spaces afford the same feelings of those controlled ones? This requires raising the awareness between urban designers to promote for this concept as feminist spaces intensify active engagement with diverse communities to understand the nature of working in the field of feminism (Oberhausers et al., 2017). Thus the study in the following part will highlight women's needs to conduct design guidelines for feminist needs in urban spaces.

4 Empirical Study

Place making in knowledge and innovation is a challenging task for urban designers due to the lack of competent conceptual frameworks and design guidelines. To address this issue, the study adopts a quantitative methodological approach to carry out an empirical investigation in a case study from New-Cairo, Egypt. Besides, when it comes to epistemology of researching the city two main tensioned strands could be identified. The first focuses on the experience of everyday life—reflected in the study through a feminist questionnaire to clarify their experiences, while the second is focusing on charting sets of power relations with the city—reflected on the study by observations and a question asking women about their needs to empower urban spaces—and its physical dimension (Fig. 2) (Cochrane, 2014). An empirical investigation is guided by a conceptual framework derived from the interdisciplinary literature review. Interviews, Visual analysis approaches targeting urban data, and various data sources are conducted among a range of women as stakeholders.

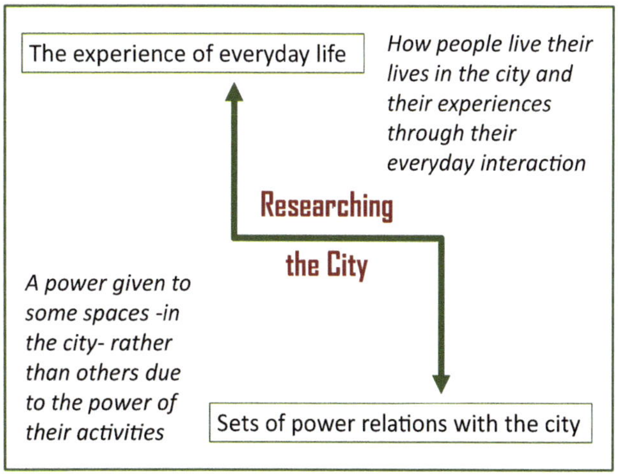

Fig. 2 Urban researching epistemology. *Source* Researcher (2019)

Thus, it is likely an important consideration when examining eighty-six social feminist diversity such as age, relationships, work, and their different activities within urban spaces in New-Cairo within various types of urban spaces such as neighborhood center, open markets, commercial paths, and heavy traffic urban zones providing opportunities to reveal attributes and considerations for a feminist functional empowerment space. The survey has been divided into parts asking about the social aspects /background affecting the female, the urban space she is mentioning with the established activities, functional satisfaction (poor till excellent), and her needs in this urban space.

The case study is chosen as New-Cairo is considered to be one of the spotted cities in Egypt nowadays, besides it targets a socio-economic community with wider ranges of activities and stakeholders, and high-class community. And as shown in Fig. 3, it is located in a strategic zone within a nearby distance from other surrounding ones. Moreover, New-Cairo carries a variety in land-uses as education, shopping streets and markets, neighborhood plazas, religious, and others.

As previously mentioned, visual analysis approaches targeting urban data took place in different urban spaces categories—commercial zones, paths, neighborhood plaza, mall entrance, and heavy traffic urban zones—at different days and times. This non-systematic visual monitoring—as shown in Table 1 helps the researcher to build up a survey in companion to the literature review. Also, it helps in the perception of the feminist empowerment in urban spaces, and their needs which should be fulfilled functionally.

5 Findings

Monitoring the spaces and interviews—designed on previous literature part—which took part in the empirical study are analyzed and concluded in two dimensions. Women were asked in the questionnaire to choose an urban space to fill about (which mostly affect her) the activity she does there, to assess it, and to recommend further functional needs she requires to fulfil her empowerment in space. In parallel, the questionnaire carries more data about her within social aspects as (married and non-married; Employed and non-employed; mothers and having no children; and within range varied from teen agers till 60 and above).

In the following, the first is a graphical relationship between women's social background, different types of urban spaces, and feminist functional quality of each type. While the second will carry different feminist needs in urban spaces—based on literature review, visual monitoring, and interviews—from which a set of main design guidelines are delivered.

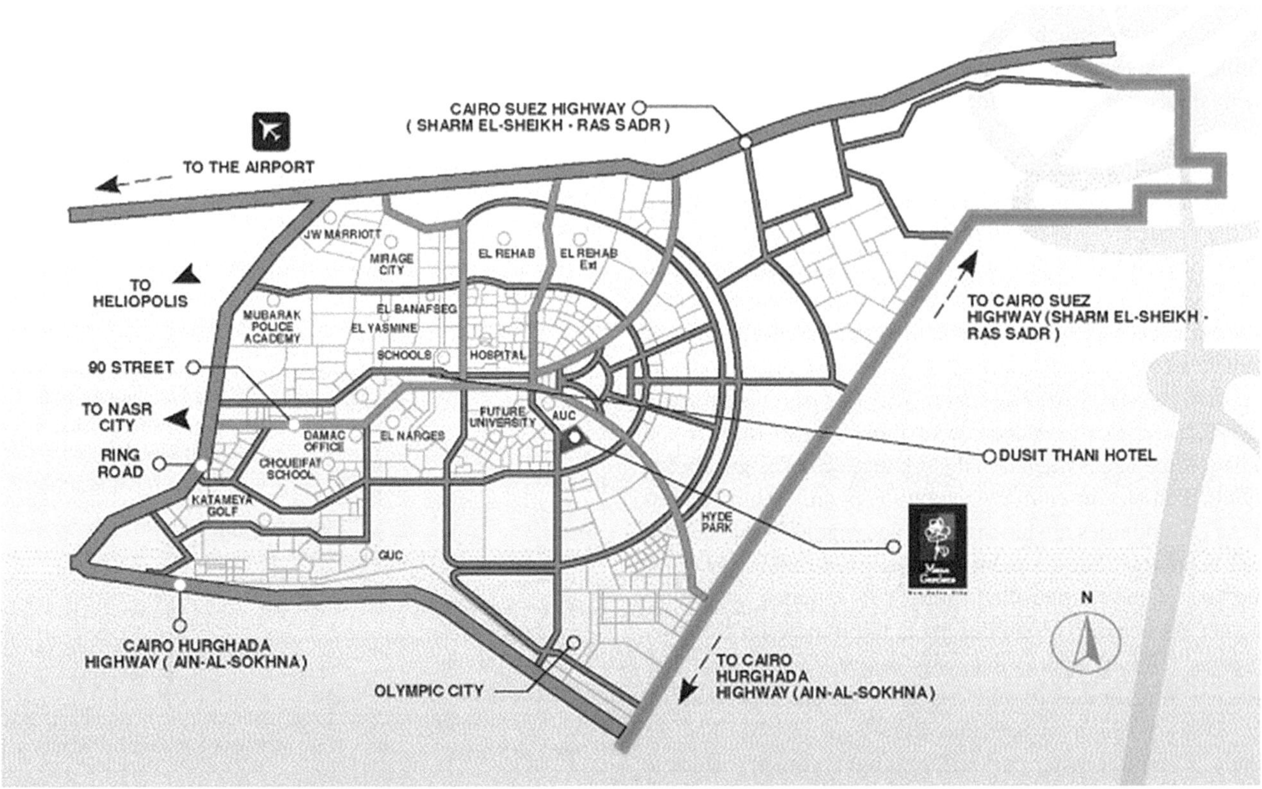

Fig. 3 New-Cairo map. *Source* Mena for touristic and real estate development

Table 1 Visual monitoring for feminist empowerment in some of the urban spaces of New-Cairo, Egypt. *Source* Researcher (2019)

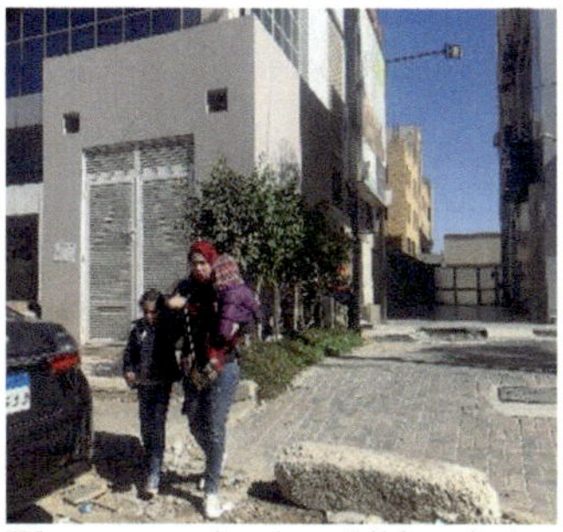 Woman carrying her children in her way to reach her car, where there is no parking zone	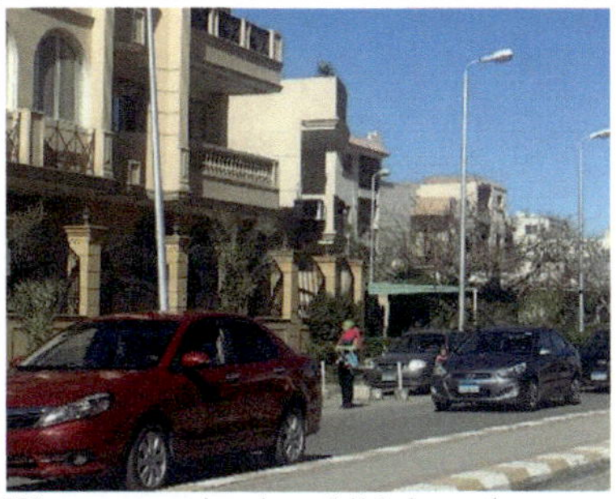 Woman carrying her child through a vehicle path to reach her destination	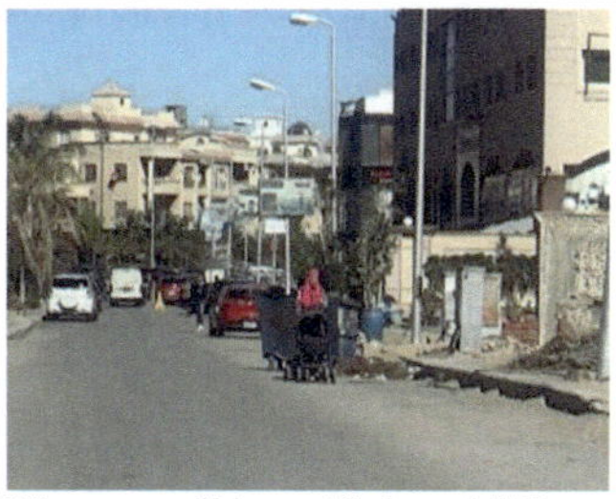 Woman walking with her child's carry kit in a non-walkable street-
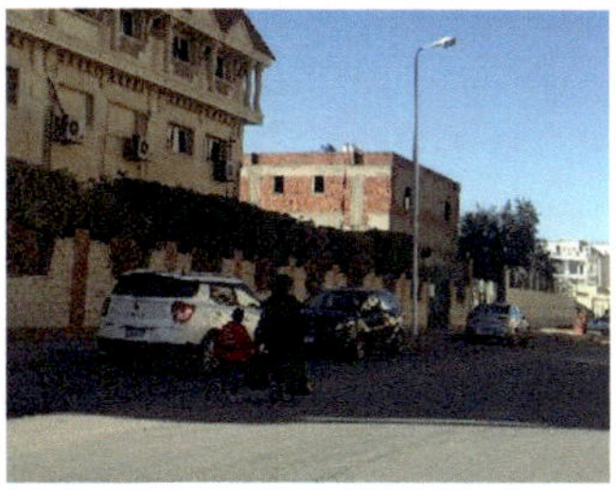 Women coming back from the market carrying goods and holding her son's hand	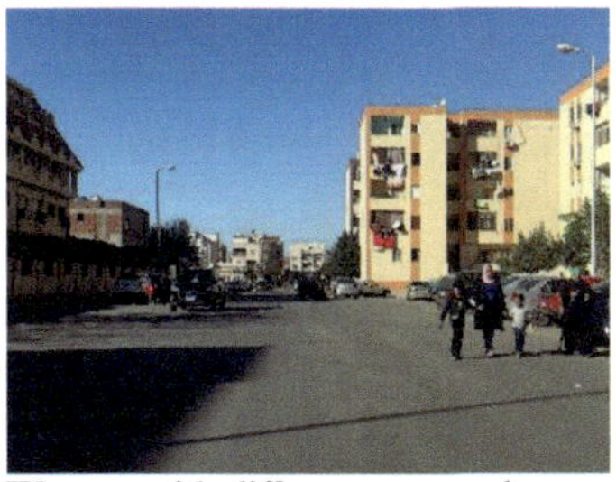 Women with different ages and purposes walking through out the mixed use path	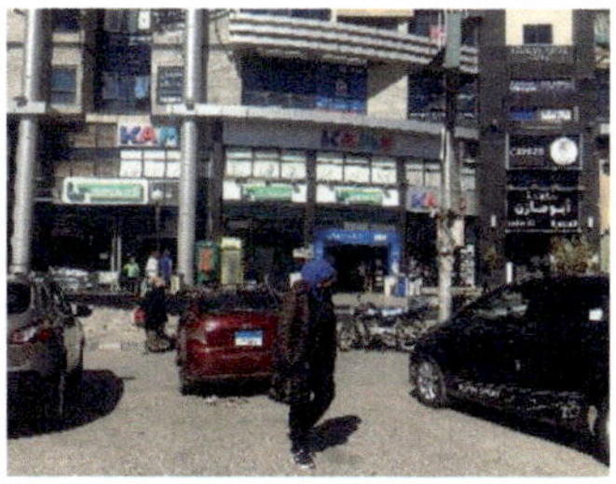 An interpenetration of spaces with mixed feminist empowerment
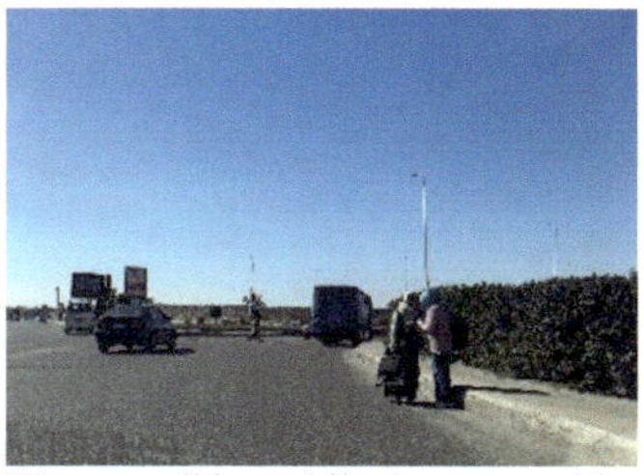 Women talking while they are waiting for a bus	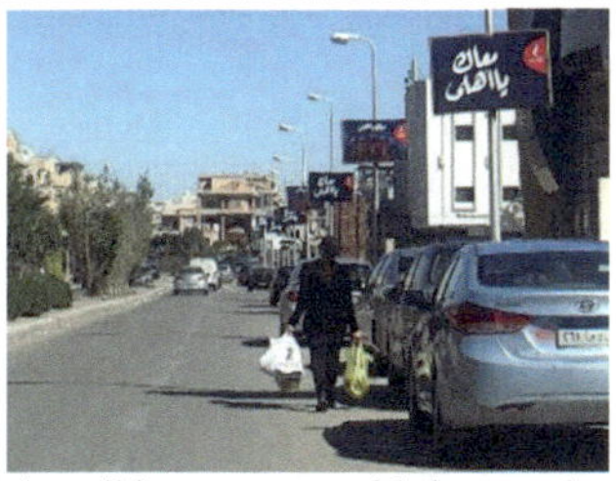 A walking woman with her market goods	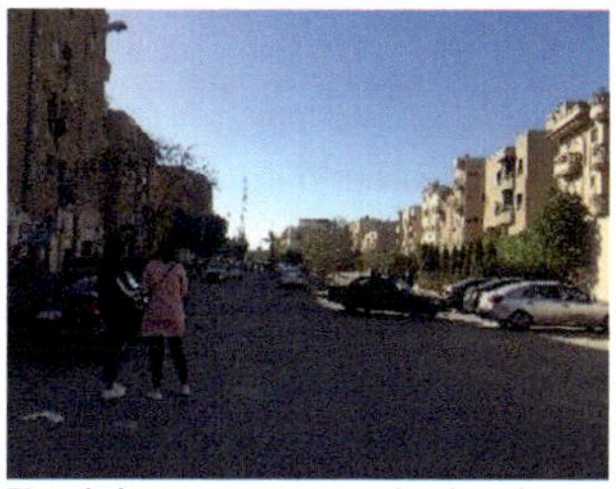 Feminist teenagers gathering in an interpenetration space (educational front space, residential, and commercial)
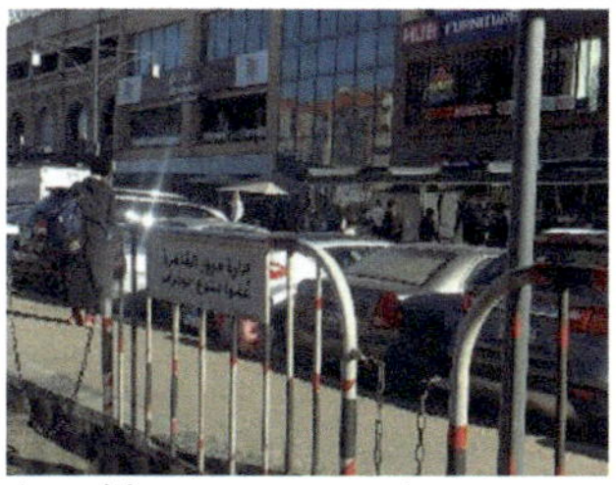 A waiting woman on the pavement with her sitting down children on the floor	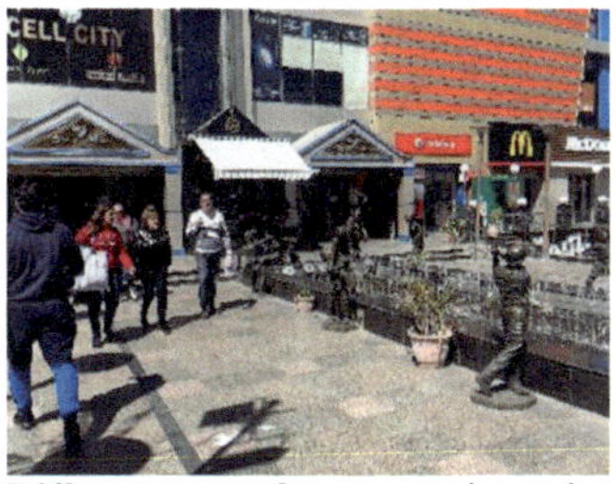 Different ages of women using urban spaces	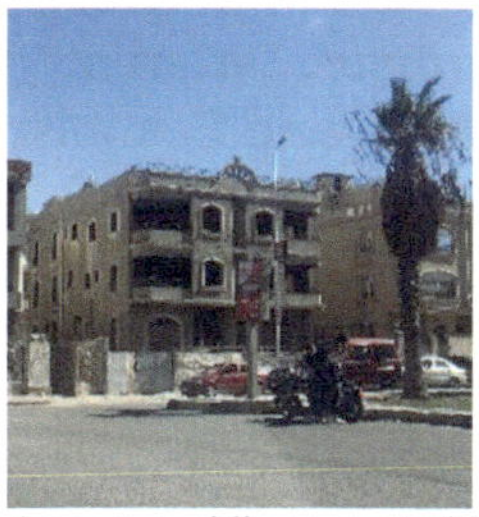 A women riding a motorbike

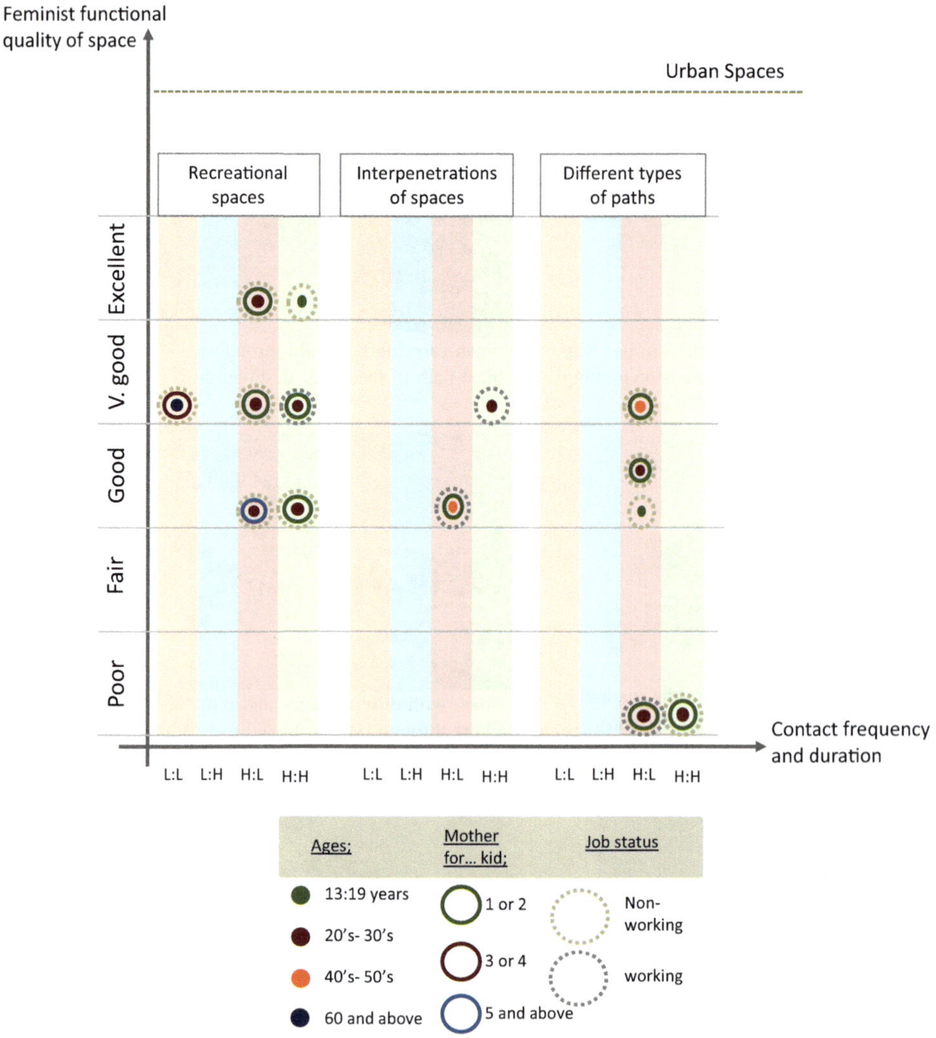

Fig. 4 Quality of different urban spaces upon feminist use. *Source* Researcher (2019)

5.1 A Graphical Relation Between women's Social Background, Different Types of Urban Spaces, and Feminist Functional Quality of Each Type (Fig. 4)

- Recreational spaces in New-Cairo offer more feminist satisfaction than paths do.
- Working women have more frequency contacts with urban spaces, and lower contact durations in urban spaces than non-working ones.
- Feminist teenagers prefer creational entertain spaces and interpenetration spaces than paths even if they were commercial.
- 20s and 30s aged working mothers do not prefer to exist in interpenetration spaces.
- 20s and 30s aged non-working mothers are having higher satisfaction toward urban spaces compared to working ones.
- The higher number of children campaign the woman, the lower frequency contact to urban spaces.
- Although women above 60s spend low durations with low frequency contacts, they have a quite fine satisfaction toward these spaces.

5.2 Feminist Needs in Urban Spaces

5.2.1 Feature and Functional Needs

- Parking:
  (Mothers usually look for a nearby parking area to move safely from cars to their target place with their children)
- Safety:
  (All humans are looking for safety, but women are seeking it even to enjoy their outings, or even to do their daily jobs. Design of urban spaces should support this feeling)
- Edged children's parks:

(Mothers are always looking for a controlled place for their children to play in, in-order to introduce joy for her kids and herself as well)

- Lightening:
  (One of the elements that may be missed in spaces at night, although it supports the concept of livability and social interaction)
- Benches and sitting elements:
  (Starting from teenagers till 60 aged and above women, comfortable sitting areas are one of the primary elements required in spaces)
- Noiseless:
  (Housewives or employed women both require a noiseless place where they can escape from the home stress and / or kids' headache in a stress relief urban space)
- Fresh air:
  (Oriented spaces and/or buffered spaces are potential for a feminist attraction)
- Private zone for women:
  (A separated private zone for women is one of the requirements for some women which may relate to their cultural or rational aspects)
- Cycling lanes:
  (Cycling in some cities is main, but for others it is one of the dreams which women need in their urban spaces)

5.2.2 Form and Image Needs

- Greenery:
  (Greens support the visual physiology of women, and are extremely required by them).
- Wideness:
  (Visually women require wide open spaces to support them physiologically).
- Clean:
  (Unclean spaces may cause stress for some women).
- Green buffer supporting semi-private sectors:
  (A separated private zone for women is one of the requirements for some women which may relate to their cultural or rational aspects).

Feature and functional feminist needs in urban spaces presents 70% of their needs, while the form and image needs represent only 30% (Fig. 5).

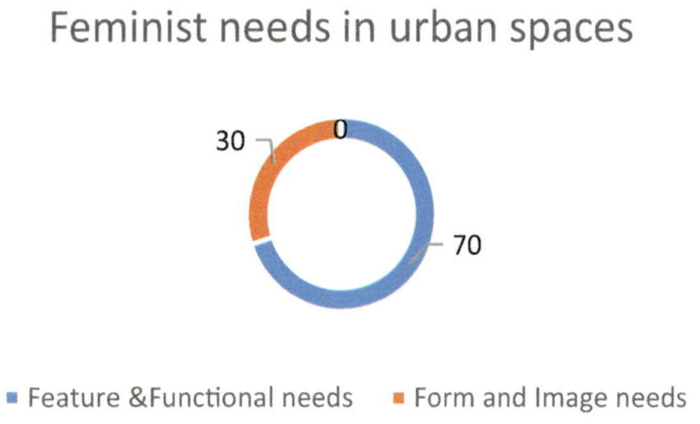

Fig. 5 Parentage of feminist needs in urban spaces. *Source* Researcher (2019)

6 Conclusion

Feminist needs in urban spaces—as for new-Cairo, Egypt—are categorized under the form and image needs, and the feature and functional needs. By which, the feature and functional needs represent 70% of femininst stakeholders' needs, while the form and image represents only 30%. According to priorities safety represents the most important need for all women using urban spaces, while having a private or a semi-public space is the least requirement for them.

Figure 6 shows each need and its magnitude within its category—where it represents a priority of elemets—forming design guidelines for a targeted feminist functional empowerment in urban spaces.

- Safety represents the most important element for women in urban spaces. Furthermore, safety elements should be integrated within different stages of designing an urban space (Shawket, 2018b).
- Creating private/semi-private sectors within different basis—through the form or functionally—is a need for women even if they are in the context of a high-class community.
- Mothers are linking their needs in urban spaces by their children's needs.
- Night-scape is important for women, thus they are searching for lighted urban spaces to support their safety aspects, and to make the space more liveable.
- Urban relief is the hidden target behind several needs.

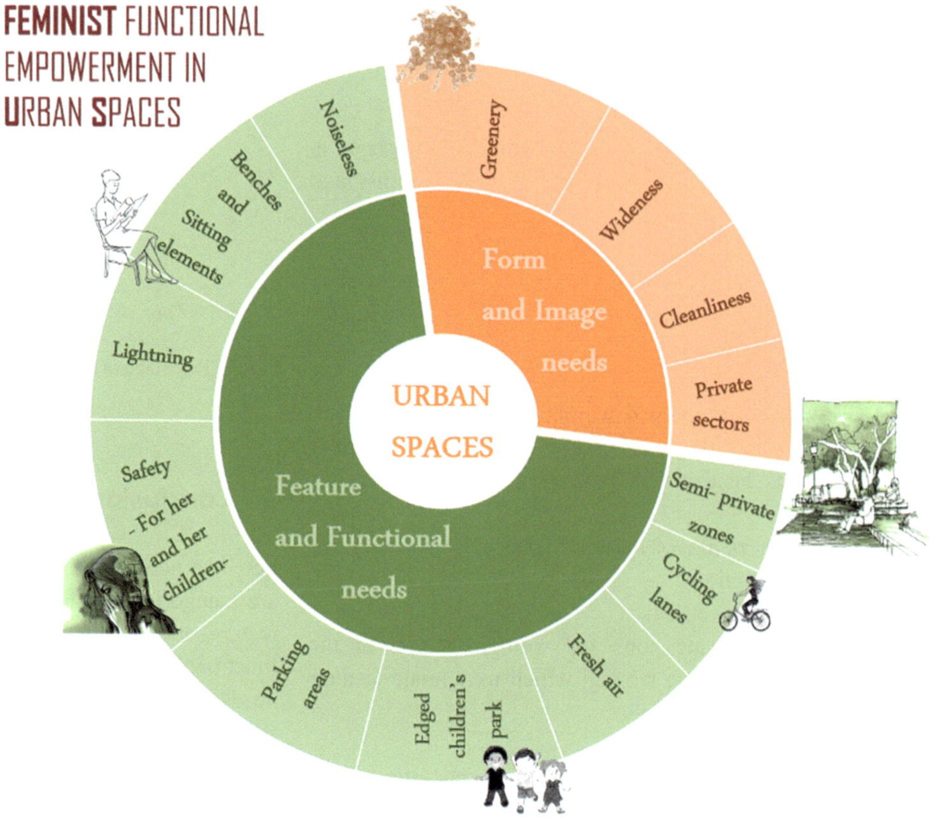

Fig. 6 Magnitude of feminist needs in urban spaces. *Source* Researcher (2019)

7 Discussion

A futher step was taken by the researcher by sketching an open space powered by previously mentioned guidelines to support feminist functional requirements, then investigating their assessment and recommendations for such a design. And this discussion could be a step by which binding further research could start from (Fig. 7).

This proposed sketch is a visual screening test for feminist functional empowerment in urban spaces and to be assessed by them. Thus, this design was assessed by 50 women—with different social aspects—and the result was as follows: 66% of them said that it is very good and they feel that they need some adjustments, none of them see that this space is poor, only 5% see that this space is fair—some of them mentioned that the private zone is not isolated enough, another said that kids area need to be more secured, 17% said it is a god place but it requires some upgrading, and only 12% of them said it was an excellent space for them (Fig. 8).

Previously statistics shows that;

- Although stakeholders are from the same city, but they still vary in needs and expectations
- Further studies should be done to identify more feminist needs in urban spaces
- A prototype of an open approved space by its stakeholders may be a good idea supporting the visual and functional identity for each city

8 Recommendations

Recommendations of the study goes on different dimensions, in which cooperation is required to reach a fulfilling urban space.

Civil society

- Support your surrounding space with allowed elements (e.g., greenery)
- Respect women's empowerment in urban spaces
- Your attitude may affect others urban relief, so try to leave your positive effect

Governmental entities

- Putting regulations organize and guaranty human—especially women rights in urban spaces
- Supporting neighborhoods with wide urban spaces

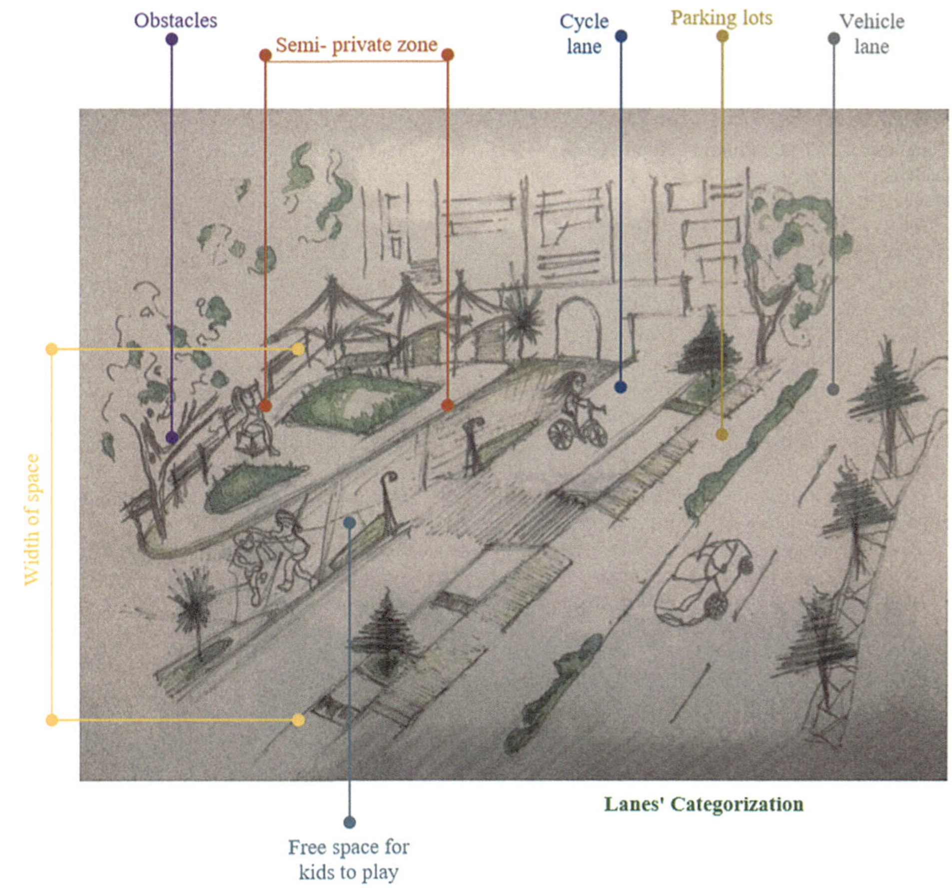

Fig. 7 Suggested urban space as a response for feminist needs in urban spaces. *Source* Researcher (2019)

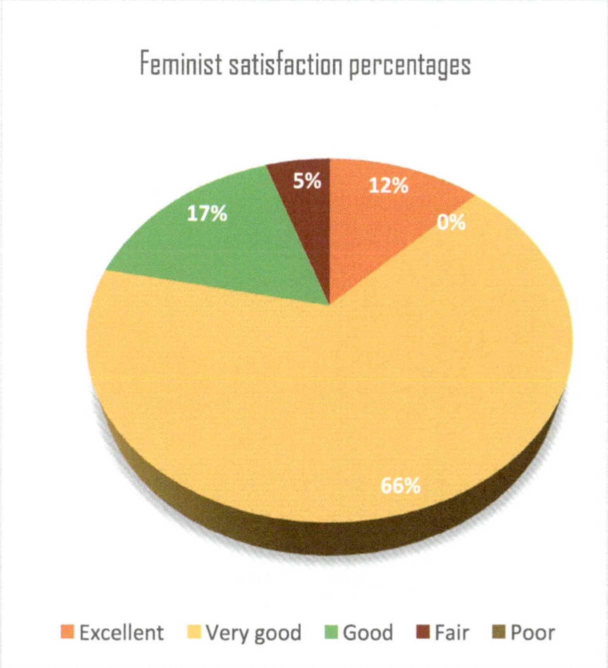

Fig. 8 Feminist satisfaction percentages for a proposed feminist functional urban space. *Source* Researcher (2019)

Urban developers

- Provide physical and physiological comfort requirements in built-environments even for those who are not high-class communities
- Support urban spaces with feminist needs according to previous guidelines
- Stakeholders—Women participation in designin process

Further studies

- It is suggested, for further research, appropriate various case studies through different communities in order to compare several results using great samples
- Building up a complete design guidelines—with many variations—that would suit feminist functional empowerment in different societies

Acknowledgements I would like to express my special thanks to all women who gave me the opportunity to do this study through helping me in gathering data. Secondly, I would also like to thank my family who supported me a lot throughout this study and helped me to finalize it within a limited time frame.

References

Andrew, C., & Milroy, B. (2011). *Life spaces: Gender*. Employment, UBC Press.

Cochrane, A. (2014). *Researching the city* (p. 38). Sage.

Corcoran, J., Zahnow, R., Wickes, R., & Hipp, J. (2018). Neighbourhood land use features, collective efficacy and local civic actions. *Urban Studies, 55*(11), 2372–2390.

Day, K. (2000). The ethic of care and women's experiences of public space. *Journal of Environmental Psychology, 20*(2), 103–124.

Francis, M. (2003). *Urban open space: Designing for user needs*. Island Press.

Gilbert, M. (1999). Place, politics, and the production of urban space: A feminist critique of the growth machine thesis. *The urban growth machine* (p. 96). Albany: State university of new your press.

Hampton, K., Livio, O., & Goulet, L. (2010). The social life of wireless urban spaces: Internet use, social networks, and the public realm. *Journal of Communication*. ISSN 0021-9916.

Hille Koskela, H. (2000). 'The gaze without eyes': Video-surveillance and the changing nature of urban space. *Progress in human geography* (Vol. 24, Issue 2). Sage.

Lillebye, E. (1996). Architectural and functional relationships in street planning: an historical view. *Landscape and urban planning* (Vol. 35, Issue 2–3). Elsevier.

Loukaitou-Sideris, A., & Banerjee, T. (1998). *Urban design downtown: Poetics and politics of form* (pp. 70–71). University of California Press.

McDowell, L. (1993). Space, place and gender relations: Part I. Feminist empiricism and the geography of social relations. *Progress in Human geography*, 157–179.

McDowell, L. (1983). Towards an understanding of the gender division of urban space. *Environment and Planning d: Society and Space, 1*, 59.

Mena for touristic and real estate development. http://www.mena.com.eg/images/concord%20gardens%20-%20location%20map.jpg.

Oberhausers, A., et al. (2017). *Feminist spaces: Gender and geography in a global context*. Routledge.

Peako, L. (1993). 'Race'and sexuality: Challenging the patriarchal structuring of urban social space. *Environment and Planning D: Socioty and Spoco, 11*, 415–432.

Santamaría, G. (2018). Planetary urbanization and its discontents: preliminary foundations for reconceptualizing urbanism. *International Journal of Current Innovations in Advanced Research, 1*(3), 1–25. ISSN: 2636-6282.

Shaftoe, H. (2012). *Convivial urban spaces: Creating effective public places* (pp. 6–7). Earthscan.

Shawket, I. (2018a). Identity in urban spaces of residential compounds: Contributing to a better environment. *HBRC Journal, 14*(2), 235–241.

Shawket, I. (2018b). Safety in real state. *New cities ans community extentions in Egypt and the Middle East: vision and challenges* (pp. 199–216). Springer.

Waller, T., Ärlemalm-Hagsér, E., Sandseter, E., Hammond, L., Lekies, K., & Wyver, S. (Eds.). (2017). *The SAGE handbook of outdoor play and learning* (p. 706). Sage.

Wei Zhai, W., Bai, X., Shi, Y., Han, Y., Peng, Z., & Gu, C. (2019). Beyond Word2vec: An approach for urban functional region extraction and identification by combining Place2vec and POIs. *Computers, Environment and Urban Systems*. Elsevier.

Yang, J., Zhang, F., & Shi, B. (2019). Analysis of open space types in urban centers based on functional features. In: *E3S Web of Conferences, 79*.

Yigitcanlar, T., Marques, J., Lorenzi, C., Bernardinetti, N., Schreiner, T., Fachinelli, A., & Wittmann, T. (2018). Towards smart Florianópolis: What does it take to transform a tourist island into an innovation capital? *Energies*, 3265.

The Main Indicators Affecting Interactive Experience Design in Contemporary Urban Spaces Using Media Interventions

Sara Tarek

Abstract

The sustained escalating development and applications of New Digitized Media and Technology is changing the settings and lives of people, in addition to their cultures and behaviour. This research addresses the impact of new media interventions and the related technology on urban spaces. It reviews its role in generating interactive experience and enhancing the urban environments. This enriches space experiences for users and strengthens their relationship with their surrounding environment. The research paper comprises an introduction, conclusions and three main parts, namely, literature review, relationship between design indicators and design considerations, verifying the indicators-consideration relationship. The research study identifies the design indicators and considerations of applying media interventions in the urban space and verifies the relationship between the indicators and considerations through experts ratings, by carrying out a questionnaire to score each indicator according to its relevance to the design consideration. The research work points out the potentials of urban media interventions' and new technologies' role in enhancing the urban experience, it also identifies the most important aspects and indicators that affects its incorporation in contemporary urban spaces, in an attempt to facilitate its design process and implementation in the urban setting.

Keywords

Media interventions • Interactive urban space • Interactive experience • Digital media technologies

S. Tarek (✉)
Department of Architecture, Faculty of Engineering, Cairo University, Cairo, Egypt
e-mail: sarat@cu.edu.eg

1 Introduction

The changes imposed by new technologies including Information and Communication Technologies (ICTs) and digital media have been influencing conceptions and practices, spanning urban space, time and cities. This resulted in the introduction and implementation of emerging concepts and determinants in urban studies and communities, hence presenting new challenges to urban designers and architects, namely, integrating the new media technologies in urban settings, to enhance the quality of its spatial constituents, elements and relationships.

The main goal of the present research work is to recognize the indicators and design considerations of the application of new media technologies in urban spaces design. The paper investigates the role of media technologies in generating interactive urban experience, which could be used in the revitalization of urban spaces. In order to achieve the stated objective, the research adopted a quantitative approach to read and understand the indicators for designing interactive spaces using media interventions.

The paper encompasses this introduction, conclusions and three main parts, namely, Part one presents a brief review of selected literature, introducing new media, its principles and impact on urban spaces and settings. It recalls and represents the three aspects of designing interactive urban spaces using new media technologies and introduces the indicators encompassed by each one of the three aspects. It also presents the design considerations of applying the interactive installations in the urban spaces. Part two synthesizes the key findings of the condensed literature review and presents the relationship between the design considerations and the indicators for the three aspects together. Part three covers the verification of the proposed relationship between the considerations and indicators through a questionnaire for experts and a statistical analysis for the responses, to form an overall view for the indicators and identify them according to their importance in the design

F. Rosso et al. (eds.), *Advances in Architecture, Engineering and Technology*, Advances in Science, Technology & Innovation, https://doi.org/10.1007/978-3-030-86913-7_3

process comprising three sections together with a concluding section combining discussion and conclusions. The presented work highlighted the reactions of the experts to urban media interventions. The quantitative investigation pointed out the possible potentials and development directives for the design indicators for this new kind of applications in the urban spaces.

2 Literature Review

Media is "*the use of computer for distribution and exhibition, rather than with production,*" and it could be seen in several forms including the Internet, virtual reality, 3D-animation and smart television. Using computer for media display or distribution differs from using it in media storage and/or production. New Media incorporates what is known as Multimedia, which could be presented as convergent images, texts and sounds in the same medium. It works by presenting programmable objects in mathematical forms which are characterized by its modular structure which varies in scale, and represented in a discrete form (pixels, voxels, scripts, etc.) (Vand Den Barthes, 1977; Boomen, 2014; Manovich, 2001). Media computerization and digitization depends on transcoding of data which allows the manipulation of models and presented data by user, and this is considered a mile stone in human–computer interaction designs (Sorapure, 2004).

Furthermore, technology is commonly defined as the everyday application of knowledge to achieve a certain purpose or to improve a certain condition (Webster, 2004). In the present discourse, some technology types need to be addressed as they are more relevant to media manipulation and its intervention in urban spaces, namely, Information and Communication Technology ICT which is based on computer information systems to provide accessibility to information through telecommunication medium. Digital Technology is the deployment of digital codes to discrete values as digital electronics (Taylor & Todd, 1995). SMART Technology refers to Self-Monitoring Analysis and Reporting Technology. It is generally used to detect different problems and malfunctions, for computers' hard disk drivers using self-monitoring and self-data gathering techniques (Nam & Pardo, 2011). Human–Computer Interaction (HCI) deals with computer systems and graphics' techniques, besides linguistics and social sciences to improve interaction between people and computers (Dix, 2009). Computer Vision is based on information elicitation from digitized images and uses it in various forms, including navigation and modelling interactions of objects; Augmented Reality (AR) is a reality that is integrated with computer simulation in which technology artifacts are developed (Aurigi & Cindio, 2008).

Designers nowadays are adopting digital media technologies for different purposes like display, communication and navigations to encourage the users of the urban space to interact, and enhance their spatial experience (Mitchell, 2003). Different opportunities for innovation in contemporary public realm are more likely available due to emergence of new media interventions and its implementations to the urban environment (Brighenti, 2011; Johnson, 2003; Townsend, 2004). Previous researches showed that sense of place is relevant to the physical properties of the place itself (McCall et al., 2005), in addition to its activities and meanings (Thompson, 2002). Applying new media technologies in urban spaces is likely to enhance users' experience and make it more interactive; Fig. 1 shows examples of applying media installations in the urban setting. Researches suggested that urban mediation is arguably considered as a new approach for making a better sense of contemporary urban life as a result of implementation of new media in the urban setting, and it lies beyond the dichotomy between "Virtual space" and "Real space" (Brighenti, 2011; Manovich, 2001).

The relevance of digital technologies does not depend on the fact that they can be touched or seen, but on relationships that they enable. They are spatial because urban space cannot be interpreted as detached from functions, relations, perceptions, and indeed from the use and role of time in the city. The dualist vision tending to keep physical and virtual, visible and invisible apart could be recognized as being at play here. This criticism can easily be extended to urban designs, where the potential of ICT in augmenting dimensions of urban space have so far been greatly overlooked (Aurigi & Cindio, 2008). Concepts like augmentation, hybridization and mixed-reality are also considered new types of the mediation process, as these concepts work by overlaying a digitized layer over real urban environment. (Bolter et al., 2006; Manovich, 2001; Ohta & Tamura, 1999).

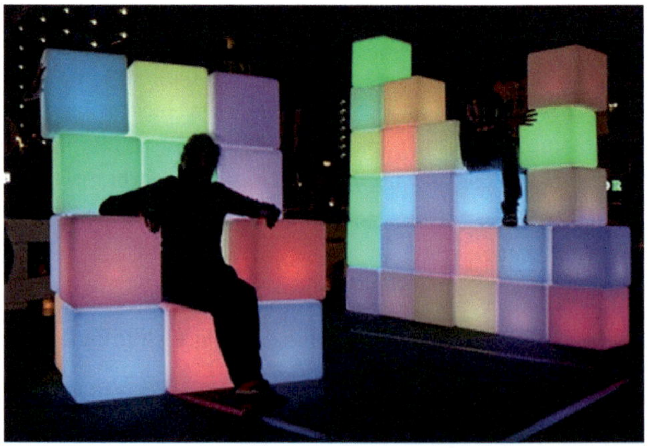

Fig. 1 Different forms of applying media interventions in the urban spaces. **a** Pixels project (interactive light cubes). Bogota, Colombia—2011, http://www.jonasvorwerk.nl. **b** Fremont Street (interactive display on a ceiling). Las Vegas, USA, http://www.vegasexperience.com. **c** Interactive Arches. Montreal, Canada—2012, http://www.atomic3.ca

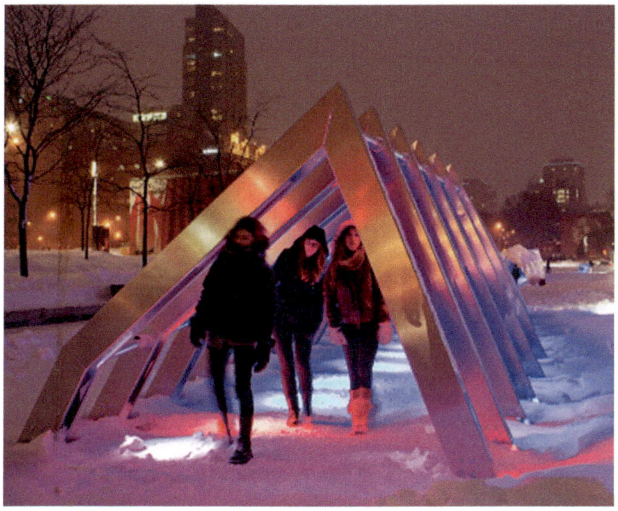

Fig. 1 (continued)

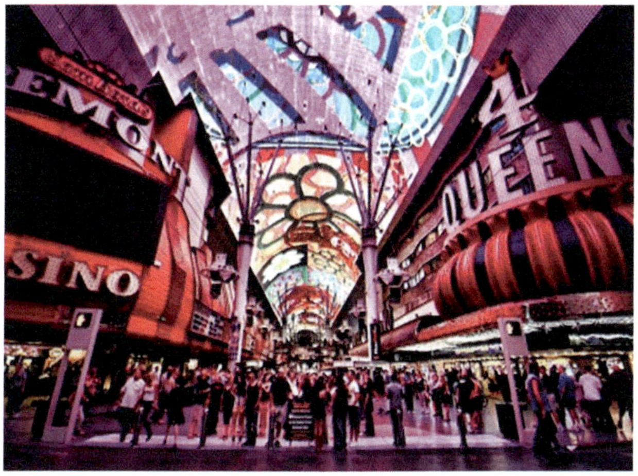

Fig. 1 (continued)

2.1 Integrating Media Interventions in Urban Spaces: Design Indicators

Formerly, studying urban spaces structure was limited to physical and social dimension only; however, due to new mediated digital applications supplemented to contemporary urban life, there is a need to add technological dimension in the structure of urban spaces. Previous researches suggest that studying the influential types of technology is essential to understand its industries, technicalities and regulations. Researchers who are concerned with mediation of urban spaces argued that urban space is becoming more "Hybrid' as it transformed to be deeply integrated with media (Aurigi & Cindio, 2008; Truijen, 2013). Consequently, in order to understand hybridization of urban spaces three aspects are proposed to be investigated, namely, Space, Users and Technology, and they form three main parameters of understanding interactive urban spaces (Foth & Sanders, 2008; Briones, 2006). Each one of the three mentioned aspects includes a set of variables that are derived from previous literature and theories about the design and experience of urban spaces.

A. Indicators for space-related aspects

Strong meaningful urban spaces are the ones that allow its users to create strong connections between the space, their lives and the larger context (Carr et al., 1992). User-interactive urban spaces offer an enriched experience according to the perception of its users.

Number of theories and literature were established to provide guidance for successful, user friendly and user interactive urban spaces. Targeting how to achieve and maintain successful urban space, a brief review for each theorist approach of urban spaces design, urban experience and achieving responsiveness and interaction in the urban environment is presented in Table 1, in addition to the relation of the presented philosophies and theories to the six dimensions of urban design (Carmona et al. 2007) morphological, perceptual, social, visual, functional and temporal.

Moreover, there are psychological aspects that are related to urban spaces' design. To understand the psychology of urban environment it is needed first to review psychological and physiological human needs as a base of functionalism concept Al bishawi and Ghadban (2011). It also helps in understanding how people react to their surrounding space and how some of reactions of users may relate to basic human needs and behaviours. On the other side, some other reactions will relate to psychological effects and interpretations (Shaftoe, 2008). Previous studies proposed that urban spaces ought to achieve five human needs, namely, *Comfort, Relaxation, Passive engagement with the environment, Active engagement with the environment and Discovery*, and those are the needs that are directly connected to space (Carr et al., 1992). Other studies proposed subsistence, *Protection, Affection, Understanding, Participation Leisure, Creation, Identity and Freedom*, as needs related to human systems and essential in the urban experience (Costanzaa et al., 2007). Designing user-interactive urban spaces and applying media interventions in contemporary urban environments requires satisfying needs of users efficiently through good control using a suitable framework.

B. Indicators for user related aspects

The term "users" refers to the audience who are experiencing interactive installation in urban space (Reid et al., 2005). Urban spaces are "the meeting ground" of benefit of many

Table 1 Different urban design pioneers concepts about urban spaces and its deduced indicators

References	Philosophy	Dimensions					
		Morphological	Perceptual	Social	Visual	Functional	Temporal
Lynch (1960)	Cognitive Mapping		•				
Lynch (1980)	City Form and Human Perception					•	•
Cullen (1961)	Townscape				•		
Kreir (1979)	Typology and morphology of Urban Open Spaces	•					
Curran (1983)	Urban experience	•					
Bentley et. al (1985)	Responsiveness of urban environments					•	•
Jacobs and Appleyard (1987)	Urban design manifesto					•	
Tibbalds (1992)	Urban design 10 commandments					•	
Moughtin (1992)	Urban design street and square		•			•	
Shaftoe (2008)	Creating convivial urban spaces		•	•		•	•

Relevance to interactive spaces design	Indicators
**	Form—function—activities
***	Exposure—vitality—accessibility—sense of control
*	Serial vision—place—content
*	Spatial typology of urban open spaces—morphology—dimensions
*	Enclosure—exposure—linear spaces—cluster spaces
***	legibility—permeability—richness—personalization—Visual appropriateness—variety—robustness
**	liveability—identity and control—imagination—joy—accessibility—communication—participation—function
*	Mixed use—walkability—human scale—legibility—sharing—complexity—joy—visual delightness—juxtaposing
**	Legibility—vitality—permeability
***	Enjoyable—human scale—accessibility—freedom—friendly—legibility—environmentally friendly

user groups. Previous studies categorized space users according to their age, gender, occupation and activities, as follows: *Age* (Children, teenagers, adults, elderly); *Gender* (male or female); *Occupation* (service, professional, unemployed/ homeless); *Activities* (recreation, eating, shopping, passive engagement, comfort) (Francis, 1989). This categorization was proposed to achieve better distinction of "publics for public space." Moreover, to recognize diversities of users and their various needs, motivations and public interactions in urban spaces, researchers added the income, social status and education to indicators they measure for the users. Thus, categorization turned out to be according to *purpose* of users in addition to *obligation and willingness* (Cooper & Francis, 1997).

Recommendations regarding users of media-scape design and applications and interactive experience design are proposed in a form of checklist that was developed to help in understanding these targeted users, and make them more comfortable, confident and in control (Reid et al., 2005), namely, *Who are the users?* Identifying the users' Age, gender and mode (Work or Leisure). *Lifestyle:* Things users are used to, who are their company (family, friends, work colleagues)? How will they be introduced to the experience (TV, radio, advertisement, oral instructions…etc. are they visitors or area residents. *Values/Likes*: identifying their likes and interests; do they like technology? Do they use smartphones, MP3 players or PDA's? Do they always need to feel in control? *Frustrations/ Stresses/Needs*: do they have enough time to enrichment activities? Or are they rushing this activity amidst many others? Do they need to be secure from any threats in the environment? *Accessibility and Provision for blind or deaf people, handicapped and people with special needs.*

Moreover, in terms of interaction design, users of urban spaces could be categorized as *Individual users*, *Group users*, *Crowd users*, *Local users* who can interact with interactive application either through specific devices or more natural interactions; *Remote users* who experience similar types of interactive behaviour through the user specific devices or device usage. Mediascape designers like to generate different forms of interaction and behaviour of users; it is their way to explore different mediums, therefore users-aspect is concerned with users' requirements for designing interactive media in urban environment. Its variables and attributes could arguably be categorized into two categories, namely, *users-location* (*local users, remote users, individual users, group users and crowd users*) and *users-type* and characteristics (Diniz et al., 2012).

C. Indicators for technology related aspects

Previous research work classified the technological application in urban space into four groups based on their functional application in urban environment, namely, *Display and expression:* represented in digital displays and LED and plasma posters besides digital facades*; Communications*: represented in wireless networks have powerful impacts on the movement and space requirements of individuals, families and workgroups; *Positioning*: represented in personal positioning systems with assistance of GPS technologies; *Documentation*: it has been greatly magnified with assistance of the GIS systems and digital location based annotation applications (Townsend, 2004). In terms of interaction design, interaction techniques application in media interventions could be classified into two types according to interactions of users, namely, *Explicit interaction*: Which is triggered by directed actions of users via some interaction techniques (voice or gesture recognition, device mediated) and *Implicit interaction*: Occurs when natural behaviour of users is used to drive the media installation (Diniz et al., 2012). In the process of interaction design it is more likely that the designer might take some indicators into considerations, namely, *Participation and engagement* by designing a non-complicated user interface, the *social protocol* which is coded and written by the designer and illustrated to the users to show them the process of interaction, the *Orientation of the installation (vertical or horizontal)* as it reflects on interaction of users with it and how they flow in space. (Briones, 2006; Rogers & Bringull, 2002).

D. Indicators selection

There are various indicators and criteria to apply media interventions in urban spaces. In order to simplify its design process and/or evaluation in urban setting, the study filtered the selected design indicators. It is proposed that the indicators and design criteria should be systematic, consistent, independent, measurable and comparable, as follows: Systemic: Essential characteristics should be reflected in the selected criteria, besides the whole performance of the system. This will help making comprehensive evaluation function of multi-indicators to obtain better results.

Consistency: indicators system should be consistent with the objectives of decision-making. Independency: alternatives and performance from different aspects should be reflected in the selected indicator. Besides it should not have "inclusion relationship at the same level criteria." Measurability: indicators should be expressed as qualitative values or measure as quantitative values. Comparability: selected indicators should be comparable; this helps in making decision-making results more rational (Ye Yc, 2006).

Based on previous literature the aspects of space comprised a set of 40 indicators coded from S01 to S40. The indicators are divided into two categories: namely, physical indicators and psychological indicators. Meanwhile aspects of users comprised a set of 18 indicators coded from U01 to U18, and were categorized according to the location and type of users. Technology aspect comprised a set of 13 indicators coded from T01 to T13 and categorized according to type of technology and interaction design.

The proposed variables are filtered according to the previous principles as shown in Table 2. Twenty-six indicators are selected as follows: 10 indicators for space aspects, 8 indicators for users-aspects, 8 indicators for the technology aspects.

2.2 Integrating Media Interventions in Urban Spaces: Design Considerations

To recombine the paradigm of mediating urban spaces with new evolving methods of urban design, previous studies presented design guidelines and considerations to help through the design process of applying media interventions in urban setting. Design considerations are based on physical factors and non-physical factors, namely, political, economic and socio-cultural factors.

The presented study categorized design considerations into four categories, namely, design considerations related to location of the installation, design considerations related to contextual impact of installation on its surroundings, design considerations related to control and management of the installation itself and its surroundings and design considerations related to the installation features, based on Fatah et al. (2009); Albrecht (2015). The Guidance on large screens of CABE (2009) and London sBorough of Bromley report (2009).

In terms of location of interactive installations in urban space the installation should

- Not be a dominant part in the space in a way that cancels its character and main features or cause a traffic disturbance or distractions.
- Be set in a flexible spot in the selected space in order to accommodate different forms of interactions and different vocabularies and proportions in the urban space.
- Be accepted by users to ensure a successful interaction process; it should be pleasant and attractive, moreover it should have a positive impact on its settings' vitality, safety and social engagement.
- Be easily accessed by all the users' types. Its location's relation to traffic, transportation and pedestrian networks should be highly considered. In addition to the exits in case of emergency and how users will escape is important to be put into consideration.

In terms of the contextual impact of the interactive installations in the urban space the installation should

- Be an appropriate part of the public realm; the space form and character should not be affected negatively, when the application is turned off or removed.
- Be added with careful consideration to make sure that the appearance of the spaces and any of its elements is not adversely affected.
- Contribute to the site permeability and legibility, in order to enhance it and turn it into an attraction point. This could be achieved by developing suitable closure and good access to the site.
- Enhance the qualities of its context.
- Be installed in a space that could accommodate the needs of disabled people in order to give them good opportunities for interaction and enjoying the interactive experience.

In terms of the control and management of the interactive installations in the urban space the installation should

- Have a strategy for the management and control of its content, noise malfunctions and handling the complaints of the users.
- Have a feasibility study of its initial and running costs which are considered with its installation in the site, maintenance, energy consumption.
- Have a strong surveillance and supervision strategy to control it and control the users' interaction together to guarantee their safety and to avoid any failure in the application's system.

In terms of the features of the interactive installations in the urban space the installation should

- Have an appropriate scale and height relevant to expected number of users.
- Have a good quality content which should not dominate the character of space and be appropriate to the background and culture of users.
- Have qualified technical criteria to validate its technical systems, its appropriateness, efficiency in the space and sustainability.

Table 2 Different urban design pioneers concepts about urban spaces and its deduced variables

Aspects		Indicators		Selection criteria					Selected
				Systematic	Consistent	Independent	Measurable	Comparable	
(a) Filtering the space indicators based on Ye Yc (2006)									
Space	Physical	Form	S01	•	•	•	•	•	**
		Exposure	S02				•	•	
		Enclosure	S03				•	•	
		Dimensions	S04				•		
		Proportions	S05	•	•	•	•	•	**
		Variety	S06	•	•	•	•	•	**
		Accessibility	S07	•	•	•	•	•	**
		Function	S08				•	•	
		Responsiveness	S09	•				•	
		Legibility	S10				•	•	
		Vitality	S11	•	•	•	•	•	**
		Robustness	S12				•	•	
		Permeability	S13	•	•	•	•	•	**
		Visual appropriateness	S14				•	•	
		Fitness	S15					•	
		Serial vision	S16		•			•	
		Juxtaposing	S17				•		
		Multiple use	S18				•	•	
		Viscosity	S19					•	
		Focal point	S20					•	
		Advantage	S21					•	
		Control	S22	•	•	•	•	•	**
		Richness	S23	•	•	•	•	•	**
		Identity	S24	•	•			•	
		Imagination	S25	•	•				
		Joy	S26		•				
		Personalization	S27	•	•	•	•	•	**
		Authenticity	S28					•	
		Walkability	S29				•	•	
		Environmentally friendly	S30		•	•		•	
	Psychological	Belonging	S31	•	•	•	•	•	**
		Participation	S32	•	•		•	•	
		Communication	S33	•	•			•	
		Pleasure	S34				•	•	
		Territoriality	S35			•	•	•	
		Social meaning	S36	•	•			•	
		Freedom	S37				•		
		Enjoyable	S38	•		•	•	•	
		Passive engagement	S39						
		Sense	S40			•		•	

(continued)

Table 2 (continued)

Aspects		Indicators		Selection criteria					Selected
				Systematic	Consistent	Independent	Measurable	Comparable	
(b) Filtering the users-indicators based on Ye Yc (2006)									
Users	Location	Local	U01	•	•	•	•	•	**
		Remote	U02	•	•	•	•	•	**
		Individuals	U03	•	•	•	•	•	**
		Group	U04	•	•	•	•	•	**
		Crowd	U05	•	•		•	•	
	Type	Age	U06	•	•	•	•	•	**
		Gender	U07	•	•	•	•	•	**
		Occupation/economic	U08	•	•	•	•	•	**
		Culture	U09		•			•	
		Special needs/handicapped	U10	•	•	•	•	•	**
		Security/stress	U11		•		•	•	
		Everyday users	U12				•		
		Visitors	U13	•	•		•		
		Passer by	U14	•	•		•		
		Visitors to events	U15	•	•		•		
		Income	U16	•	•		•		
		Social status	U18	•	•		•		
(c) Filtering the technology indicators based on Ye Yc (2006)									
Technology	Purpose	Display and expression	T01	•	•	•	•	•	**
		Communication	T02	•	•	•	•	•	**
		Positioning	T03	•	•	•	•	•	**
		Documentation	T04	•	•	•	•	•	**
	Interaction Design	Sound design	T05		•			•	
		Image design	T06		•			•	
		Explicit	T07	•	•	•	•	•	**
		Implicit	T08	•	•	•	•	•	**
		Social protocol	T09				•	•	
		Networks	T10		•			•	
		Location technologies	T11	•			•		
		Participation and engagement	T12	•	•	•	•	•	**
		Orientation (vertical/horizontal based)	T13	•	•	•	•	•	**

- Have a good strategic plan for its forms and times of construction according to the selected space and its programming and technical issues.

3 Relationship Between Design Indicators and Design Consideration

It is possible in the light of the reviewed literature to outline the likely relationship between the design considerations and the selected design indicators related to the application of media interventions in the urban spaces. The proposed interrelationship may help in dealing with the complexities of the design of interactive urban spaces using this kind of new interventions.

The matrix presented in Table 3 combines the set of 16 design considerations (coded from C01 to C16) with the closely relevant indicators.

The proposed interrelationship may be outlined as follows: The installation dominance on the space (C01) depends on the form, scale and proportions of urban space altogether with the variety of space elements; it is also related to interactions of users in addition to technicality and framework, either it is explicit or implicit in addition to orientation of the installation in the space. The flexibility for the location of installation scale (C02) is found to be related to space form, proportion and installation's orientation. Users' acceptance (C03) for the installation is related to the accessibility and permeability of space and the installation itself, in addition to the quality of the content and safety and security control besides interaction type (explicit or implicit), also participation and engagement of users in the process is essential for their acceptance to the installation. Installation's accessibility (C04) is related to the form of space and permeability, and they are key controllers in this process; it also includes accessibility of disabled people and their participation in the interaction process. Moreover, it is related to the interaction of users (individual or group) in addition to the role played by the installation whether it is for display and expression, communication, positioning or documentation.

The impact of the installation on public realm (C05) is related to the form and proportions of space in addition to variety of elements and activities in space. In addition to its vitality and richness. Interference of the installation with space elements and features (C06) is likely to be related to variety of space elements and its vitality, as well as its relation to the interaction type (explicit or implicit) and the orientation of the installation itself. The contribution of media installation to permeability and legibility of the space (C07) arguably depends on the form, proportions, accessibility and permeability of space, in addition to interaction location of users (local or remote). Consequently, interaction of users, whether it is individual or group is also an important indicator. Design quality of the media installation and its space enhancement (C08) is likely depending on the form, proportions and richness of space. As a social aspect Media installations and special needs design (C09) is related to accessibility of space in addition to interaction of users whether it is local or remote, besides individual or group interaction. It also depends on special needs requirements and technical criteria, in addition to the role played by the installation, type of interaction and the orientation of the installation.

Management and control of media installation (C10) is likely to be related to space control and richness, in addition to interaction of users (individuals or group) besides age, gender and special needs requirements. Costs of media installation (C11) depends to both space control and richness. Location surveillance and supervision (C12) is about securing the intervention itself and its users so it is more likely to depend on the form, proportions and control of space. Also, interaction type of users, regarding their safety, whether it is implicit or explicit, in addition to participation of users and installations orientation.

Form and content of media installation (C13) is likely to depend on form, proportions and accessibility of space, in addition to personalization of users to his/her interaction space and sense of belonging of space; interaction of users (individual or group) are important indicators to this design consideration, in addition to gender of users and special needs requirements. Media installation's form and content (C14) is related to space's form and proportions, in addition to control and richness of space. It is also related to interactions of users whether they are local or remote, besides their age and the special needs requirements, in addition to type of interaction (explicit or implicit), participation and engagement of users and orientation of installation. Technical criteria and specification considerations (C15) are related to the role played by the installation (display and expression, communication, positioning or documentation). Form and process of construction of the media installation (C16) consideration is related to the form of space and participation and engagement of users.

4 Methods and Procedures: Verifying the Relationship the Indicators and Considerations

This research study aims to answer explanatory questions regarding applying urban media interventions in contemporary urban spaces. The empirical study is conducted to

Table 3 The relationship between the design considerations and the design indicators

Aspects	Indicators	Design considerations															
		Location				Contextual impact				Control and management				Installation features			
		Dominance of the installation on the space	Location of the installation scale flexibility	Users acceptance for the installation	Accessibility of the installation	Impact of the installation on public realm	Elements and features of the urban space	Contribution of the installation to the urban space's permeability	Design quality and enhancement of the urban space	Special needs design requirements of the installation	Management and Control of the installation	Costs of the installation	Surveillance and Supervision of the installation	Scale and Height of the installation in relation to users' scale	Form and Content of the installation	Technical criteria of the installation	Forms, process and time of construction of the installation
		C01	C02	C03	C04	C05	C06	C07	C08	C09	C10	C11	C12	C13	C14	C15	C16
Space	S01	•	•		•	•		•	•			•	•	•	•		•
	S05	•	•			•		•	•				•	•	•		
	S06	•				•	•										
	S07			•	•			•		•				•			
	S11					•	•										
	S13	•		•	•			•									
	S22			•							•		•		•		
	S23	•				•	•		•			•			•		
	S27													•	•		
	S31													•			
Users	U01				•			•		•	•			•	•		
	U02				•			•		•	•			•	•		
	U03	•			•			•		•	•	•		•			
	U04	•			•			•		•	•	•		•			
	U06			•						•	•	•		•	•		
	U07			•							•			•			
	U08			•													
	U10				•					•		•		•	•		
Technology	T01				•						•	•				•	
	T02				•						•	•				•	
	T03				•						•	•				•	
	T04				•						•	•				•	
	T07	•		•			•			•		•	•		•	•	
	T08	•		•			•			•		•	•		•	•	
	T12				•			•		•	•		•		•	•	
	T13	•	•	•	•		•	•		•	•	•	•		•	•	•

understand the relation between indicators and design considerations regarding the mentioned issue, through a statistical proof for the influence on each other. Standardized questionnaire is considered an effective method for reliable data gathering as it uses direct and uniform questions in addition to regular data presentation (Haralambos and Holborn 2000). It is characterized by its ability to track more than one study indicator. It also can predict the relations between the indicators and each others by identifying the mean rating given to the relationship between design indicators and considerations. Thus, the research study conducted a questionnaire targeting a sample of experts to read and understand their reactions and evaluations regarding urban media interventions.

4.1 Sampling and Participants

Since the questionnaire is directed to architects and urban design experts, stratified sample was rigorously chosen as sampling method. It shows a proportional representation of the population in terms of stratifying criteria. Whereas users sampling method was selected to be simple random sample. This make sure that each unit of the population has the identical chance to be involved within the sample (Bryman, 2012). The sample included (30) architects, urban designers, planners and landscape architects/designers, with seven years (or more) professional experience. The selected experts sample size was thought to be adequate at this phase of investigation, to give reliable indicators and support preliminary conclusions regarding the research objectives. The majority of the sample were 10–15 years of experience (18 = 60%), and around 27% were more than 15 years of experience (27%), and around 13% were from 7 to 10 years of experience (13%).

4.2 Stimuli

The questionnaire was divided into three parts, namely, introduction to the research study, general information about the participant and indicators ratings. Each questionnaire had an introductory statement outlining the purpose and scope of the study, in addition to a 50 s video showing how media interventions work in its different patterns, and then the introductory part ends by a brief definition for the indicators and the design considerations to help participants to understand the questions and answer it. Then participants are asked to give a brief about their background: Age, Gender, Years of experience and Nationality. Finally using Likert scale (from 1 to 5), participants are asked to rate the indicators given in each question in relation to the design consideration relevant to it. The questionnaire was designed using Google forms, and was distributed via e-mail to the participants.

5 Results and Discussion

The questionnaire results were analyzed using the SPSS software to calculate the indicators relevance score to each one of the design considerations. The relevance score is categorized into five categories: Very Strong (from 5.0 to 4.0), Strong (from 4.0 to 3.0), Moderate (from 3.0 to 2.0), Weak (from 2.0 to 1.0) and Very Weak from (1.0 to 0.0).

Participants answers came out to show that the strongest relevance score is between special needs design consideration (C09) and special needs user interaction technicalities and procedures indicator (U10) that scored mean rating = 4.9, in addition to relevance score between acceptance of users design consideration (C03) with the age of users (U06) came in the second place that scored a mean rating = 4.7. The relevance score between the dominance of the installation on space design consideration (C01) and space form (S01) and space proportions (S05), the impact of the installation on the public realm (C05) and the spaces vitality (S11), the contribution of the installation to permeability of urban space consideration (C07) and the space form (S01), in addition to the form and content of the installation design consideration (C14) and participation and engagement in interaction design (T12) all scored a mean rating = 4.5. The relevance score between users acceptance of the installation (C03) and explicit interaction (T07), the accessibility of the installation (C04) and local interaction of users (U01), the special needs design consideration (C09) and accessibility of spaces, surveillance and supervision of the installation (C12) with space form (S01) scored a mean rating = 4.4. The contribution of the installation to permeability of urban space consideration (C07) and its accessibility (S07), design quality and enhancement of urban space (C08) and proportions of space (S05) and technical criteria of the installation (C15) and display and expression technology (T01) scored a mean rating = 4.3. The relevance score between the users-acceptance design consideration (C03) and the space accessibility (S07), the impact of the installation on the public realm (C05) and the space form (S01), the design quality and enhancement of the urban space (C08) and the space form (S01) as well, surveillance and supervision of the installation (C12) with the space control (S22) scored a mean rating = 4.2. The users-acceptance design consideration (C03) with the implicit interaction technology (T08), the special needs design consideration (C09) with the participation and engagement technicalities (T12), the scale and height of the

Table 4 Relevance score (mean ratings) between the design considerations and the design indicators

ASPECTS	INDICATORS	DESIGN CONSIDERATIONS															
		LOCATION				CONTEXTUAL IMPACT				CONTROL & MANAGEMENT				INSTALLATION FEATURES			
		C01	C02	C03	C04	C05	C06	C07	C08	C09	C10	C11	C12	C13	C14	C15	C16
SPACE	S01	4.5	3.6		3.8	4.2		4.5	4.2			3.7	4.4	4.1	4.1		3.8
	S05	4.5	4.0			3.8		4.0	4.3				3.8	3.5	3.9		
	S06	3.7				3.7	4.2										
	S07			4.2	2.0			4.3		4.4				3.2			
	S11					4.5	4.0										
	S13	3.5		3.5	3.9			4.2									
	S22			2.9							4.0		4.2		3.2		
	S23	3.3				3.5	3.7		3.8			3.0			3.8		
	S27													3.5	3.9		
	S31													3.8			
USERS	U01				4.4			3.5		3.2	3.4			3.6	3.8		
	U02				3.5			3.5		3.8	3.2			3.5	3.5		
	U03	3.0			3.7			2.9		3.3	3.5	2.8		3.2			
	U04	3.0			3.6			3.8		3.1	3.5	3.0		3.5			
	U06			4.7						3.8	3.8	3.0		3.5	3.2		
	U07			3.1							3.5			3.5			
	U08			2.7													
	U10				3.7					4.9		3.6		4.0	4.1		
TECHNOLOGY	T01				3.2						3.2	3.7				4.3	
	T02				3.1						3.1	3.7				4.0	
	T03				3.4						2.3	3.5				1.9	
	T04				3.5						3.1	3.6				4.0	
	T07	3.5		4.4			3.1			3.5		3.6	3.3		3.3	3.7	
	T08	3.5		4.1			3.3			3.1		3.4	3.5		3.1	4.0	
	T12				3.6			3.9		4.1	3.4		3.4		4.5	3.6	
	T13	3.9	3.7	3.0	3.1		3.0	3.8		4.0	2.8	2.9	2.5		3.8	3.2	2.4

V. Strong Strong Moderate Weak V. Weak

V. strong, strong, moderate, weak, V. weak

installation in relation to scale of users (C13) with the space form (S01), the content of the installation design consideration (C14) with both space form (S01), and special needs (S10) all scored a relevance of mean rating = 4.1. The rest of the indicators showed a strong and moderate mean ratings to their related design considerations, as illustrated in Table 4. The results showed only one weak relation between the technical criteria of the installation (C15) with the positioning technologies (T03).

The questionnaires result and mean ratings pointed out a very strong consideration–indicator relationship which is the special needs design requirements in the interaction process of the installation itself and the urban space preparations for this process to this type of users. Experts emphasized the importance of the space form as an indicator for the design process in addition to the accessibility, permeability and vitality of space. Moreover, the importance of taking the age of users, form and content of the installation in addition to theinteraction design into consideration supported the notion that "integrating interactive media installations in urban-spaces enriches the experience of users and enhances the quality of space" (Tarek et. al., 2019).

6 Conclusions

The present research work identified the likely impact of new media technologies on the design of urban spaces. It investigated promising contributions of media in generating interactive urban experience. It critically reviewed the selected literature on new media and different types of technologies that are integrated with its implementation in urban spaces. It also investigated new media contributions to quality of urban environment. The research identified and deployed key indicators for designing interactive urban spaces through reviewing previous literature concerning three major aspects that are proposed to be essential in the process of mediating the experience of urban spaces, namely, space, users and technology. It also recalled design guidelines and considerations for applying media interventions in urban spaces.

The research synthesized key indicators with the reviewed design considerations and identified the likely relationship between them for a better deployment of these considerations in interactive urban spaces design process. The proposed relationship between the design considerations and indicators was verified through a questionnaire targeting a stratified sample of experts, and a statistical analysis for their responses. Mean ratings to the relevance score between the suggested relationships between 26 indicators and 16 design considerations highlighted and showed up the relative weight of importance for some indicators than the others, and it identified the strongest indicators and the weakest ones in terms of their relevance to design considerations. This is proposed to contribute in the design process of interactive spaces. The various aspects of the presented research work supported the notion that "Applying media interventions in the urban spaces; enhances the quality of space by providing a mediated interactive experience." Furthermore, it pointed out key indicators for the design process of interactive spaces using media interventions. This may serve in the decision-making process of designing these new spaces, incorporating guidelines of experts to integrate new media interventions in urban settings.

The research findings and conclusions point out directives for future research, namely, to further develop a proposed decision-making tool that incorporate the considerations–indicators relationship presented in this research work, and support it with more factors to read and measure the generated interactive experience and set an interactivity index that evaluate level of interactivity in the urban setting.

Acknowledgements The author would like to acknowledge Professor Dr. Sayed Ettouney, Professor Dr. Nasamat Abdel Kader and Professor Dr. Ahmed Amin for their guidance and consistent encouragement, which without it this research wouldn't have been completed.

References

Albrecht, J. (2015). Digital media façades for lively public spaces: Promoting dialogue, participation and social innovation in urban environments. In: *Proceedings of 8th Making Cities Liveable Conference, Melbourne*.

Al bishawi, M., & Ghadban, S., (2011). *A methodological approach for reading urban open space*. Archnet-IJAR. *5*. 73–85. https://doi.org/10.26687/archnet-ijar.v5i1.219.

Aurigi, A., & De Cindio, F. (2008). *Augmented urban spaces, articulating the physical and electronic city*. Ashgate Publishing.

Barthes, R. (1977). *Elements of semiology*. Macmillan.

Bentley, I., Alcock, A., Murrain, P., Mcglynn, S., & Smith, G. (1985). *Responsive environments*. Elsevier Ltd.

Bolter, J., MacIntyre, B., Gandy, M., & Schweitzer, P. (2006). New media and the permanent crisis of aura. *Convergence*.

Van den Boomen, M. V. T.(2014). *Transcoding the digital: How metaphors matter in new media*. Institute of Network Cultures.

Brightenti, A. M. (2011). *New media and urban motilities: A terrtoriologic point of view*. Urban Studies.

Briones, C. (2006). *LED's urban carpet: A portable interactive installation for urban environments*. University College London.

Bryman, A. (2012). *Social research methods*. Oxford University Press Inc.

Carmona, M., & Tiesdell, S. (2007). *urban design reader*. Architectural Press, Elsevier.

Carr, S., Francis, M., Rivilin, G., & L. and Stone M A. (1992). *Public spaces*. Cambridge University Press.

Cooper, C., & Francis, C. (1997). *People places: Design guidelines for urban open spaces*. Wiley Publishers.

Cullen, G. (1961). *Townscape*. Reinhold Pub. Corp.

Curran, R. (1983). *Architecture and the urban experience*. Van Nostrand Reinhold Company.

Development Control Committee. (2009). *Large digital screens in public spaces*. Report. London Borough of Bromley.

Diniz, N., Duarte, C. A., & Guimarães, N. M. (2012). Mapping interaction onto media façades. In: *Proceedings of the 2012 International Symposium on Pervasive Displays*.

Dix, A. (2009). *Human-computer Interaction* (3rd ed.). Springer.

Fatah gen Schieck, A., Boddington, G., & Fink, P. (2009). *Framework for the implementation of urban big screens in the public space*. Research report. Bartlett School of Graduate Studies, University College London, London, UK.

Foth, M., & Sanders, S. P. (2008). Impacts of social computing on the architecture of urban space. In A. Aurigi & D. F. Cindio (Eds.), *Augmented urban spaces: Articulating the physical electronic city* (pp. 73–91). Ashgate publishing.

Francis, M. (1989). *Control as a dimension of public space quality*. In: I. Altman & E. Zube (Edis.), *Public places and spaces*. New York: Plenum Press.

Haralambos, M., & Holborn, M. (2000). *Sociology: Themes and perspectives*. Collins.

English Heritage and Commission for Architecture and the Built Environment. (2009). *large digital screens in public spaces*. Joint guidance from English Heritage and CABE.

Jacobs, A., & Appleyard, D. (1987). Toward an urban design. Manifesto. *Journal of the American Planning Association*.

Johnson, S. (2003). *Wireless communications and the future of cities*. Report of Taub Urban Research Center.

Kreir, R. (1979). *Urban space*. Academy Editions.

Lynch, K. (1960). *The image of the city*. The Technology Press & Harvard University Press.

Lynch, K. (1980). *A theory of good city form*. MIT Press.

Manovich, L. (2001). *The language of new media* (1st ed.). MIT Press.

McCall, R., O'Neill, S., Carroll, F., & Smyth, M. (2005). Responsive environments: Place and presence. *Psychology Journal*.

Mitchell, W.(2003). *ME++: The cyborg self and the networked city*. MIT Press.

Moughtin, C. (1992). *Urban design: Street and square*. Oxford: Butterworth Architecture.

Nam, T., & Pardo, T. A. (2011). Conceptualizing smart city with dimensions of technology, people and institutions. In: *Proceedings of the 12th annual International Digital Government Research Conference: Digital Government Innovation In Challenging Times*. ACM.

Ohta, Y., & Tamura, H. (1999). *Mixed reality: Mergiing real and virtual worlds* (p. 1999). Springer.

Reid, J., Cater, K., Fleuriot, C., & Hull, R. (2005). Experience design guidelines for creating situated mediascapes. *Mobile and Media Systems Laboratory*.

Rogers, Y., & Bringull, H. (2002). *Subtle ice-breaking: Encouraging socializing and interaction around a large public display*. Workshop on Public Community and Situated Displays.

Shaftoe, H. (2008). *Convivial urban spaces: Creating effective public places*. Earthscan in Association with the International Institute for Environment and Development.

Sorapure, M. (2004). *Five principles of new media: Or playing Lev Manovich*. Kairos.

Tarek, S., Amin, A., Abdel-Kader, N., & Ettouney, S. (2019). Urban media interventions—Designing urban spaces for interactive experience. *Journal of Engineering and Applied Sciences, 66*(5).

Taylor, S., & Todd P. A. (1995). Understanding information technology usage: A test of competing models. *Information Systems Research, 6*.

Thompson, C. W. (2002). *Urban open space in the 21st century*. Landscape and urban planning.

Tibbalds, F. (1992). *Making people-friendly towns*. UK: Longman Group UK.

Townsend, A. (2004). *Digitally mediated urban space: New lessons for design*. Praxis.

Trujin, K. (2013). *Urban dialogues: Media interventions as assembling devices*. MA Thesis. New Media and Digital Cultures University of Amsterdam.

Webster, M. (2004). *Merriam-Webster's collegiate dictionary*.

Ye, Y. K., & L. (2006). *System synthetical evaluation technology and its application*. Metallurgical Industry Press.

Modern, Environmentally Friendly Electric Transport as a Step Toward Improving City Welfare

Mehad Sayed Emara

Abstract

The role of public transportation is important as one of the foremost modes of utilities in cities, in achieving user comfort and welfare. Transport would help to transfer people from one place to another, and with the high population density, the demand for transportation, whether cars or buses, has increased. Thus, it has become a major source of pollution in the world; in general, and particularly, in the Arab Republic of Egypt. Hence, the researchers, from developing countries, have discussed the problem extensively within the academic literature, contemplating innovative and smart solutions to address such problems by working to reduce greenhouse gas emissions. This paper aims to provide practice for recording, illustrating, and discussing the tardiest technologies that are considered a change or revolution in public transportation sector systems. The role of modern electric transportation, which is environmentally friendly low-carbon means, is adopted by many developed countries, including the Arab Republic of Egypt as part of social sustainability. As a consequence, the researcher emphasizes the importance of governments, authorities, designers, and planners' role in building up a framework for the design of environmentally friendly charging centers within the Arab Republic of Egypt to operate electric public transit without any further cost. This could be accomplished through the usage of solar energy panels, PV, by designing modern architectural methods to connect PV panels with charging centers and work them into sustainable centers, where many developed countries and some Arab countries also have now proceeded to sweep up this concept, trying to use it in order to improve city welfare. ©2016 The Authors. Published by Elsevier Ltd. Peer-review under responsibility of IEREK, International experts for Research Enrichment and Knowledge Exchange.

Keywords

Public transport • Electrical transport • PV panels • Environmental pollution • Friendly environmental charging centers

1 Introduction

Ministry of Transport Greater Cairo Urban Transport Master Plan (2003) states that energy is considered as one of the human fundamental necessities, as its degree sources diversity and determines the way of social life in addition to its development level. There is no development without sources of energy, so there is the inevitability of finding renewable energy sources such as solar cells and photovoltaic panels as an alternative solution to the environmental pollution problem within the Arab Republic of Egypt because public transportation is one of the major sources of pollution. Founded on that government adopted the concept of using electric transport, which has lately become a recent trend for environmental sustainability in Egypt.

The government is charged with additional duties to design sustainable charging centers and establish a framework that invests solar energy conducted from photovoltaic panels that can operate public transportation without using oil or natural gasoline.

The transport sector in specific is a major source of diverse kinds of polluter, including "carcinogens and smog-forming pollutants; fine particulate matter (PM2.5), nitrogen oxides (NOx) and sulfur oxides, hydrocarbons, soot, etc. Furthermore, from the economic viewpoint, the phasing out of fuel subsidies is part of Egypt's plan for economic reform, implying steady increase in fuel prices and thus a pressing need to explore fuel-saving solutions to offer

M. S. Emara (✉)
Modern Academy for Engineering and Technology, Elhadaba El-Wosta—Elmokattam, Cairo, Egypt

F. Rosso et al. (eds.), *Advances in Architecture, Engineering and Technology*, Advances in Science, Technology & Innovation, https://doi.org/10.1007/978-3-030-86913-7_4

citizens. Even with the foreseen phasing out (or rationalization) of subsidies, the government will still need to reduce fuel consumption in passenger transport in order to save fuel that can be used more profitably elsewhere, such as in export, in industry, or in petroleum-based products." (Fig. 1) (Centre for Environment and Development in the Arab Region and Europe, 2016).

1.1 Research Questions

1. Is it possible to achieve a framework that contributes to overcoming the problem of environmental pollution caused by public transportation in Egypt?
2. Is it possible to bring down the cost and burden on the state through sustainable means of transportation to improve the standard of living in the future?

1.2 Hypothesis

A carbonless future can be achieved by integrating public transportation into the public transit system, which is one of the most significant sources of pollution that threatens the future of the environment.

1.3 Research Objectives

According to the statistics provided by the competent authorities, it became clear that there is a continuous increase in the population rates within the Republic, where the population problem expresses the imbalance between the number of population, resources and services. The level of the world's countries in terms of population, which has led to an increase in the demand for transport vehicles, and consequently, an

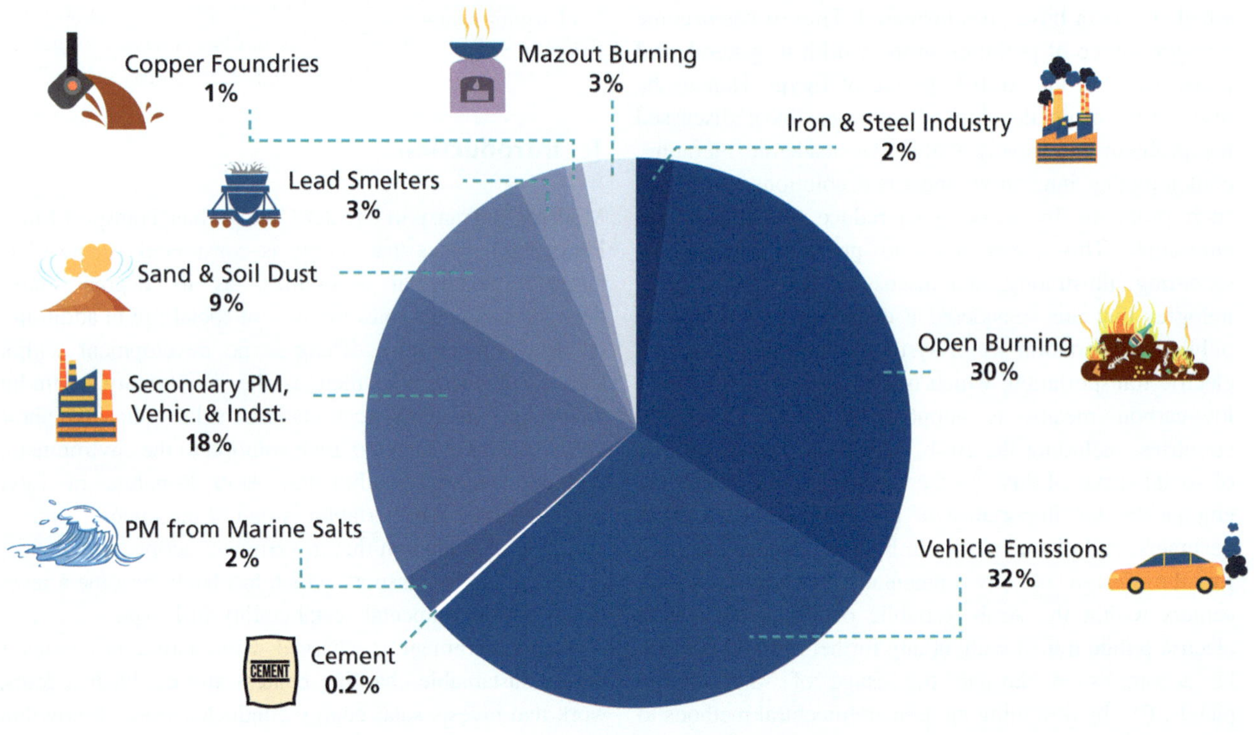

Fig. 1 Source-attribution of PM10 air pollution in Cairo dominated by vehicle emissions. *Source* USAID, edited

increase in greenhouse gas emissions and environmental pollution rates. Resulting in influencing various aspects of environmental, economic and social life.

The objectives are to design a framework to integrate modern electrical transport systems, how to employ solar energy to produce the electricity required to bring down energy consumption, and then identify a plan which addresses pollution problems caused by using traditional transportation, thus addressing negative effects and damages caused consequently.

In addition, address innovative framework to design environmentally friendly charging stations to operate modern electrical transport with no additional cost.

1.4 Research Methodology

To attain the objective and hypothesis of the study to establish its validity, the study adopted the theoretical and analytical approaches as follows.

Theoretical approach: This is the phase of documentation and monitoring of causes/factors of increasing the environmental pollution problems caused by means of public transportations. Add-on, the negative effects of greenhouse gas emissions, besides, by documenting the modern, environmentally friendly electrical transport role and how to utilize the solar energy to operate vehicles through generating electricity by using photovoltaic panels.

Analytical approach: From the previous phases, this approach is a concluded framework with how to integrate the role of transportation environmentally friendly electricity with solar energy through designing sustainable (charging stations) eco-friendly to operate the means of electrical transport with no additional cost. It monitors the causes of environmental pollution problems caused by public transportation.

Haug et al. (2001) mentioned that cars, trucks, and buses within the Republic of Egypt are considered as the main sources of air pollutants, particularly in suburbs with heavy traffic. As the numbers of vehicles are continuously increasing, these vehicles are playing a prominent role in generating major air pollutants with many health and environmental risks resulting from several types of exhausts as gasoline or diesel. Due to these harmful gases coming out of burning fuel, numerous studies conducted around the world proofing that major cities suffer from a high percentage of air pollution, transportation is considered the main source (by nearly 60%).

WHO (2001) explained that many governments in both developed and emerging nations have taken measures and enacted laws to decrease such pollutants. Thus, the level of pollutants (especially lead) has decreased from the air in numerous European, American, and some Arab cities, including the Arab Republic of Egypt (Figs. 2 and 3).

"EGP 110 bn had been allocated to subsidize petroleum products in 2017, highlighting the substantial fiscal burden of fuel consumption. 3 Subsidy rationalization is however underway as part of Egypt's reform policies, and is felt in the steady increases in fuel prices. Diesel fuel in specific is a higher burden than gasoline as it remains largely imported" (Fig. 4) (Environmental Pollution Centers, 2017).

1.5 Disadvantages of Transport Emissions

The transport sector in specific is a key source (no. 1) of different cases of pollution, including carbon monoxide, lead, nitrogen oxides, atmospheric and volatile organic compounds, as mentioned previously by the Environmental Protection Agency (EPA). That is why the EPA has determined transportation as the most important cause of air pollution in the world now and is expected to worsen in the coming years (Fig. 5) (C40 Cities, 2017).

In the meantime, the Ministry of Planning in Egypt referred that "air quality has been estimated by the World Bank to cost Egypt EGP 3.3–9.6 bn (up to 3.2% of GDP) in environmental and health damage. 4 Consequent impact on tourism is also substantial. Movement towards increased use of public transport is underway, while informal transport continues to fill a large gap in mobility needs. In other market segments, transport network companies (ride hailing services) are also expanding, with the leading player already exceeding [150,000] registered cars alone" (2016).

1.6 Factors and Reasons for the Increasing Problem of Environmental Pollution Resulting from Transport

In fact, improvements in engines and transport exhaust technologies that are negatively confronted by the rapid increase in vehicle numbers. Furthermore, old vehicles are heavily traded in poor and emerging countries, as in the case of Egypt.

According to the latest data maintained by the Statistics Authority, Egypt is acquiring an annual gain in car numbers on road, by a norm of 700 thousand vehicles each year, and the number of vehicles in traffics during 2014 was counted by more than 7.5 million vehicles. It raised to 8.6 million vehicles in 2015, with a continuous rise to reach 9.3 million vehicles in Great Cairo roads by the end of 2016 (Fig. 6) (African Development Bank, 2017).

Accordingly, Green Climate Fund (2018) mentioned that the effects of emissions from the transport sector vary in their chemical composition and the extent to which they are concentrated in the air on human health and the environment. The most important factors are:

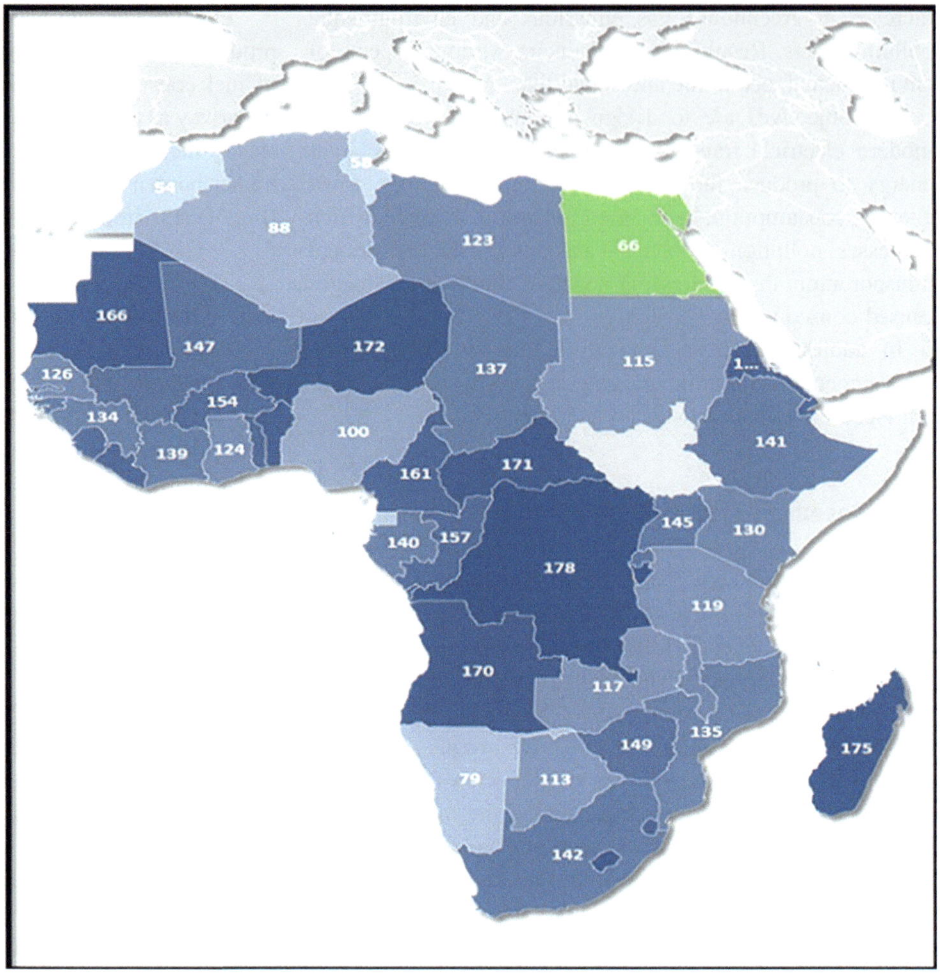

Fig. 2 Egypt ranks 66th out of a total of 180 countries. *Source* http://www.nrea.gov.eg/About/Strategy

- Transportation plays a major part in increasing global warming and air pollution, regardless of its technical advancement.
- Increase in heartbeat rates, which can lead to death.
- Poor eyesight and deafness.
- Adverse impact on both concentration and the nervous system.
- Increasing the lead concentration from 10 μg/deciliter to 30 μg/deciliters leads to minimizing the children's IQ rates by close to 4.4–5.3 points.

The U.S. Environmental Protection Agency (EPA) has stated six essential air-polluting elements on the planet: carbon monoxide, lead, nitrogen oxides, volatile organic compounds, sulfur dioxide, and solid residues. Nevertheless, pollution out of transportation is the worst, and recently classified as a major cause of air pollutants according to EPA information (Wong et al., 2002).

Tie et al. (2014) confirm that one of the most significant challenges Egypt is facing is the low quality of diesel fuel, which contains sulfur rates more than 100 times from international standards, where it exceeds 5000 parts per million, and the consumption of diesel fuel is still increasing rapidly with economic growth and increasing the numbers of public bus fleets, and microbes, so fuel consumption on public buses in Greater Cairo has doubled in the last 10 years (Fig. 7).

2 Environmentally Friendly Electrical Transport

Means of transport that works using electric power and replacing the internal fuel combustion engine with an electric one instead had been considered the easiest ways to switch from fuel to electricity while maintaining the other components of vehicles, as engines are supplied with the necessary energy sources from the storage batteries (Jorgensen 2008).

Hence, the electric car designs are based on electric motors, electrical control system, and a strong battery that can be recharged while maintaining its light weight. This

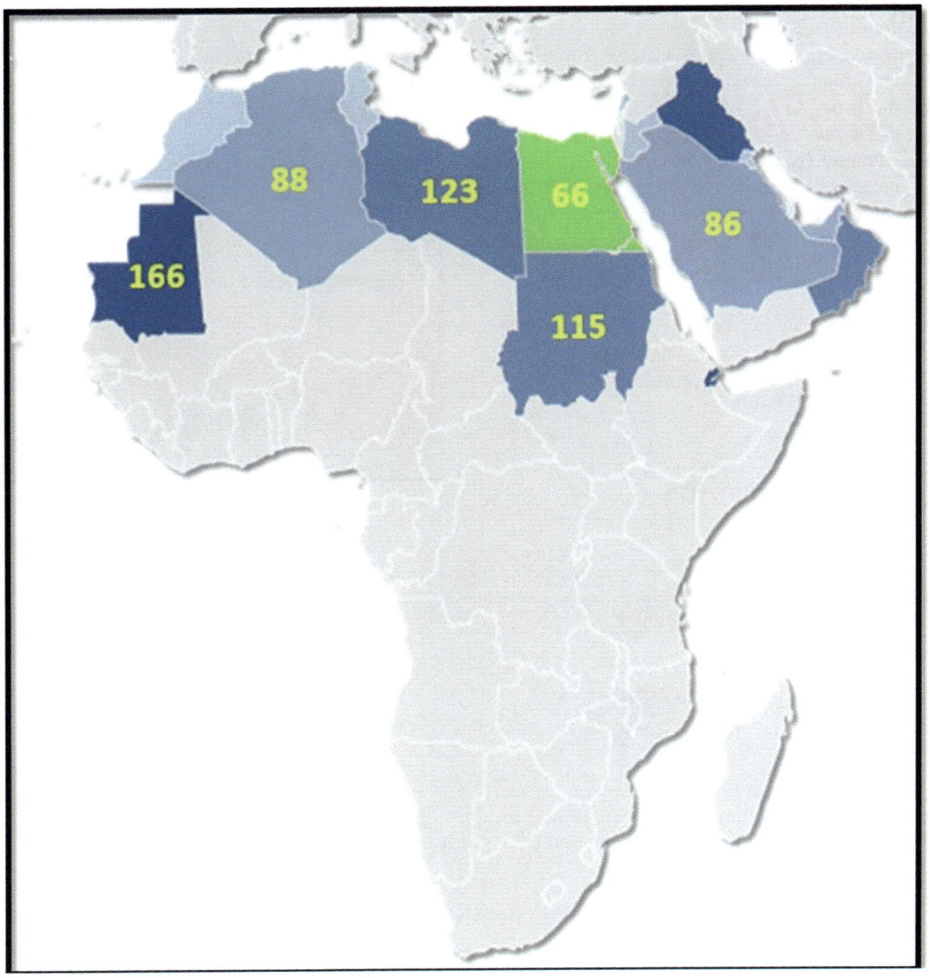

Fig. 3 Egypt ranks 4th in Africa, out of a total of 51 countries. *Source* http://www.nrea.gov.eg/About/Strategy

type of electrical transport is more suitable than other combustion engines in terms of environmental protection where it does not make any harmful waste to the environment.

2.1 Electrical Transportation System Features

The extent of these experimental methods is even below 200 km but it is becoming serious in large transport factories with the encouragement and financial funding of governments all around the world to develop lithium (LI-Tec.) batteries, operating for a total distance of 300,000 km (CEDARE, 2016).

The current difficulty is increasing the car range of above 200 km, with a single battery charge and also reducing recharge time as it currently takes around 8 h to charge during car park, where it needs around 650 electrical voltages (Sioshansi & Miller, 2011).

2.2 Electric Transport Development in the Global Market

According to Renewable Energy Policy Network (Statistics R.C., 2017) estimates by the twenty-first century, more than three million electric vehicles are currently traveling on the world's roads, accounting for 1% of the international automotive marketplace. Sales rate in 2017 was more than 1.2 million electric vehicles, an increase of over 70% from 2016 (International Renewable Energy Agency, 2018).

China, then Europe, especially Norway, Netherlands, Germany and France, the United States, and Japan are the countries that widely used electric vehicles, where the percentage of electric vehicles in markets such as Norway has reached 40% and targets 100% by 2025, while the traditional car sales banned.

In the Netherlands, the share of electric cars exceeds 10% of the market size, in France and Sweden close to 5% with expectations for banned of traditional auto sales by 2040. By

(a) (b)

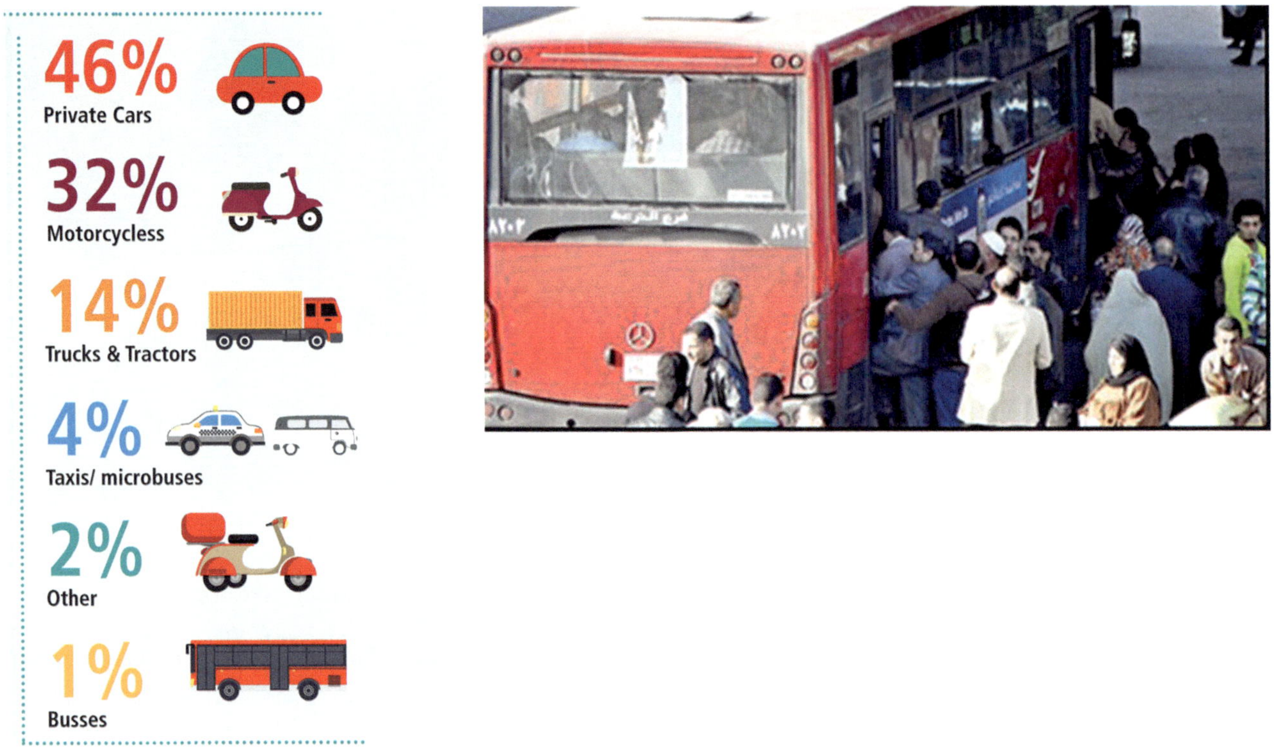

Fig. 4 **a**, **b** Egyptians suffer from transportation. *Source* http://www.environmentalpollutioncenters.org

(a) (b)

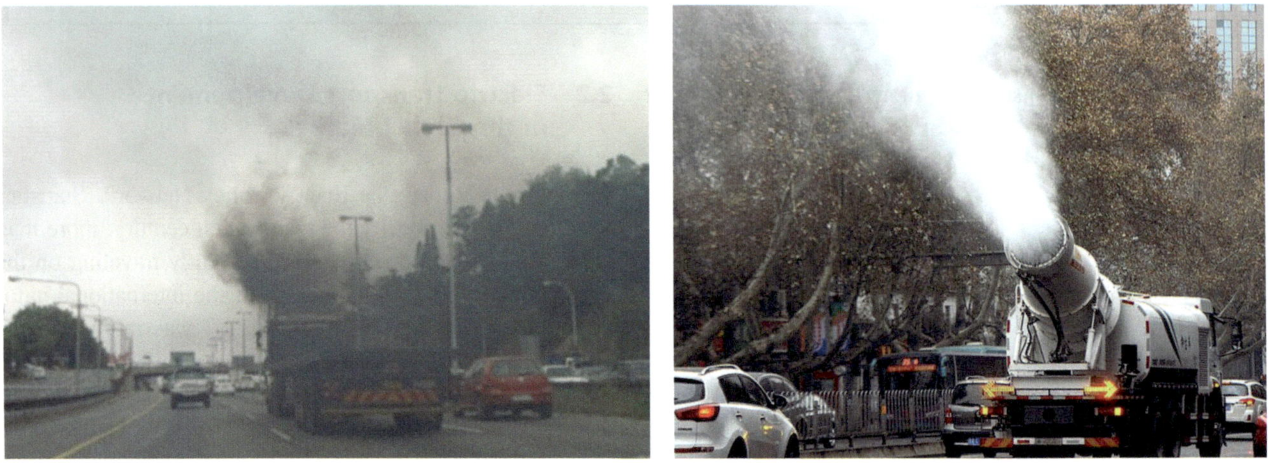

Fig. 5 **a**, **b** Pollution resulting from traditional means of transport. *Source* Ammonia poisoning, http://www.environmentalpollutioncenters.org

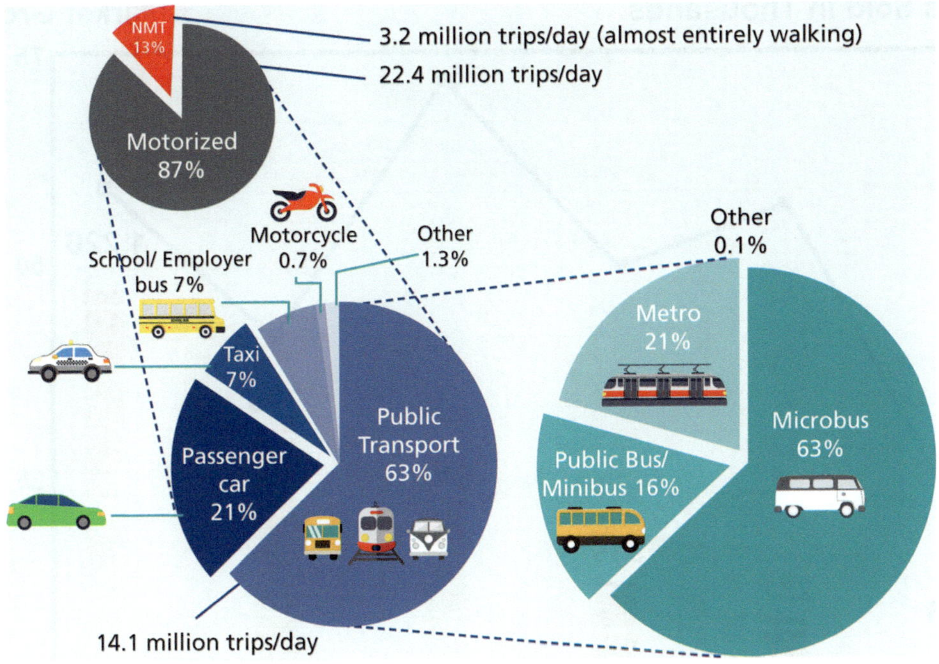

Fig. 6 Shows various means of transport distribution on the 26 million daily trips in Greater Cairo in 2014. *Source* Data update based on the type of means of transport, the Egyptian Center for Transport Excellence in the Ministry of Transport citing 2002 Updated 2014

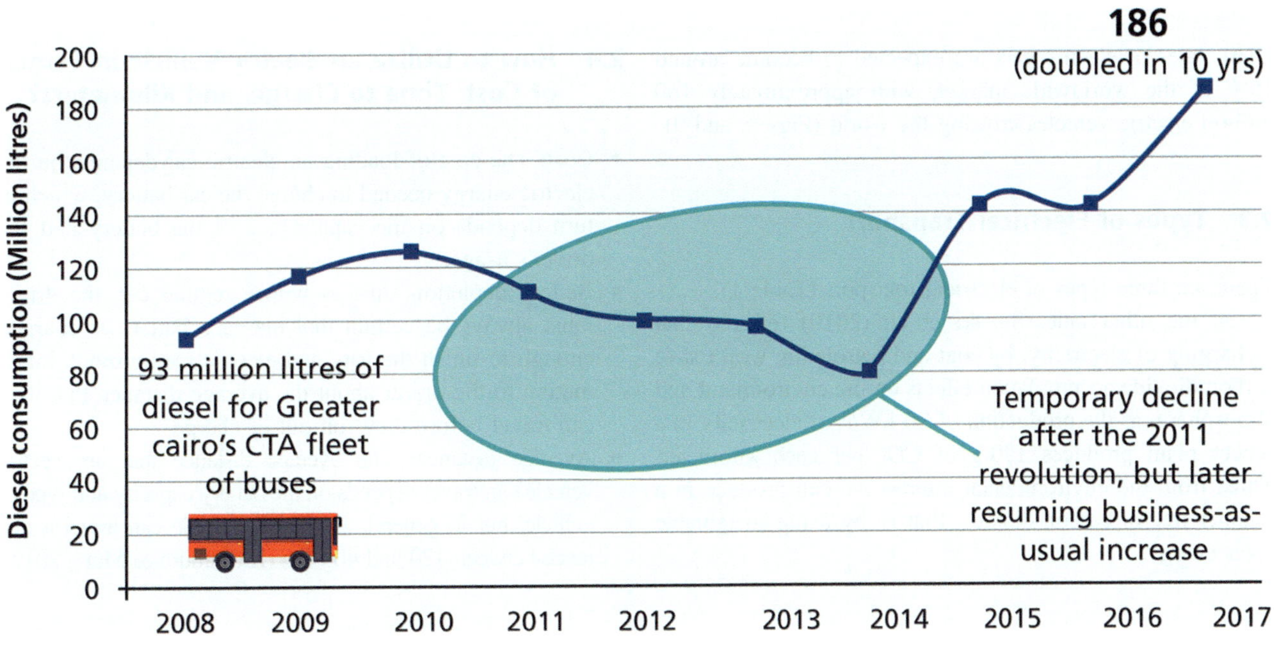

Fig. 7 Diesel consumption for Cairo Transit Authority (CTA) fleet doubling in the past 10 years. *Source* 2018 data, Ministry of Finance

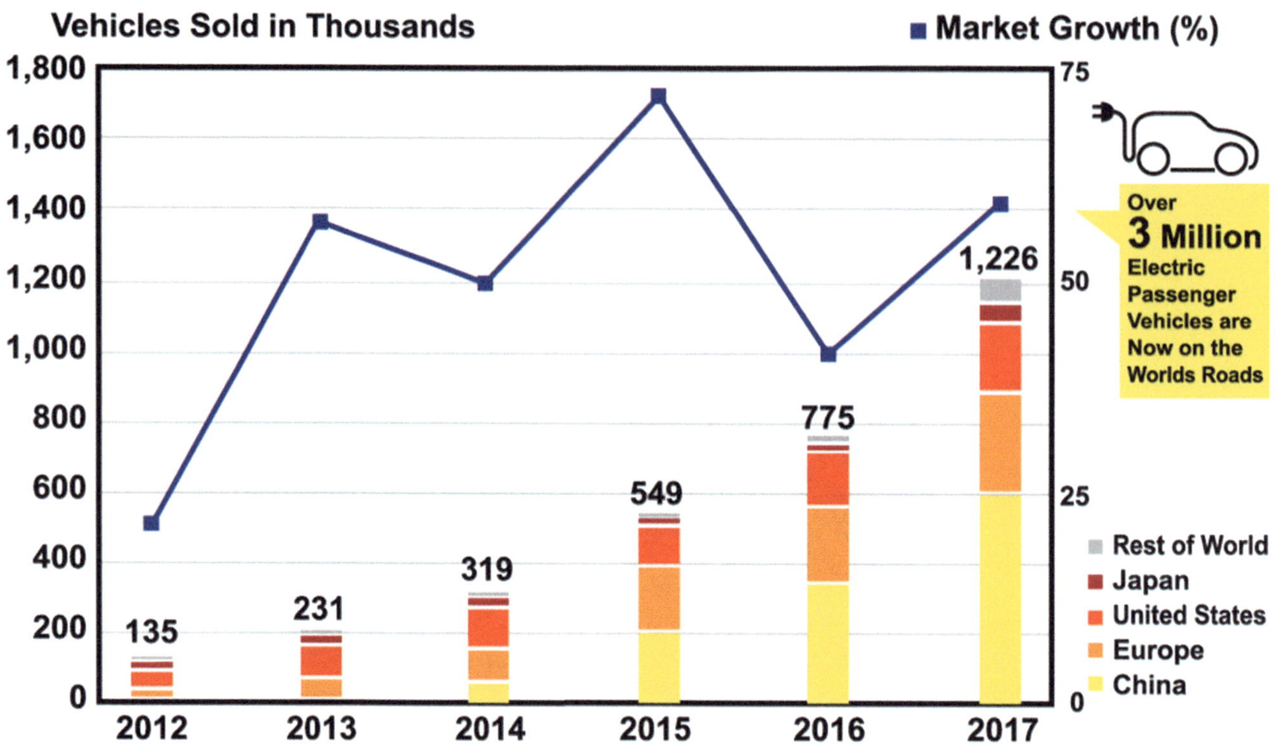

Fig. 8 Growth of global electric car stock throughout 2013–2017. *Source* International Renewable Energy Agency (IRENA), International Energy Agency (IEA), Perspectives for the Energy Transition (Abu Dhabi: March 2017)

2030, sales of electric cars are expected to become around 15% of the worldwide market, with approximately 160 million electric vehicles cruising the world (Figs. 8 and 9).

2.3 Types of Electrical Transport

There are three types of electrical transport (Table 1).

On the other side, Suzuki et al. (2010) referred that recharging of electricity, by coal and petroleum, won't save carbon dioxide accumulation effects on the environment and atmosphere, as the production of 20 kW/h of electricity in a power plant produces 120 g of CO_2 per each kilometer. Thus, from the environmental interest we can produce in a manner free of carbon dioxide, that is, by using sustainable solar energy.

2.4 How to Utilize an Electric Vehicle in Terms of Cost, Time to Charge, and Kilometers?

- Cost: The cost of loading an electric car depends on the electric energy needed to charge the car battery, which in turn depends on the capacitance of the battery and the driver's usage.
- Battery depletion: Just as with a regular car, the driver must always be certain that his car's battery is charged enough to finish the trip, and electric cars provide information to the driver about the average distance that they can travel before they run out of charge.
- Average distance: The average distance that an electric vehicle can travel depends on the battery capacity and type of vehicle, but in general, an electric vehicle can travel a distance between 120 and 400 km (Kawamura & Muta, 2012).

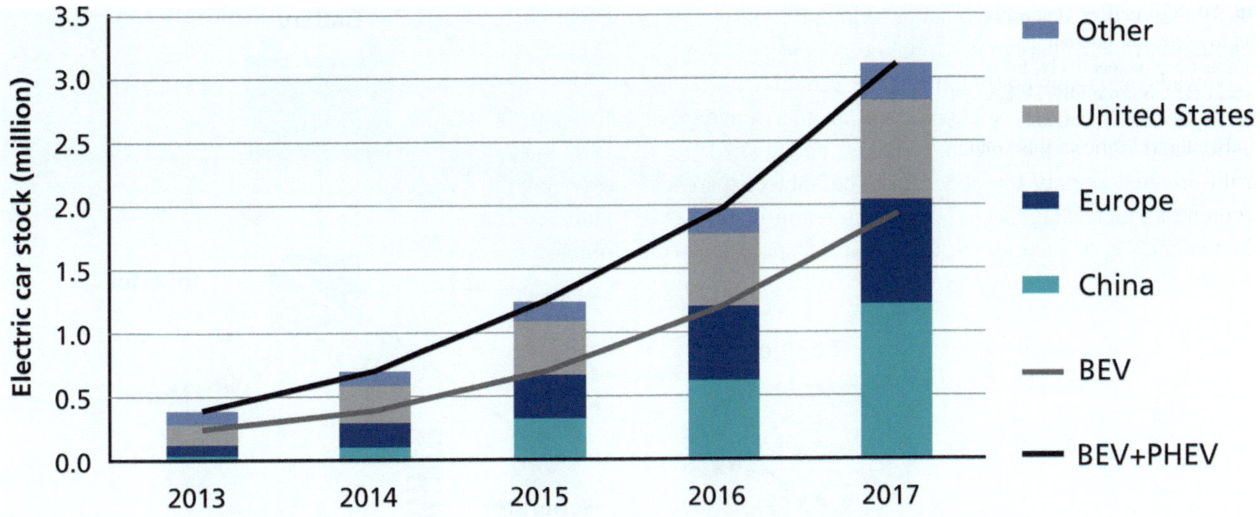

Notes: The electric car stock shown is primarily estimated on the basis of cumulative sales since 2005. Where available, stock numbers from official national statistics have been used (provided that the data can be shown to be consistent with sales evolutions).

Sources: IEA analysis based o country submission, complemented by ACEA (2018); EAFO (2018a).

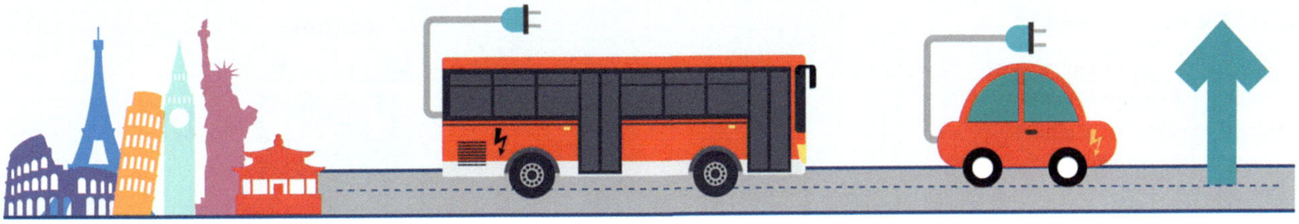

Fig. 9 Growth of global electric car stock throughout 2013–2017. *Source* The IEA analysis, based on data provided by each country, supported by European Association data for car manufacturers (2018), European Alternative Fuel Observatory (2018)

Table 1 Types of electrical transport. *Source* US Department of Energy. Plug-in electric vehicle handbook for public charging station hosts. Clean cities, US department of energy

Hybrid electric vehicle	Plug-in hybrid electric vehicle	Fully electric vehicle battery
It has a system of two engines, one electric, and the other is a regular internal combustion engine. The battery is charged by a braking system that converts the car's engine into a generator during brake pressure, permitting the batteries to charge, which is called regenerative braking, and there is a friction braking system. Backup works if the system is down (Figs. 10 and 11)	It is a system of two engines, one electric, and the other a regular internal combustion engine. Batteries are charged by external charging units as well as the brake system. The two engines are connected by three alternatives: parallel, series, series, and parallel (Fig. 12)	This type of car derives its energy only through electricity. It is a fully electric car. Batteries are charged by plugging into electricity from an external source through charging points, where the car consists of batteries, switching systems electrical control, charging plug, engine, transmission systems for mechanical (Fig. 13)

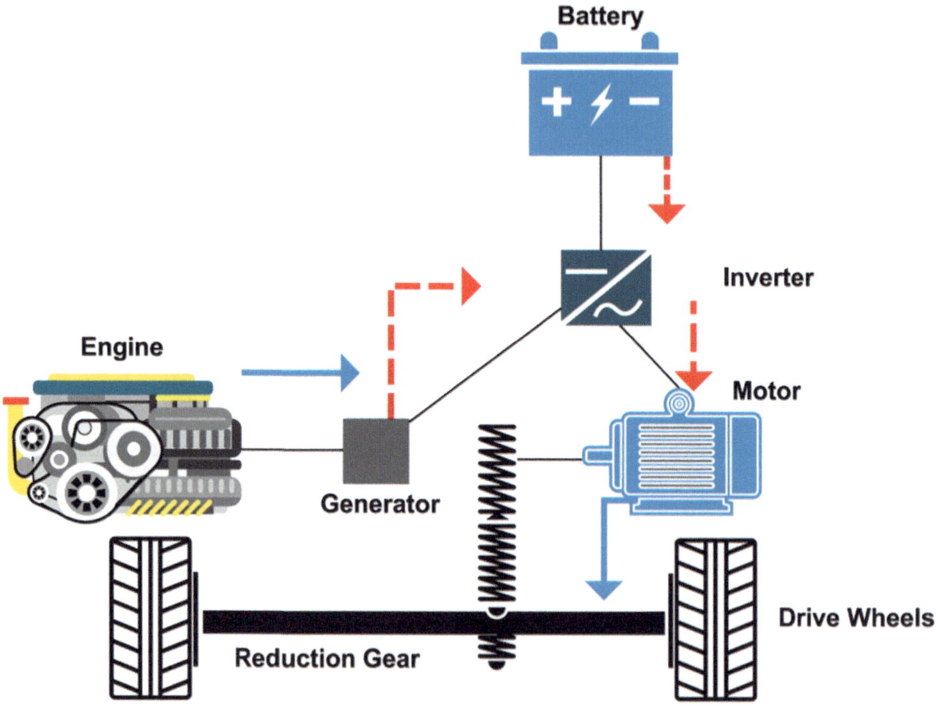

Fig. 10 The idea of connecting engines (electric internal combustion), respectively, in hybrid cars. *Source* OECD/IEA (2018). Global EV outlook 2018/million electric cars: beyond one

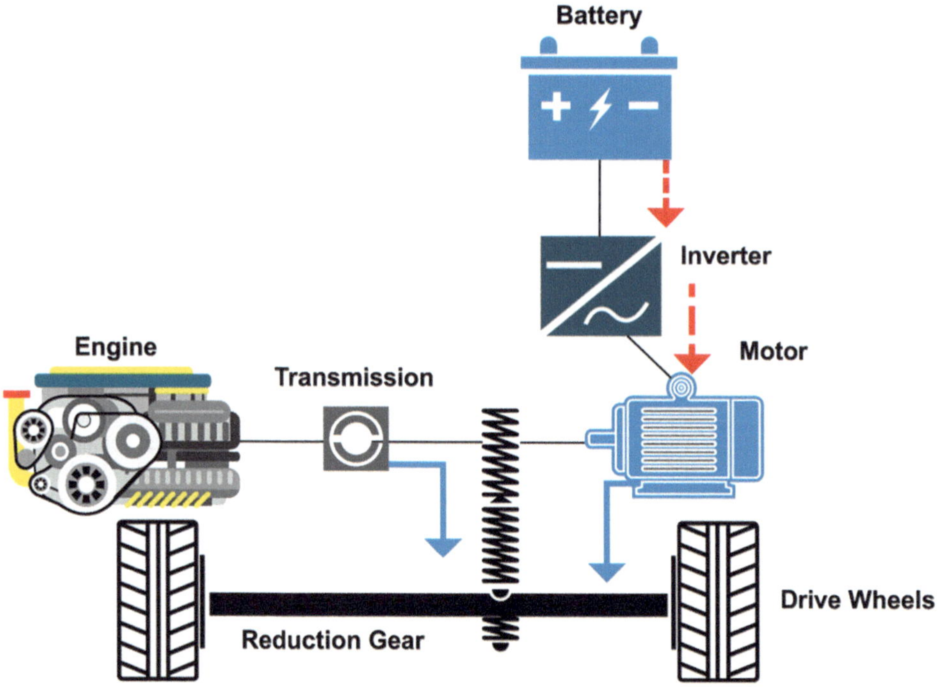

Fig. 11 The idea of connecting motors (electric internal combustion) in parallel in hybrid cars. *Source* OECD/IEA (2018). Global EV outlook 2018/million electric cars: beyond One

2.5 The Importance of Electric and Hybrid Transport Charging Centers

Probst and Roecker (2007) agreed that electric and hybrid transports need electricity to charge their batteries. As mentioned previously, it saves for an estimated distance of 10–200 km or more depending on the amount of savings that exist as well as hybrid transportation. For long-distance journeys and remote places, situations must have a place or source to charge energy for the rest of the journey; most importantly, they need to fix this energy from a sustainable and renewable source such as solar energy.

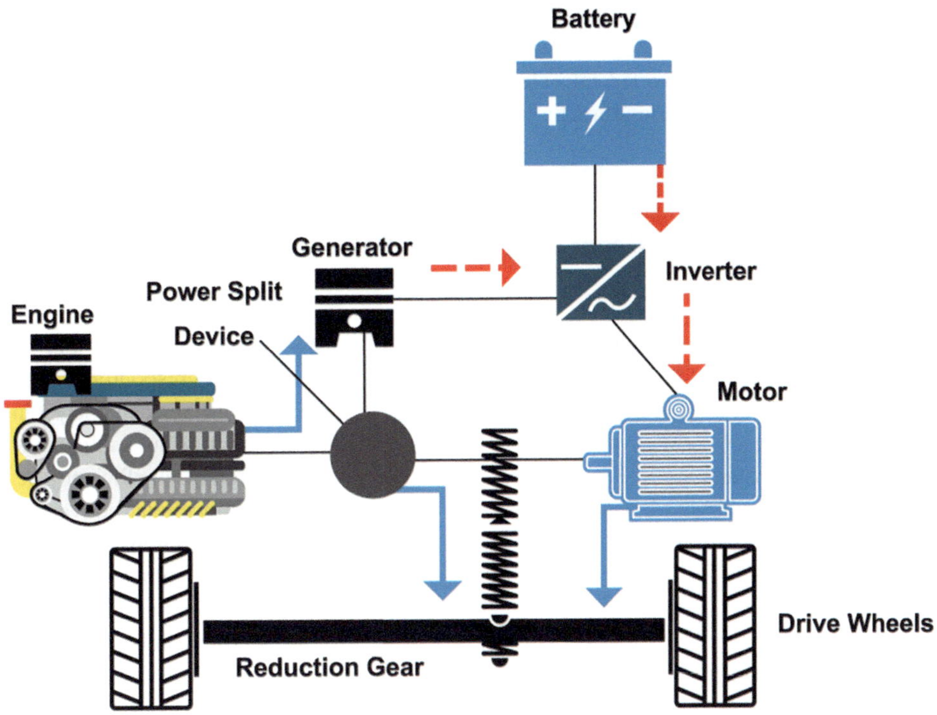

Fig. 12 The main components of the electric car. *Source* OECD/IEA (2018). Global EV outlook 2018/million electric cars: beyond one

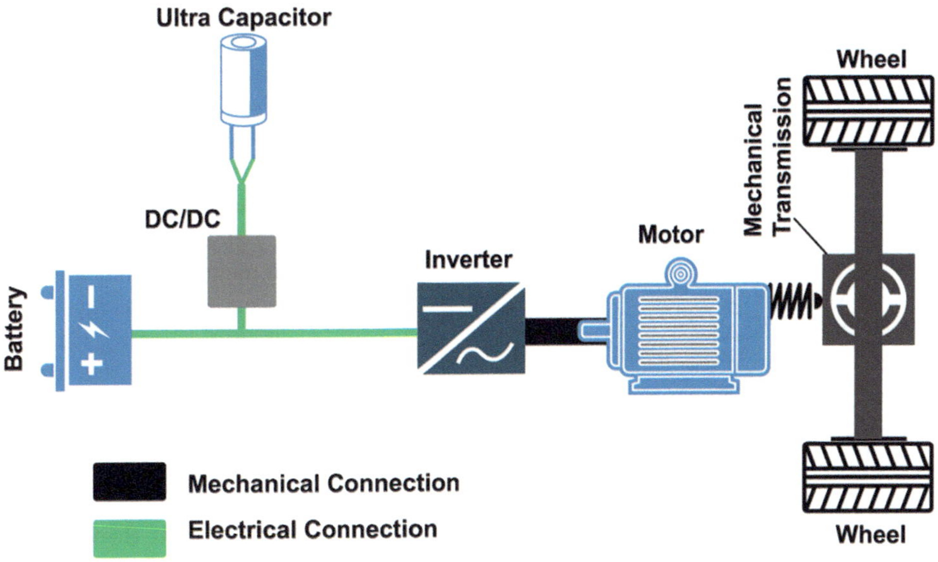

Fig. 13 The idea of connecting the engines (internal combustion electric) between successive and parallel hybrid cars. *Source* OECD/IEA (2018). Global EV outlook 2018/million electric cars: beyond one

3 Use of Solar Energy to Generate Electricity

Solar energy can be converted into electric power by converting solar or light radiation directly into electric power by photovoltaic solar cells.

3.1 Solar System for Generating Electric Power Components

The solar system for generating electrical power consists of four main items (Fig. 14), which are:

- PV photovoltaics
- Charger controllers
- Batteries
- Power inverters.

3.2 Solar Panels

It is the visible component of the solar system (Hall et al., 2017) (Fig. 15), which is installed on the surface of Pergolas, generates electric power, and consists of the following:

- Solar cell
- Solar module
- Solar array.

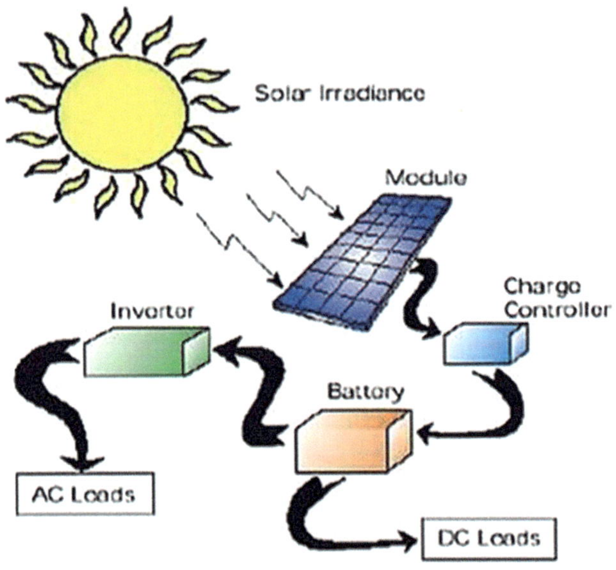

Fig. 14 The solar system components to generate electricity. *Source* https://www.greenclimate.fund/500m.

Thus, the solar panel is a solar cell that is grouped together, used to run or store some equipment, or saved in rechargeable batteries and used more than one time. The effectiveness of these cells is measured in watt units. There are small plates ranging from 5 W or 15 W, which can be grouped to achieve billions of watts.

3.3 Solar Cell (Photocells)

It is the principal and the smallest component of the solar system, responding to direct and indirect solar radiation converting such energy into an electric one. The solar panels benefit from sunlight, which activates electrons to create current. Cell efficiency depends on two factors: the efficiency of intracellular conversion and the power of the solar cell to absorb photons.

Photovoltaic cells consist of semiconductors, mostly of silicon, which is compressed into a treated chip to form an electrical field, positive on one end and negative on the other. When the light energy reaches the cell, electrons are released from atoms in the semi-conductors material as sunlight photons. It stimulates electrons to a higher state of energy to generate electricity. Electrons are collected in the form of electric current if electrical transmissions are connected to the negative and positive sides, so is the form of an electric current, if electrical conductors are connected to the negative and positive ends. The generated energy can then be stored in batteries of different capacities so that they can be used during the sun's demise.

Moreover, Lechner (2014) is worth mentioning that many foreign and Arab countries have moved to take the idea of using solar energy to create electrical power for smart transportation by 2018, as it is helping people for changing their lifestyles and working to create an eco-system that includes mobility solutions along with other friendly solutions for the environment.

3.4 Solar Radiation in Egypt

El-Metwally (2004) discussed that Egypt is located geographically between latitude 22 and 31.5 north, and thus Egypt is at the center of the global solar belt, and therefore it is one of the most productive nations in the world with solar energy. Available data from the meteorological station and the addition of some new stations and advanced measuring equipment have resulted in the formulation of an atlas of solar energy in Egypt. The accompanying figures, Figs. 16 and 17, show the monthly averages of the hours of sunshine at the sites of monitoring stations in Egypt.

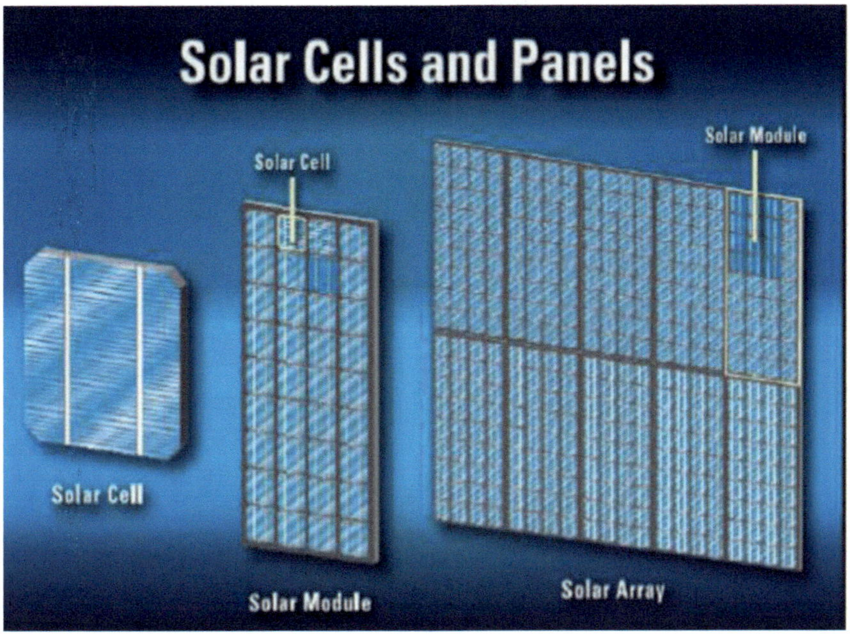

Fig. 15 Solar cells and panels. *Source* https://www.greenclimate.fund/500m

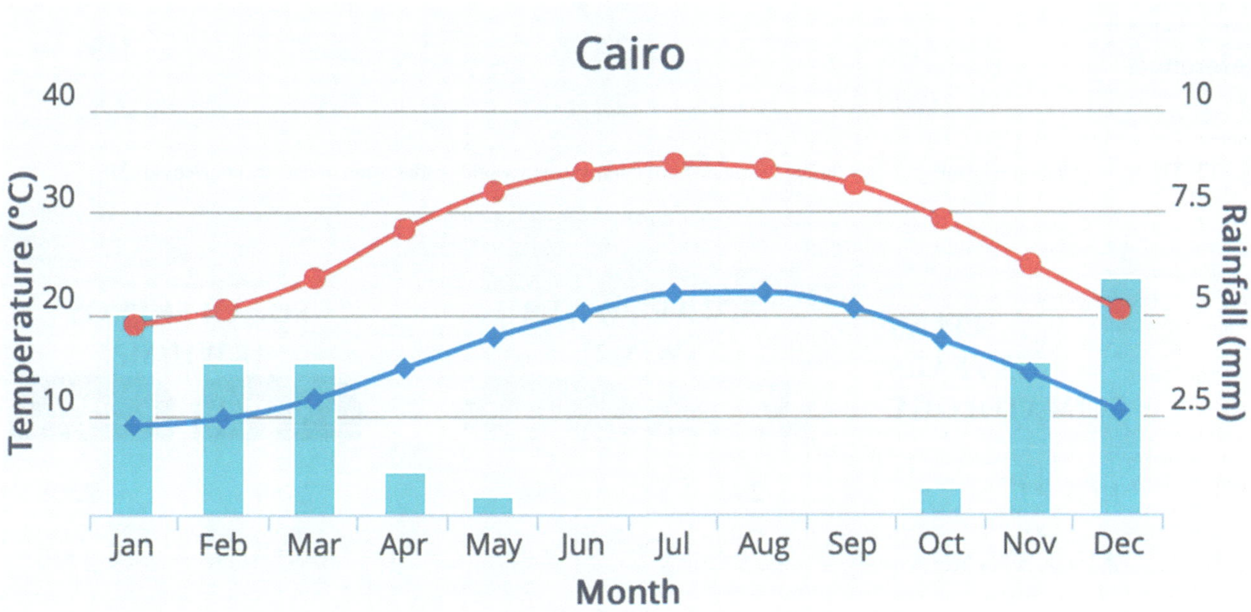

Fig. 16 Thermal radiation in Egypt. *Source* https://www.worldweather.wmo.int/en/city.html?cityId=248

Consequently, the use of solar energy to generate energy through solar panels in Egypt is considered to be highly successful, depending on the level of large solar radiation in most months of the year. As per the current uses in Egypt, solar cells are done at an experimental level and include the generation of electricity necessary to pump water, salinity displacement, cooling, and communications. Figure 18 shows the rates of power generation using solar panels in Egypt (Eissa, et al., 2015).

Month	Mean Daily Minimum Temperature (°C)	Mean Daily Maximum Temperature (°C)	Mean Total Rainfall (mm)	Mean Number of Rain Days
Jan	9.0	18.9	5.0	3.5
Feb	9.7	20.4	3.8	2.7
Mar	11.6	23.5	3.8	1.9
Apr	14.6	28.3	1.1	0.9
May	17.7	32.0	0.5	0.5
Jun	20.1	33.9	0.1	0.1
Jul	22.0	34.7	0.0	0.0
Aug	22.1	34.2	NaN	0.0
Sep	20.5	32.6	NaN	0.0
Oct	17.4	29.2	0.7	0.5
Nov	14.1	24.8	3.8	1.3
Dec	10.4	20.3	5.9	2.8

Fig. 17 The solar radiation schedule in Egypt during the year. *Source* https://www.worldweather.wmo.int/en/city.html?cityId=248

ATLAS 15 YEARS CLIMATOLOGY	SOLAR POWER (W/M2)		SOLAR ENERGY (KWH/M2)	
	DNI	GHI	CSP	PV
EGYPT	292	252	2554	2208
ALEXANDRIA	294	255	2572	2230
CAIRO	328	279	2875	2447
SOUTHERN	315	269	2756	2357

Fig. 18 Mean solar power and energy for Egypt and its three big cities. *Source* http://www.nrea.gov.eg/Content/files/SOLAR%20ATLAS%202018%20digital1.pdf

Table 2 A proposal to employ solar panels in operating electrical transport in Egypt. *Source* researcher with reference to New Generation Motors Corporation (2002)

	Overview	Benefits
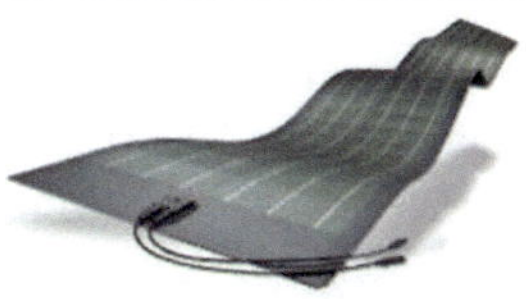	New levels of monitoring	The savings rate will increase with the addition of new and renewable energy sources to the energy mix used in the electricity sector. The reduction in local pollutants will be greater, as an effect of disposal of old vehicles that use low-quality fuel
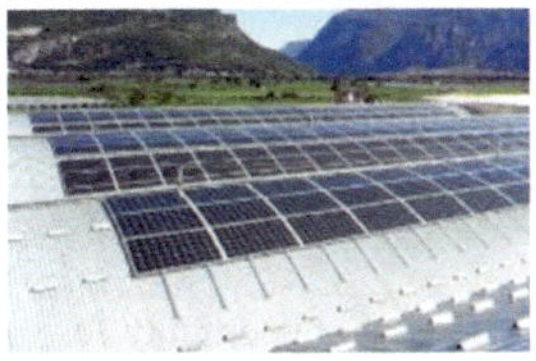	Used batteries in the network to improve power quality	By studying the power of solar panels to produce close to 2208 kWh, it can be achieved to provide many charging stations, as one charge per vehicle is sufficient to travel from 120 to 400 km
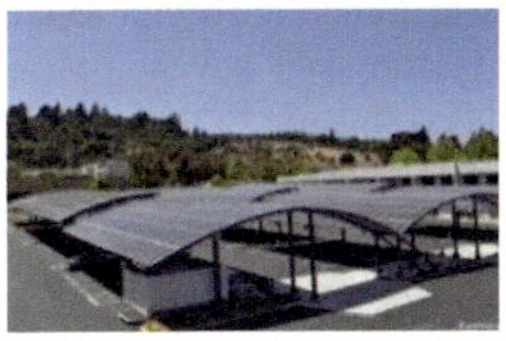	Lower cost	The figures indicate that energy saving for long-distance buses is 67%; that is, it is a huge decrease if the vehicle depreciated only Yet, when counting the energy consumption of the electricity network as well, the saving rate decreases due to the dependence of most power plants today, in Egypt, on fossil fuels, as the savings rate goes down to 12% when calculating the most comprehensive consumption

3.5 A Proposal About Using Solar Panels to Operate Electric Transportation in Egypt for Electric Buses

Through studying the foregoing, a proposal to employ solar panels in operating electrical transport in Egypt can be reached (Table 2).

4 International Models Adopting the Idea of Using Solar Energy to Produce Electricity for Environmentally Friendly Transport (Table 3)

Table 3 International models adopting the idea of using solar energy to produce electric power for environmentally friendly transport. *Source* https://www.testbook.com/blog/indian-power-plants-gk-notes-pdf/german-sustainable-mobility.de

Country	Partnership sector	Project concept	Challenges
Germany (German partnership for sustainable mobility)	European union for urban energy TKI urban energy (Public and private sectors)	Project name: sustainable mobility The first station opened in Germany. The express Fright station is located along the Limburg highway and has several chargers up to 350 kW. This allows electric drivers to charge up to 100 times faster than home, making it possible to travel long distances It is aimed to establish a European network of 1,000 fast-charging stations in Germany. The iconic stations are in crowded locations along highways and urban centers, giving electric drivers the possibility to drive across Europe	Switching as quickly as possible to electric driving, along with renewable energy, can prevent climate change, making our cities cleaner and less noisy This is by building fast-charging stations where all electric vehicles can be charged using renewable energy from the sun and the wind
India (Deb et al., 2019)	(EAI) is India's leading business catalyst for clean energy, environment, and sustainability sectors. (public and private sectors)	Project name: smart charging station More than 100 units of private and public electric transport chargers' station installed in India's major cities. It is an innovative product developed for the development of the electric transport industry	Electronic traffic can play a key role in the evolution of the Indian economy, helping to bring down fuel costs, government expenses, lower operating, and maintenance costs

5 Domestic Models to Adopt the Idea of Using Solar Energy to Produce Electricity for Environmentally Friendly Transport (Table 4)

As a result, through presenting the above international and domestic models that adopt the idea of using solar energy, to produce electric power for smart transports, the researcher was able to accomplish a set of controls and guidelines through a framework for designing environmentally friendly charging centers and electric transport used within Egypt (Tables 5 and 6).

Thus, the proposed model was reached by presenting a group of pictures of the current situation (Figs. 19, 20, 21, 22 and 23) and the proposal (Figs. 24, 25, 26, 27 and 28) how much is clear in Table 7.

Table 4 Local models to adopt the idea of using solar energy to produce electricity for environmentally friendly transport. *Source* https://www.envirocitiesmag.com/articles/pdf/envirocities-article10.pdf

Country	Partnership sector	Project concept	Challenges
Jordan (Wamda, 2019)	Public and private sectors, which include a range of local and international partners	Project name: solar electric car charging project The project aims to spread 100,000 electric vehicles in Jordan by its completion in 2020, supported by a network of 10,000–20,000 electric vehicle charging stations. It will be equipped with advanced lithium-ion batteries with a capacitance of almost 2 GWh to isolate the charging stations of electric vehicles from the power grid to avoid pressure on that grid providing any additional storage capacity that can be used during peak hours as part of the WAN System Strategy. In addition, a solar PV system was installed and connected directly to the grid compensating electricity used to charge electric vehicles	The project offered an opportunity for potential users in Jordan to recognize the technical and performance advantages of electric vehicles compared to traditional gasoline vehicles, as considerably as the obvious environmental impact and economic benefits of reduced fuel consumption (electric driving costs a quarter of gasoline driving). Electric vehicles have a better moment with fewer spare parts, which brings down the cost of maintenance and increases the vehicle average life
UAE (Abdelsalam, et al., 2019)	Sharjah electricity and water authority	Project name: electric car charging station The Authority plans to increase its electric vehicle number. The first phase of which includes the establishment of 100 electric car charging stations supported by solar energy in commercial centers, hospitals, government service centers, and main roads in various areas of the Emirate of Sharjah. Then expanding electric car charging infrastructure projects in all regions of the Emirate of Sharjah. Adding the construction of the first fully environmentally friendly charging station supported by solar panels installed on the roofs of the parking lot, which will save 9,500 kWh of solar energy per year	Raising awareness of the importance of electric vehicles for environmental preservation, highlighting the role played by environmentally friendly situations, including electric car battery recharging stations, in promoting good environmental practices
UAE (Dubai Electricity and Water Authority, 2019)	Dubai electricity and water authority	Project name: green charger In October 2018, Dubai Electricity and Water Authority (DEWA) announced the completion of the second phase of the Green Charger initiative, which included the installation of 100 new green charger stations to charge electric vehicles across the Emirate of Dubai, doubling the total number of green charger stations in Dubai from 100 to 200, as part of the Authority's efforts to encourage the public to acquire this type of eco-friendly vehicle, which contributes to reducing carbon emissions and supporting sustainable transport in Dubai	Connecting plants to solar energy technology that converts solar energy into electricity for charging, thus helping to significantly reduce greenhouse gas emissions. Excess energy can be stored and then transferred to the country's power grid overnight

Table 5 Framework methodology for designing the environmentally friendly electric transportation centers within the Arab Republic of Egypt

Designing a framework methodology for the environmentally friendly electric transportation centers within the Arab Republic of Egypt		
The problem	Many health and environmental risks resulting from car exhausts of various car types	The ever-growing population density
		Increased demand for transportation
	Many of the usual obstacles associated with the introduction of any modern technology	Doubts about technical and financial feasibility, legal and regulatory prerequisites, market responsiveness, and impact on employment local industry
Problem analysis	Transportation is considered the first main source of global warming	High emission of carbon monoxide, lead, nitrogen oxides, and atmospheric organic compounds within the Republic
		Increase in heartbeat rates that can lead to fainting and death
		Poor eyesight and deafness
		Adverse impact on the nervous system
		Adverse impact on concentration
Information Gathering	Use of environmentally friendly electrical transportations	Working by using electrical energy, by replacing the original engine of the car and laying an electric motor instead
	Different types of charging stations for electric transport	Identify their types and how they operated
		Usage of solar panels to generate electricity
	Use of solar energy to charge electric transport batteries	Identify their benefits, methods of use and how can be used to reduce greenhouse gas emissions, load onto the electricity grid, thus cutting the cost
Assumptions	Start a plan to design charging centers using photovoltaic panels and convert them into sustainable charging centers to reduce harmful emissions by at least 50% by 2030	Strengthening public-private sectors partnership also, aside from private institutions
		Transit Authority, the Electricity and Energy Authority, Solar Energy Authority, Atomic Energy Authority, private companies
		Awareness programs
		• How to use eco-friendly transport • What are the types of charging stations for electric transportation • How to use electric charging stations • How to get assistance on how to use • Times of charging stations required for operation • How to buy an electric car, and identify its cost • Average distance the vehicle can move • What happens if the battery is giving out of charge before reaching the charging station • The expected distance between each station and the other • How to get an electric transport recharge card
		Sharing experiences
		Share experiences with countries/cities that have reached different levels of development in this field of study
Goal	Work on consolidating the concept of a smart city, and achieve the strategy of reducing carbon emissions by incorporating environmentally friendly transport means with renewable sources of energy	
Getting the perfect solution	Update the list of environmentally friendly charging stations and their locations in the Arab Republic of Egypt on a regular basis. Initiate an agreement with private companies to produce electric transport to promote their products within the Republic	

Table 6 Outlines the framework for the design of environmentally friendly electric transport centers within the Arab Republic of Egypt

A framework for designing environmentally friendly electric transportation stations within the Arab Republic of Egypt		
How the charging center works: (charging hybrid or electric cars and the state of the storage system (savings))		
Step 1: The charging centers; charge hybrid or electric vehicles and the storage unit secured	The charging system charges transport by energy generated from solar panels and energy stored in savers. When there is no sunlight, the energy is secured from the savers. Savings can secure the energy from its stock until the smaller voltage is reached so that the system stops. Charging prevents savings from increasing discharge; when this voltage is reached, the energy flow of the savings stopped and the system switches to the second step	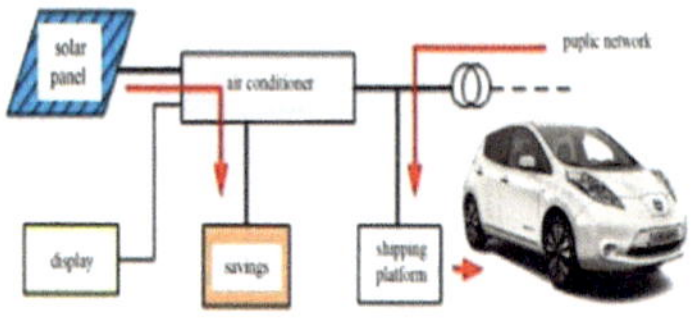
Step 2: The charging center; charge hybrid; or electric transport, but the storage unit does not supply power	The public grid charges the means of transport, while savings are unable to give energy when the smaller effort reaches. The energy generated from the solar panels charges the savings to prevent it from being damaged	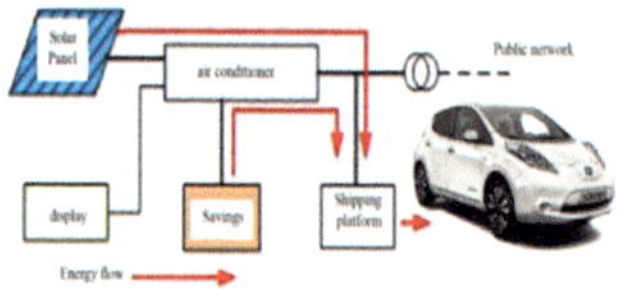
Step 3: The charging center does not charge hybrid or electric transport and the solar panels charge the storage unit	The charging system: charge savings from solar panels, and protect savings from increased charging. The system prompts to type 4 when the maximum saving effort is reached	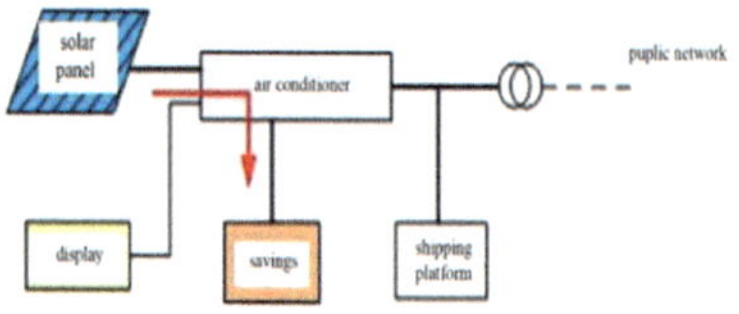
Step 4: The charging center does not charge hybrid and electric transport; solar panels does not charge the storage unit	In this type, the system gives electricity to the public grid so that electricity is supplied to some nearby appliances or sold to the electricity company, and savings in case of suspension or floating charging	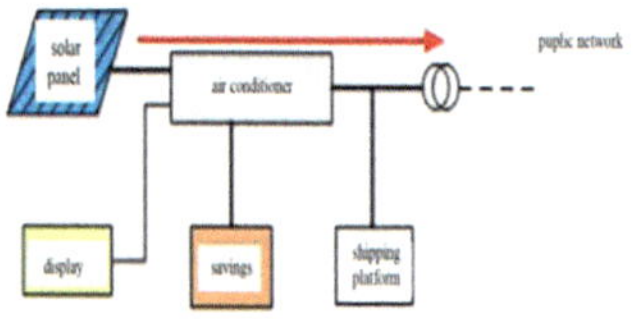

(continued)

Table 6 (continued)

A framework for designing environmentally friendly electric transportation stations within the Arab Republic of Egypt	
Challenges	
Urban density	The urban environment in Egyptian cities is characterized by its highly dense population, which supports the utilization of electric vehicles and reduces necessary investment costs of infrastructure and gas station space requirements
The nature of traffic (traffic and frequent parking in traffic)	Driving in Egyptian cities is characterized by slow drive with frequent traffic jams (stop-and-go), which is linked up to the congestion and urban density, which increases the proportional benefits of using electric vehicles compared to conventional vehicles in this scenario compared to other settings characterized by smoother driving cycles and high use of main roads
Target results	
Local production, Local energy consumption	By storing the energy storage unit (savings) to obtain localized production (generation) of energy, thus local consumption. As the power generated by solar cells is delivered to electric and hybrid vehicles charged, and when the charging platform is not connected for car chargers, the energy generated is stored within the energy storage unit (savings) for charging the future vehicle
Not affected by weather conditions	By plugging into the system to a public network, therefore charging platform can be used even in cases of energy depletion of savings. Caused by difficult weather conditions of permanent rain or cold, this system is equipped with two methods of protection. The first is the power conditioner which cannot charge savings from the public network through its circuit. Secondly, the electric savings voltage is about 320 V compared to the 220 V effort which is the general grid voltage
Providing emergency energy during disasters	By using PV panels in general to get continuous, regular output in emergencies. Even so, it is difficult to secure a stable output due to changes in solar radiation, and since the electro-solar system includes a volume of energy storage (savings) the energy stored in it can compensate for solar radiation fluctuations. This electro-solar system is able to offer a stable electricity supply of 220 V maximum output of 1.9 kW ac

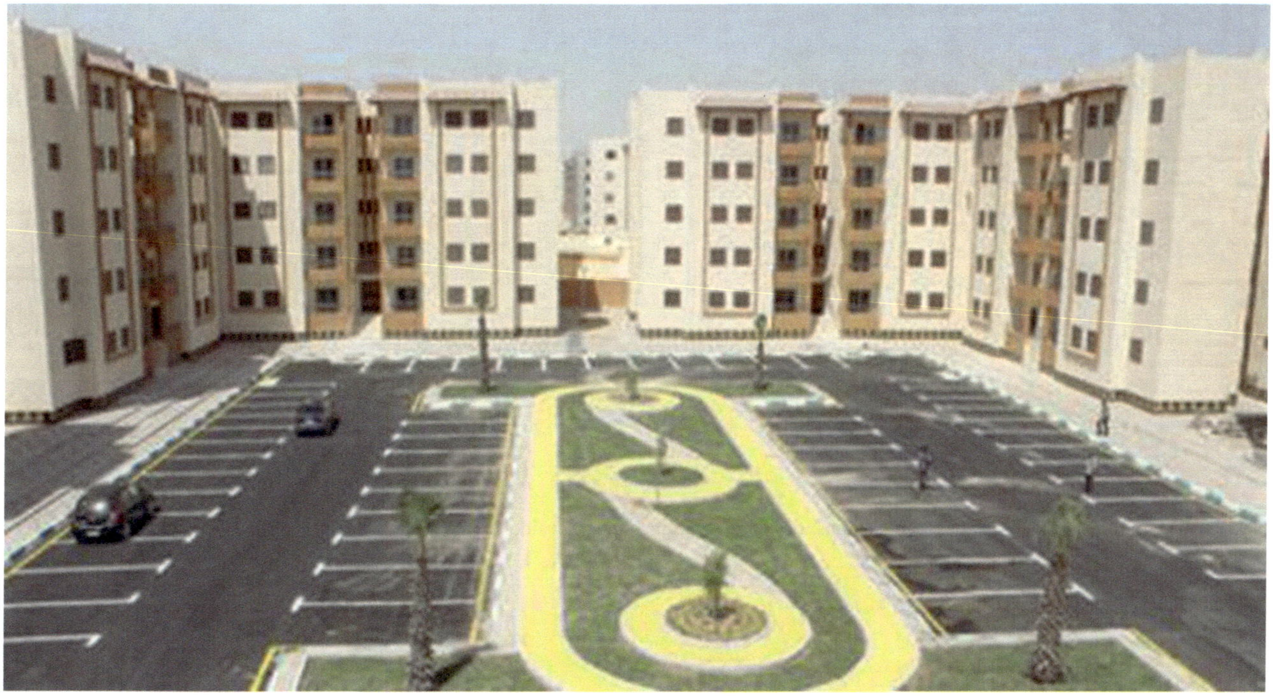

Fig. 19 The current situation for charging centers: design independent charging centers within residential neighborhoods

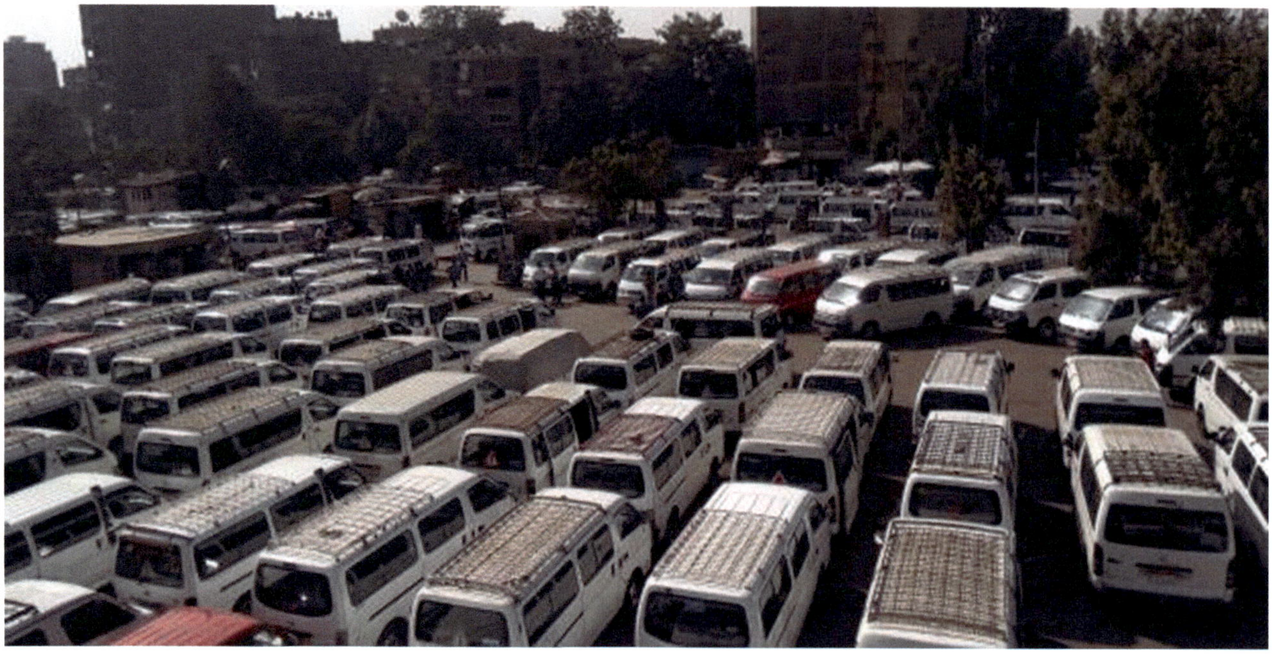

Fig. 20 The current situation for charging stations: design charging centers inside (private) car parking lots

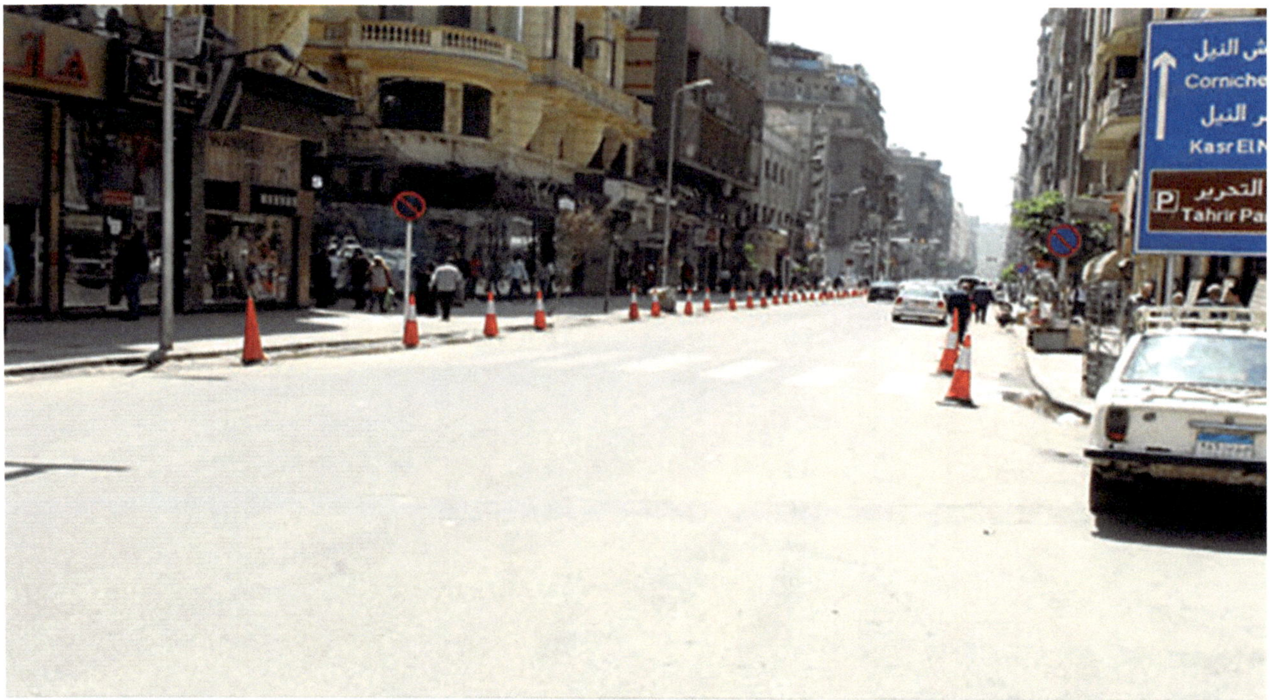

Fig. 21 The current situation for charging stations: design of charging centers inside (public) parking lots

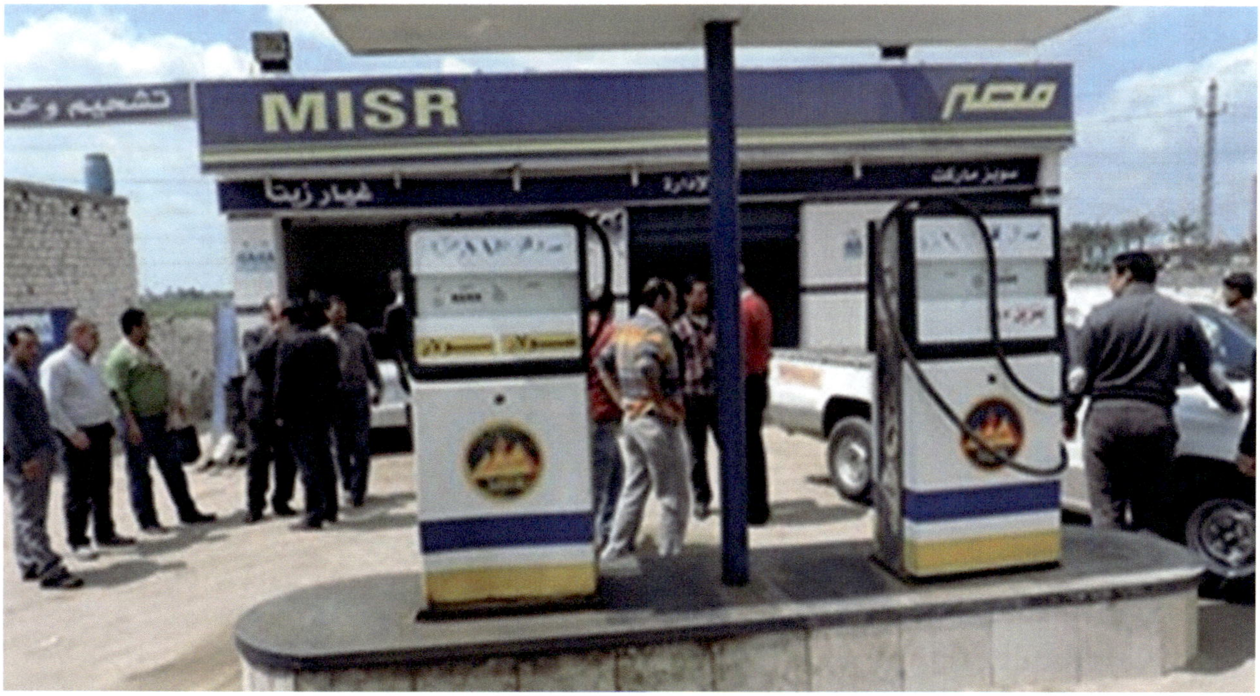

Fig. 22 The current situation for charging stations: design of public/general charging stations

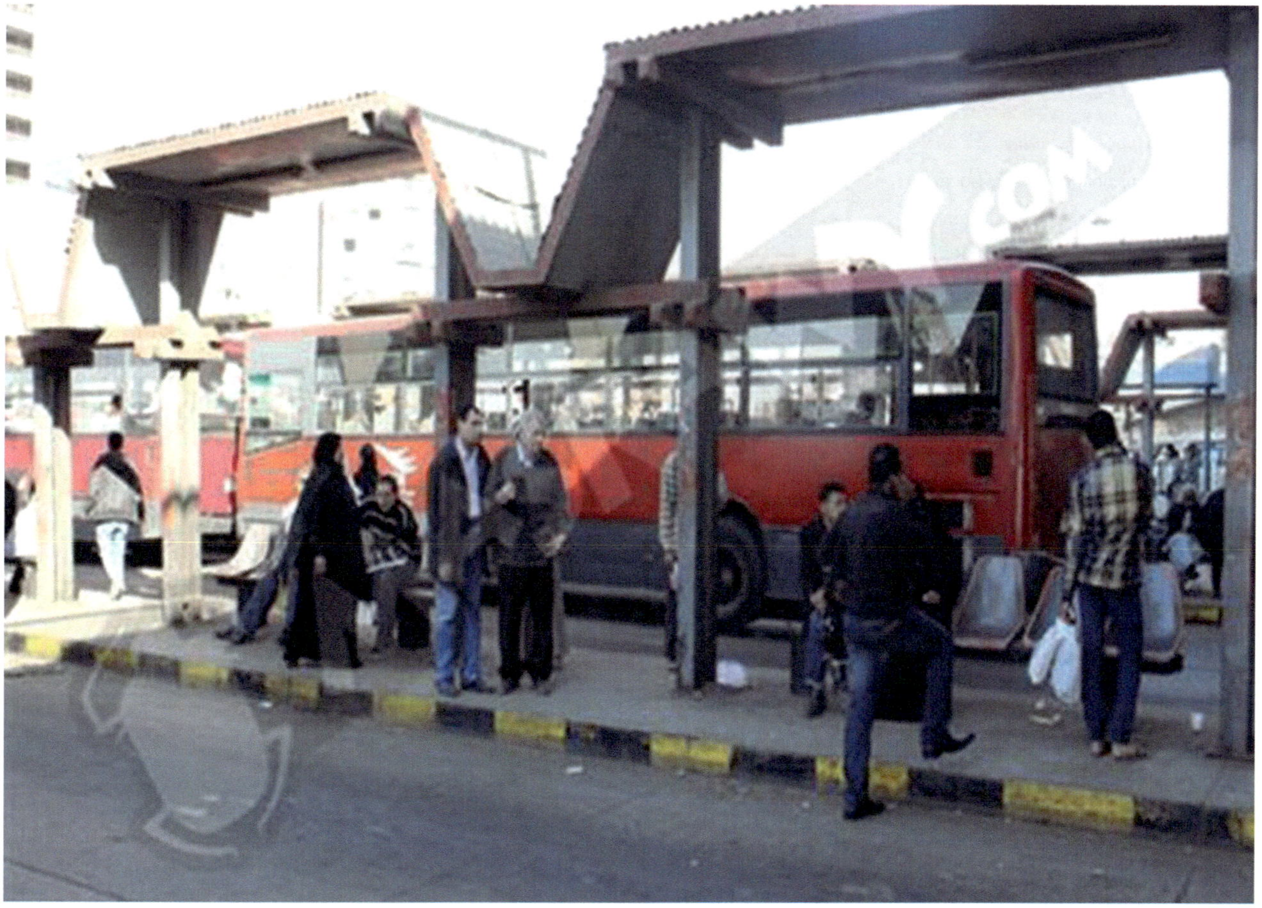

Fig. 23 The current situation for charging centers: design charges centers in public transport waiting stations

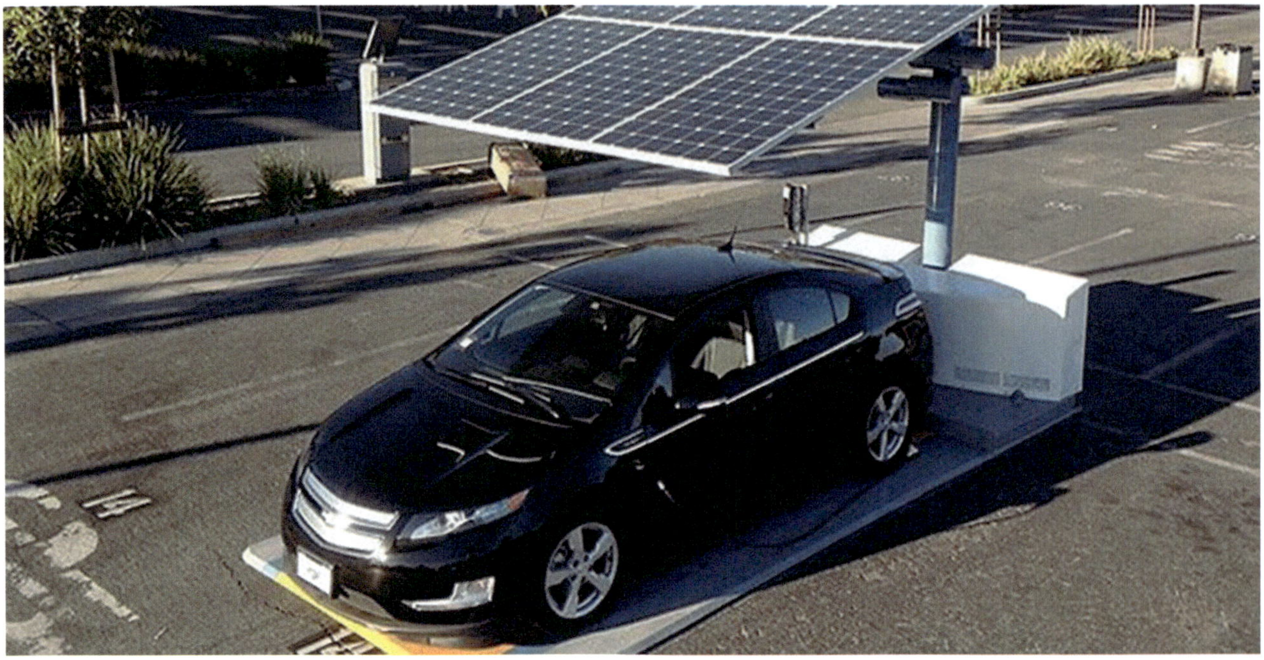

Fig. 24 The proposal for charging centers: design independent charging centers within residential neighborhoods

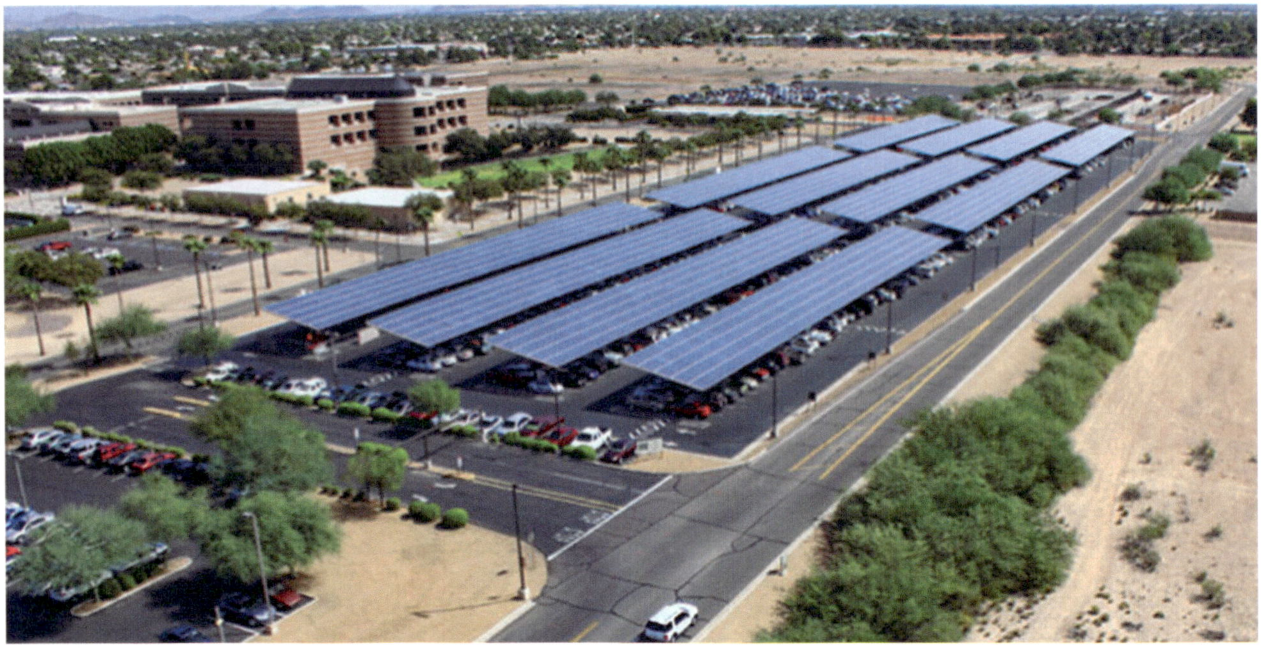

Fig. 25 The proposal for charging stations: design charging centers inside (private) car parking lots

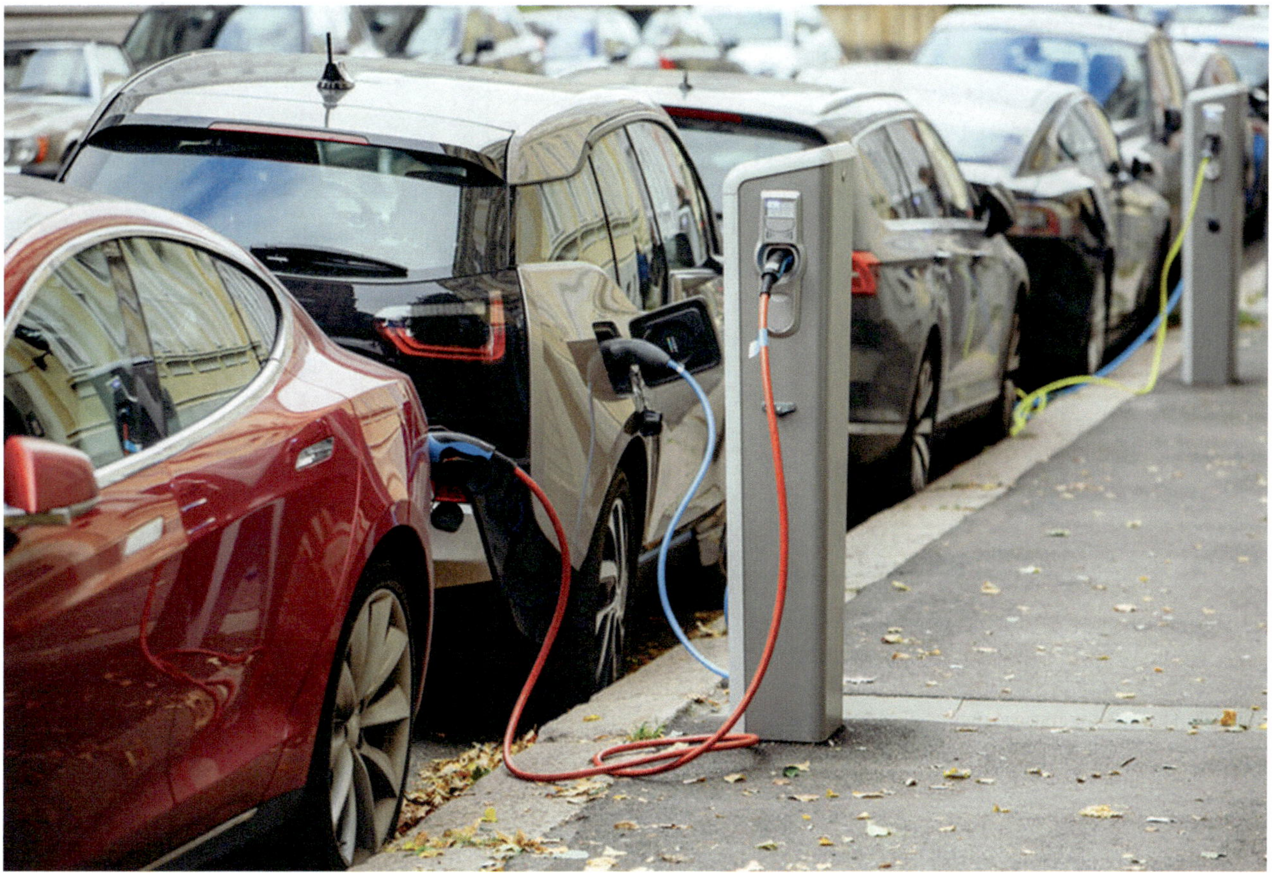

Fig. 26 The proposal for charging stations: design of charging centers inside (public) parking lots

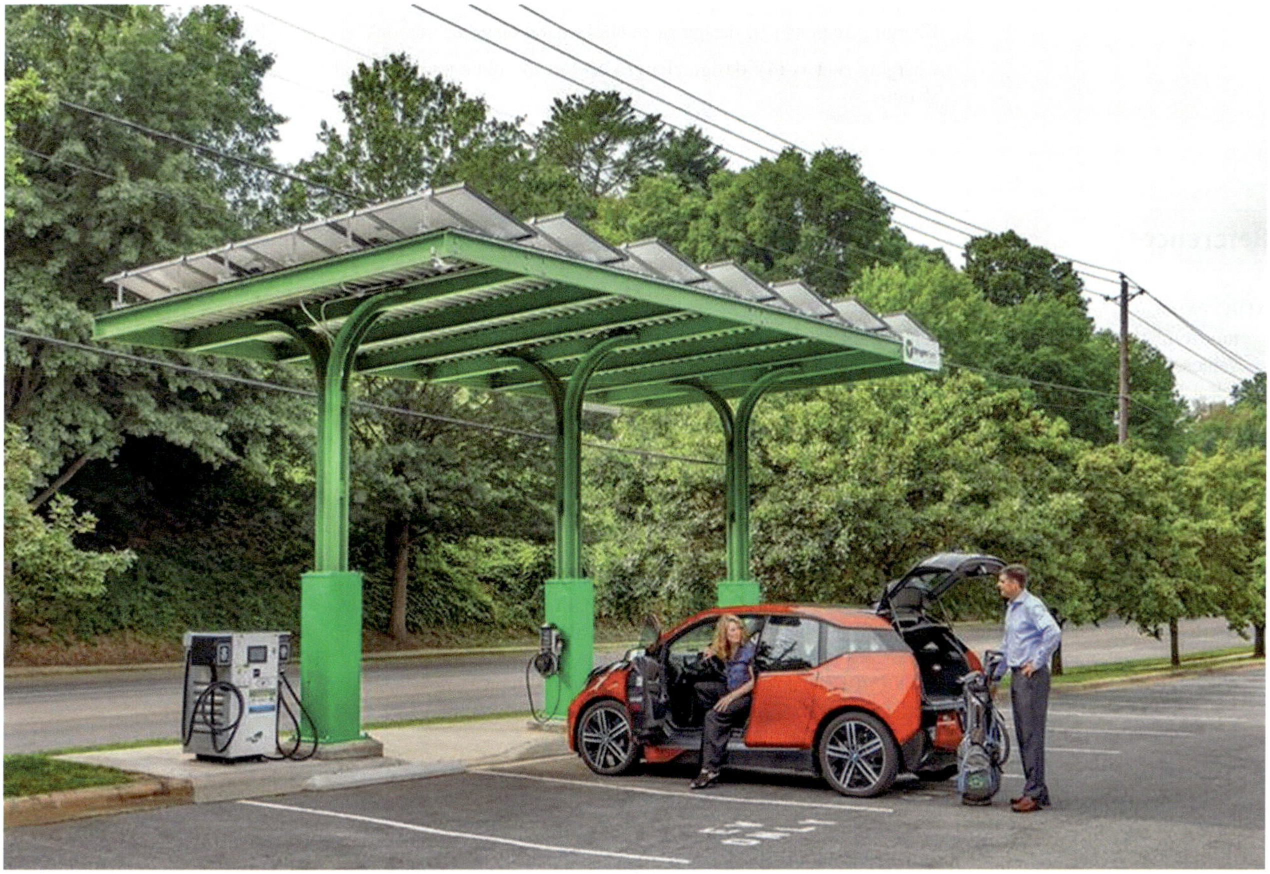

Fig. 27 The proposal for charging stations: design of public/general charging stations

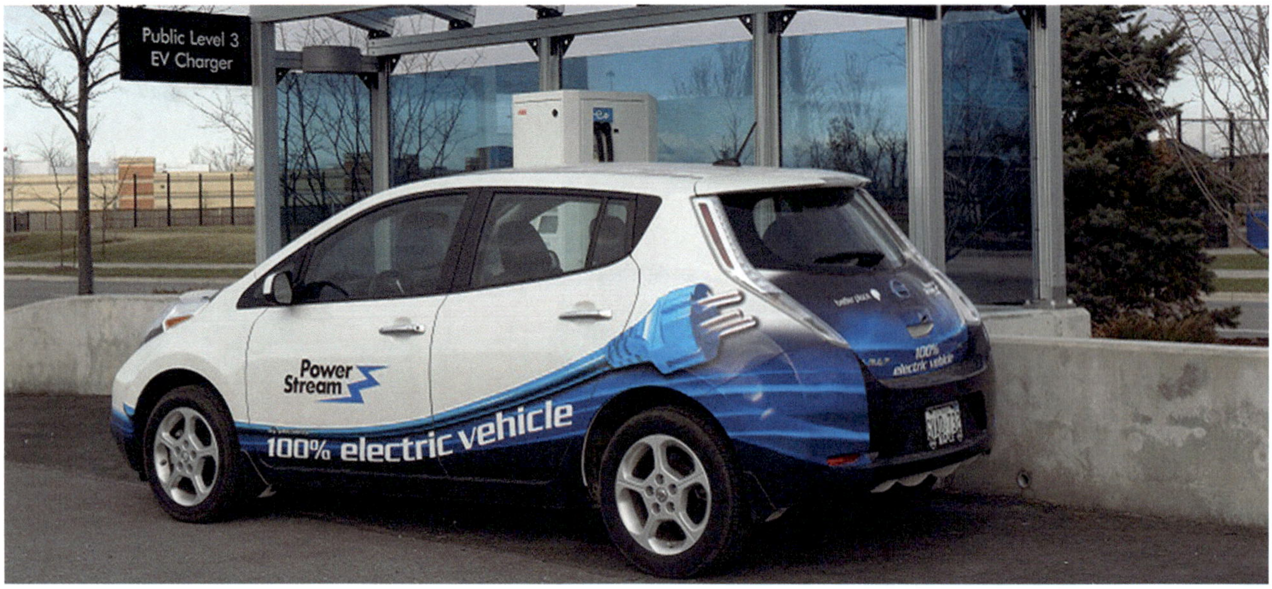

Fig. 28 The proposal for charging centers: design charges centers in public transport waiting stations

Table 7 Outlines the framework of the current situation and the proposal

Proposed models	The current situation	Proposal
Charging centers (1): design independent charging centers within residential neighborhoods	Figure 19	Figure 24
Charging stations (2): design charging centers inside (private) car parking lots	Figure 20	Figure 25
Charging stations (2): design of charging centers inside (public) parking lots	Figure 21	Figure 26
Charging stations (3): design of public/general charging stations	Figure 22	Figure 27
Charging centers (4): design charges centers in public transport waiting stations	Figure 23	Figure 28

References

ADB (African Development Bank). (2017). Philippines: Market transformation through introduction of energy-efficient electric vehicles project. https://www.adb.org/projects/43207-013/main

Abdelsalam, M. A., Obaid, W., Ahmad, F. F., Hamid, A. K., & Ghenai, C. (2019). Modeling and simulation of a solar powered golf cart charging station in Sharjah. In: *2019 Advances in Science and Engineering Technology International Conferences (ASET)* (pp. 1–4). IEEE.

C40 Cities: Green and healthy streets. Fossil-fuel-free streets declaration - planned actions to deliver commitments. https://www.crow.nl/downloads/documents/kpvv-kennisdocumenten/1426_greenandhealthystreets_c40-original.

CEDARE. (2016). Fuel economy policies and labeling for new cars: Improving fuel efficiency and CO_2 emissions in Egypt. http://www.pharos.cedare.org/wpcontent/uploads/2017/01/Handout-Fuel-Economy-forCars_Policy_Brief_EGYPT-GFEI.pdf.

Deb, S., Tammi, K., Kalita, K., & Mahanta, P. (2019). Charging station placement for electric vehicles: A case study of Guwahati city, India. *IEEE Access, 7*, 100270–100282.

Environmental Pollution Centers. (2017). http://www.environmentalpollutioncenters.org.

El-Metwally, M. (2004). Simple new methods to estimate global solar radiation based on meteorological data in Egypt. *Atmospheric Research, 69*(3–4), 217–239.

Eissa, Y., Korany, M., Aoun, Y., Boraiy, M., Abdel Wahab, M. M., Alfaro, S. C., Blanc, P., El-Metwally, M., Ghedira, H., Hungershoefer, K., & Wald, L. (2015). Validation of the surface downwelling solar irradiance estimates of the HelioClim-3 database in Egypt. *Remote Sensing, 7*(7), 9269–9291.

GCF. (2018). GCF in brief: A simplified approval process. Green Climate Fund (GCF). https://www.greenclimate.fund/sites/default/files/document/gcf-brief-simplified-approval-process_0.pdf.

German partnership for sustainable mobility. "Sustainable Mobility—Made in Germany". http://www.german-sustainable-mobility.de. Accessed 20 Apr 2019.

The Emarits. Goverment of Dubai, Dubai Electricity and Water Authority. https://www.dewa.gov.ae/en/builder/useful-tools/faq—consultants-and-contractors. Accessed 10 Apr 2019.
Hall, D., Moultak, M., & Lutsey, N. (2017). Electric vehicle capitals of the world. In: *ICCT White Paper.*
Haug, T., Søstrand, P., Løvik, M., & Langård, S. (2001). Public health and air pollution. *The Lancet, 357*(9249), 70–71.
WHO. (2001). Air quality guidelines for Europe 2000, World Health Organisation.
IRENA. (2018), Corporate sourcing of renewables. In: *Market and Industry Trends* - REmade Index 2018. International Renewable Energy Agency, Abu Dhabi.
Jorgensen, K. (2008). Technologies for electric, hybrid and hydrogen vehicles: Electricity from renewable energy sources in transport. *Utilities Policy, 16*(2), 72–79.
Kawamura, N., & Muta, M. (2012). Development of solar charging system for plug-in hybrid electric vehicles and electric vehicles. In: *2012 International Conference on Renewable Energy Research and Applications* (ICRERA) (pp. 1–5). IEEE.
Lechner, N. (2014). Heating, cooling, lighting. In: *Sustainable design methods for architects.* John Wiley & Sons.
Ministry of Transport Greater Cairo Urban Transport Master Plan, CREATS. (2003). Updated, (2014).
Ministry of Planning of Egypt. (2016). Sustainable Development Strategy: Egypt Vision 2030.
New Generation Motors Corporation. (2002). Impact of EV and HEV Applications in the Reduction of Greenhouse Gas Emissions in Egypt, and Socio Economic Analysis. *Social Fund for Development.*(SFD), Cairo, Egypt
Probst, M. M., & Roecker, C. (2007). Towards an improved architectural quality of building integrated solar thermal systems (BIST). *Solar Energy, 81*(9), 1104–1116.
Sioshansi, R., & Miller, J. (2011). Plug-in hybrid electric vehicles can be clean and economical in dirty power systems. *Energy Policy, 39* (10), 6151–6161.
Statistics, R. C. (2017). International Renewable Energy Agency (IRENA).
Suzuki, H., Hyogo, T., Hyodo, M., & Saito, E. (2010). Development of charging station for plug-in hybrid vehicle. *Toyota Industries Technical Review, 59*, 54–57.
Tie, C. H., Gan, C. K., & Ibrahim, K. A. (2014). May. The impact of electric vehicle charging on a residential low voltage distribution network in Malaysia. In: *2014 IEEE Innovative Smart Grid Technologies-Asia (ISGT ASIA)* (pp. 272–277). IEEE.
Wamda. Reasons behind the mushrooming of electric cars on Jordan's roads [Q&A]. https://www.wamda.com/memakersge/2017/10/reasons-mushrooming-electric-cars-jordans-roads-qa. Accessed 15 Apr 2019.
Wong, C. M., Atkinson, R. W., Anderson, H. R., Hedley, A. J., Ma, S., Chau, P. Y. K., & Lam, T. H. (2002). A tale of two cities: effects of air pollution on hospital admissions in Hong Kong and London compared. *Environmental Health Perspectives, 110*(1), 67–77.

Towards Greener Neighborhoods: A Case Study for Street Renovating Solutions in Cairo

Zeina Mohamed Zedan, Germin El-Gohary, and Sherouk Mohamed

Abstract

A sustainable "green" city outlined while thinking of the social, economic, and environmental effects. It considered energetic living space for the existing residents while considering the impact of these sites on the upcoming generations and the population growth, fulfilling the resident's needs without ignoring the future society's capacity to meet their needs. The continuous increase in population is imposing challenges globally, leading to urban poverty and limited resources. Having overpopulated cities leads to deficiencies in infrastructure and residents' quality of life. Handle the city's problems and needs, and it is necessary to organize and thoughtful, considering solving the city's desires without distorting other vital factors of people's living situations. This specific paper will focus on the green mobility network and its impact on Cairo's quality of life. The Capital of Egypt. The purpose of this paper is to implement good street designs from a public space perspective. The way was to review neighborhood hierarchy, street classifications, and design. Introducing the Green Mobility Network that combines bicycles and pedestrians (Officials, C. T. (2019). Lane Width. 1–11.). Two international case studies networks were analyzed. Introducing findings and recommendations by proposing a project in Cairo's problematic streets in Nasr city to find solutions for Cairo's neighborhood streets will help make it more efficient and sustainable.

Keywords

Green mobility • Road design • Overpopulation • Streets • Renovation

Z. M. Zedan · G. El-Gohary (✉) · S. Mohamed
Ain Shams University, Cairo, Egypt
e-mail: germin_elgohary@eng.asu.edu.eg

Z. M. Zedan · S. Mohamed
British University in Egypt, Cairo, Egypt

1 Introduction

Cairo's streets are always a challenge for people due to the overcrowding of streets with cars and other mobility problems. Roads used for different purposes beyond directly linking destinations act as shared spaces for multiple users, including different genders and ages, where they engage in social, cultural, political, and economic activities (Tobergte & Curtis, 2013). Therefore, streets signify the main factor of public space that embodies different users with several interests.

Investigating sustainable neighborhoods and their green dynamics in general and Cairo's streets in specific is a multidisciplinary topic. It attracts Egyptian urban designers, city planners, environmentalists, public health authorities, and city managers. UN Habitat (2013) mentioned that streets designed for different activities, like; walking, cycling, and public transportation, rather than being car-oriented streets, would result in fewer crowds. If it intended to encourage this, people could cycle or walk rather than taking cars or public transportation. As a result, it cuts down the need for fossil fuels and fewer carbon emissions, which is more ecofriendly and will lead to more sustainable urban development (Ewing & Handy, 2009). These streets can inspire social variety and connectivity; therefore, having neighborhoods that are more unified and lively is more attractive for the residents and the visitors (Kheir-El-Din et al., 2012).

2 Problem Statement

Cities that are considered more developed are the ones that recognize the importance of having proper public spaces. In contrast, cities with lower productivity and quality of life often fail to have adequate street designs and functions (Tobergte & Curtis, 2013). Cairo is the most populated city in Egypt, with highly congested neighborhoods and streets. According to the world population review in 2018, Cairo has

F. Rosso et al. (eds.), *Advances in Architecture, Engineering and Technology*, Advances in Science, Technology & Innovation, https://doi.org/10.1007/978-3-030-86913-7_5

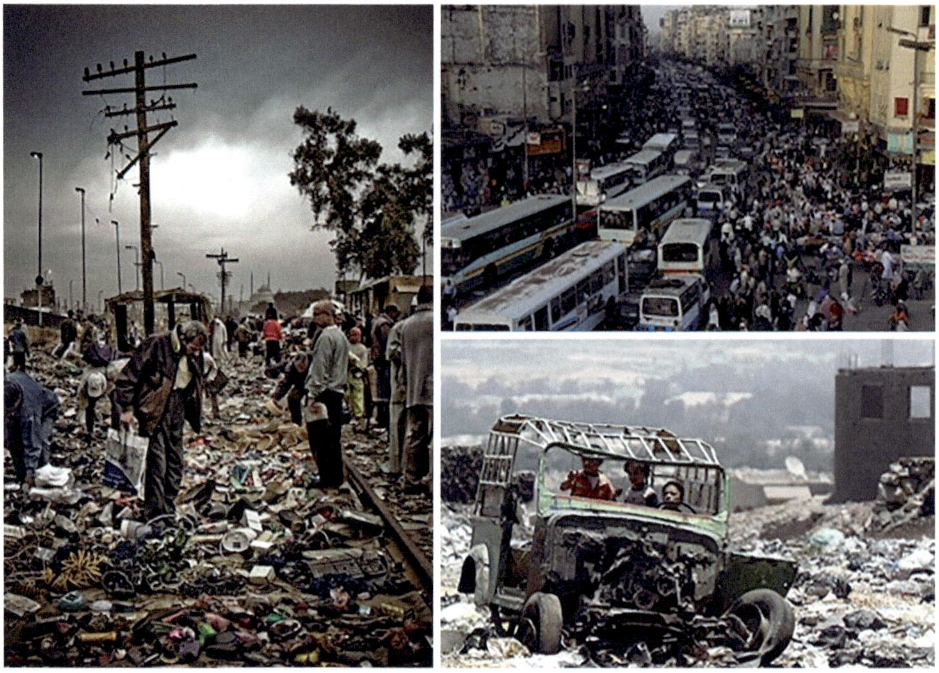

Fig. 1 Cities problems

a population of 16.4 million, with an expectation to reach 24 million by 2027 compared to the second-largest city in Egypt, Alexandria, with a population of 5.71 million. Accordingly, this statistic indicates the severity of overpopulation in Cairo's neighborhoods and its outcome on Cairo's streets. Our roads are jammed and filled with problems to be solved regarding design and planning, as shown in Fig. 1. We are trying to address Cairo's street problem as a public space, making people more comfortable while walking, driving, cycling, or other physical activities (Jan Gehl, 2011).

3 Overpopulation Street Impact

A. Population Growth has Harmed the Surroundings

For living purposes, people use more land; thus, the surrounding environment changes. The population grows in certain cities, so they use more resources to the realm of the population's well-being (Weber & Sciubba, 2018). "On November 17, 2016, the world population clock displayed the existing world population: 7, 465, 023, 315, and increasing quickly. World widely, overpopulation threatens the world. According to the UN's medium plan, the growing population of seven billion expects to reach at least nine billion. Every year, the world population grows by 1.2%. Although this number may not sound shocking, the global population is doubling in less than 50 years. Every second of every day, the birth of nine lives into the world. Every 10 s, 44 people are born, which counts around 140 million people per year" (Doris, 2017).

Overpopulation is one of the biggest problems on streets and living standards, as shown in Fig. 2 in Cairo's streets, the capital of the 14th largest country in population. Its purpose is to have better street designs, more green spaces, a healthier lifestyle, and more appropriate ways to solve Cairo's street problems for a better quality of life. Our cities should grow without fearing the future; low environmental quality, economic problems, and cultural breakdown-threatened city life (Rudofsky, 1969).

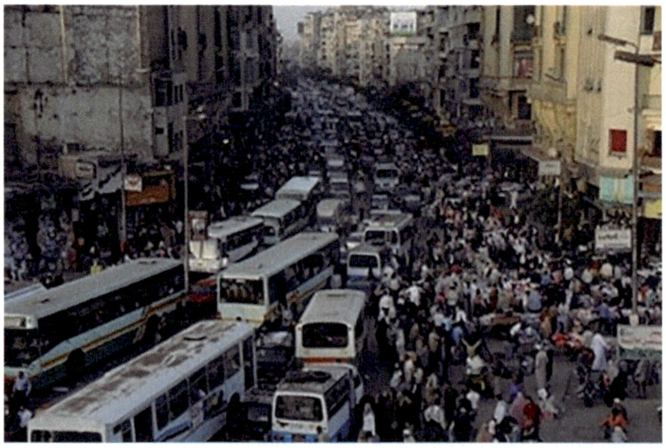

Fig. 2 Streets of Cairo, Egypt

B. Impact of Population Growth on Infrastructure

Various perceptions have been progressed with an end goal to realize the connection between infrastructure request and improvement. To begin with, the supply of infrastructure prompts social, financial improvement, and besides, the improvement potential makes interest for better infrastructure. Along these lines, infrastructure facilities are imperative to the social development of urban and rural regions through direct and indirect advantages (Asoka, THUO, & Bunyasi, 2013). The exponential growth in population and the rising improvements have put difficult stress on the essential framework in the neighborhood. This incline that requires an extended or upgraded foundation to have the capacity to successfully carry important and basic infrastructure needs. For example, the place to stay, water supply, security, instruction, health & waste management components, and road networks. All these are considered the main public spaces of neighborhoods to address the necessities of the expanded population (Asoka et al., 2013).

C. Impact of Population Growth on Road Network

Street structures and transportation systems form the spatial organization of any city. The quality of the roads and streets is important in improving accessibility and connectivity. Considering a street as a main public space, people should feel comfortable in. They should be relaxed while transferring from one place to another by; driving, walking, or cycling in the street (Asoka et al., 2013). However, the increase in population may affect street productivity. As a result, road traffic congestion resulted from poor street conditions. According to a 2007 report by the World Bank, 20% of the streets are in a good state but suffer from low service quality and are at risk of early disaster. It can be rescued if there is a considerable rise in maintenance actions (IFC, 2007). The report also stressed that the area is loaded with infrastructural inefficiencies and old-established constructions. These harmfully affect the movement of trade, the effectiveness of Egyptian manufacturing, and road safety.

D. Overpopulation in Egypt

Overpopulation in Egypt remains a threat to the country's resources and sustainable growth, according to Egypt's improvement toward Vision 2030. "Throughout the 1st gathering of the 4th National world Youth Conference in Alexandria on July 24, 2017, President Abdel Fatah al-Sisi said that overpopulation and terrorism are Egypt's two real

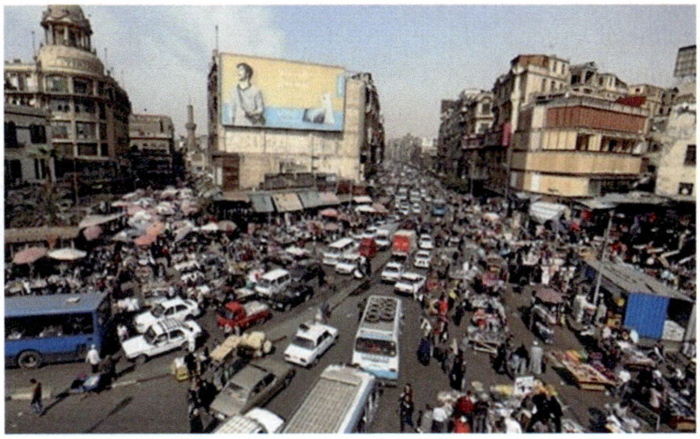

Fig. 3 Cairo traffic and street problems

dangers. Later in October 2017, the Central Agency for Public Mobilization and Statistics (CAPMAS) stated that the population in Egypt has doubled over the last 30 years as it increased from 48 million in 1986 to 95 million in 2016, that increased the traffic specially in Cairo as shown in Fig. 3. Furthermore, according to CAPMAS, Egypt drops approximately 60,000 acres per year as a result of soil erosion and construction to meet population needs, including housing" (Al-Wahaidy, 2005).

E. Egypt's Attempts for Managing Overpopulation

Egypt announced its first population target to back off the population growth rate and speed up financial development in the mid-1960s. In 2013, the human population's effect anticipating the personal satisfaction and economic improvement stated at the National Population and Development Conference. It created an interministerial group organized by the National Population Council (NPC) to build up a complete National Population Strategy (NPS) 2015–2030 to confront overpopulation (Asoka et al., 2013). The National Population technique defines enhancing population control and all Egyptians' satisfaction. It understands administrations and endeavors, including government, private and public segment, media foundations, and volunteers (Al-Wahaidy, 2005).

F. Negative Human Behavior in Egyptian Streets

There are a lot of factors affecting the streets other than its design; the truth is that traffic congestion is caused by multiple causes. One of them is human behavior, which is a very important factor (Rosen, 2013). Egyptian visitor's first reaction is its national hymn: vehicle sirens, the famous

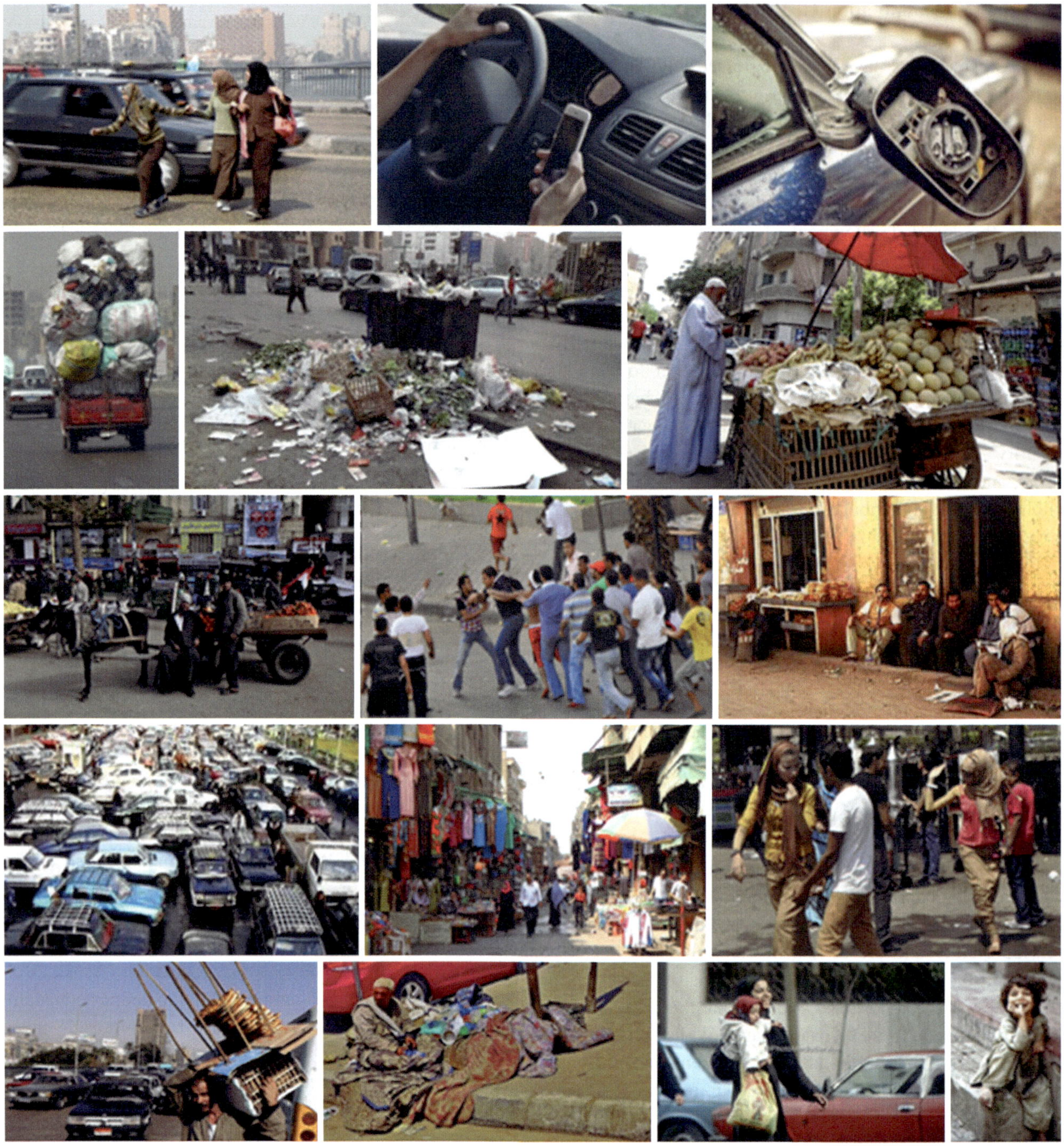

Fig. 4 Disorder on Egyptian roads (*Source* The author

Egyptian traffic. Car horns say "Hello" for the warning, "Watch out I am mergin into your lane whether you like it or not," shouting (EFGHERNES, 2013). Egyptian disorder roads impact, leads to more than 10,000 deaths per year, as shown in Fig. 4. Accidents range from crushed citizens by lorries while crossing the street to crash into busses. For example, Assiut's school bus disaster, which killed 51 children in 2012 (EFGHERMES, 2012).

The malicious Human Behavior in Egyptian streets is collected in Fig. 5, which shows around 14 specific negative behaviors.

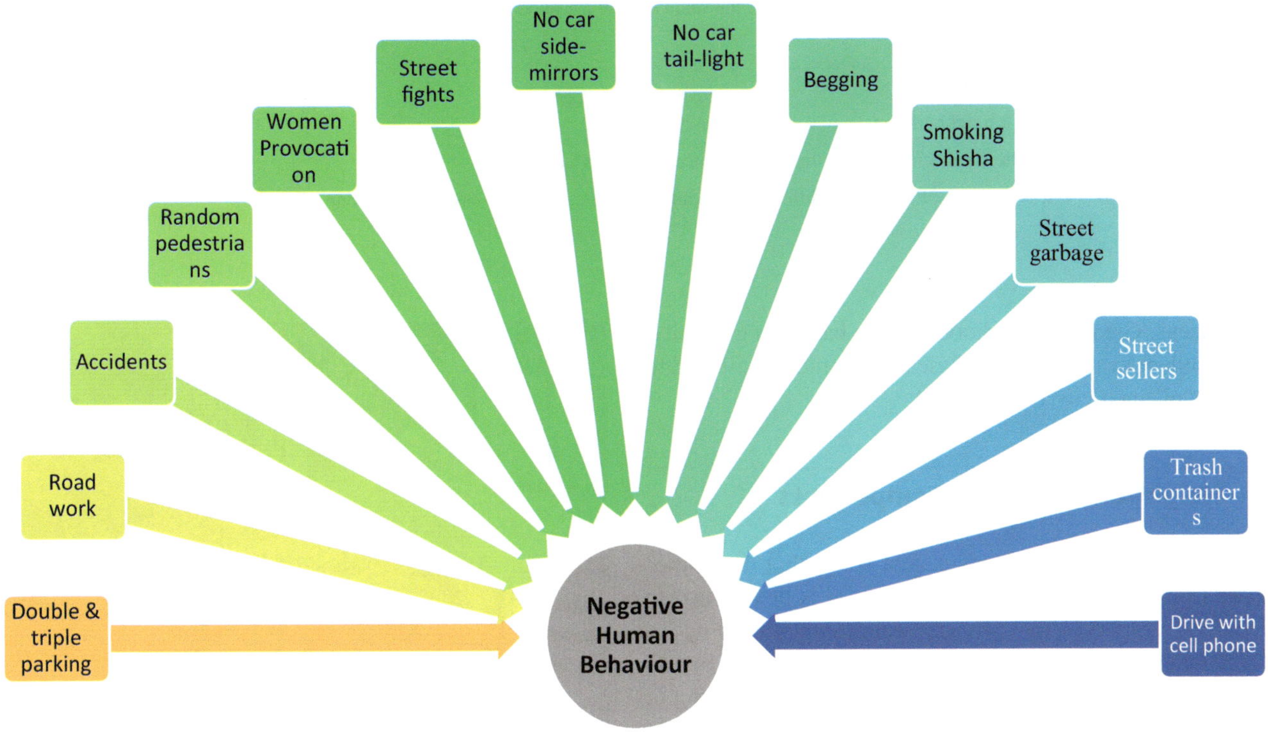

Fig. 5 Negative human behavior in Egyptian streets (*Source* The author)

4 Neighborhoods and Its Importance

A neighborhood is a local community within a larger city town, district, or country. To understand what a neighborhood is and how it was created, we must realize the urban hierarchy to form a city (Fig. 6) Urban Hierarchy in the creation of the City (Table 1): This hierarchy is concluded by the author; 2 District. City.

Residential group. → Neighborhood. → District. → City.

Specifically discussing neighborhoods, it is defined as a specific physical area and functionally as a social network. Neighborhoods are also a place where face-to-face social communications happen, the personal situations to recognize conventional morals, socialize youth, and keep real social control. Good communities result from careful planning and thoughtful designs that generate sustainable, vibrant, walk-able, livable, and social, which increases the quality of life

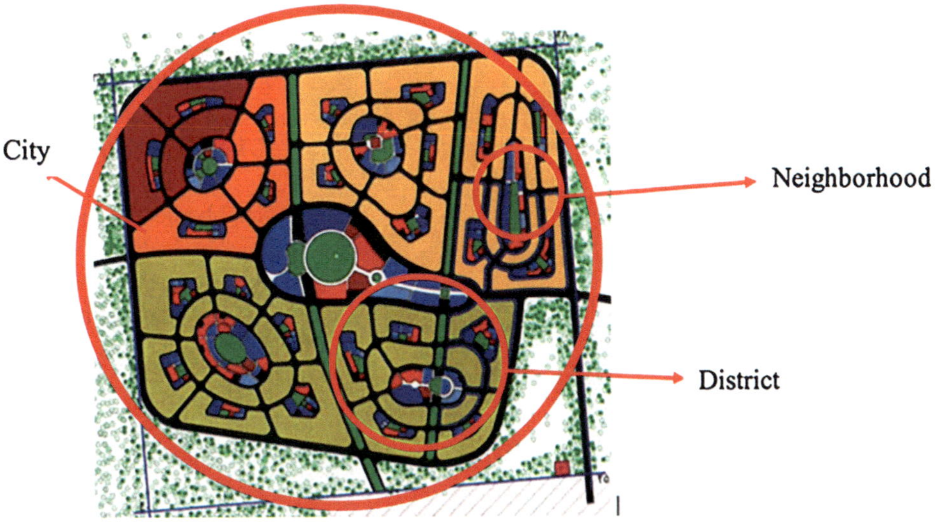

Fig. 6 Urban city hierarchy (*Source* The author)

Table 1 Urban hierarchy in Neighborhood importance as the main base for a good urban city (*Source* Deer 2013)

Hierarchy	Components	Number of residents
City	4 or 8 districts	150 000–300 000 resident
District	4 or 8 neighborhoods	20 000–25 000 resident
Neighborhood	Smallest residential space that offers educational services	5000–7000 resident
Residential groups	It is a group of similar or different residential buildings that are assembled in one place	_

for citizens. Great neighborhoods are the base of a great city. An excellent healthy planned base neighborhood will lead to a better-planned city (Deer, 2013).

5 Neighborhood Main Components

It is the smallest scale of residential space that accommodates from 5000 to 7000 residents. Its size is determined to provide a school for primary education see (Fig. 7). Depending on the number of children at the age of five to eleven. There are certain circumstances that make a living in the neighborhood great; lifestyle match, great schools, pride in ownership. A sense of ease and calm come from low crime rates, outdoor activities, stepping back in time, medical care access, family-friendly, close to public transportation, nearby shopping and restaurants, nightlife and entertainment, and walkability (Trulia, 2014). Therefore designing the outdoor space and street network is very important for living a high quality of life in the neighborhood (Fig. 8).

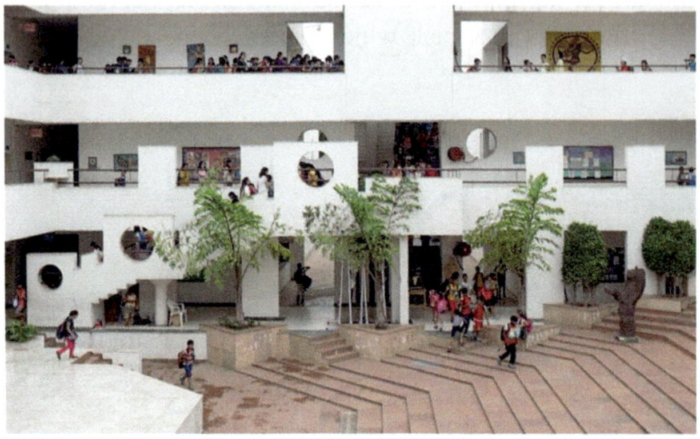

Fig. 7 Primary school in a neighborhood

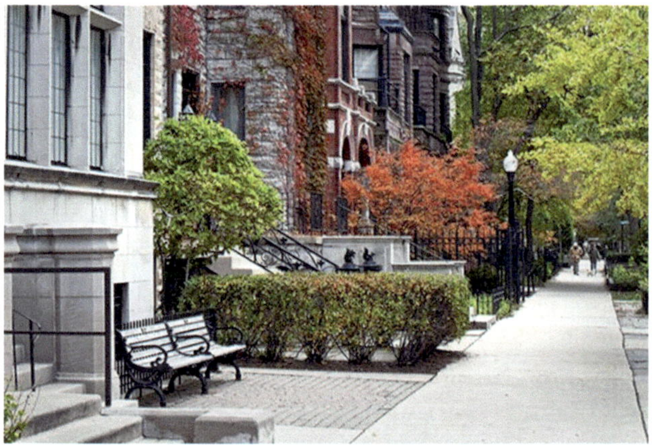

Fig. 8 Good neighborhood (Trulia, 2014)

6 Residential Categories

Residential land is divided into three categories based on the research findings see (Fig. 9), as shown in Table 2 and listed in the following;

A. 1st category, small area.
B. 2nd category, medium area.
C. 3rd category, large area.

7 Analysis of Findings

7.1 Neighborhood Streets Classification

Neighborhood street classification is derived from the relationship findings between all neighborhood categories. All three categories of areas in the neighborhood must be connected by streets. Different kinds of streets float in any neighborhood.

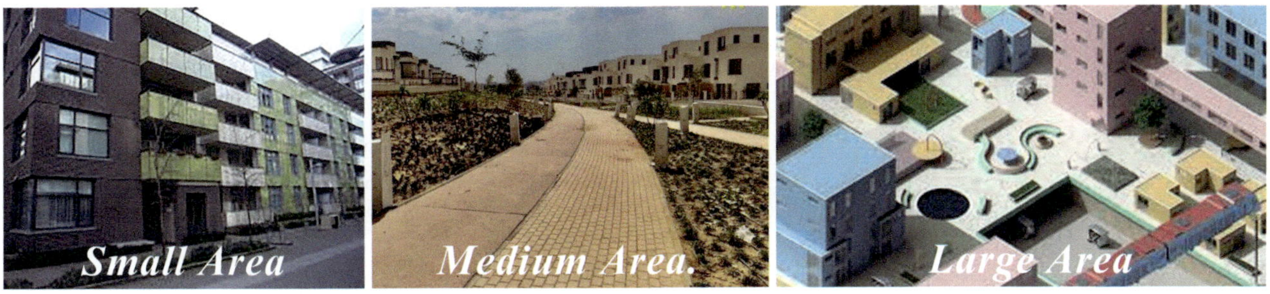

Fig. 9 Residential categories analyzed by the (author)

Table 2 Residential program (Bevan & Croucher, 2011)

Category	Description	Area	Street width
Small area	The distance between the two successive streets is sufficient for the existence of two pieces of land (back-to-back), and this distance is approximately 36 m	150 m^2 or less	9 m
Medium area	The distance between the two successive streets is sufficient for the existence of two pieces of land (back-to-back), and this distance is approximately 54 m	150–400 m^2	12 m
Large area	The distance between the two successive streets is sufficient for the existence of two pieces of land (back-to-back), and this distance is approximately 72 m	400–700 m^2	12 m

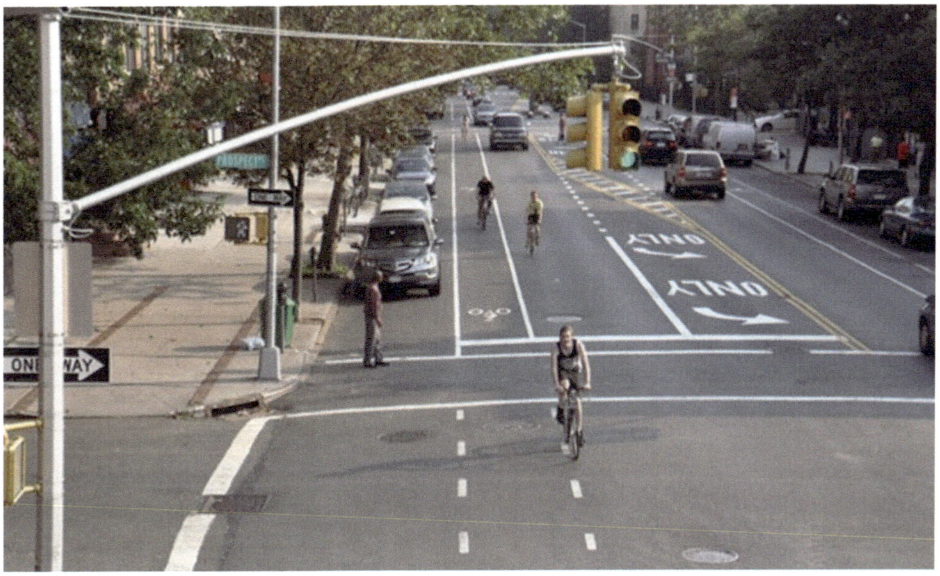

Fig. 10 Neighborhood main street

A. **Neighborhood Main Streets**

The neighborhood life connection is the main neighborhood streets with high pedestrian capacities. Main streets should limit traffic speeds and create a well-designed pedestrian crossing. Roads should be from four to three lanes of car travel lanes, with bike lanes and a median see (Fig. 10). The main streets should contain the following elements shown in Table 3.

7.2 Neighborhood Local Streets

Native streets in residential neighborhoods should be safe and pleasant walking areas with direct access to different zoning, as shown in Fig. 20. They can be 10–20 m wide (McCutchan, 2013) Its primary function is serving the building since most of the building facades and building entrances overlook this type of road, as shown in Fig. 21. Neighborhood streets differ in design types according to

Table 3 Main streets elements (*Source* The author)

Element		
1. Lane reduction or road recanalization	Road diets increase traffic flow and decrease clashes with turning cars or any vehicle while improving the road's efficiency In a 3-lane arrangement, the conflicts eliminated	Figures 11 and 12
2. Turn lanes	It is a single lane in the center of a road that is used for turning cars. It allows cars to turn left without disturbing the traffic movement. See Fig. 13	Figure 14
3. Bike boxes	It's a bike sign on the car lanes that allows bikes to turn left or right in front of traffic at the red light. See Fig. 15	Figure 16
4. Parklets	It is a sidewalk extension that offers more space for people using the street. Usually installed on parking lanes. Parklets generally range out from the sidewalk at the same level to the width of the adjacent parking space (NACTO, 2013). See Fig. 17	Figures 18 and 19

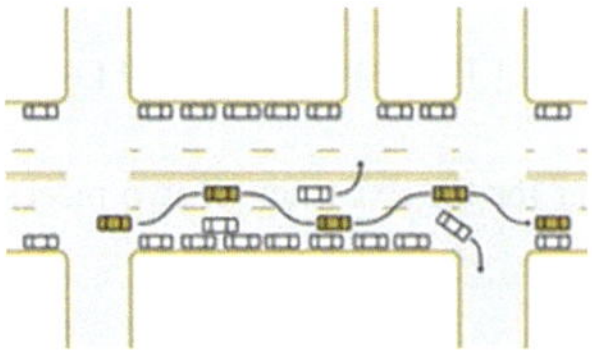

Fig. 11 Road diets

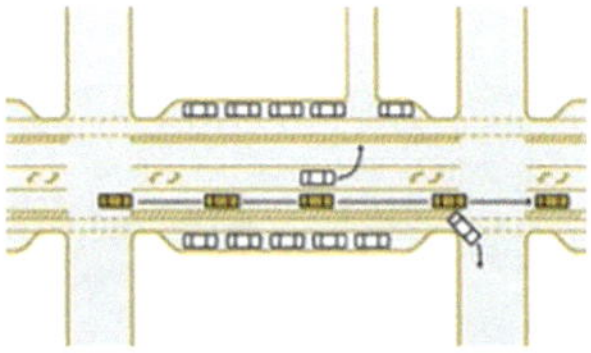

Fig. 12 Three lanes arrangement

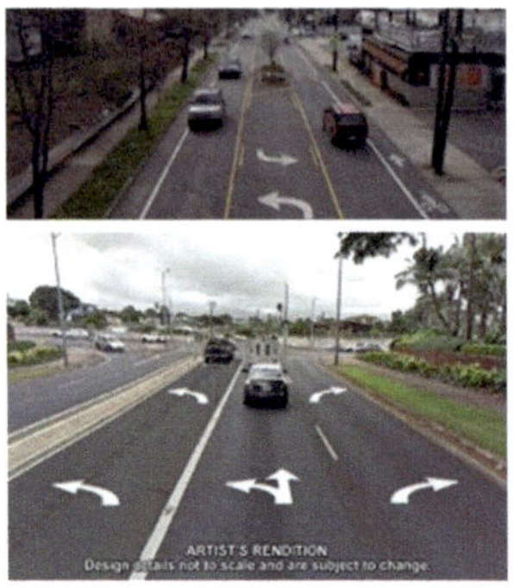

Fig. 13 Turn lane

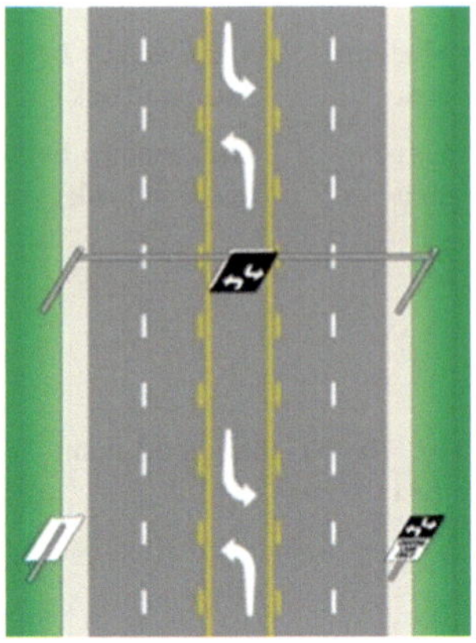

Fig. 14 Turn single lane

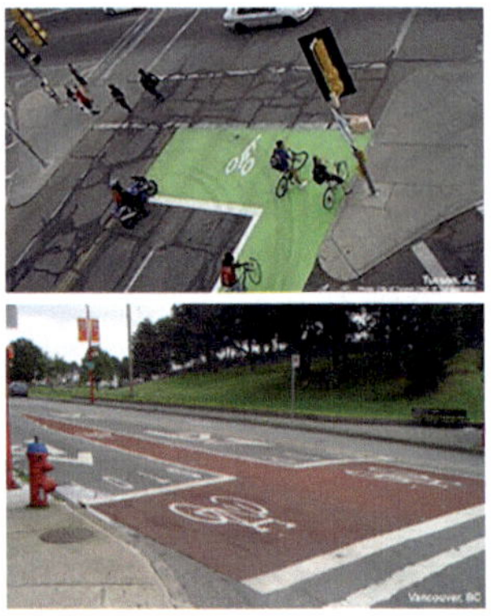

Fig. 15 Bike boxes in Vancover

their described speed, parking, signals, lanes, and sidewalks density; for example, high-speed streets design don\t allow parking lots while urban streets allow it and so on, as shown in Table 4.

7.3 Street Elements

Street elements are the components that create the city roads including; sidewalks, drive lanes to see (Fig. 22), curb extensions creating safe crossing, gateways, bus bulbs, pinch points, and chicane (NACTO, 2013).

7.3.1 Road Lanes

The width assigned to lanes (Fig. 23) for bikes, motorists, trucks, buses, and parked cars is a critical feature of any street design. Each lane width should be well-thought-out after understanding the traffic goals of each street. Travel lanes are lined to describe the planned path of travel for

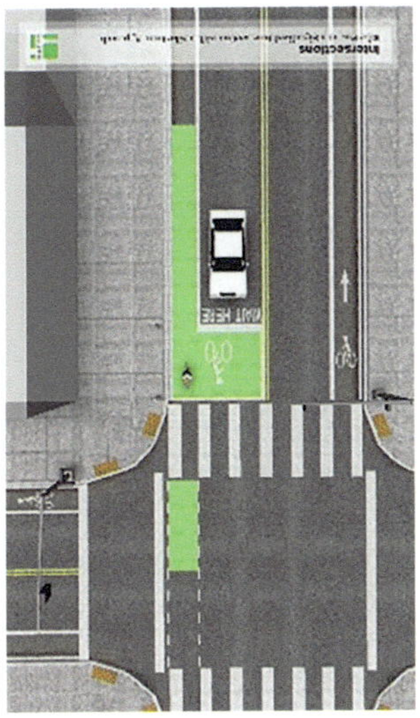

Fig. 16 Bike box

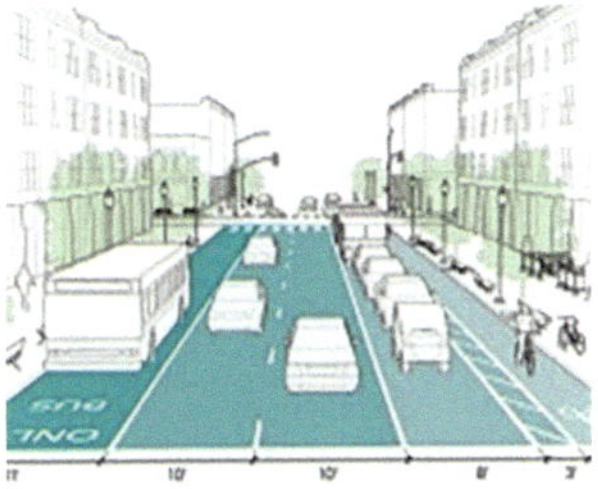

Fig. 21 Local streets (McCutchan, 2013)

automobiles. Generally, broader travel lanes (3–4 m) were preferred to make a more forgiving barrier to drivers, particularly in high-speed streets where thin paths or roads may feel uncomfortable or lead to sideswipe crashes (NACTO, 2013).

A. Car Lanes

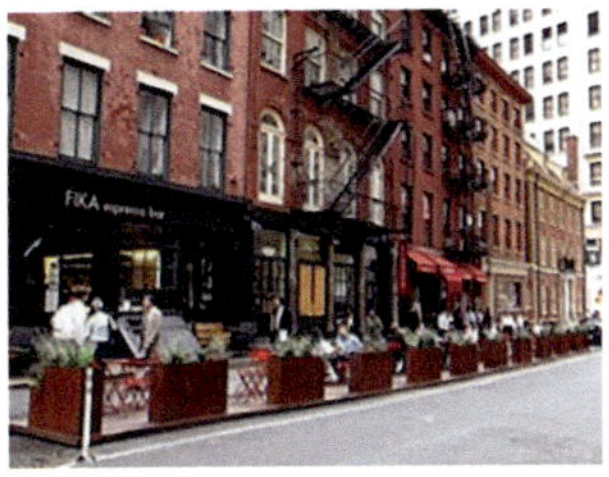

Fig. 17 Parklets

The sizes of car lanes usually range from 9 to 15 ft. or 2.7 to 4.6 m. Lane widths are often finer on low-volume roads and broader on higher volume roads. There's no proof that more wide paths are secure. At the point when trails turn out to be very wide, they turn out to be less protected. Crash frequencies increase once path width exceeds 3.4 m, and have notable more significant amounts of accidents at 3.7 m or more. One possible explanation behind this is drivers may increase the speed with more wide paths (McCutchan, 2013).

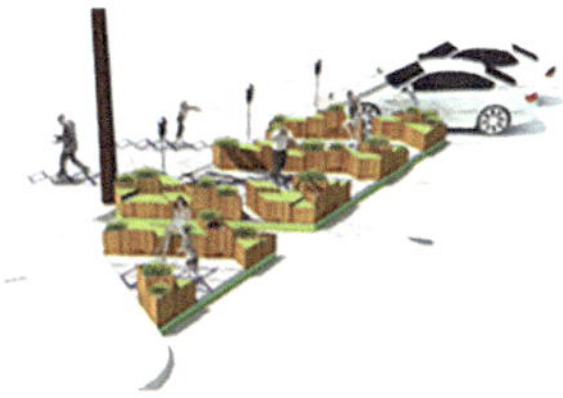

Fig. 18 Parklets

B. Bike Lanes

A Bike lane definition is; a well-defined space of the road labeled by signage and pavement markings for bicyclists. It is logical, to begin with, basic concepts and characteristics. (Wisconsin Department of Transportation, 2009).

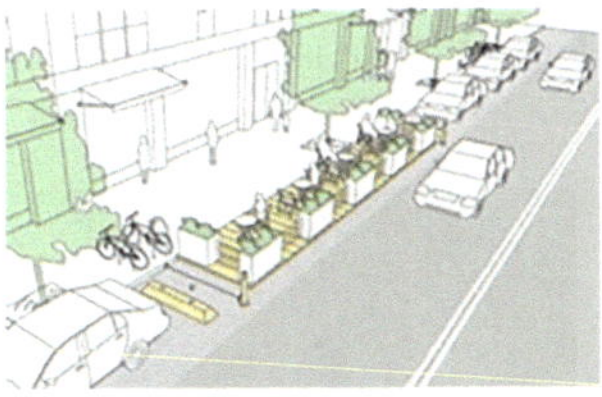

Fig. 19 Parklets

The minimum width of a bike lane should be 1.2 m in "No curb" roadways. If parking is allowed, the bike lane should be placed between the parking area and the travel lane and have a minimum width of 1.5 m (Fig. 24) (Flowers, D., Warne, T., & Pyers, 1999) (Fig. 25).

- If no curb/no parking: 1.2 m.
- If curb and gutter/no parking: 1.5 m from the curb face.
- If parking: generally 1.5 m (Flowers, D., Warne, T., & Pyers, 1999).

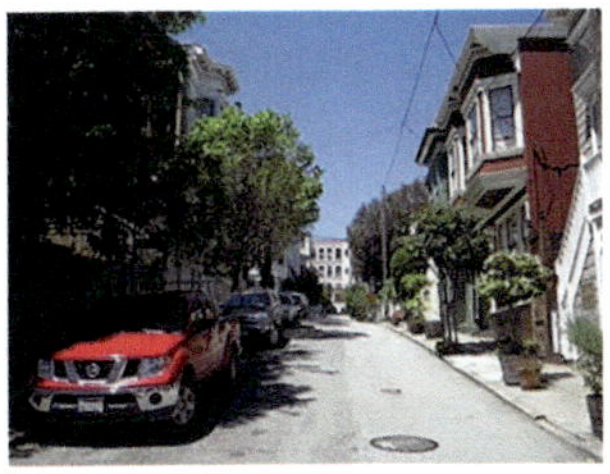

Fig. 20 Local streets

Bicycles are between 0.60 and 0.9 m wide. An adult tricycle or a bicycle trailer is approximately 0.80–1.1 m wide. The length of a bicycle is about 1.5–1.8 m. The longitudinal distance grows from 2.6 to 2.9 m see Figs. 26 and 27 (Wisconsin Department of Transportation, 2009).

Table 4 Design types and elements (*Source* The author)

Design types	Description	Parking	Signals	Lanes	Sidewalks
High-speed streets	These are the streets with very low driveway	No parking is allowed	Signals are rare, spread out at long distances	Maybe multilane	A moderate-high density roadside expansion
Suburban streets	They signify streets of low driveway	Parking is allowed in some parts	Moderate signals	Maybe multilane	A moderate-high roadside expansion, unlike suburban streets
Intermediate design streets	They signify urban streets with reasonable driveway	Parking is allowed in some parts	Moderate signals	Maybe multilane	High roadside expansion, unlike roads on suburban streets
Urban streets	They signify urban streets with high access point density	They usually provided with roadside parking	More signals	Maybe multilane	Highest roadside expansion density between all previous types

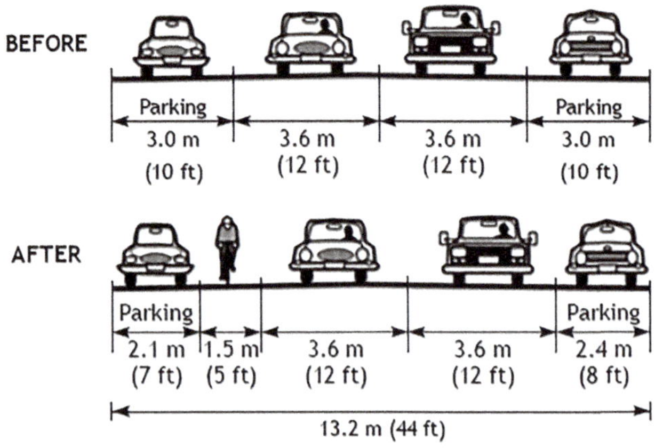

Fig. 22 Street elements

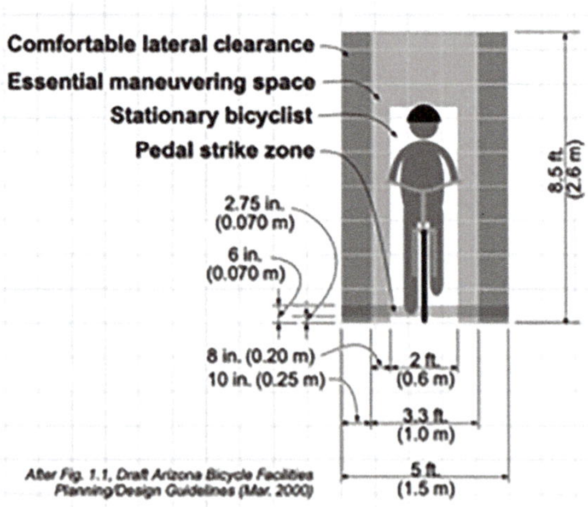

Fig. 24 Bike lane dimensions

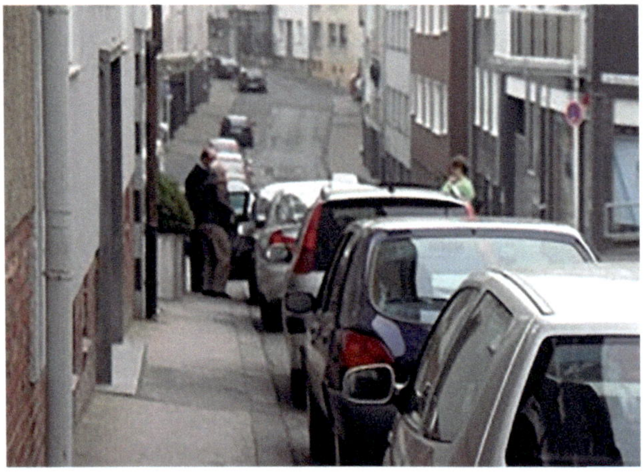

Fig. 23 Parallel parking

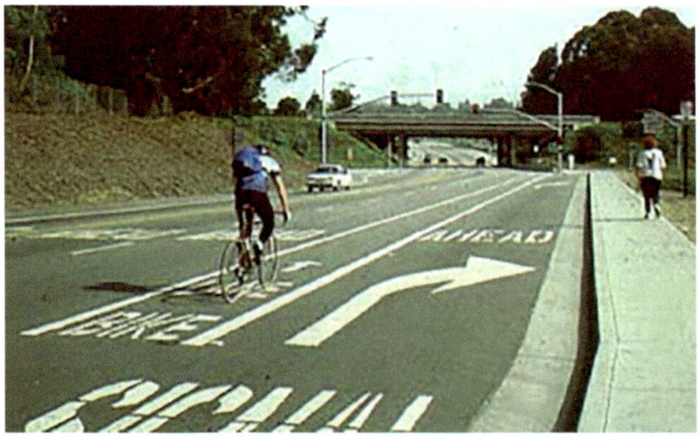

Fig. 25 Bicycle lane at Austin Texas (Smith et al., 2019)

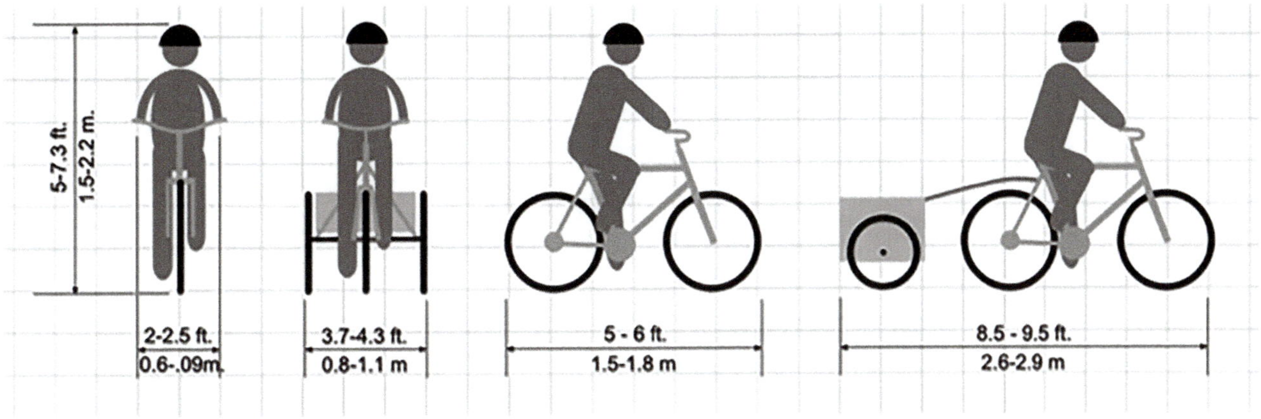

Fig. 26 Common dimensions for bicycles, tricycles, and bikes with trailers (*Source* Wisconsin Department of Transportation 2009)

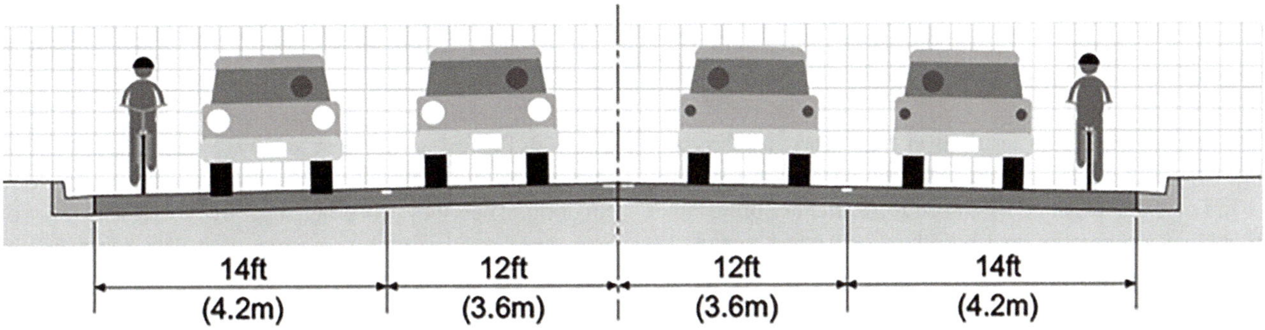

Fig. 27 Standard "wide outside lane" configuration showing a (4.2 m) outside lane and a 12ft (3.6 m) inside lane

7.3.2 Conventional Bike Lanes

They are bike lanes that use pavement markings and signage. The bike lane is adjacent to car travel lanes, and in the same direction, it flows as shown in Fig. 28). hey are usually on the right side of the streets, between the travel way and the road edge, or the parking lane. It can be located on the left side when used in a one-way street. This type increases comfort for cyclists on busy roads and creates a separation between both cyclists and motors for more safety. Bike lanes can be applied on streets with 3000 fewer motor vehicle traffic daily, on roads with high transit vehicle volume, streets with high traffic volume, and regular truck traffic as shown in Fig. 29 (Cart, 2019).

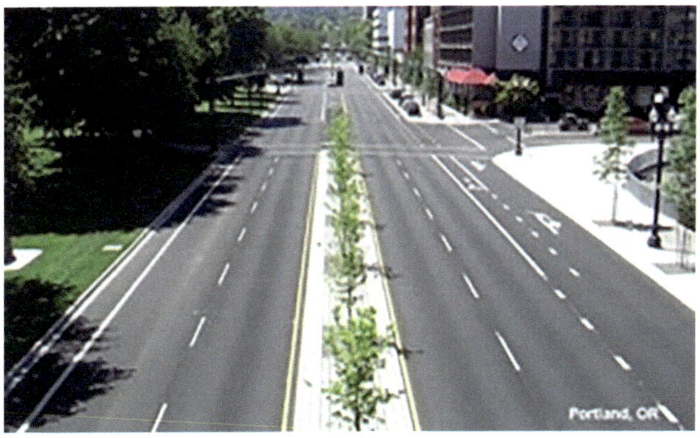

Fig. 29 Bike Lane, Portland, OR (*Source* NACTO 2013)

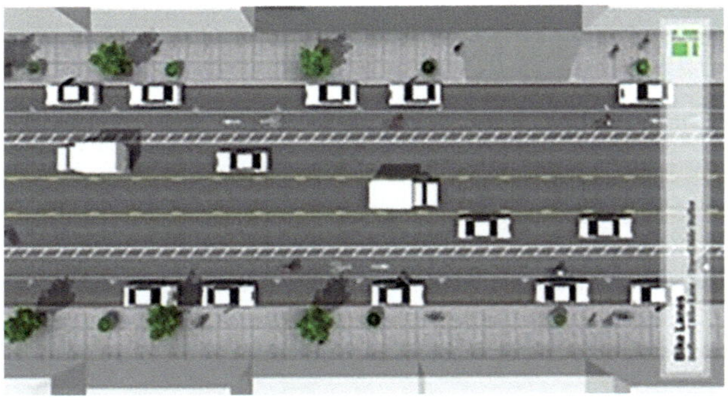

Fig. 28 Conventional bike lanes

7.3.3 Buffered Bike Lanes

It is a conventional bike lane that is joined by a buffer space, separating between the cyclists and motor vehicles or parking lanes, as shown in Fig. 30. It allows extra space created in between the cycling lane and travel lanes (Cart, 2019) A buffered

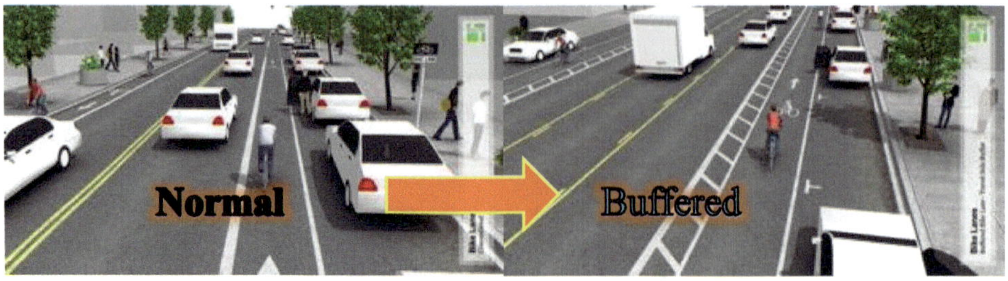

Fig. 30 Buffered bike lane (*Source* The author)

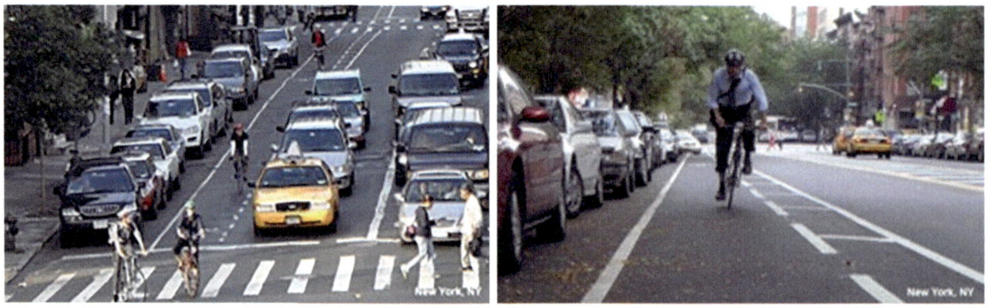

Fig. 31 New York buffered bike lanes (*Source* NACTO 2013)

bike lane trial has been installed in New York, as shown in Fig. 31, creating a safe zone for bikes in crowded streets.

In Los Angeles, there was an additive trial in Spring street by adding a green buffered bike lane, as shown in Fig. 32. The city's first green lane was created first by the buffered bike lane; bike counts were taken before and after showing a 52% increase in cyclists on Spring Street. Figure 33 indicates a 250% increase in cyclists on weekends and a 161% increase in female riders. His evidence encourages the city to increase the investment in; bike lanes, bike paths, and separated or buffered infrastructure (Newton, 2012).

7.3.4 Contraflow Bicycle Lanes

Contraflow bicycle lanes' design allowed cyclists to ride in the opposite direction to the travel lane. They provide access to bicyclists traveling in both directions and connectivity, as shown in Fig. 34, to decrease sidewalk riding. In the streets, large numbers of bicyclists use them to ride in the wrong direction. Thus they work better at low speed and low volume streets unless the provision of buffer separation or physical protection to see (Fig. 35) (Cart, 2019).

7.3.5 Left-Side Bike Lanes

They are conventional bike lanes located on the left side of one-way streets or two-way streets with medians. Left-side bike lanes offer advantages along roads with heavy transit use and frequent parking on the right side. It avoids possible right-side bike lane clashes on streets, reduces door zone clashes next to parking because of rarer door openings on cars' passenger side. In Commonwealth Avenue, in Boston. The bike lane is located either; to the left of the vehicle road or the park's right see (Fig. 36). It is easy for a bike to travel faster

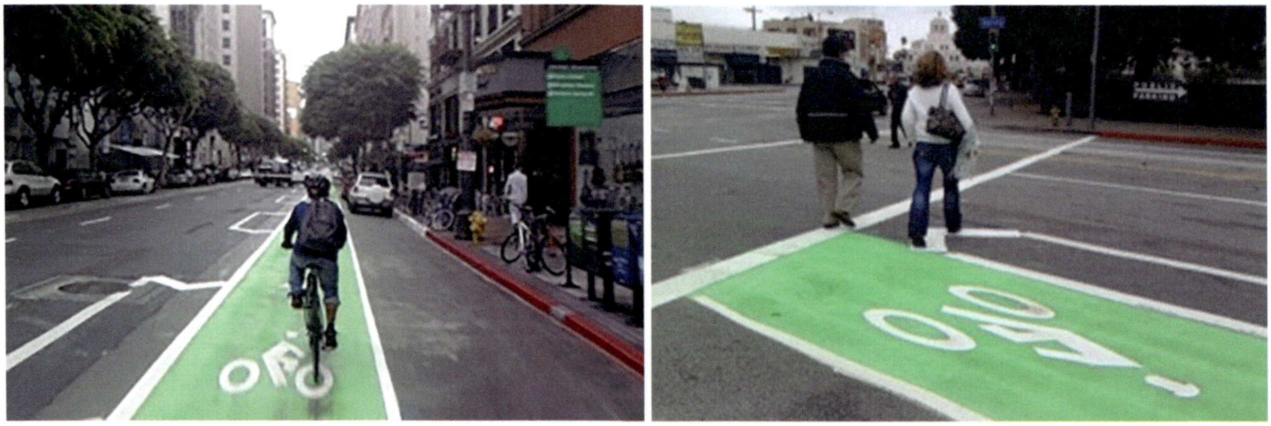

Fig. 32 Spring street (*Source* Linton et al., 2020)

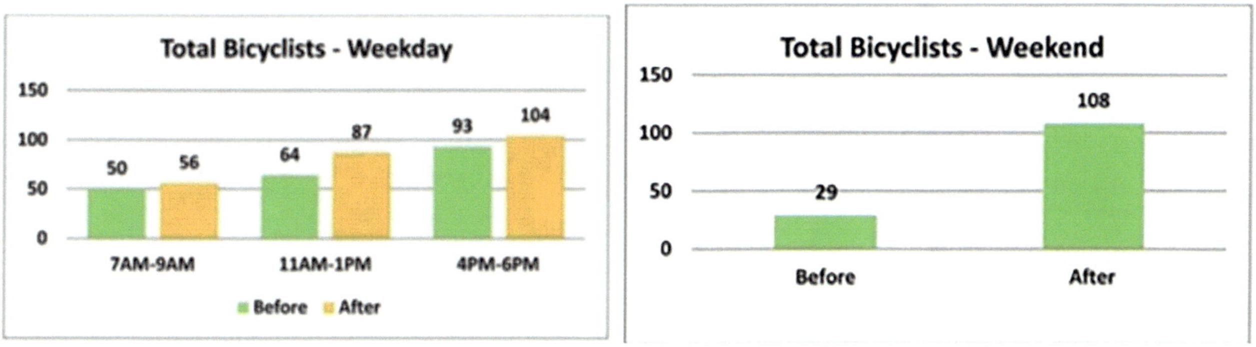

Fig. 33 The stats from Spring street (*Source* Newton, 2012)

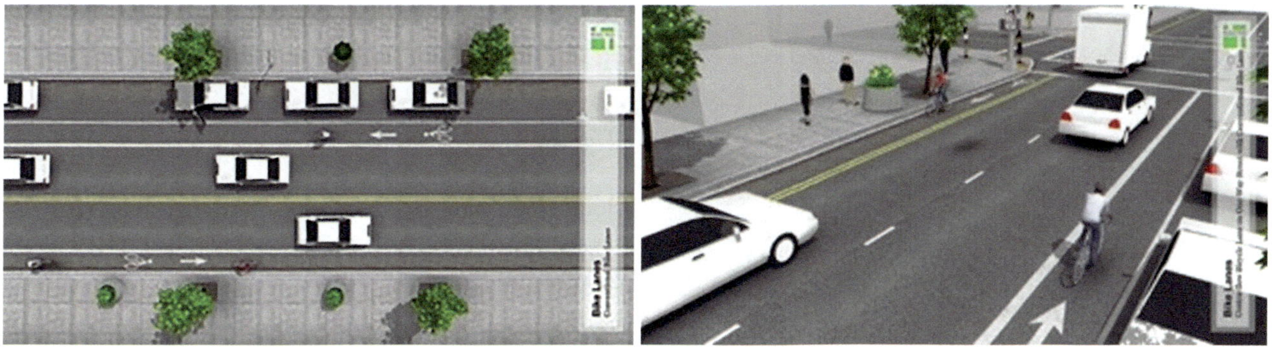

Fig. 34 Contraflow bike lanes

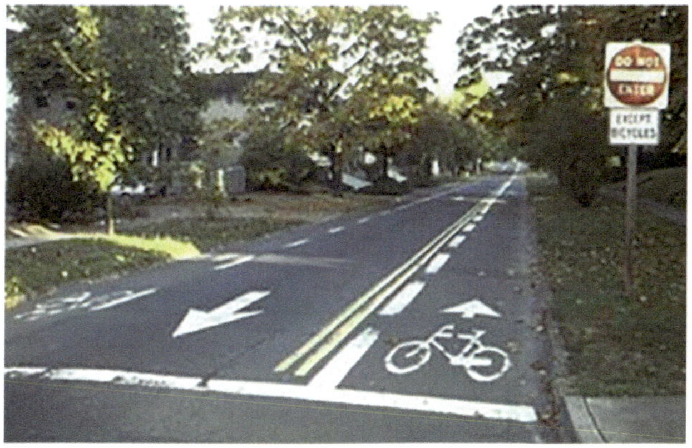

Fig. 35 Contraflow bike lanes can provide direct access to high-use destinations

than a car in heavy traffic, it is not next to any parked cars, and so there is no risk of dooring see (Fig. 37) (Almy, 2012).

7.3.6 Sidewalks

Walkways fill in as the visible strides to the city as open spaces, socially and financially activating roads. Sidewalks see (Fig. 38) play an active part in city life. They increase their connections, thus encouraging walking (NACTO, 2013).

Sidewalk zones are (NACTO, 2013);

A. Frontage zone
   The front zone includes the walkway piece that serves as an extension of the building, either entrances and doors or walkway cafes. The frontage zone contains both the structure and the building's facade facing the street; space ends to end to the building.
B. Pedestrian sector
   The pedestrian zone is the leading open path that goes equivalent to the street. The pedestrian region confirms that walkers have a safe place to walk and should be 1.5–2 m wide in residential locations and 2.5–3 m wide in commercial areas.
C. Street furniture
   Space of street furniture is well-defined as the pedestrian sidewalk zone in which street furniture, such as benches, lighting posts, newspaper booths, trees, and bicycle parks. It also includes green substructure elements, such as rain gardens or flow-through planters.
D. Buffer zone
   Buffer zone directly follows the walkway that might include a range of different elements as; curb extension parklets, bike racks, parking, stormwater management features, cycle paths, and bike-share stations see (Fig. 39).

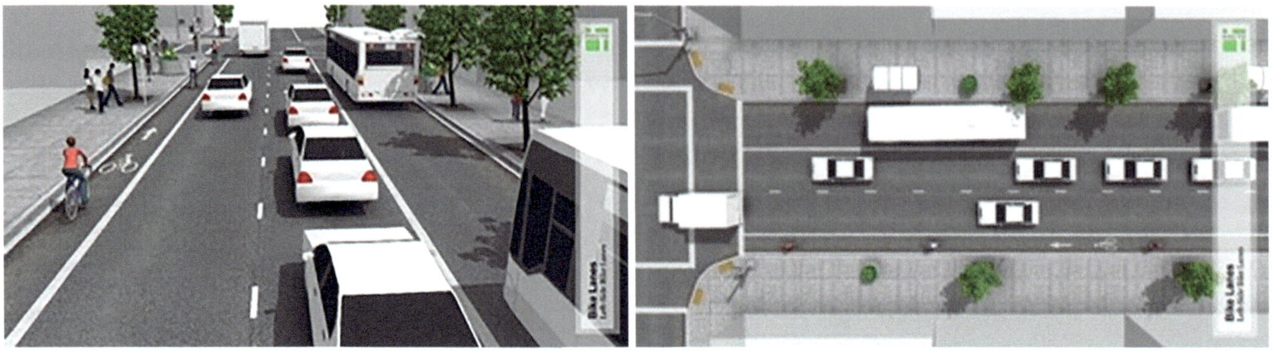

Fig. 36 Left-side bike lane (*Source* NACTO, 2013)

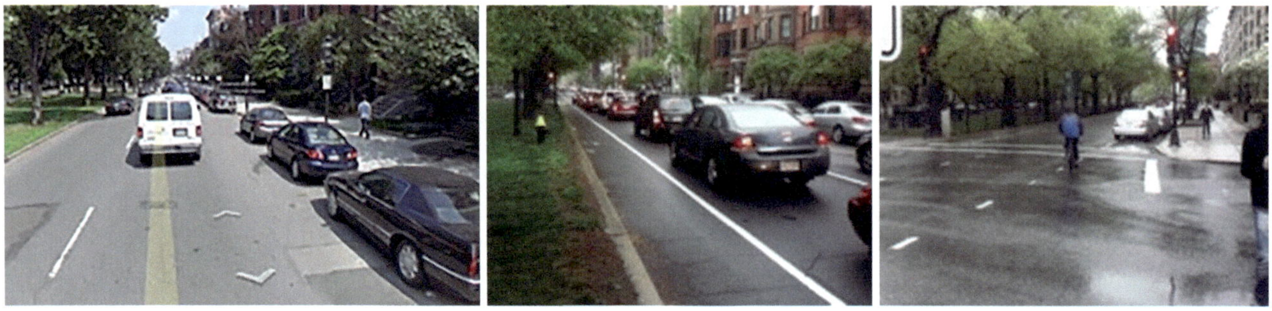

Fig. 37 Commonwealth Avenue in Boston with left-side bike lane (*Source* Almy, 2012)

Fig. 38 Sidewalks

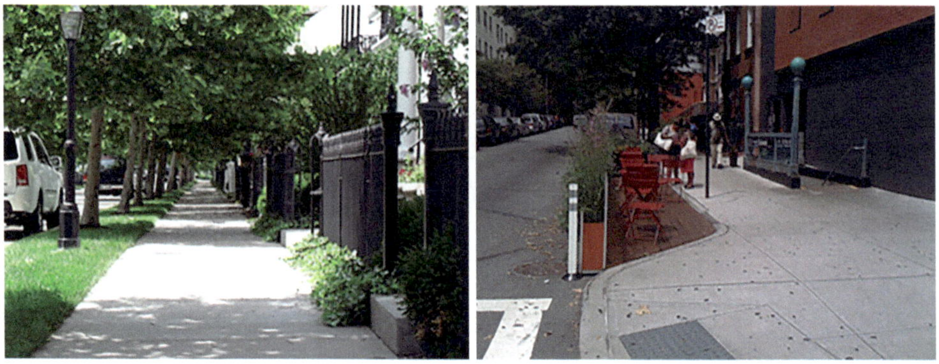

7.3.7 Curb Extensions

Curb extensions narrow the roadway (Fig. 40), creating safe and small walking passages while expanding space for street furniture, benches, vegetation. Also, they reduce the roadway's width and serve as a visual signal to drivers entering a neighborhood street. Its length should be at least equivalent to the width of the crosswalk. The four types of curb excisions is shown in Table 5.

Gateway. B. Pinch point. C. Chicane. D. Bus bulbs.

A. **Sidewalk eight principles of design see** Fig. 45 (Maneola, 2015);
   1. Proper sizing.
   2. Quality surface.
   3. Efficient drainage.
   4. Universal accessibility.
   5. Secure connections.
   6. Attractive spaces.
   7. Permanent security.
   8. Clear signage.

B. **Sidewalks widths**

The recommended sidewalk width for each street type is shown in Table 6. Sidewalks below the minimum width for widening the relevant street type as opportunities allow, as shown in Figs. 46 and 47.

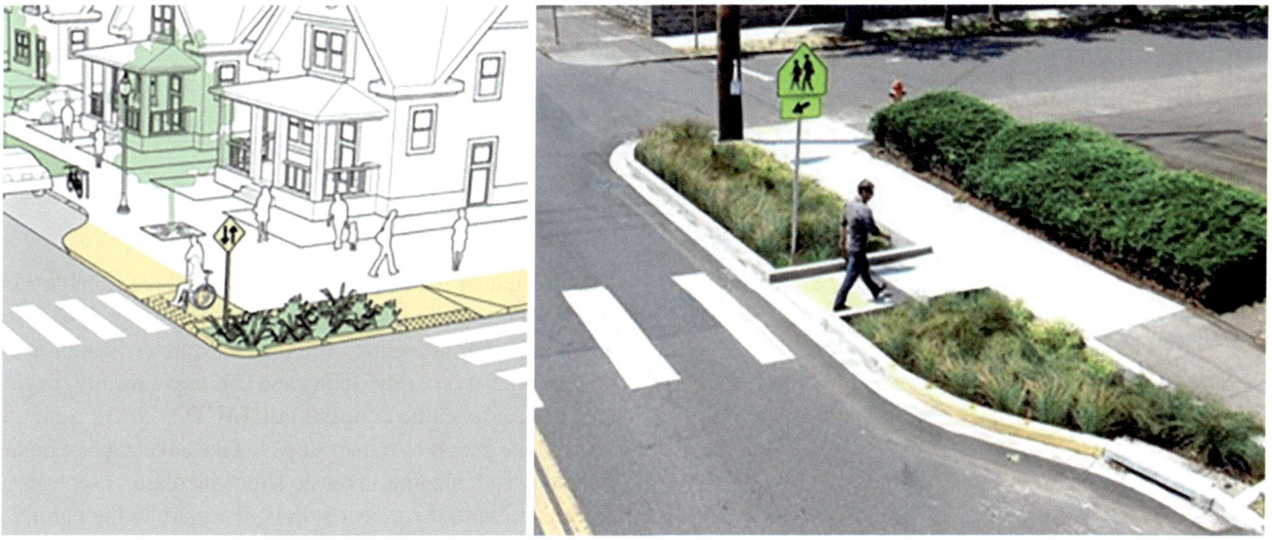

Fig. 39 Buffer zone of sidewalks (*Source* NACTO, 2013)

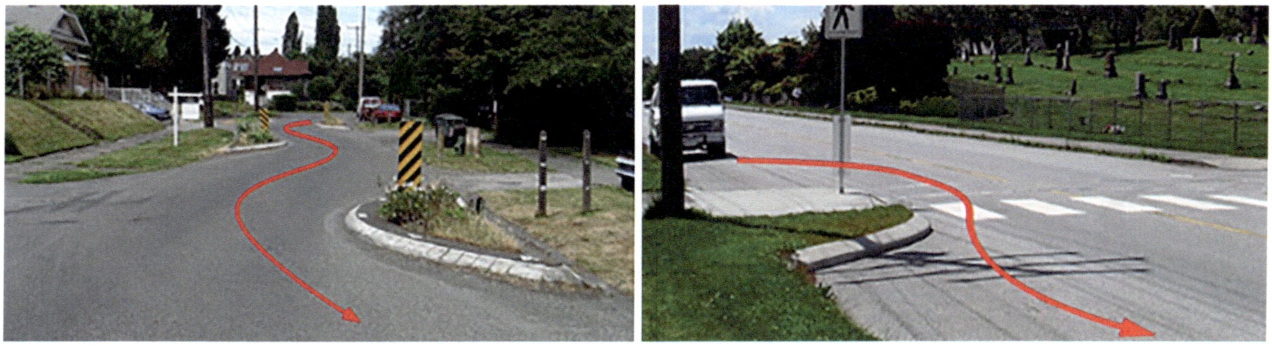

Fig. 40 Curb extension in Vancouver, Canada (*Source* Thomas Elli, Serwicka Ilona, 2015)

Table 5 Types of curbs (*Source* The author)

Curb extensions types	Description	Pictures
Gateway	We are using them at the connections and entrances of less speed streets. It is planned to spot the shift in streets	Figure 41
Pinch point	Curb extensions located at midblock for traffic slow down and public space addition	Figure 42t
Chicane	This type creates a chicane in low-traffic streets that significantly reduces traffic speeds. It increases the area of public space and street furniture	Figure 43
Bus bulbs	Bus bulbs' design is to line up the bus stop with the parking road, letting buses stop and board people without leaving the travel lane. It helps buses transfer quicker and more consistently by reducing the total time lost when merging in and out of traffic	Figure 44

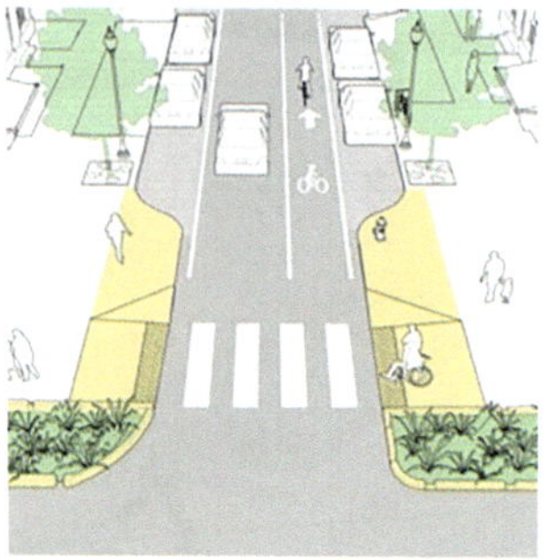

Fig. 41 Gateway

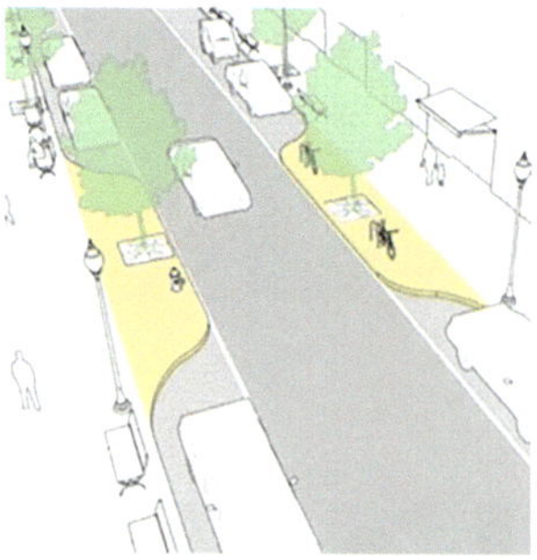

Fig. 42 Pinch poin

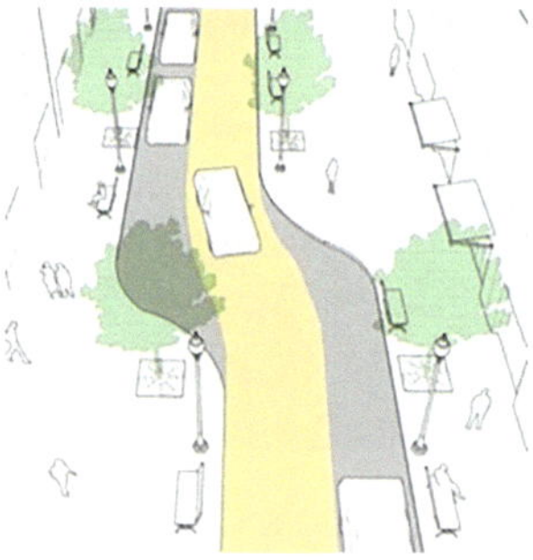

Fig. 43 Chicane

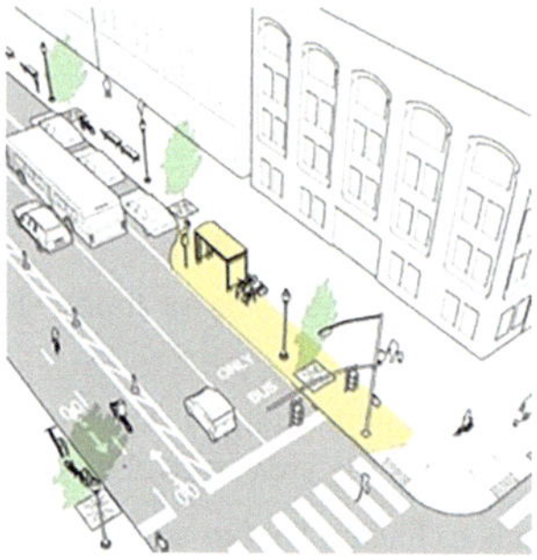

Fig. 44 Bus bulbs

8 Vertical Speed Control Elements

Vertical speed control elements manage traffic speed and strengthen pedestrian-friendly rates. It may be suitable in a variety of street types but mainly used in merchandise traffic of the neighborhood, residential, or low-speed streets see Table 7. Vertical speed control elements include three types (Namee & Witchayangkoon, 2011);

A. Speed bumps to see Figs. 48 and 49.
B. Speed table to see Figs. 50 and 51.
C. Speed cushion to see Figs. 52 and 53.

9 Transit Streets

Devoted transit roads, as shown in Fig. 54, are suitable base signal controls and operational activity upgrades. They guarantee that traveling vehicles encounter insignificant holdup time at crossing points and can move openly, paying less respect to traffic congestion (NACTO, 2013). Safe and appropriate access to transit stops to take advantage of higher ridership and income. Transit street increases user safety, secure access to the transit stop is necessary to the public. If pedestrians do not feel safe and protected, they will not walk to the bus stop. All transit customers are pedestrians for some part of the trip. Transit street includes the walk from one's origin to the halt, better access to transit leads to better conditions for other walking journeys. The more direct a transit route is, the less time and cost it is necessary to offer a certain level of service (SUDS-RP-UD, 2012).

A. Dedicated Offset Bus Lanes

A bus lane shown in Fig. 55 is a lane made only for buses and is usually used to hurry up public transportation that would be else held up by traffic overcrowding. Bus lanes are a vital component of improving bus travel speeds by decreasing interruption caused by other traffic. A bus lane can take place as a part of a road, which also has lanes serving other motorized transport. In New York City, The 34th Street Select Bus Service design, is based on an analysis of the traffic, transit, and curbside access needs on the corridor. By community participation techniques and after many discussions with residents, businesses, institutions, and rigorous traffic analysis, as shown in Fig. 56 (New York City, 2010).

B. Median Bus Lanes

Median bus lanes are useful on main roads with numerous movements, as shown in Fig. 57. Using a median bus lane parallel to the centerline of a multilane roadway combined with nearby transit stops. In South Korea, they are implementing Median lanes, where they use alternative or public transportation more often. In Midtown Seoul, median bus lanes cut through heavy traffic and reduce pollution, as shown in Fig. 58 (Jang, 2014).

The following Table 8 is comparing the street elements and their description and show the importance of each component.

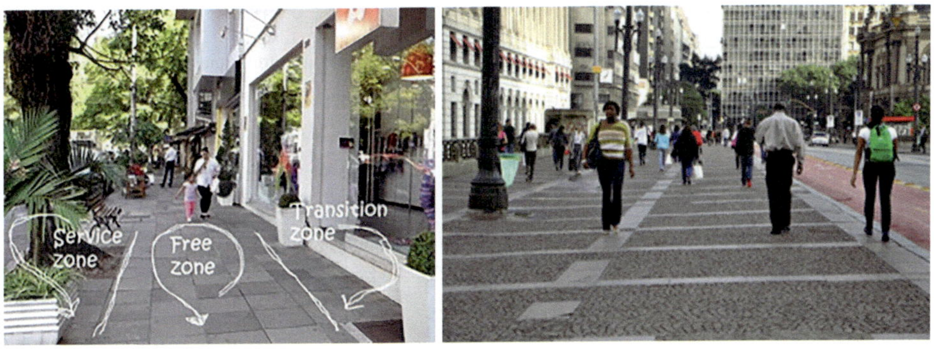

Fig. 45 Sidewalk best design (*Source* Maneola, 2015)

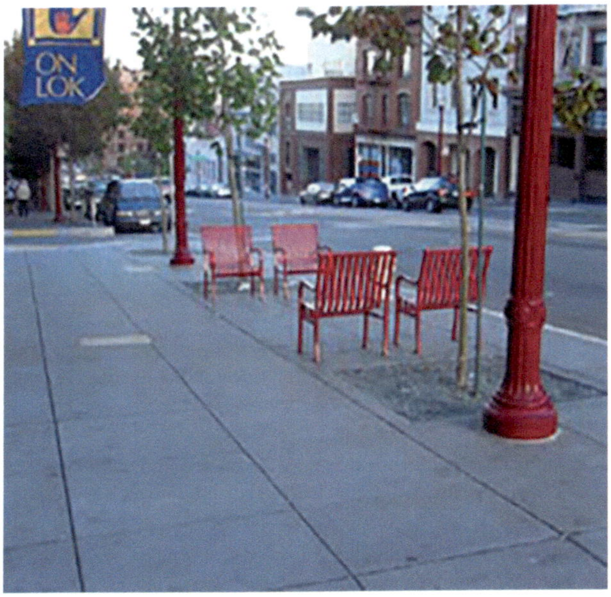

Fig. 46 Sidewalks

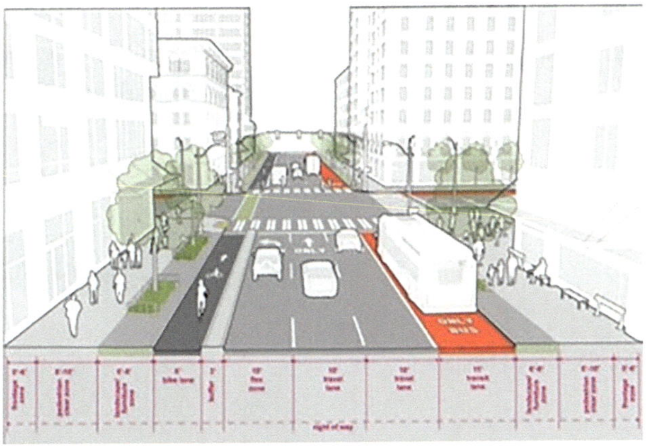

Fig. 47 Sidewalks and road way zones

9.1 Middle Islands in Streets and Their Importance

A middle island is a part of the roads that separate the two directions of the street; it's defined as an area between the two sides of the street that controls traffic movements. It may be low, raised, or the same as the road's surface height. Raised islands offer space for pedestrian safety features, traffic control devices, services, landscaping, and stormwater management.

9.2 Median Refuge Island

They are safe areas in the middle of streets, where people may securely wait while crossing a street. Pedestrian refuge islands are mainly helpful as relaxing areas for elders, people with disabilities, kids, and anyone who cannot pass the road at one stage. They offer a protected space for bicyclists and pedestrians to pause for an acceptable gap in traffic, decreasing the complete crossing length to automobile movement for a bicyclist or pedestrian see Fig. 63 (Corner, 2019).

9.3 Median Island Placement

Islands should be well-thought-out under the following situations:

A. Streets with high bicycles and pedestrian movement.
B. They are crossing spaces of eighteen meters or more.
C. Close and inside neighborhood selling areas, public and institutional uses, and schools.
D. Places with many transmissions between transportation lines (Corner, 2019).

Table 6 Sidewalks width (*Source* SFPD, 2010)

	Street type	Minimum width	Recommended width
Commercial	Downtown commercial	Per downtown streetscape plan	
	Commercial throughway	12′	15′
	Neighborhood commercial	12′	15′
Residential	Downtown residential	12′	15′
	Residential throughway	12′	15′
	Neighborhood residential	10′	12′
Other	Industrial	8′	10′
	Mixed-use	12′	15′
Special	Parkway	12′	17′
	Park edge	12′	24′
	Multiway boulevard	12′	15′
	Ceremonial	Varies	Varies
	Alley	6′	9′
	Shared public way	NA	NA
	Paseo	Varies	Varies

Table 7 Vertical speed elements (*Source* The author)

Vertical speed control elements	Description	Pictures
Speed bumps	Devices that custom vertical bend to slow motor vehicle traffic to increase safety conditions (Namee & Witchayangkoon, 2011). See Fig. 48	Figure 49
Speed tables	Speed tables are traffic calming devices that raise an automobile's whole wheelbase. They decrease their traffic haste. They are Longer than speed bumps and flat-topped (Ewing, 1999). See Fig. 50	Figure 51
Speed cushions	Speed cushions are considered speed bumps or speed tables that contain wheel cutouts, for significant automobiles pass unaffected while decreasing average car speeds (Namee & Witchayangkoon, 2011). See Fig. 52	Figure 53

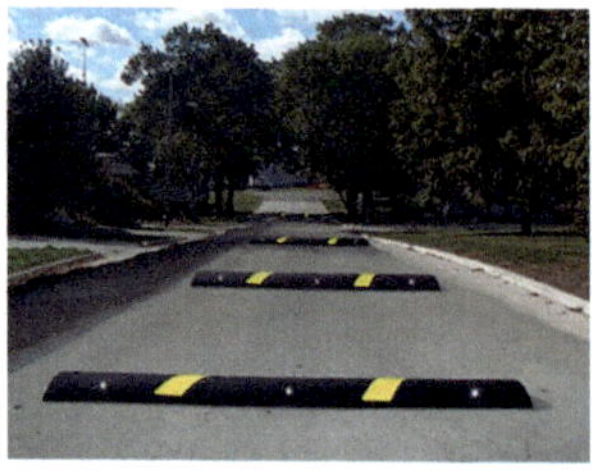

Fig. 48 Speed bumps

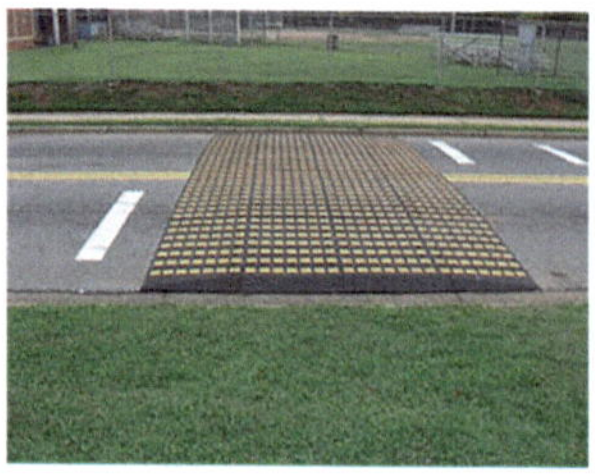

Fig. 50 Speed tables

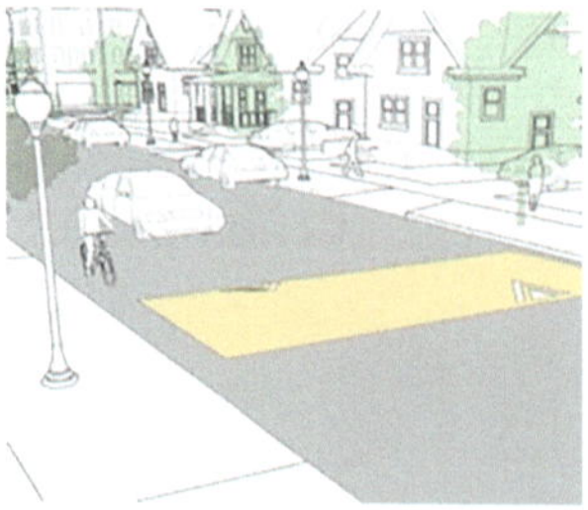

Fig. 49 Speed bumps

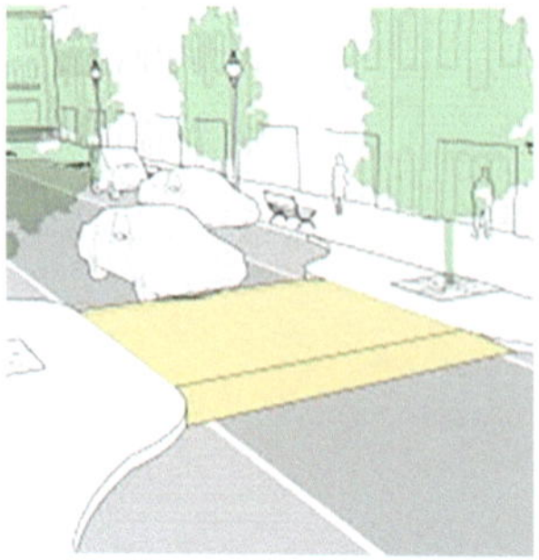

Fig. 51 Speed tables

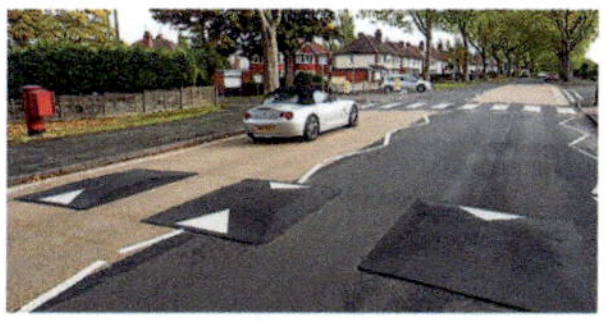

Fig. 52 Speed cushions

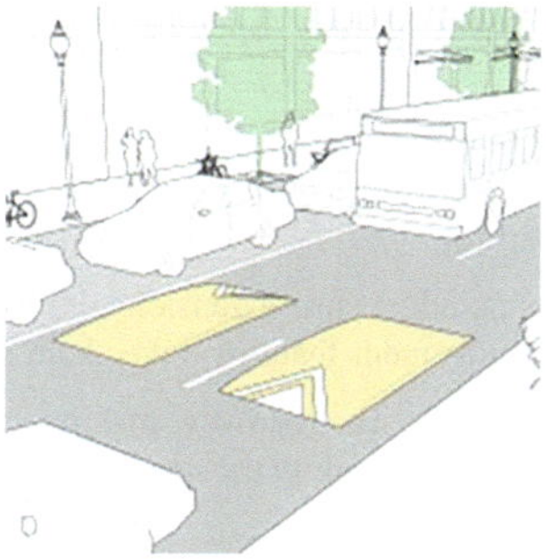

Fig. 53 Speed cushions

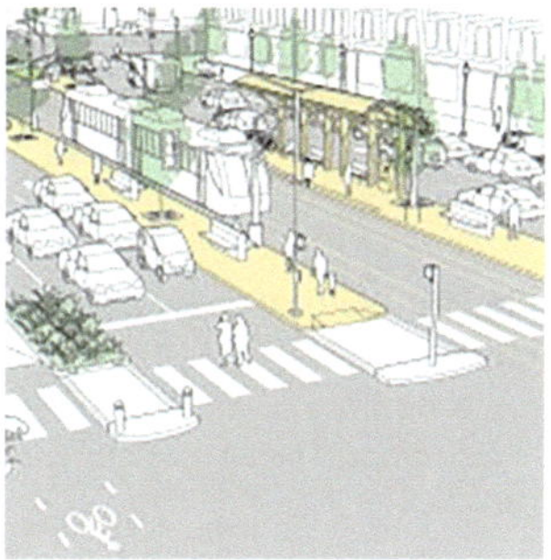

Fig. 54 Transit streets (*Source* NACTO, 2013)

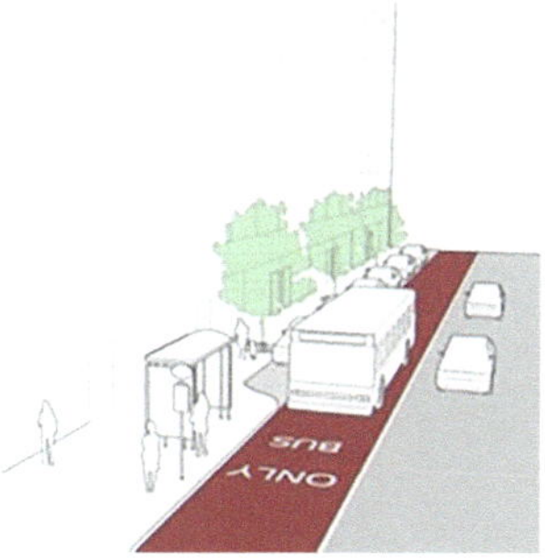

Fig. 55 Bus lanes (*Source* NACTO, 2013)

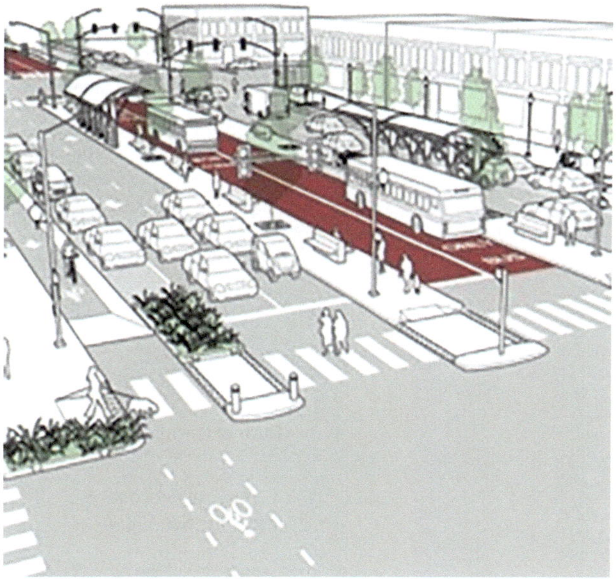

Fig. 57 Median bus lanes (*Source* NACTO, 2013)

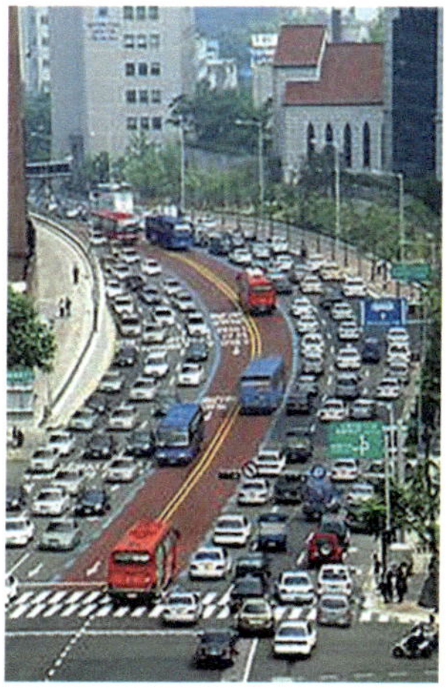

Fig. 58 Bus lanes in downtown Seoul (Jang, 2014)

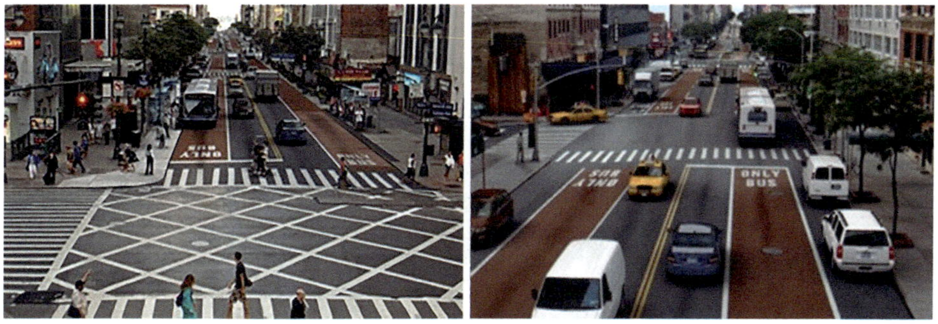

Fig. 56 34th Street New York (*Source* New York City, 2010)

Table 8 Shows the street elements that can be found in any street (*Source* The author)

Street elements		Description
Lane width		Width assigned to bike lanes, trucks, motorbikes, buses, and car parking
Sidewalks. See Fig. 59	Sidewalks	They are paths for pedestrian access, and increase their connections and encourage walking
	Frontage zone	It describes the walkway's piece that purposes as an extension lead of the building, either entrances and doors or walkway cafes
	Pedestrian zone	It is the primary, open path that goes equivalent to the street
	Furniture zone	It is well-defined as the sidewalk and the pedestrian zone where street furniture as; lighting, seats, newspaper booths, trees depth, and bicycle parking
	Buffer zone	It includes; curb extensions, bike racks, parking, stormwater management features, bike share stations, and cycle paths
Curb extensions. See Fig. 60		It narrows the roadway, making safer and smaller passages for walkers while allowing extra space for street furniture
Vertical speed control elements. See Fig. 61		Manage traffic speeds and strengthen pedestrian, fast, safe speeds. It may be suitable for various street types and broadly used in the neighborhood, residential, or low-speed streets
Transit streets. See Fig. 62		They guarantee that the travel vehicle encounters can move openly, paying less respect to traffic congestion

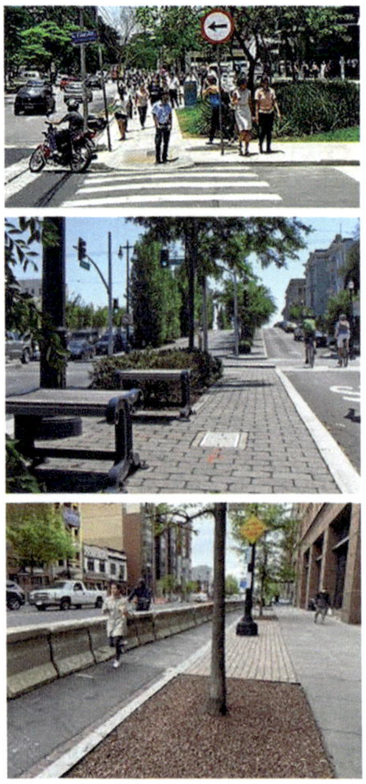

Fig. 59 Sidewalks

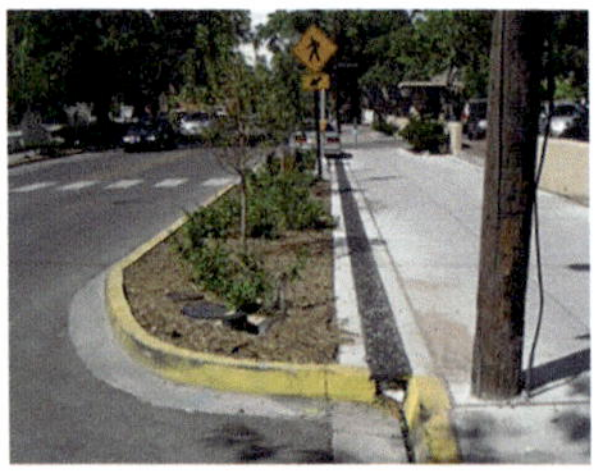

Fig. 60 Curb extensions

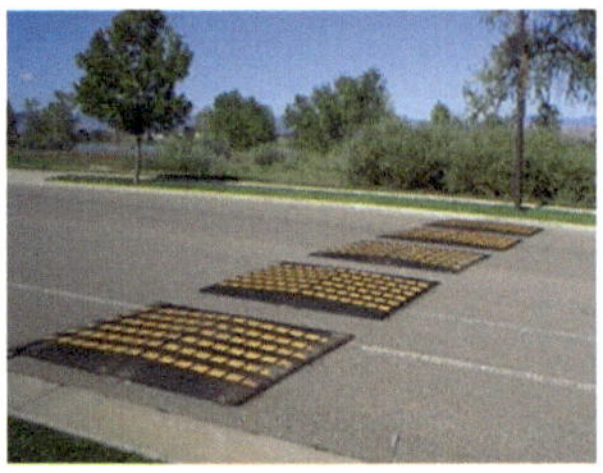

Fig. 61 Vertical speed control elements

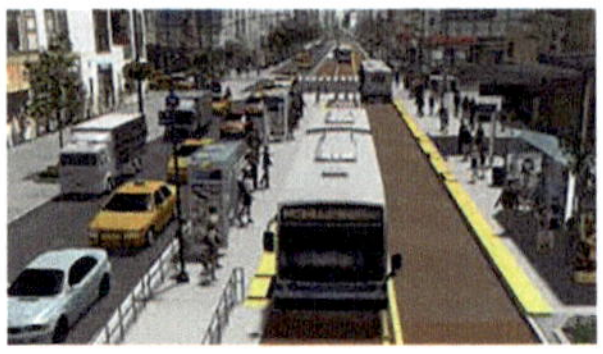

Fig. 62 Transit streets

Streetscape Elements

Trees, lighting, pedestrian furnishing, pavement, and other elements fill the places of the streetscape with life and make a street more comfortable and usable space for people.

Trees. B. Lighting. C. Benches. D. Bicycle racks. E. Trash receptacles. F. Bus shelters.

A. Street Trees

Shown road trees in Fig. 64, and other landscaping, used to give unique characteristics to specific streets and neighborhoods. Trees and gardens need structure in coherence with

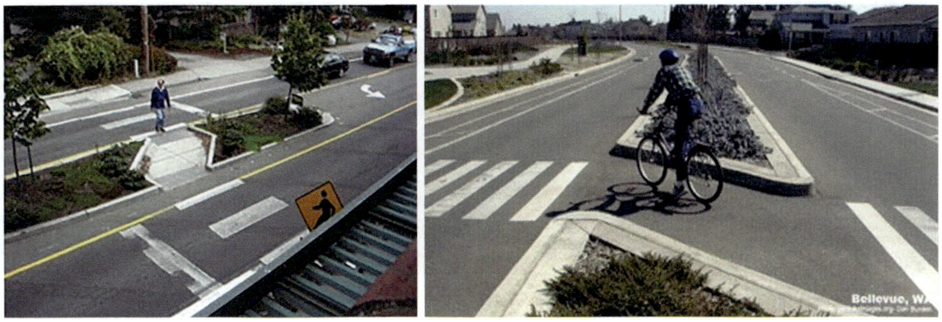

Fig. 63 Median Island (NACTO, 2013)

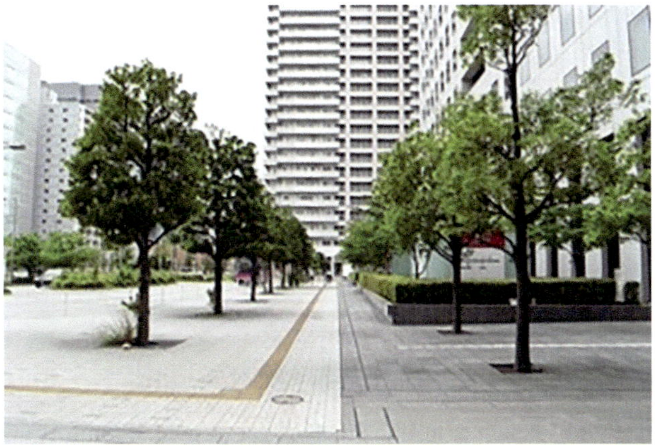

Fig. 64 Road trees in Japan (*Source* Images, 2015)

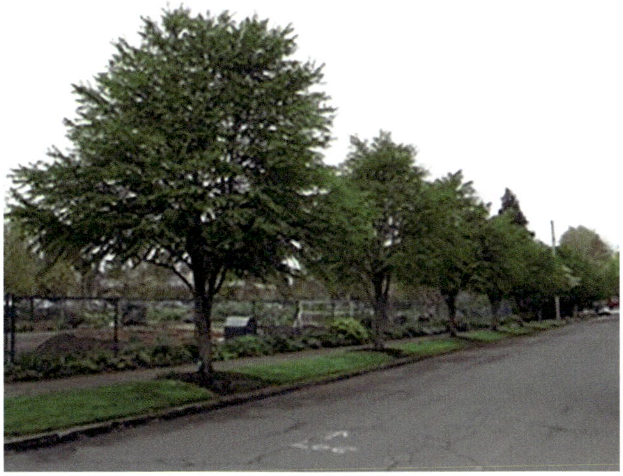

Fig. 65 Street trees (Street, 2020)

road lighting, walkway services, and building locations (SFPD, 2010). Planting trees in urban areas is for the sake of the planet; thus, it is essential to allocate plants in the right place, to avoid removing that valuable tree, later on, to see Fig. 65 (Street, 2020).

10 Tree Placement

Classically planting trees were in basins or sidewalk cut-outs in sidewalks. It is not essential to have tree basins specifically. In some conditions, the placement of trees could be above ground. Tree locations and spacings are determined according to the mature scope of the tree. Small trees (<6 m in diameter) were planted at four and a half meters in the center.

- Medium trees (six to ten and a half meters in diameter) were planted at 7.5 m in the center.
- Tall trees (>10.5 m in diameter) were planted at ten and a half meters in the center.

In general, trees with smaller diameters (less than 6 m) should be planted at closer space, while trees with larger diameters (>12 m) should be afforded wider spacing.

We should think in a futuristic term when it comes to planting trees. Incorrect trees placement can lead to untold damage to hard surfaces and infrastructures (Street, 2020).

B. Street Lighting

Street lighting is one of the main street features; this includes roadway and pedestrian paths lightings. Lighting supports the night driving and any other night activities. It shows the street quality for traffic safety and pedestrians' security (Fig. 66).

11 Lighting Placement

Locating street lighting should be as the following:

- **Entrances**: lighting around residential building entrances doorways make it safer.
- **Edges**: the boundaries of any place, such as a park, that help in identifying the space. Even buildings placed on its

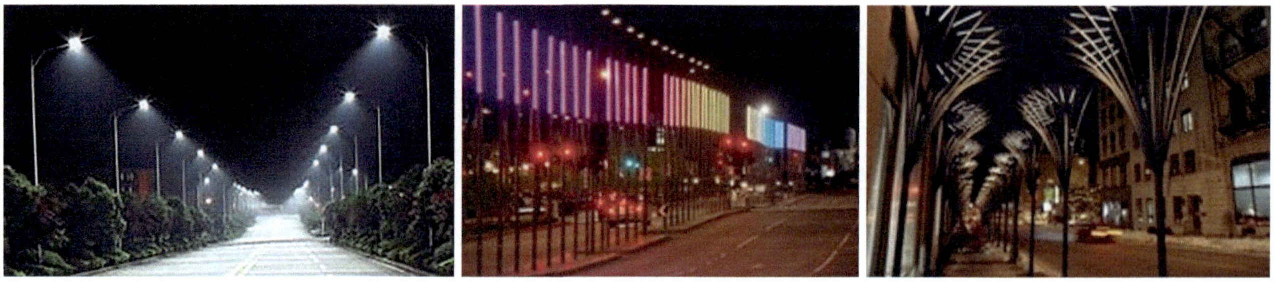

Fig. 66 Street lights

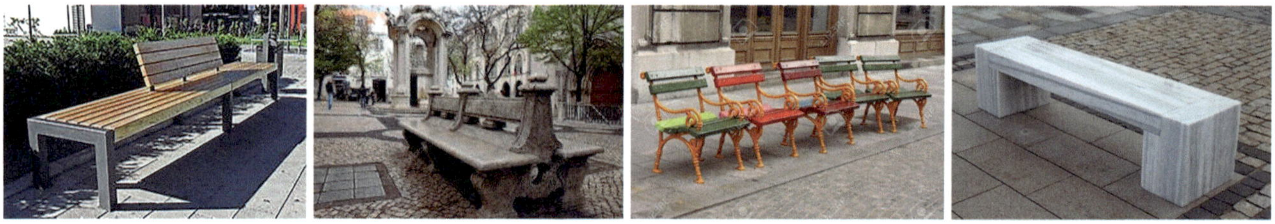

Fig. 67 Street benches (Description, 2020)

side can also have seasonal lights conveying attention to the more prominent district outside the park.

- **Selling places**: light even if stores are locked. This method helps raise the number of people on the street, which is a significant donor to safety.
- **Architectural details**: grazing technique in lighting at entrances, cornices, arches, columns attracts attention toward a specific building detail.
- **Signage**: lighted signage and maps should be visible for all street users, both driving or pedestrians, to correctly identify those signs.
- **Traffic-calming device**: the difference between a pedestrian-lit street and a highly illumined road directly leads drivers to enter a new zone. It also forces them to slow down their speed.
- **Streets with high pedestrian volumes**: with small sizes, considering their safety and security. For example, backstreets, muses, parks, pathways, and pedestrian paths (Cityscape Institute, 2008).

C. **Benches**

Installing benches in the streets is for people's comfort and relaxing. The preferred materials in street benches are; wood, stone, granite, concrete, and metals, as shown in Fig. 67.

D. **Bicycle Racks**

The term bicycle rack or bike rack refers to a bicycle carrier. Bicycles are mounted for transport, as shown in Fig. 68. It is considered a parking rack, a stationary fixture to which a bike can be secured by using a bicycle lock.

E. **Trash Receptacles**

All streets must have trash receptacles for hygiene and waste disposal. Nowadays, there must be four trash receptacles for recycling purposes, as shown in Fig. 69.

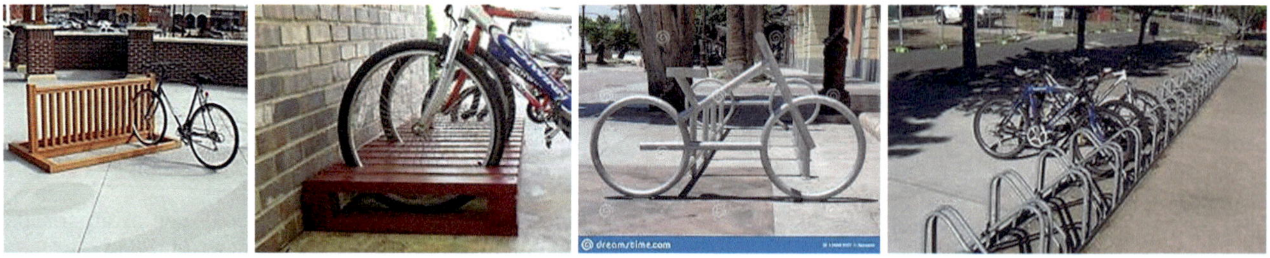

Fig. 68 Bicycle racks

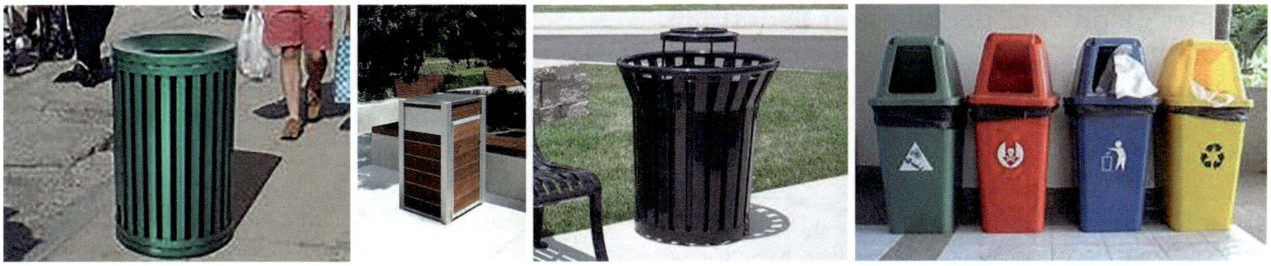

Fig. 69 Trash receptacles

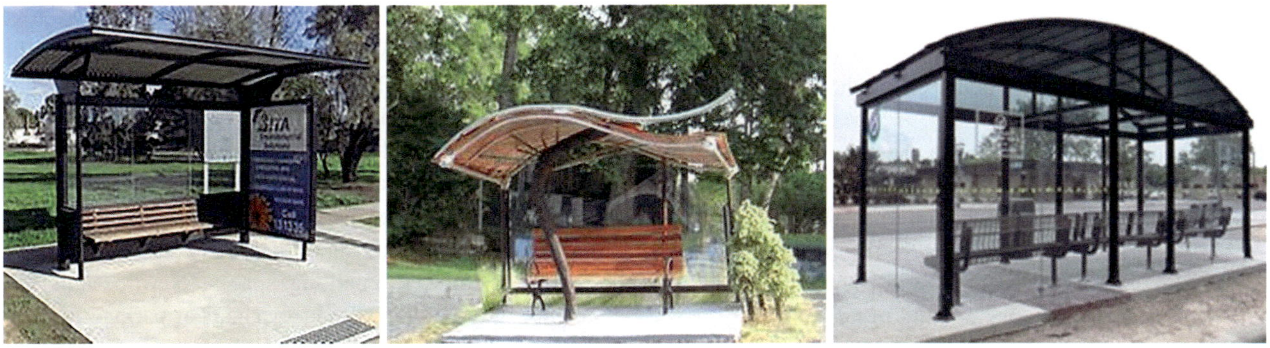

Fig. 70 Different bus shelters (Newman, 2020)

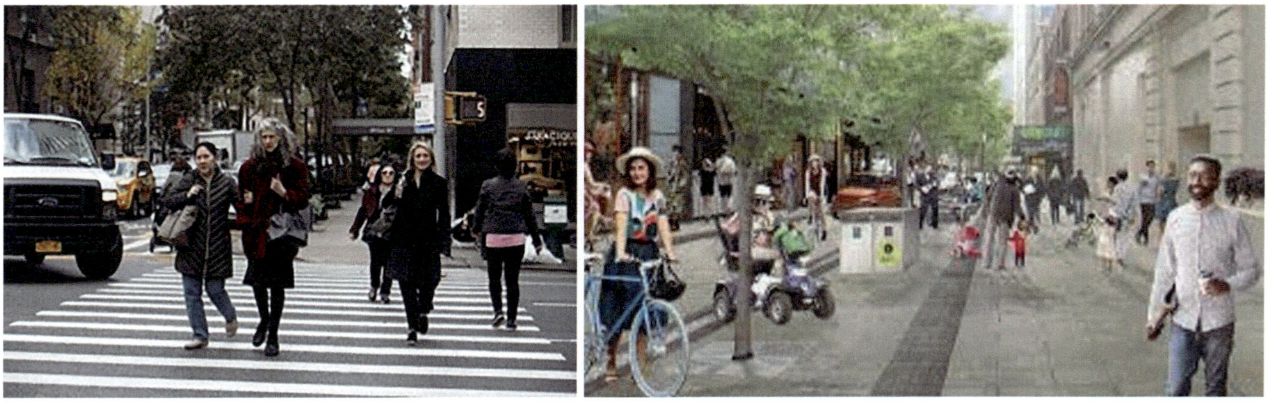

Fig. 71 Pedestrian-friendly streets

F. **Bus Shelters**

Waiting for a bus needs a shaded area for users' comfort with embedded seats. They can vary in design and materials for giving a unique esthetic ambiance to the street see Fig. 70.

Pedestrian-Friendly Streets

Pedestrians, generally, are people physically walking instead of traveling in a car. Pedestrians can also be using skateboards, roller skates, scooters, wheelchairs, or other mobility aids. The automobile power in the streets has had a harmful effect on the deterioration of public life. They are improving urban environments into an energetic space rather than to destroy essential connections (Appleyard, 1981). Accordingly, numerous urban design ideas came out. "Traffic-free zone" term is applied to a wide range of urban spaces as plazas, squares, promenades, esplanades, and parks. On the other hand, this expression describes a more detailed concept, representing urban areas with forbidden private vehicles, and priority to pedestrian movement and public transportation Brambilla & Longo, 1977). The priority to pedestrian and vehicle access is banned totally. If vehicles are allowable, their speeds are limited to have a protected, excellent, and safe environment, as shown in Fig. 71.

12 Designing Considerations for Pedestrian Paths

A. The design follows function: the pedestrian paths are planned upon the following;
   - Spaces you want to access.
   - The residents' age and physical abilities differ.

B. Safety for pedestrian should be applied: to smear protection for pedestrian paths, this can be done by;
   - The separation between pedestrian paths and vehicle roads.
   - Intersection studies.
   - Study pedestrian paths that meet vehicle roads.
   - Façade that sees open paths should allow social interaction.

C. Comfort for the pedestrian: the user has to reach his target without feeling tired;
   - Users must avoid difficulties to ease movement.
   - Use of un risky materials for the path finishing, not sliding, and will not cause feet pain.

D. Pedestrian path and its relation to the journey purpose: the number of pedestrians using the road;
   - Users are either children (nursery) or senior (shops and terminals)
   - There are special conditions, such as the desire for quick access in the case of going to work or carrying bags in a travel case.
   - The goal is, either sport or recreational.
   - It can take two paths to the same goal, the first primary and short, and the second secondary and more extended.

E. Automated service streets and its relation with pedestrian paths.

F. Street furniture details.

13 Recommendation on Solutions for Cairo's Street Problem

The followings are research findings recommendation to resolve the issue on Cairo's Urban Street:

A. **Smart Parking**

Smart parking, or multistory smart parking lots, is a leaving technique that combines innovation and human advancement with an end goal to use few resources. For example, fuel, to accomplish quicker, less demanding. Denser leaving of vehicles for most of the time, they stay inert. It combines innovation, functionality, and economy (Offer, Implementations, Company, Parking, & European, 2019). It is a vertical structure that can take up to 16 cars in the space of just two cars. An automatic multistoried car parking system reduces and minimizes parking areas needed in urban neighborhoods, as shown in Fig. 72. It offers extra flexibility of optimum parking solutions like; speedy tower, rotary carousel, optima, multi, and lift & side parking, as shown in Fig. 73. Vertical and horizontal movements simultaneously ensure fast parking and retrieval times, so it helps in time-saving (Of & Parking, 2019). High-rise buildings are proof of old architecture; It is cheaper to build up than out; the same is valid for parking spaces. New designs now construct fully automated and mechanical parking systems. This technological design will encourage using robotic car parking systems in areas where land is scarce and expensive —also, sites with problematic congestion, zoning, and crime.

The first multistory smart park in Egypt was opened on Al-Gomhouria Street in El-Mansoura, Dakahlia. It is the first attempt to reduce traffic congestion in the city. El-Mansoura governorate indicates that the city is suffering from a horrific traffic problem and then will repeat the trial to build more 20 multi-story smart parking to solve the parking problems see Fig. 74 (Ramadan, 2019).

A. **Carpooling**

They share car rides to travel in a car with more than one person. Having more individuals using one vehicle reduces every individual's movement costs—for example, fuel costs, tolls, and the worry of driving. It is a sustainable and eco-friendly traveling way. They are sharing rides decreases, air pollution, carbon emanations, traffic blockage on the streets, and the requirement for parking spots. Someone can share

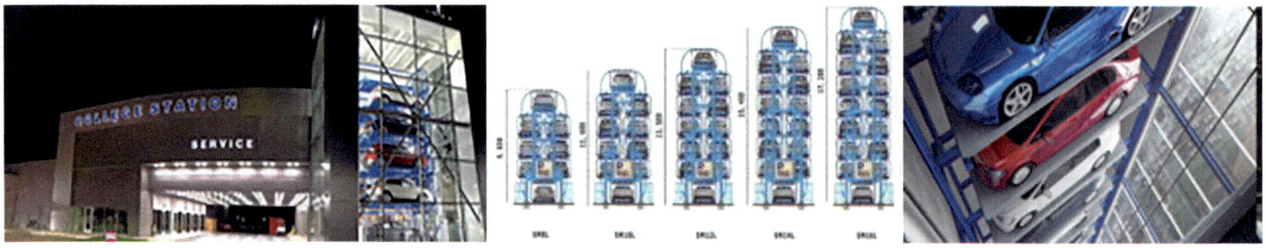

Fig. 72 Automated and mechanical parking systems (Solutions, 2020)

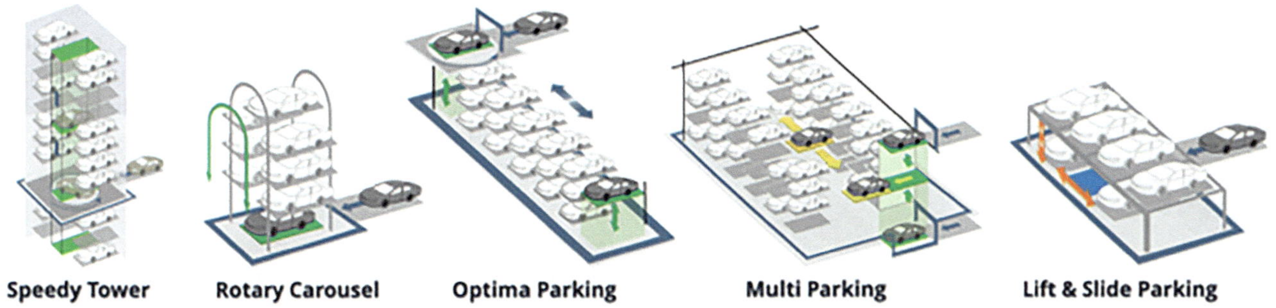

Fig. 73 Automated and mechanical parking systems (Solutions, 2020)

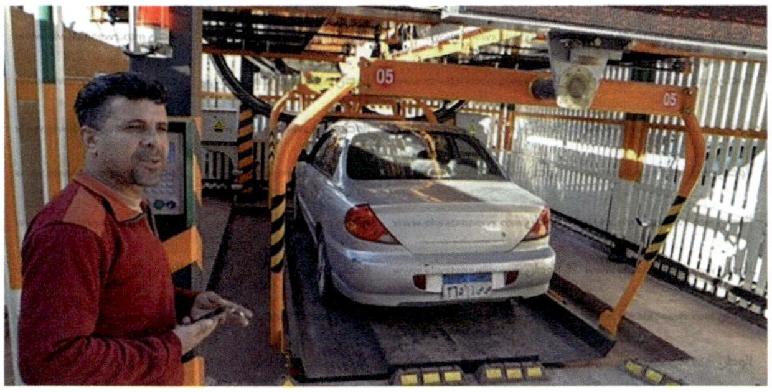

Fig. 74 The first multi-story Smart Park in Egypt (Ramadan, 2019)

his neighbors who work at different companies located only a short distance apart and have the same work hours see Fig. 75. Runner passengers can carpool to and from stations together. Green cities are leading now to shift people to public transport and restrict single-occupancy vehicles in urban centers (Forecasting, 2019).

14 Research Methodology: Case Studies

Two case studies will be analyzed by using the specific matrix to reach a comparative analysis between them.

A. **Downtown Kenton Redevelopment and Denver Ave Streetscape Plan Washington**

In North Portland, the downtown Kenton District area is a historical, commercial area. It has been struggling to maintain and attract a viable economic base for thirty years now. They suffered from high storefront vacancy rates, dilapidated buildings, commercial services not complementary to a neighborhood commercial core, and a high crime rate. For future development planning, hiring a professional team to perform development opportunity studies for six volunteer sites in the corridor for assisting property owners. Concepts included building and façade improvements and recommendations for new mixed-use and transit-oriented buildings, as shown in Figs. 76 and 77 (Downtown, District, Portland, & Avenue, 2008).

For the construction improvements to the project, BergerABAM is an organization for event management services. Creating a strategic communications plan; drafted key speaking points; arranged for on-site support, such as tents and sound equipment; and pitched the event to local media as shown in Fig. 78 (BergerABAM, 2020).

B. **The Better Block Project** (2010)**, Norfolk, Virginia**

Norfolk is an independent city in the state of Virginia in the United States. It's residents decided to follow a new project for three months period, as shown in Fig. 79, to refresh a single commercial block in an underused neighborhood corridor. A group of community planners, neighbors, and property owners gathered together servicing the project for the better block. These residents realized their ability to be assets that can help make their community economically generative. They used digital fabrication, CNC routers, laser cutters, and 3D printers to adapt to the built environment

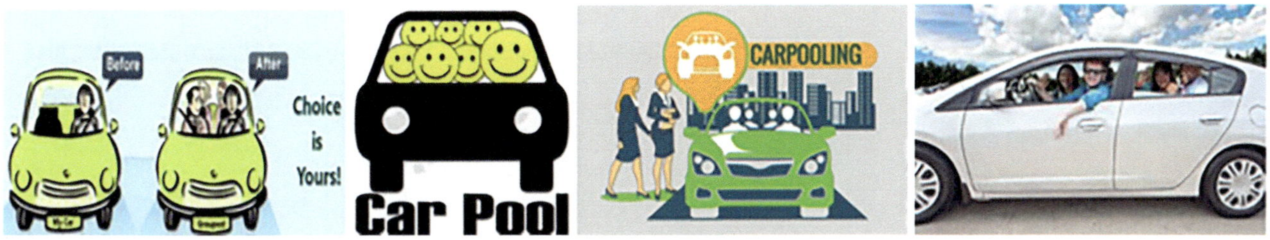

Fig. 75 Carpooling (*Source* The author)

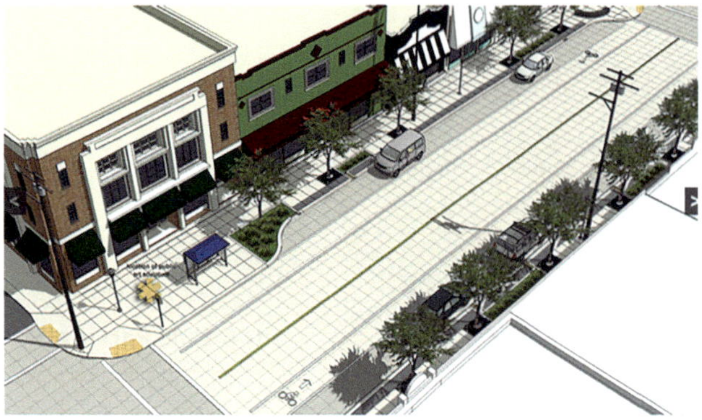

Fig. 76 Denver renovation cityscape project (Downtown et al., 2008)

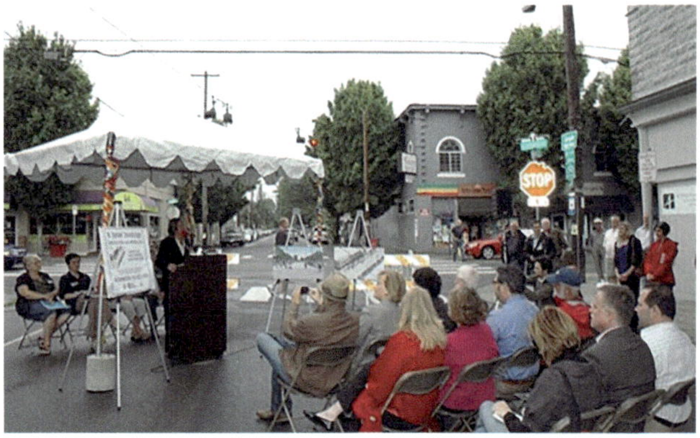

Fig. 77 North Denver avenue streetscape improvements Berger ABAM (Berger ABAM, 2020)

rapidly. The project incubated temporary businesses with residents, testing market viability for creating permanent businesses. By creating rapid-prototyping events and marketplaces, street safety had increased too (Roberts, 2020). Residents participated to have a better quality of life as shown in Fig. 80.

15 Research Recommendations

A. A Proposed Project at Hassan Maemoon, Cairo, Egypt

The street is a secondary street, deprived of Mostafa El Nahas road in Nasr City, Cairo. It is full of residential buildings in a two-way road 44 m wide, with 5 m median in-between. The median had been transformed into a garbage area and a relaxing sitting area for people. The street's condition was riddled with cracks, with no trees, no parking lots, no street furniture, no leading signs at all, and animals in the streets as shown in Fig. 81 (Alyoum, 2016).

The researcher proposed a project to renovate the street quality shown in Fig. 82, adding bike lanes, bus lanes, parking lots, shading trees, and proper sidewalks as shown in Fig. 83. The proposal should add better conditions for the whole area and elevate the quality of residents' lives.

Fig. 78 A section in the renovated street (Downtown et al., 2008)

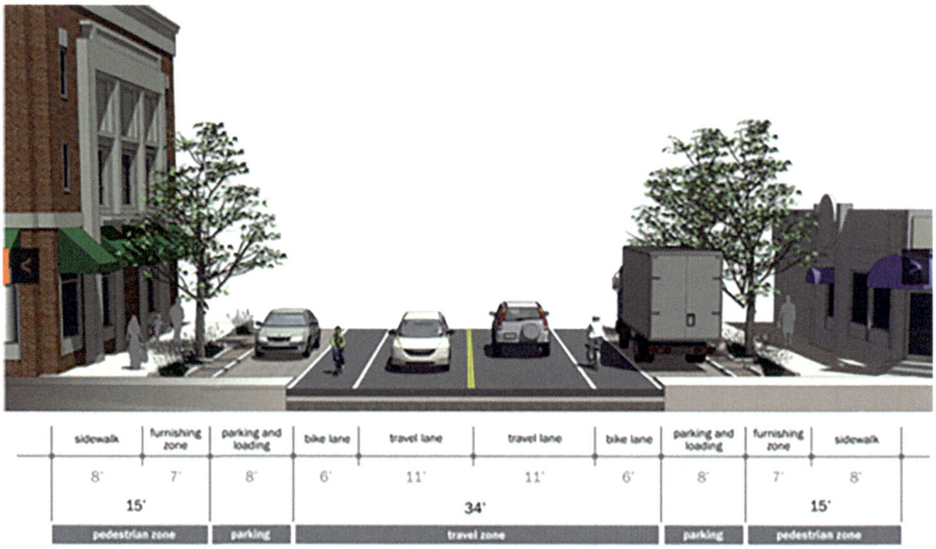

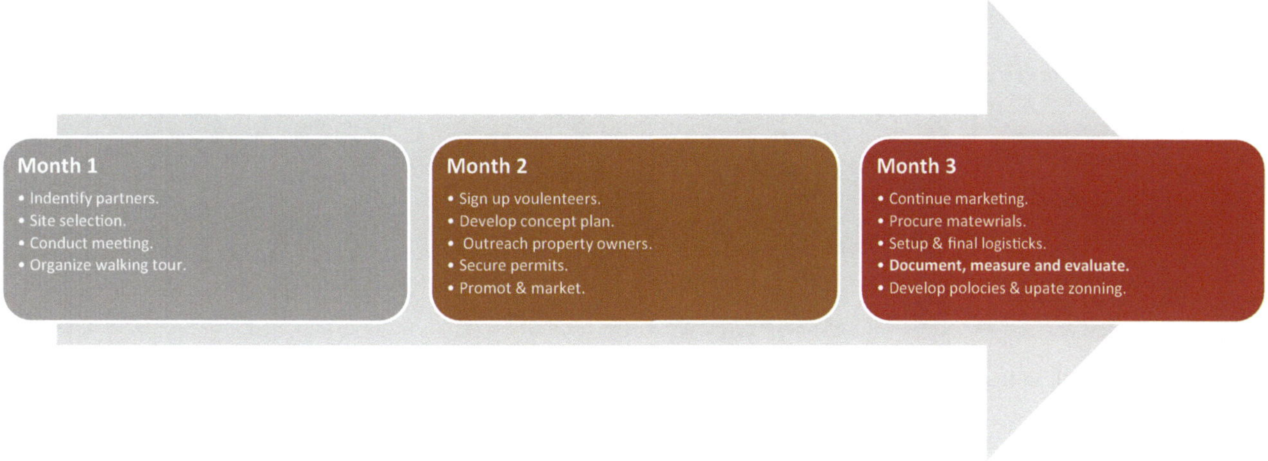

Fig. 79 Norfolk Better Block project time line (*Source* The author)

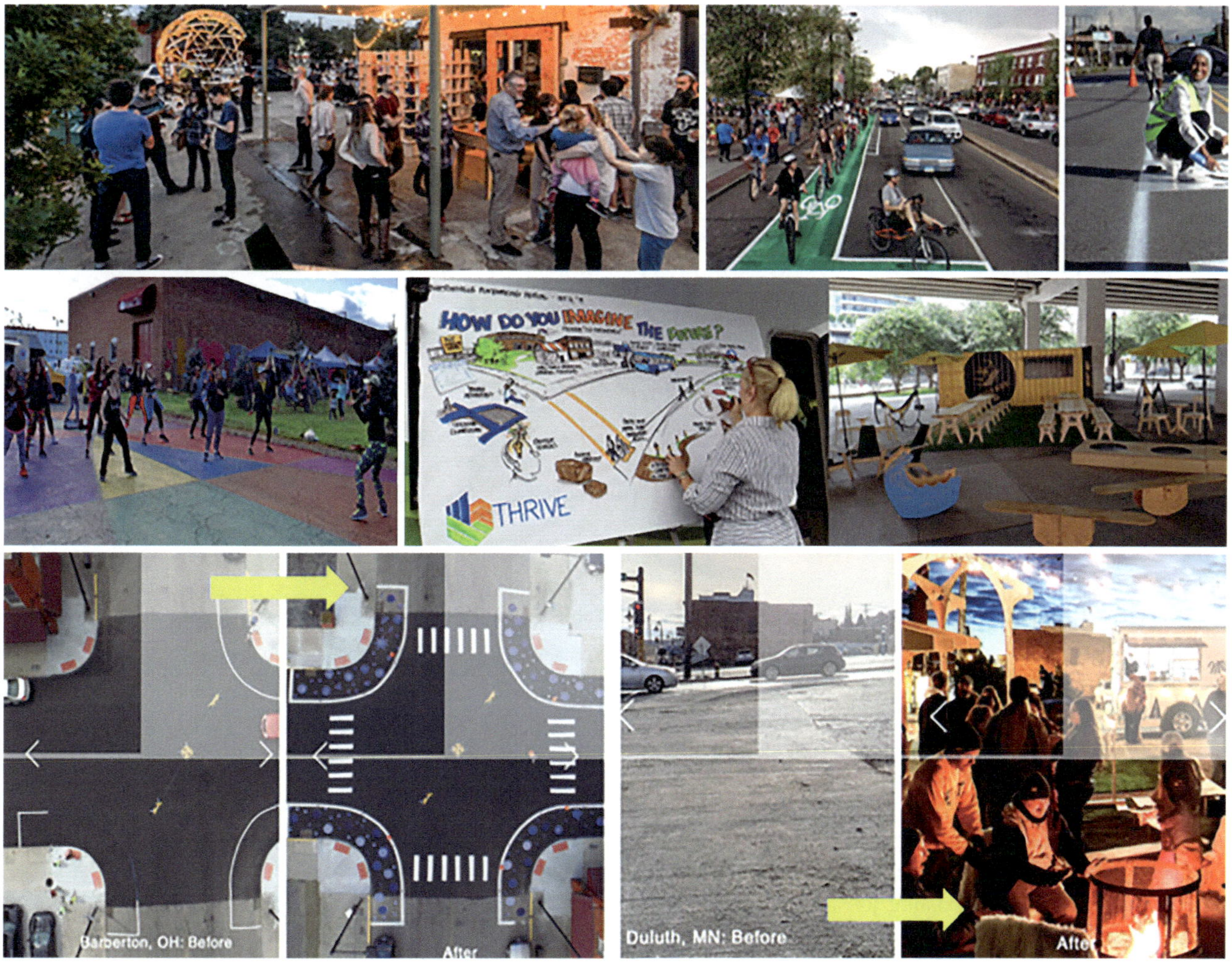

Fig. 80 Community participation development in "Better Block Project" (Edited by the author: *Source* Block, Donate, Mission, & Our, 2020)

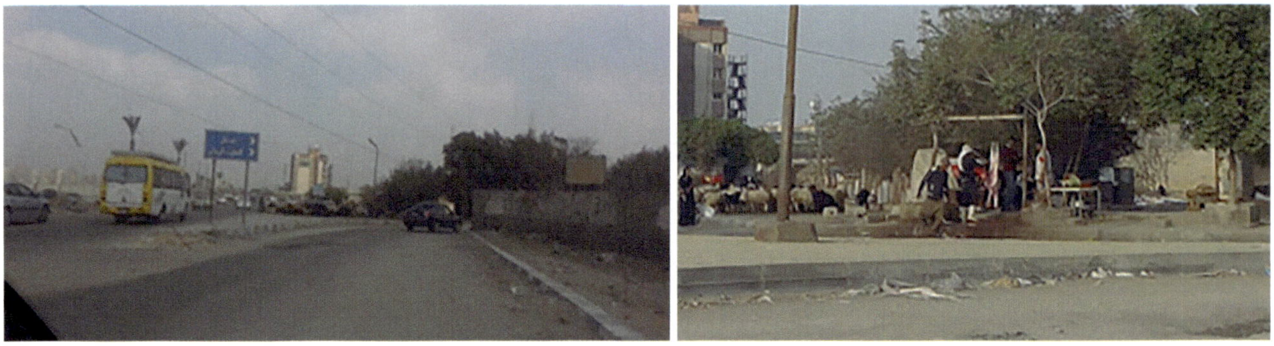

Fig. 81 Raising animals in Hassan Maomoon street (Alyoum, 2016)

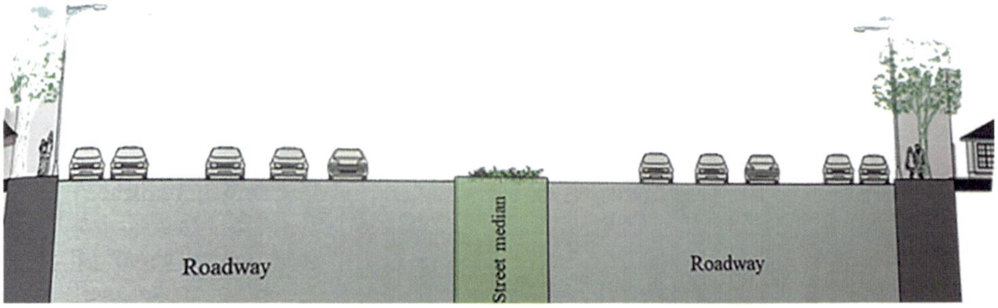

Fig. 82 Existing road situation of street Hassan Maemoon (*Source* The author)

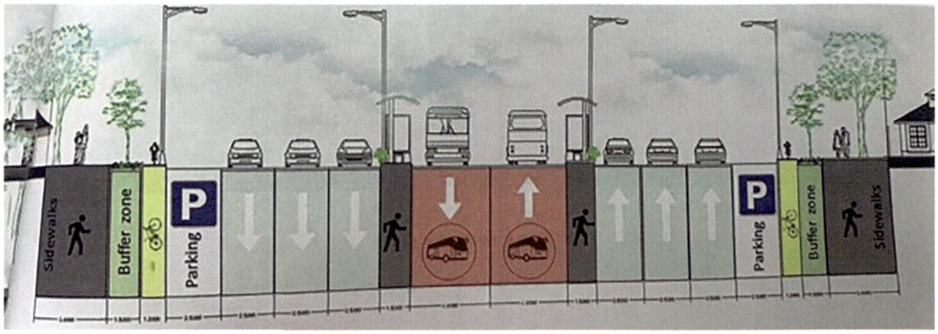

Fig. 83 Proposed project for renovation of Hassan Maemoon street (*Source* The author)

B. Services Program

Neighborhoods offer a mix of housing, selling, and employment usages that are always located in their centers. The feature of lifetime neighborhoods has the potential to allow local authorities and residents to meet a variety of similar aims, such as; striving toward sustainability and helping to decrease harmful impacts on the environment. Residents working in the service sector are 20–30% of all residents and 18–24% of existing jobs. Each service has a minimum and a maximum load of capacity, in which the residents can use the service with the maximum possible efficiency (Bevan & Croucher, 2011). Services that need to be in a neighborhood; are a primary school, three to four nurseries, a mosque, and commercial facilities. Table 9 shows the percentages of services according to the areas of urban hierarchy (Thomas Elli, Serwicka Ilona, 2015).

Table 9 Services percentages (Thomas Elli, Serwicka Ilona, 2015)

Centers locations	Percentage of services
City center	12–15% from the whole city area
District center	8–11% from the whole district area
Neighborhood center	13–18% from the whole neighborhood area

Land Use

Great neighborhoods hold a combination of land uses, as shown in Fig. 84 with a recognizable commercial center. These areas provide goods and services to meet the resident's daily needs and are an essential community assembly place. Neighborhood nodes may also be locations of recreation and transportation. Nodes can differ in size depending on the context; they provide options for living, learning,

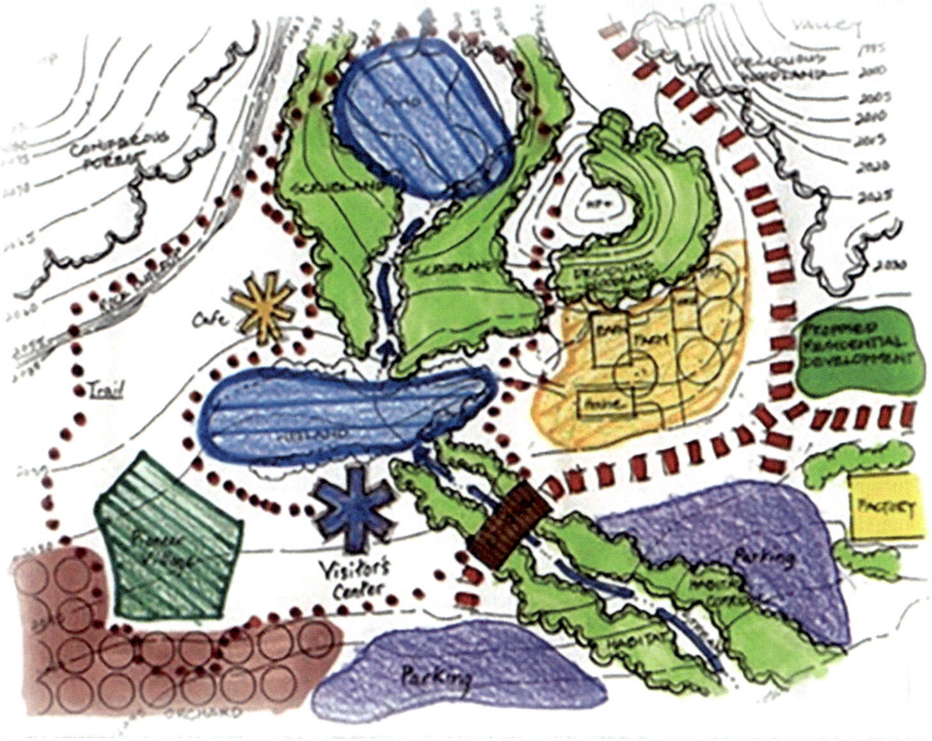

Fig. 84 Land uses (Deer, 2013)

working, and playing. More concentrated land uses are linked and focused around transportation, other transportation styles, and parks. All citizens can easily access daily shopping and recreational needs in their neighborhoods.

The best neighborhood nodes include: A mix of uses, combined higher density residential housing, a pedestrian public land, and are within a short walking space of most residents in a neighborhood (Deer, 2013). The prosperous neighborhood should include both; buildings as a mix of residential, services, and open areas.

16 Road Networks

Each neighborhood offers mobility choices for residents to travel to, from, and within the neighborhood. Its relationship with the nearest entrance depends on the neighborhood's location and different ways to access services. Streets and paths are well connected to inspire active modes of travel. Traffic and parking are managed and do not control the neighborhood (Bevan & Croucher, 2011). Road types in neighborhoods vary as the following.

Neighborhood Road types

1. Substreets that occur from high streets.
2. Main streets get along with the central services of the neighborhood.
3. Secondary streets drive to residential buildings.

Healthy Neighborhoods Regarding Streets

This research concentrates on renovating Cairo streets, starting with the neighborhoods as they are the main base for a healthy city. Roads and highways must provide multiple transport choices and better access for people of all ages and physical mobility levels. They should target different users, not only links between destinations for drivers, as roads are considered a public space for all people. Many activities take place in the streets beyond driving, such as cycling, shopping, and walking. So, the road participants are drivers and pedestrians, disabled citizens, motorcyclists, and cyclists. According to the city scale, road classification is either arterial or central, local, or collector (secondary) roads. Understanding roads and considering people's needs and health should have good street designs and

Table 10 Street elements comparison matrix (*Source* The author)

Points of comparison							
Neighborhood residential groups			Percentage of surrounding land use		Road network		
Small area	Medium area	Large area	Bbuildings	Open spaces	Sub-streets	Main streets	Secondary streets
Points of comparison							
Street elements							
Lane width				Sidewalks			Curb extensions
Cars		Bikes	Buses	Frontage zone	Pedestrian zone	Street furniture	Car way
2–3 m	*3–4.6 m*						
Points of comparison							
Street elements							
Curb extensions			Vertical speed control elements			Transit streets	
Pinch point	Chicane	Bus bulb	Speed pumps	Speed tables	Speed cushions	Offset bus lanes	Mediun bus lane
Points of comparison							
Streetscape elements							
Trees	Light post	Benches	Benches	Trash receptacles	Bus shelters		

thoughts that will lead to a healthy, green, and sustainable neighborhood. Finally, traffic engineers undertake a big responsibility of providing safe traffic movements to the road users and ensure their safety (Mathew & Bombay, 2014). Adding accessibility to streets for all types of users and create more ecofriendly cities. Drivers of urban prosperity are contribution to cities' success. This task integrates roads into five dimensions of well-being; productivity, infrastructure development, ecofriendly, quality of life, and equity. All sizes detangled the quality of the street pattern (Tobergte & Curtis, 2013).

Creating an efficient road transportation system, serve effectively different land use in an urban area, and guarantee network improvement. It is vital to set up a pathways system separated into frameworks. Each one serves a specific capacity or specific reason. It improved city extreme road classification with a specific transportation benefit capacity (Mathew & Bombay, 2014).

17 Conclusion

It concludes the previous work in a collective matrix (Table 10), for all streets and landscape elements in neighborhoods, facing different road networks of different residential groups. The research helps in analyzing any neighborhood retrieving problems to focus on adding elements.

Cairo's streets are facing a huge problem; unplanned streets, no correct designs, garbage in every corner, and overpopulation, and this impacts the whole environment and harms the streets as a public space.

Having a green and sustainable city is outlined with thought for social, economic, and environmental effects and healthy living space for the current population, taking into attention the effect of these neighborhoods and cities on the upcoming generations and gross in populations.

Egypt's problem lies in that the streets were designed many years ago and cannot be used by the new generations and the generations to come by new environmental and technological standards, as it had not planned for such from the beginning. That is why street design is an essential element in our daily lives; it is a public space we interact with every day. Therefore, it should be planned well and accurately for the safety of people and their importance.

References

Almy, I. (2012). Calm streets boston student reviews of traffic calming and bike-pedestrian treatments in the Boston area. Left-side bike lane on commonwealth but not including transition from right to left. https://calmstreetsboston.blogspot.com/2012/04/left-side-bike-lane-on-commonwealth.html.

Al-Wahaidy, F. (2005). T a t a. *Building*, (January 2018) (pp. 172–190). https://doi.org/10.5433/2176-6665.2012v17n1p172.

Alyoum, A. (2016). سكان حسن مأمون يشكون انتشار الزرائب وتربية الحيوانات داخل الكتلة السكنية. *AlMasri Alyoum* (pp. 1–3). https://sharek.almasryalyoum.com/cities/cairo/500158/.

Appleyard, D. (1981). *Livable streets/Donald Appleyard, with M. Sue Gerson and Mark Lintell*. In: M.S. Gerson & M. Lintell (Eds.). Berkeley: University of California Press.

Asoka, G., & THUO, A., & Bunyasi, M. (2013). Effects of population growth on urban infrastructure and services: A case of eastleigh neighborhood Nairobi, Kenya. *Journal of Anthropology & Archaeology, 1*, 41–56.

BergerABAM. (2020). North Denver Avenue Streetscape Improvements. https://www.abam.com/portfolio/project/436.

Bevan, M., & Croucher, K. (2011). Lifetime neighbourhoods. In: *Department for communities and local government*. https://www.gov.uk/government/uploads/system/uploads/attachment_data/file/6248/2044122.pdf.

Block, B., Donate, V., Mission, O., & Our, V. (2020). Using design to activate spaces and bring people together. https://www.betterblock.org/.

Brambilla, R., & Longo, G. (1977). *For pedestrians only: Planning, design, and management of traffic-free zones*. https://www.books.google.com.eg/books?id=6i9SAAAAMAAJ.

New York City. (2010). Bus rapid transit—34th street select bus service. *Bus Rapid Transit*, (3).

Cart, S. (2019). Urban bikeway design guide (2nd edn). *ISLANDPRESS* (pp. 1–3).

Corner, M. (2019). Medians and Islands. *SF Better Streets*, 1–5.

Deer, R. (2013). *Neighbourhood planning & design standards*. Red Deer City.

Description, P. (2020). *Vario-bench Leander*. https://www.stylepark.com/en/union-public-street-furniture/vario-bench-leander.

Doris, B. (2017). Overpopulation and the Impact on the Environment. *CUNY Academics Works, 60*. https://www.academicworks.cuny.edu/gc_etds/1906?utm_source=academicworks.cuny.edu%2Fgc_etds%2F1906&utm_medium=PDF&utm_campaign=PDFCoverPages.

Downtown, T., District, K., Portland, N., & Avenue, N. D. (2008). Downtown Kenton Redevelopment & Denver Ave Streetscape Plan. https://www.seradesign.com/projects/kenton-redevelopment-denver-avenue-streetscape-plan/.

EFGHERMES. (2012). *Assiut train disaster: 51 Egyptian children killed*. https://www.egyptianstreets.com/2012/11/18/assiut-train-disaster-51-egyptian-children-killed/.

EFGHERNES. (2013). *Egyptian roads: A trip through hell*. https://www.egyptianstreets.com/2013/01/24/egyptian-roads-a-trip-through-hell/. Accessed 30 June 2019.

Thomas, E., Serwicka, I., & Paul, S. (2015). Why people live where they do. In: *Citylab*. https://doi.org/10.13140/RG.2.1.1053.8965.

Ewing, R., & Handy, S. (2009). Measuring the unmeasurable: Urban design qualities related to walkability. *Journal of Urban Design, 14* (1), 65–84. https://doi.org/10.1080/13574800802451155

Ewing, R. (1999). *Traffic calming state of practice*. https://www.safety.fhwa.dot.gov/speedmgt/ref_mats/fhwasa09028/resources/TrafficCalming-stateofthepracticeSLIDESHOW.pdf.

Ramadan, S. (2019). □□□ "□□□□ □□□□□"□□□□□ □□□ □□□□□□□□ □□ □□□□□ □□□□. https://www.elwatannews.com/news/details/4053404.

Flowers, D., Warne, T., & Pyers, C. (1999). *Guide for the development of bicycle facilities*. 86. http://www.azmag.gov/Documents/pdf/cms.resource/PWB-RBTF_2008_Guide-for-Development-of-Bicycle-Facilities83606.pdf.

Forecasting, W. (2019). *Can carpool reduce emissions and congestion in Indian cities?|WRI INDIA*, 1–6. http://www.wri-india.org/blog/can-carpool-reduce-emissions-and-congestion-indian-cities.

Gehl, J. (2011). Life between buildings: Using public space. *Landscape Journal, 8*. https://doi.org/10.3368/lj.8.1.54.

IFC. (2007). *An integrated report*. https://www.openknowledge.worldbank.org/handle/10986/11849.

Images, R. (2015). *Explore Log in*. https://www.pixabay.com/photos/japan-shinagawa-mall-street-trees-828429/.

Cityscape Institute. (2008). Lighting use & design. *PPS.Org*, pp. 1–7. https://www.pps.org/article/streetlights.

Jang, Y. (2014). CityNet blog global warming from a Korean perspective. https://www.citynetmembers.wordpress.com/2014/01/14/global-warming-from-a-korean-perspective/.

Kheir-El-Din, H., El Ehwany, N., Ragab, A., & Fouad, H. (2012). Roads and highways in Egypt: Reform for enhancing efficiency. *Toward More Efficient Services in Egypt Reforming Tourism, Construction, Information Technology, Wholesale and Retail, Roads, and Banking Services, 152*, 217–254. https://doi.org/10.5743/cairo/9789774164941.003.0007

Linton, J. O. E., Street, S., Street, S., Angeles, D. L., Avenue, C. C., & Angeles, L. (2020). Los Angeles eco-village reinventing how we live in the city beautiful new buffered green bike lane on spring street. https://www.laecovillage.wordpress.com/.

Paula Maneola. (2015). Os 8 princípios da calçada. https://www.wribrasil.org.br/pt/blog/2015/04/os-8-principios-da-calcada.

Mathew, T. V., & Bombay, I. (2014). Transportation systems engineering: accidents studies. *Transportation Systems Engineering*, 1–28. https://nptel.ac.in/courses/105101008/downloads/cete_42.pdf.

McCutchan, S. (2013). How Wide Should a Neighborhood Street Be? – Part 1. *Planning Commissioners & PlannersWeb.Com, September*, 1–3. Retrieved from http://plannersweb.com/2013/09/wide-neighborhood-street-part-1/

NACTO. (2013). Street design elements. *Urban Street Design Guide, Island Pre* (New York) (pp. 31–70). https://doi.org/10.5822/978-1-61091-534-2_2.

Namee, S., & Witchayangkoon, B. (2011). Crossroads vertical speed control devices: Suggestion from observation. *International Transaction Journal of Engineering, Management, & Applied Sciences & Technologies, 2*(2), 161–171.

Newman, N. (2020). How can I find a list of companies that make bus shelters? https://www.quora.com/How-can-I-find-a-list-of-companies-that-make-bus-shelters.

Newton, D. (2012). Add a green buffered bike lane and number of cyclists explode—Streetsblog Los Angeles. http://www.la.streetsblog.org/2012/05/11/add-a-green-buffered-bike-lane-and-number-of-cyclists-explode/.

Of, A., & Parking, M. C. A. R. (2019). Multi-storey car parking system —Project.

Offer, O. U. R., Implementations, P., Company, W., Parking, S., & European, T. (2019). *Smart parking systems*. http://www.smartparking-systems.com/.

Officials, C. T. (2019). *Lane width* (pp. 1–11).

Roberts, J. (2020). Better block. http://www.visitfinland.com/about-us/.

Rosen, A. (2013). What really causes traffic congestion? https://www.bklyner.com/what-really-causes-traffic-congestion-sheepshead-bay/.

Rudofsky, B. (1969). Streets for people: A primer for Americans. https://www.books.google.com.eg/books?id=2uJUAAAAMAAJ.

SFPD. (2010). *San Francisco better streets plan*. https://www.sfplanning.org/sites/default/files/archives/BetterStreets/docs/Guide_to_BSP.pdf.

Smith, A., Zucker, S., Lladó-Farrulla, M., Friedman, J., Guidry, C., McGrew, P., & Duchesne, J. (2019). Bicycle lanes. *Journal of Trauma and Acute Care Surgery, 87*(1), 76–81. https://doi.org/10.1097/ta.0000000000002328

Solutions, R. P. (2020). *Commercial and residential parking solutions rotary display system automated multi parking lifts speedy automated parking rotary carousel parking lift and slide puzzle parking parkmatic provides automated and mechanical click for digital catalog free Quot* (pp. 3–5).

Street, A. (2020). Addressing street and front yard tree issues in cities. https://www.goodnewstree.com/2019/08/29/addressing-street-and-front-yard-tree-issues-in-cities/.

SUDS-RP-UD, A. (2012). Design of on-street transit stops and access from surrounding areas. *APTA Sustainability and Urban Design*

Program. http://www.citeseerx.ist.psu.edu/viewdoc/download?doi=10.1.1.278.4944&rep=rep1&type=pdf.

Tobergte, D. R., & Curtis, S. (2013). Streets as public spaces and drivers of urban prosperity. *UN habitat* (pp. 1–152). https://doi.org/10.1017/CBO9781107415324.004.

Trulia. (2014). 12 Things that make a neighborhood truly great. *Forbes*, 1–5. https://www.forbes.com/sites/trulia/2014/11/29/12-things-that-make-a-neighborhood-truly-great/#25d3d1735f6a.

Weber, H., & Sciubba, J. D. (2018). the effect of population growth on the environment: Evidence from European regions. *European Journal of Population = Revue Europeenne de Demographie*, *35* (2), 379–402. https://doi.org/10.1007/s10680-018-9486-0.

Wisconsin Department of Transportation. (2009). *Wisconsin bicycle facility design handbook*.

Social Housing Design: A Guideline for Enhancing Dwellers' Livelihood in Egypt Through Sociocultural Aspects

Youssef Baseet, Gehan Nagy, and Dalya Maguid

Abstract

Since the 1950s, the Egyptian government has proposed and constructed several low-income housing prototypes in Cairo. However, to date, many of the constructed prototypes have failed to meet their occupants' sociocultural needs. As policies, rules, and regulations defined by the government do not consider aspects of quality of life for the occupants. As a result, occupants tend to reappropriate their living spaces through transformations and territorial extensions into the surrounding urban space. This research aims to propose guidelines for the design of low-income housing that adheres to occupants' sociocultural needs. The method applied includes an in-depth investigation of earlier literature concerned with low-income housing design considerations. In addition, sociocultural activities, and their impact on the quality of life are investigated.

Keywords

Sociocultural aspects • Housing design • Design guidelines • Livelihood

1 Sociocultural Design Aspects of Housing

The research gives an understanding of the sociocultural design aspects which define the difference between housing and home. furthermore, it highlights the impact of implementing social–cultural aspects on the livelihood of the dwellers. Besides, the research explains diverse types of social housing and its related sustainable values with a focus on sociocultural aspects and its relationship with distinctive design levels aiming to create a design matrix that involves sociocultural values on each design level.

1.1 Housing and Home

Housing is a sizable subject that corresponds not only to the records of architecture since its beginnings but to the very history of humankind. The beginning of housing would be through a man who shelters himself from the rain and the solar through protecting oneself. The primal definition of housing would be a refuge, while the dictionary of French architecture of 1854 states that "*The idea of safety is a clear measurement of the idea of housing*" on the other hand, the Architect Viollet-Le-Duc says: "*In the art of architecture, the house is what excellently characterizes the customs, the tastes, and behavior of a population. A house isn't a historical report, it is the expression of a geological and social historical past*" another definition is, "*A house is an expression of a geological and social historical past and not just a historic report*" (Ortelli, 2018).

Housing is defined by being a physical structure through, which a family lives in a place where fundamental human activities take place (Després, 1991), while Cambridge dictionary explains the concept of housing "*that it's a building where people live in*," on the other hand, Oxford dictionary defines it as; "*a casing for enclosure and protection*." Similar definitions are provided in the Webster dictionary by being a dwelling provided for people.

However, a home is defined through critical behavioral interpretations identified through the general categorization of its occupiers those categories include security, control, and safety. The home also reflects one's ideas and values showing financial, physical, and emotional involvement with the owners of the dwelling unit. Moreover, a home is a place where someone experiences belonging through durability

Y. Baseet (✉) · G. Nagy · D. Maguid
Faculty of Engineering, British University in Egypt, Cairo, Egypt
e-mail: Youssef.baseet@bue.edu.eg

G. Nagy
e-mail: Gehan.nagy@bue.edu.eg

D. Maguid
e-mail: Dalya.maguid@bue.edu.eg

F. Rosso et al. (eds.), *Advances in Architecture, Engineering and Technology*, Advances in Science, Technology & Innovation, https://doi.org/10.1007/978-3-030-86913-7_6

and continuity creating roots for people living in a certain place. A home Also stands for the opportunity for having relationships with family and friends through multiple practices and activities. Furthermore, a home displays personal status and social repute by supplying a place to own which gives a stable foundation to family lifestyle and pleasure (Després, 1991).

1.2 Types of Social Housing

There are multiple types of social housing programs including family housing, senior-specific, and housing with adaptation to the disabled. These types of housing have various sources for funding it might be governmental or aid driven from other organizations.

Family housing is the most common between housing types as different age groups are eligible to dwell in it since it is not very user-specific (communitylegalaid.org, 2017). The number of rooms for such type is determined according to the number of family members. On the other hand, senior housing is affordable for seniors, such type of funding depends on the cut of senior's wages during their employment and its design should consider their needs for safety due to their possibility of being harmed by creating a safer design through carefully installing more lighting, bars and creating a safer environment for them (homeforallsmc.org, 2014). Moreover, the special needs housing identified through having a medium to deal with difficulties in their habitual environment (Government AU, 2016).

The modified social housing aims to fulfill the requirements and needs of living, being user-specific through understanding their health, security, safety, and needed services. In addition to, responding to their needs by multiple design considerations and location, the idea of integrating the disabled among the society is vital. Such type of housing should consider also visual tactility to the hearing impaired for alerts, accessibility, and circulation measures of any entrance access point, baths, toilets, and basins through certain applicable design features and guides (Osman, 2013).

1.3 Sustainable Aspects

The aspects of sustainable design include environmental, economic, and sociocultural aspects which create a housing development that suits the dweller's needs and future wants.

Environmental sustainability in housing may be executed through addressing useful resource limits of the environment through efficient consumption of non-renewable resources, minimizing the impact of waste substances, and pollution through utilizing appropriate technologies (Nair, 2005). On the other hand, the most crucial financial sources are the capacity financial savings of the population. Economic sustainability or affordability of housing daily are embedded in a monetary development approach, which strengthens the economic self-reliance of household participants. Meanwhile, the poor frequently can only manage pay of public housing through assistance for financial sustainability or affordability of the schemes (Glaeser & Tiwari, 2001). Lastly, sociocultural sustainability in housing entails numerous dimensions including adaptability, equality, integration of facilities and services, self-help housing or beneficiary participation, and network involvement.

Sustainable housing daily depends on the sociocultural practices of the households and groups (Nair, 2005). Through, targeted housing improvement that promotes social interplay of individuals and cultural enrichment of the community and is aimed every day to lessen the inequality of housing among communities. meanwhile, it accelerates the improvement in social development, family members, and interactions (Nair, 2005). The multiple sustainable aspects entailed defining how social housing should fulfill multiple dimensions for the dwellers to feel more satisfied in the living experience they part take in.

The aspects of sustainable housing design include environmental, economic, and sociocultural aspects. considering different aspects throughout the design process would enhance housing livability.

Table 1 Shows the difference between housing and home

House	Home
–Physical structure –A place for fundamental human activities –A building where people live –Enclosure and protection	–Security and control –Reflections of one's ideas and values –Applying modification to the unit –The ability to continue to live and use the dwelling unit create memory and roots –Family and friend's interpersonal relationships –Center for activities –Personal status indicator –A place to own

1.4 Sociocultural Aspects Affecting Housing Design in the Middle East

A community is a locality about the size of a suburb or a town that houses people with similar sociocultural, socioeconomic, and sociopolitical characteristics. Accordingly, the community has a clear spatial dimension which is represented as geographic concentrations of people with similar characteristics, sharing the locality and interpersonal ties with one another, with varying levels of intimacy. Through a series of local social relations, such as those between households local voluntary organizations, formal organizations, and local government, a community culture covering the rules, customs, laws, norms, etc., is formed. That state of how individuals and groups who form the community should behave housing is the key material resource for individuals and households making up the community with multiple factors taken into consideration as aspects of social culture (Reingold, 1995).

1.4.1 Safety

Safety is a state of being protected from a perception concerning the neighborhood, social, and physical dimensions forming its context. Approaching the idea of conforming social organization forming social ties creating residential stability, which aids in the decrease of emotionally perceiving crime.

The level of safety of neighborhoods is affected by the socioeconomic composition of its residents, leading to networking which creates social ties and bridges gaps with the outer world. Identifying who relates to a certain place is correlated to knowing people on the streets that mark the territory of such place decreasing fear of crime (Clampet-Lundquist, 2010). Besides, the benefits that come from the social ties could also be securing a job from a neighbor's recommendation which aids in safety in economic status or even a cooked meal that increases the quality of life (Clampet-Lundquis, 2010).

Approaching theories on security and crime prevention through either social control, space encloser and defensible space, through having shops that give a reason for the people to use the streets decreasing unfamiliar people's physical contact with parts of the community and Defensible space might also be the solution through increasing the safety by households surveillance their dwelling units (Newman, 1996).

Safety is a key part of everyday living and through understanding key theories in housing design such as social control, space enclosure, and defensible space it could be accomplished. Through, the aid of community ties and kin the cornerstone of the design shall aid to achieve and have a perception of social stability.

1.4.2 Privacy

Privacy is defined through one's ability to govern acoustic-visible and olfactory interplay with other people (Lang, 1987) leading to the conclusion of how social interaction and privacy closely intertwine in a relationship on everyday activities shall be conducted without being gawked and stalked at or even have a fear of being observed (GhaniSalleh, 2008). Moreover, privacy defines the potential control of social interplay and being capable to choose the social interaction desired (Schwartz, 1968).

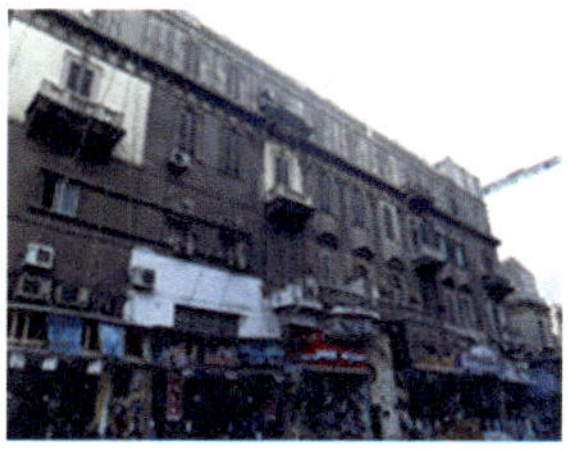

Fig. 2 showing social control factor of the shops creating safer environment by surveillance.

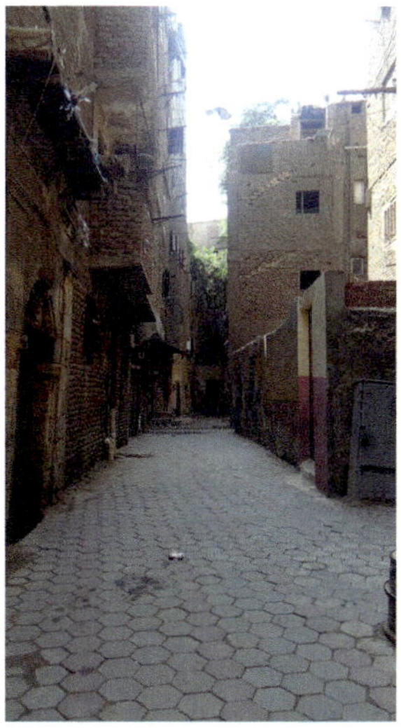

Fig. 1 Showing space enclosure in Khalifa neighborhood, khedive Cairo

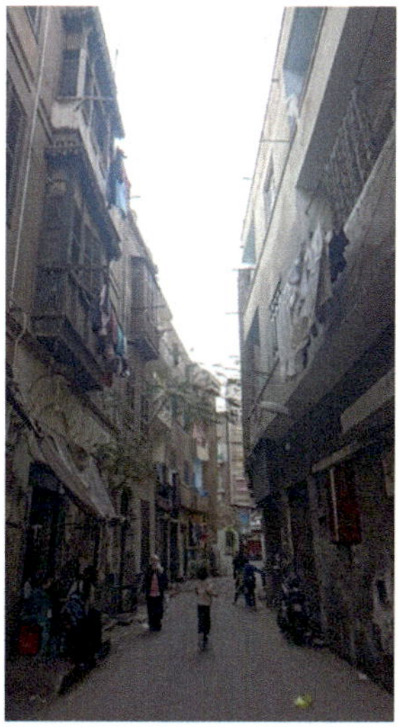

Fig. 3 Showing mashrabiaya as a transformative addition for privacy in Qawawir neighborhood

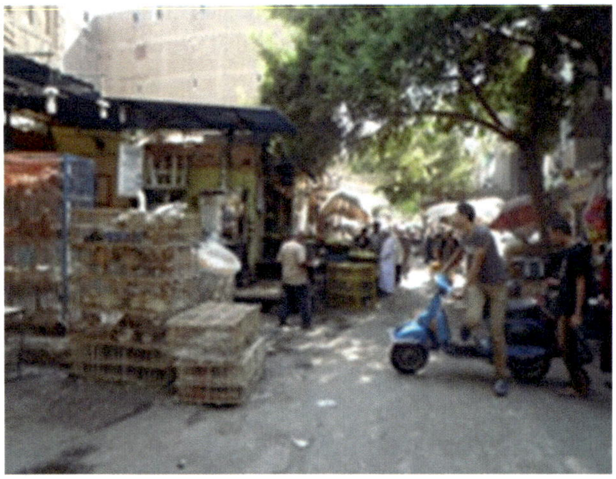

Fig. 4 Showing daily social interaction in residential areas of Cairo

Spaces with indefinite boundaries of public and private spaces have a lower amount of social interaction, defining the need for physical privacy for social acts, enabling personal choices of a wider range to occur (Rapaport, 2015). Through avoiding contact with others and creating spatial territorial control over a certain space by clear boundaries both will lead to privacy, however, the rate with defined and clear boundaries in the interior or exterior of the building and in open public areas affecting people's satisfaction with the homes they live in (Opoku, 2010).

Traditional Middle Eastern cities have an association with their specific societies' culture, organically formed or not the manifestation of their Islamic beliefs accordingly to certain principles that define character traits that include the principle of privacy (Opoku, 2010). Through which, cultural perception of privateness becomes one of the determinant elements inside the system which remains obvious in the hierarchy of areas in context (Othman, 2015). Through, separating males that are not related to the family and females, privacy becomes vital to the everyday life of a Muslim satisfying needs of psychological relation.

Creating inwardly positioned and oriented layouts and environments, using Mashrabiyas, and positioning windows in higher levels cultivate the concept of privacy in home designs offering a more psychological and religious satisfaction in the middle eastern sociocultural context since Islam is the widely followed religion (Al-kodmany, 1999).

1.4.3 Social Interaction

Physical and mental well-being are strongly correlated to social networking and interaction, which also plays a role in the fabrication of social capital, giving inhabitants of a town or city a coping mechanism against scarcity of resources, hardships, and being poor (Wandersman, 1985). This social interaction can be considered the building block of a community, through simple greetings and visual contact one shall feel a sense of familiarity of being at home. Moreover, senior adults are the ones who lack the most social interaction which could lead to their isolation for the multiple challenges that they usually are facing from decreased mobility, illness, disability, and bad health effects, showing how crucial human interaction can be. However, young adults spend most of their time distant from their homes commuting to and from their work, having a low amount of time to spend at home decreasing the amount of social interaction practiced in their neighborhood (Farshidi, 2016).

Through two types of social interaction within residential communities, the informal and formal beholding varied strength of weak interactions such as spontaneous corridor talks and strong interactions including neighborhood organizations participation and communal service work, communities get an indicative assessment on how much they are bonded and cohesive (Farshidi, 2016). A general increase in social interaction comes with homeownership over renting since homeowners have a higher general sense of community since they are more interested in maintaining their property and neighborhood taking part and participating in communal acts. Also, more encounters in local shops such as a bakery or grocery store increase neighborhood social interaction and the sense of community (Gehl, 1996).

Social interaction can get affected by design through physical attributes of building elements affecting communal behavior through space creating opportunities for interaction and encounter. Those physical attributes affect certain qualities such as safety and privacy impacting the environments which hold interactions (Farshidi, 2016). Moreover, the design quality defining the character of space plays an important role in cultivating social interaction through the legibility of being a clear design, accessible, and affordable (Ewing & Handy's, 2009). It can be argued that design features are impartial and may be gauged with the aid of measuring bodily attributes, however, consumers' perception is completely subjective and can be measured best through asking for users' reviews. Defining design qualities through their effect on the interaction among dwellers, including Accessibility, bodily proximity of communal areas and layout immediately impacts motion patterns. Accordingly, growing more opportunities for passive touch, increasing the use of communal areas and public spaces by making them legible of multiple activities and visually connected yet not lacking privacy and creating visually attractive environments increasing the probability of meetups and interaction making communal areas more inviting (Sherrod & Downs, 1974). Through creating affordance, space offers to be flexible with multiple functionalities according to dweller's wants and needs (Gehl, 1996).

To sum up, social interaction is extremely influential on all levels since it is the building block of the community, it is needed for a healthy, non-isolating living environment. Also, social interaction achieves certain design qualities including; physical proximity by having main access route proximity and dwelling units proximity, main access route, and dwelling units through visual connectivity. Moreover, creating attractive visuals through the use of view, color, light, and greeneries through having clear boundaries that create affordance.

1.4.4 Lifestyle

According to the Cambridge dictionary lifestyle is defined by, "*someone's way of living: the things that a person or a particular group of people usually do.*" Investigating the everyday practices and daily life of people of certain social groups would only give the understanding of their lifestyle, also by analyzing certain districts' consumption of designs and symbols an impression will typically be left. Information about value could be analyzed through people's selection of goods bought, even though what has a certain high value at a certain area or district might be considered of low value in another (Salama, 2006).

Reaching an idea of how lifestyle alters one's home through how people work and spend the rest of their time living. Although, it may change it usually has basic forms that a home can be adaptable to accommodate different lifestyles; self-employment, where the home plays a part of the production of one living, using their homes for representative purposes and wage earners who live separating their work from home spending their leisure time only at home. Considering, the ability of the individual to alter his home to adapt to his needs or to find a home that suits his life mode would create a much more satisfactory experience for the dweller's lifestyle (ÆrØ, 2006).

1.4.5 Family Structure

A family is defined through its household consisting of two or more, the family might include children, or be a single-parent family (insee, 2016). Also, a family is dependent on the number of members it beholds. Upon the number of members, the number of rooms in their homes should be suited. In instances of having children of different sexes a three-bedroom apartment would be more suitable and in case a senior citizen needs someone to accompany him a two-bedroom apartment would suit his/her needs (communitylegalaid.org, 2017).

Families being assigned to housing suiting their family structure are more satisfied (communitylegalaid.org, 2017). The incapability of paying for the extra space is one of the main reasons why families do not opt for bigger space and other reasons such as the unavailability of three-bedroom apartments in social housing is also a very critical reason of such deficiency in how family structure is not considered by having less available types of units for families with more members (SeattlePlanningCommission, 2014).

1.4.6 Hospitality

Acceptance and receptiveness to others are how hospitality can be defined as a trait, playing a role that is defined as being of value in the Islamic traditional society that is heavily associated with middle eastern culture (Shraim, 2000). Moreover, compassionate treatment of strangers has an association with how hospitality is perceived, and the importance of sharing is a general idea and that the culture of Islam prospers (Sobh & Belk, 2013).

The domain of public hospitality in the Islamic culture creates comfortable surroundings and setting as an environment to host males and females. However, they get hosted separately since men get hosted in the "majlis" which is translated to a place to be seated being directly accessible from the street and women get hosted in salons where it usually more elegant in the use of materials and décor in comparison to the place of seating that men use (Sobh & Belk, 2013).

In conclusion hospitality of the home needs separateness of females and males to take place in the Islamic culture. However, the home is seen as a feminine domain, a place to practice religious activities, a place to extend the relationships with one's family, and to be open to.

Table 2 Shows the sociocultural aspects and give example on relations on influence

Sociocultural aspects	Housing influence relation
1. Safety	–Social control (Jacobs)—space enclosure—Defensible space (Newman)
2. Privacy	–Acoustic privacy—Visual privacy—Clear boundaries between neighbors
3. Life style	–The ability to repurpose and reconfigure space
4. Social Interaction	–Accessibility and layout—visual connectivity—visually attractive space –The affordance of space having multiple functions
5. Family structure	–Spaces suiting the need for space according to family size and need
6. Hospitality	–Different spaces for males and females—Approaching theories on security and crime prevention through either social control, space encloser, and defensible space. and more decorated spaces for receiving strangers

2 Affordable Housing Design Considerations

The design consideration's chapter on housing design trys to define a formative checklist to identify key troubles of designing a social housing project with defined sequential development procedures that are taken for the design of such projects.

2.1 Site Selection

The site selection process is one of the vital jobs that are first taken by experts to create and build residential housing for multiple considerations which includes site selection with good transport facilities, near civic services, and good soil. The selected site should have enough space to accommodate the needed sum of units and parking required by the project (Anupoju, 2016).

2.1.1 Site Location

Choosing a projects site location is one of the very first steps taken into creating a housing project and a lot of research is taken into consideration including resident's socioeconomic status concerning housing suitable accessibility to amenities and services and the ease that dwellers shall find into using transit hubs that need to be located nearby to reach out the work, or the city core (Collier, 2014). Moreover, the site entry should be taken into consideration being emphasized to dwellers and non-residents alike for the definition of the territory of home from the greatest scale onwards to a smaller scale to create a feeling of safety (Advisor, 2004).

2.1.2 Parking

Parking is a hard issue to clear up in low-cost housing improvement projects. vehicle/pedestrian interaction must be carefully analyzed, with a focus on minimizing conflicts. Keeping it away from letting it dominate belongings or the streetscape (Rashid, 2017). However, properly dealt with parking can fade into the historical past and nonetheless continue to be fully useful, reachable, and secure. Different approaches to a parking location should be used—for play or automobile maintenance. If parking is provided underneath a podium, the podium itself must be designed as a public open area (authority, 2017).

2.2 Urban Scale

The urban scale development consists of open public, open private spaces. The connectivity of the buildings to the open spaces and the layout to how the tenants of these areas interact with the built environment and that is vital for such areas to be more successful for dwelling.

2.2.1 Public Open Space

Public open areas—shared outside areas intended for use of citizens—must be as thoughtfully designed as another "space" in an improvement (AucklandCouncil, 2019). It is far helpful to think about open spaces as outdoor rooms and to design and supply them with the identical care put in any room in your own home. Such rooms should be easy to access from any living unit in a development. and they must have clear limitations so that citizens and traffic apprehend what is public and what is non-public (Dadkhah, 2015). Surveillance is likewise critical: as many devices as feasible must have visibility of open spaces, especially play areas.

Finally, public open regions must be designed to be used at night time in addition to during the day. Intelligently designed middle of the night lighting fixtures will ensure that public spaces are attractive and safe after sunset (Advisor, 2004). Leading to, delicately designed public open space that turns an amazing development right into a superb one, supplying an enduring facility for residents and outcomes alike.

2.2.2 Private Open Space

Non-public open space—outdoor regions in which residents can enjoy the sun and sky with privacy is vital for life for all people. Patios, porches, decks, balconies, and yard needs to also be of good enough size with smooth access from each living unit. unique care must be taken whilst designing balconies to efficaciously balance the want for light and look at with safety and privacy issues (authority, 2017).

A nicely designed, correctly sized private accessible area will improve the high quality of any affordable housing improvement and should be considered a need rather than a comfort (Collier, 2014).

2.2.3 Landscaping

Landscaping—too regularly dealt with as a secondary consideration or removed altogether because of the value of their use. Paved areas are essential and need to be designed as a part of the panorama. Edges between paved and planted areas should be designed so the two nation-states work nicely personally and collectively. Paths and door seating must suit the overall panorama plan and taken into consideration how and while residents will use them (Advisor, 2004).

Landscaping has a big effect on any undertaking as it supplies and enhances an undertaking of its neighborhoods. If it was not completed or carried out poorly the exceptional of a task suffers drastically.

2.3 Building Allocation and Design

The building location and layout are initially well thought of taking into consideration multiple design aspects, then the building's shape is designed taking into consideration the community that would be living so that it reflects their identity and be visually appealing, respecting the size and form.

2.3.1 Building Location

Constructed dwellings are placed on a site have a powerful effect on the perception of the dwellers. The site entry—where it is and the way it appears—is vital to the public photo of an improvement. Likewise, the improvement's setback sample can affect the public belief of the venture. Finally, a construction's placement on a site will have an impact on how, and what kind of, the weather will affect the construction (Advisor, 2004). How a building is positioned on a domain is one of the "big" early choices in the design manner. Getting it right makes each later step exceptionally easier (QueenslandGovernament, 2015).

2.3.2 Building Layout

How well construction is laid out will decide how well it "works" for residents, a team of workers, and site visitors. Entries should be defined, welcoming, and comfortable. Central spaces and common rooms must be clean for all residents to get to and use. Likewise, help and service areas must be easy for management personnel to use. Stairs, elevators, and admission to corridors serve as crucial public gathering locations and must be designed accordingly (authority, 2017). A building's designed layout and circulation should result in a secure to live in and less complicated premises.

2.3.3 Building Shape

The general size and form of new construction will have a big effect on the encircling neighborhood and on how improvement is perceived through the community wherein its placed (Collier, 2014). The height of new construction is extraordinarily vital—too excessive and it can overwhelm its associates; too low and it can create a gap in the physical "cloth" of a neighborhood. the overall scale and massing of a brand-new building ought to also try to shape that of the encompassing neighborhood. Moreover, the overall shape of a brand-new construction must include as much variety as workable and keep away from large expenses (Advisor, 2004).

creating a construction whose size and shape complement the scale and form of surrounding homes will pass a long way toward making a brand-new improvement acceptable to its acquaintances. at equal time it will fortify the notion amongst citizens that their housing is "much like every person elses."

2.3.4 Building Appearance

How a dwelling looks is crucial to its acceptance inside a community. Since, the home windows, façade, roof form, length, a rhythm of openings, trim, materials, and color should be commonly well matched with the surrounding community. Finally, unique interest must be paid to the layout of front doorways because they bring such sturdy messages about the great development and its citizens (authority, 2017).

2.3.5 Unit Layout

A large mix of unit kinds, supplying variety and versatility, must be taken into consideration (QueenslandGovernament, 2015). The relationships between rooms inside each unit type must additionally be carefully analyzed, since every room layout should be based totally on how that room could be used, and every layout must be examined to make certain it may be supplied in at least two exceedingly different methods (AucklandCouncil, 2019). Also, entries, eating rooms, toilets, and storage areas should get hold of special attention, as a need for the views to be had from the windows (Collier, 2014). Moreover, access to daylight hours and air flow ought to be maximized. for max gain to the citizens, substances, and home equipment/mechanical systems should be selected primarily based on lengthy life and ease of renovation (Advisor, 2004).

The residing unit—greater than the overall apartment building or the open area—is what residents will name their home. every effort must be made to create dwellings that might be a delight to live in.

2.4 Conclusion

The conclusion matrix shows the relations between sociocultural aspects in both forms tangible and non-tangible concerning all design guideline aspect points including the site location, building location, parking, public spaces, private spaces, landscaping, building appearance, building layout, and unit layout. Also, given the clear understanding of how each of these aspects gets affected by these designs, considerations lead to a deeper statistical analysis with defined percentages of each of the design guidelines' effect on each sociocultural aspect. Moreover, the statistics were carried out by counting how many points each aspect got in

Table 3 Shows the sociocultural aspects and their relation with housing design aspects

Design Aspects		1-Safety	2- Privacy	3-Life style	4-Social Interaction	5-Family structure and size	6- Hospitality
Site Location	1. Housing location suitably accessible to services and amenities			X		X	
	2. Transit services nearby			X			
	3. Site entry emphasizing main entry	X		X			
Building Location	1. Solar access, natural ventilation & views access		X	X			
	2. Setbacks being respected	X			X		
Parking	1. Avoid letting parking dominate street scape	X		X	X		X
	2. Provide parking location minimizing walking distance	X		X		X	
	3. Minimize conflicts between vehichles and pedestrians	X					
	4. Secure entries and well lit parking	X					
	5. Site furniture and adequate landscaping			X	X	X	
Public Open Spaces	1. Outdoor spaces for both residents and non-residents			X	X	X	X
	2. Accessability from the dwelling units	X		X	X	X	X
	3. Clear boundries between public and private controlled spaces		X		X		
	4. Surveillance through viral access to shared open spaces	X			X	X	
	5. Play areas for children	X			X	X	
	6. Night time lighting	X			X		
Private Open Spaces	1. Provide each household a certain form of open space [balcony/yard]			X		X	
	2. Adequate size			X		X	X
	3. Using fending/screening for balaconies	X	X				
Landscaping	1. Variety of plantation				X		
	2. Paved areas/nodes	X			X		
	3. Planted edges				X		
	4. Outdoor seating/furniture		X		X	X	
	5. Pedestrian paths	X			X		
	6. Storage for landscape equipments						
Building Shape	1. Building heightS	X	X				
	2. Building scale and massing						X
	3. Building form [unbox like form]				X		X
Building Apperance	1. Building image reflecting the community				X		X
	2. Visual complexity with variations						X
	3. Maximizing window number and size	X	X				
	4. Front door quality	X					X
	5. Façade relating to the character of the region		X				X
	6. Roof shape relating to the character of the region						X
	7. Trim and details adding to the character of the building				X		X
	8. Material and colour adding variety and visual interest				X		X
	9. Indvidual identity to break large repetative structures						X
Building Layout	1. Entries prominent and visible	X			X		
	2. Facilities and common rooms available	X			X	X	
	3. Stairs should be durable	X				X	
	4. Elevators available and adequate space in front	X		X		X	
	5. Access corridors avoiding exessive length	X					
	6. Security of units by the ability of residents surveillance	X	X				
Unit Layout	1. Unit entry providing indvidual identity		X				X
	2. Room relationships creating private and less private living areas		X			X	X
	3. Room design [Tested arrangement of furniture arrangement]			X		X	X
	4. Multiple variety of unit types			X		X	
	5. Dinning room having enough space				X	X	X
	6. Bathroom being visually screened		X			X	
	7. Day lighting and ventilation	X				X	
	8. Storage space	X		X		X	
	9. Window views	X	X				
	10. Materials used should be easily maintained by residents	X		X			

(Non-Tangible Activities/Aspect: 1-Safety, 2- Privacy, 3-Life style, 4-Social Interaction, 5-Family structure and size, 6- Hospitality)

the design criteria and turning it into a percentage chart. Resulting in, a conclusion aided in defining what the social interaction has held the highest percentage of charts creating the conclusion that it should be most thought of during the design procedure of social aspects and then comes lifestyle and safety.

Tables 6.4, 6.5, and 6.6 later show the final concluded housing guidelines per the three phases site location, urban design phase from the noted design aspects and building, and unit design phase.

2.5 Recommendations and Further Research

Further research is needed in the gap between how municipalities manage social housing and the process of finding dwellers to units that suits their needs, which plays a big part in their satisfaction and creates a more socioculture-friendly atmosphere by not cutting down dwellers motility with their old neighbors and acquaintance which made them more capable beforehand especially in cases were whole neighborhoods get reconciled and people knowing each other beforehand.

Table 4 Showing urban design as the second phase of the design aspects

Site Location		
Design aspects	Site location	Building location
	•The housing location should be suitably accessible to services and amenities since it affects the family's lifestyle and affects families for their size and structure •The housing project should be closer to nearby transit services •The project should emphasize site entry for safety and lifestyle	•Building location should have good solar access, natural ventilation, and view access as it affects their privacy and lifestyle •Setbacks must be respected for the safety and social interaction of the residences

Table 5 Showing building and unit design phase as the third phase of the design aspects

Urban design phase			
Parking	Public open spaces	Private open spaces	Landscape
• Avoid parking dominating streetscape for safety, respecting resident's lifestyle, increasing social interaction, and hospitality • Provide parking locations close by the housing minimize walking distance for safety, adaptability to family lifestyles and family structure, and size • Minimize conflicts between vehicles and pedestrians for safety purposes • Secure entries and well-lit parking should be set up for the safety of residents • site furniture should be added for lifestyle, increased social interaction, and family structure and size	• add outdoor open spaces for residents and non-residents because of resident's lifestyle, social interaction, family structure, and hospitality • The open spaces should be easily accessible for safety purposes, lifestyle, social interaction, family structure, and hospitality • Boundaries should be created between private and public areas for privacy and social interaction • Surveillance through viral access should be set up for safety, family structure, size, and social interaction • Adequate play areas should be created for safety, family structure, and social interaction • Pubic open spaces night time lighting should be added for safety social interaction	• Each house should be supplied a certain form of private open space a balcony or a yard for the life style and family structure • The private open space should be adequate in size for adaptation to family lifestyle, family structure and size, and hospitality • Fencing and screening should be used for safety and privacy	Plantations should be varied aiding an increase in social interaction Areas and nodes should be paved for safety and increased social interaction Edges should be planted increasing social interaction Outdoor furniture should be added for increased social interaction and privacy depending on the type of furniture Pedestrian paths should be created for the safety and increased social interaction

Table 6 Showing Building and Unit design phase as the Third phase of the design aspects

Building and unit design phase			
Building shape	Building appearance	Building layout	Unit layout
• Building height should be according to the regulations for safety and privacy • Building scale and massing should be well thought out and design since it affects hospitality • Building design should be well thought out and not be in a box-like form for social interaction and hospitality	• Building images should reflect the community for an increase in social interaction and hospitality • Visual complexity should be integrated since it increases hospitality • Window sizes should be inadequately maximized number and size taking into consideration safety and privacy • Front door quality should be of good quality for safety and hospitability • Facades and roofs should be relating to the character of the region for the sense of hospitability and privacy • Trim, details, colors and individual building identity breaking large repetitive structures should be adding to the character of the building affecting the hospitality and social interaction	• Entries must be prominent for safety and social interaction • Common rooms and facilities should be available for safety, added lifestyle, and social interaction • Stairs and elevators should be durable with enough space in front of safety, adaptability t lifestyle, and social interaction • Access corridors should avoid excess length for safety • Surveillance for the units should be available for the sense	• Unit entry should supply an individual identity for creating a sense of privacy and hospitality • Defined Room relationships should be created into private and less private spaces also affecting hospitality and family structure and size • Room design with tested management of furniture affecting lifestyle, family structure, and size, and hospitality • Multiple varieties of unit types must be offered affecting lifestyle and family structure and size • A dining room should have enough space for social interaction, family structure and • Visually screening for bathroom affecting privacy • Storage spaces must be offered for lifestyle adaptations •Windows views considerations being affected by privacy and safety •Materials of easy maintenance should be used for safety purposes and lifestyle considerations

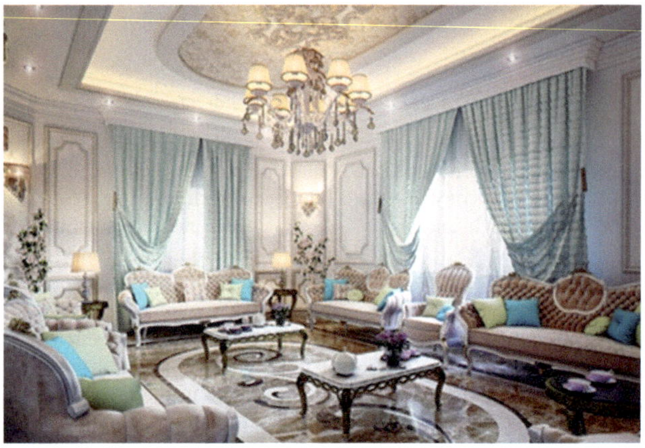

Fig. 5 showing enclosed living spaces heavily decorated and embroidered for women

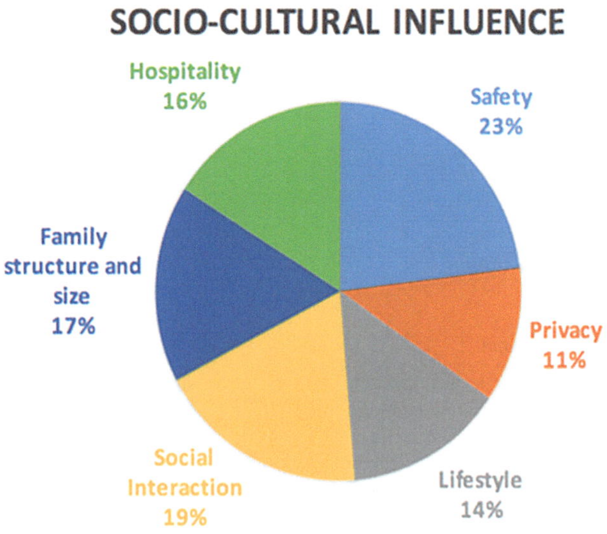

Fig. 6 shows the sociocultural aspects and their influence on design percentage.

References

Advisor, A. H. (2004). *Affordable housing design advisor*. Retrieved from https://web.njit.edu/abs/frameset.html?https://web.njit.edu/abs/check/check_buildloc_01.html

ÆrØ, v. T. (2006). Residential choice from a lifestyle perspective. *Housing, Theory, and Society*.

Al-Kodmany, K. (1999). Residential visual privacy: Traditional and modern architecture and urban design. *Journal of Urban Design, 4* (3), 283–311.

AucklandCouncil. (2019). *apartment building design*. Retrieved 5 2, 2019, from aucklanddesignmanual: http://www.aucklanddesignmanual.co.nz/sites-and-buildings/apartments/guidance/placing-the-building/designing-for-privacy#/sites-and-buildings/apartments/guidance/placing-the-building/designing-for-privacy

Authority, S. H. (2017). Form of Housing 1.1 house design guide. 30.

Clampet-Lundquist, S. (2010). "Everyone Had Your Back": Social Ties, Perceived Safety, and Public Housing Relocation. *City & Community*.

Collier, P. (2014). *Urban gateway*. Retrieved from http://www.urbangateway.org/news/importance-housing-africa%E2%80%99s-economic-development

Dadkhah, M. (2015). Apartment-Housing Effects on People's Social Interaction Case study: Abbas Abad District of Isfahan. *Current World Environment, 10*(1), 294–301.

Després, C. (1991). The meaning of home: literature review and directions for future research and theoretical development. *Journal of Architectural and Planning Research, Vol. 8, No. 2, Special Issue: The Meaning and Use of Home* , 96–115.

Government, Q. (2016). . Retrieved from Queensland government: https://www.qld.gov.au/housing/public-community-housing/eligibility-applying-for-housing/housing-for-people-with-disability

homeforallsmc.org. (2014). *Impact of Affordable Housing on Families and Communities.* Enterprise .

insee. (2016). *insee*. Retrieved from insee: https://www.insee.fr/en/metadonnees/definition/c1626

Lang, J. (1987). Privacy, tterritoriality,and personal space- proxemic theory .

Luca Ortelli, C. W. (2018). *What is housing?* EPFL, architecture. EPFL.

Nair, D. (Deepa.G. Nair). A CONCEPTUAL FRAMEWORK FOR SUSTAINABLE – AFFORDABLE HOUSING FOR THE RURAL POOR IN LESS DEVELOPED ECONOMIES. *The 2005 World Sustainable Building Conference.*

Opoku, R. A.-M. (2010). Housing preferences and attribute importance among low-income consumers in Saudi Arabia. . *Habitat International.*

Osman, H. (2013). *Housing for Special Needs: Physical Interior Design to Accommodate Special Needs Presenters* .

Othman, Z. A. (2015). Privacy, modesty, hospitality, and the design of Muslim homes: A literature review. Frontiers of Architectural Research, .

QueenslandGovernament. (2015). *Design Standards for New Construction Social Housing: Houses and Apartments.* Queensland Governament. Queensland: QueenslandGovernament.

Rapaport, a. (2015). *Using "Culture" in Housing Design.*

Rashid, S. A. (2017). Safety of street: The role of street design. *AIP Conference*, (p. 7).

Salama A. (2006). A life style theories approach for affordable housing research in saudi arabia . *Emirates Journal for Engineering Research,*, 10.

Schwartz, b. (1968). *The Social Psychology of Privacy.*

SeattlePlanningCommission. (2014). *Family-Sized Housing An Essential Ingredient to Attract and Retain Families with Children in Seattle.* SeattlePlanningCommission. Seattle Planning Commission.

Shraim. (2000). Hospitality and visibility in domestic space: An analysis of visual separation between the men's and women's domains of domestic space in Riyadh. (Doctor of Philosophy / Thesis). Georgia Institute of Technology.

Sobh, R. B. (2013). *slamic Arab hospitality and multiculturalism Marketing Theory.*

Efficient Designs: High Performance, Energy- and Resource-Efficient Buildings

Futuristic Interior Design Concept Through the Evolution of Biotechnology: Towards a New Model of Bio-sustainable Space

Eman Ahmed El Sayed Mahmoud Al Akaby,
Mahmoud Hassan Ibrahim Almallah,
and Marwa Khaled Ibrahim Mahfouz

Abstract

The research presents an interdisciplinary study focusing on the integration between biotechnology and interior architecture design. Over the past decades, a new model of interior space has emerged. However, the extent of the biotechnological impact on the transformation of the interior architecture design concept needs to be clarified. Furthermore, it is necessary to reach an optimal solution for interior design that can develop automatically and adapt to the surrounding environment by utilizing the creative solutions offered by biotechnology to address a new bio-design. Accordingly, the purpose of this study is to propose and test an integrative model of bio-interior space by performing a descriptive analysis of individual experiences of biological treatments and tracking the evolution of synthetic biology to reveal their contribution. It helps in solving environmental problems and developing biologically sustainable interior spaces. Additionally, the research aims at studying the impact of vital tissue applications on the interior design concept, which affects the future of interior design and its livability in terms of form and function using bio-tools. Moreover, it presents a design assumption for semi-living interior architecture as a proposal that clarifies the new design thinking criteria for building the shape of the interior space and function. This model shows the selection of biomaterials and biological treatments by presenting an experimental design for a group of furniture units and individual elements, then demonstrating how they fit together within the surrounding environment. Finally, the model proposed in this study sheds light on the impact of incorporating biomaterials in interior spaces which encourages new further interdisciplinary studies to search for the possibility of formulating a new complete sustainable interior design.

Keywords

Bio-architecture • Sustainable spaces • Biotechnology in interior design • Vital tissue

E. A. E. S. M. Al Akaby (✉)
Lecturer, Faculty of Fine Arts, Decor Department, Interior Architecture, Alexandria University, Alexandria, Egypt
e-mail: Eman.A.Akaby@alexu.edu.eg

M. H. Ibrahim Almallah
Professor, Faculty of Fine Arts, Decor Department, Interior Architecture, Alexandria University, Alexandria, Egypt

M. K. Ibrahim Mahfouz
Associate Professor, Faculty of Fine Arts, Decor Department, Interior Architecture, Alexandria University, Alexandria, Egypt
e-mail: Marwa_Mahfouz@alexu.edu.eg

1 Introduction

Biotechnology has left its footprint in the architectural design field through its many achievements. It is the third most influential technology, after Nanotechnology and Cyberspace, that led to the collapse of the old ages of interior architecture design. It is a revolutionary change in the concept of traditional thought of interior design, which considering nature is the real source of raw materials and architecture.

> We overlook only too often the fact that a living being may also be regarded as raw material, as something plastic, something that may be shaped and altered, Herbert George Wells said. (HG Wells, 1975).

1.1 Research Problem

The research problem lies within several points:

- The lack of clarity on the impact of the biotechnological evolution on the interior design concept and the design of its components.
- The integration of the two fields is limited to individual attempts, some of which are theoretical or experimental treatments applied in small scales or on prototypes, and

F. Rosso et al. (eds.), *Advances in Architecture, Engineering and Technology*, Advances in Science, Technology & Innovation, https://doi.org/10.1007/978-3-030-86913-7_7

do not come together to show an integrated result that serves the field of interior spaces' design.

- The need to transform the space surrounding its users from a traditional one to one that develops, adapts, and interacts automatically with the surrounding environment and becomes on with its system without harming it, rather encouraging its sustainability.
- The lack of clarity on the new role of the interior designer when creating a vital or semi-living interior space design.

1.2 Objectives

The main *aim* of the research is to clarify how to achieve a new model of bio-interior space that can complement the surrounding environment's sustainability. Another remarkable goal of the research lays in pinpointing and further clarifying the impact of biotechnology's evolution on the transformation of the interior design concept through monitoring theoretical ideas or actual experiments that produce biological materials or design treatments, and trying to classify them and define their intellectual trends. All of which accomplished by the achievement of the rest of the *objectives*, which are shown in the following points:

- Recognize Biotechnology's new role in achieving the sustainability of interior spaces.
- Identify the methods and methodologies that have been used beneficially in creating biomaterials that can be used in the design of interior architecture and its components.
- Explore what biotechnology has added to the field of interior design in terms of changing the general concept of its form, function, use of material, and design treatment.
- Attempt to devise a new role for the interior architecture designer and strengthen his relationship with specialists in the field of biotechnology. Moreover, propose new standards for the future of sustainable bio-interior architecture.
- Attempt to suggest an experimental design for vital furniture units, as well as an integrated interior space that can be considered a semi-living interior space.

2 Methodology

The study follows a descriptive-analytical approach through monitoring individual experimental attempts to create a new model of biological materials and design treatments, trying to classify them, and recognize their impact on the creation of new functions as well as change in the concept of form and function. On the other hand, the research attempts to suggest new criteria for an updated type of semi-living interior space.

The study also shows the need to know of similar design trends that paved the way for research in creating inhabitable interior architecture. Also, to recognize the role of biotechnology in achieving sustainability by monitoring and extrapolating previous experimental ideas to infuse the vitality of interior architecture design through the following points:

- The research conducts an integrative review of the interaction of biotechnology in the field of design with description and analysis using the following points.
- Analyzing past experiences to point out some of the environmental problems for which biological materials and treatments have provided solutions.
- Analyzing and classifying the main methods which used biotechnology to create these biological materials and treatments.
- Analyzing and classifying new functions that have vital characteristics added to interior design and that improve the quality of users' life.
- Finally, the paper provides an added proposal for completing what was not taken care of in the individual bio-experiments. It advocates the intellectual, ethical, and functional standards of design, especially within the new role of the interior designer. Finally, it proposes an experimental design suggestion for some furnishing units and a complete semi-living interior space. This method demonstrates the integrity of their work and does not harm the surrounding environment.

3 A Review of the Interference of Biotechnology in Interior Architecture Design Concepts

The biological model of interior architecture design proved that it is not about relying on metaphors of form or biomimicry but about understanding and considering the interior design as a form of life and an integral part of the living environment; they have a longstanding relationship.

Nature is a complex system. Interior architecture is a complex system, as nature, where self-organization is the biological evolution of living organisms. The changes in structures like their cells and tissue arrangements through a series of complex processes can be analyzed by new engineering systems to build envelopes, components, and find biotreatments (Pazzaglini, 2015; Abo alkhier, 2008).

Nature has been a role model for designers and architects for a long time. The significance of this phenomenon lies in the emergence of the Biomimicry movement, which considers nature as the ultimate evolution to learn from, shedding light on the concept behind biological simulation, response (Connell, 2009) and finding out the ways in which biotechnology interferes with space skin and components as a kind of relating biology with technology. (Braham, 2003) (Fig. 1).

An existing literature review conveys that it was a dream for architects and designers since the 1950s, as an experimental design led by "William Katavolos" in 1961 imagines the origin of built architecture and interior space as seeds (Fig. 2).

The Katovolos's concept developed, and many designers followed it such as the 1992 attempts of "Nat Chard" to express space through anatomical drawings of human organs (Spiller, 2006). And over the past years and decades, emerged some grown spaces with their components, such as the "Molecular House", "The Multistorey Apartment" and portable spatial shelter like "The Spatium Gelatin" model series from 2004 to 2007 (Lally, May–June 2009) (Figs. 3, 4, 5).

4 Bio-architecture is the Way Towards Sustainability

Of the problems challenging the natural environment leading to losses in its ideal balance, climate change is on top of the list. This is where the role of biotechnology in sustainability comes in and offering new and developed sustainable materials and treatments as an agreement with the well-known view "from cradle to cradle". (Spiller and Armstrong, 2011). Biotechnology derives evolution from the surrounding nature and living creatures to produce and recycle all types of materials, making living sustainable. Biological processes can replace polluting chemical methods, allowing the increase of efficiency and improving the act of creating new materials with less pollution, all while optimizing the use of resources such as water, land, and energy. (Clara Rodríguez Fernández—22/10/2019 7 min—Tops, 2020, October 14).

The simplest definition of sustainability in the dictionary is the quality of being able to exist constantly. The use of natural resources to create sustainable biological materials and raw materials supports the management and improvement of the functions and contents of the interior architecture's skin. It also guarantees its continuation and renovation, based on criteria in the following points:

- Establish a reciprocal relationship with the surrounding environment.
- Create an energy producer instead of an energy consumer.
- Be able to merge into the surrounding ecosystem.
- Not produce toxic chemicals.
- Be responsive to climate changes such as moisture and thermal changes.
- Be able to regenerate, endure, and be self-sufficient.

Biotechnology, in collaboration with nanotechnology, has enabled the production of alternative materials for currently available materials. These replacements may be cheaper and more energy-efficient while being abundant in nature. Some of them can be explained and analyzed below.

4.1 Biomaterials

4.1.1 The Crystals of the Cellulose

As a result of a series of experimental research at the University of Purdue University in Indiana, a crystalline material interspersed with plant cellulose, which causes the shape of the plants treated with nanotechnology to produce the crystals of the cellulose (Fig. 6), was revealed. It can be used in structure and interior space design to build a whole city.

This material has some benefits as it is highly available in plants, algae, some marine organisms, and bacteria. While it may appear fragile, it has very high rigidity. Also, it can be a potential alternative to carbon nanotubes treated with polymers and concrete. Finally, it mimics the behavior and structural performance of plants.

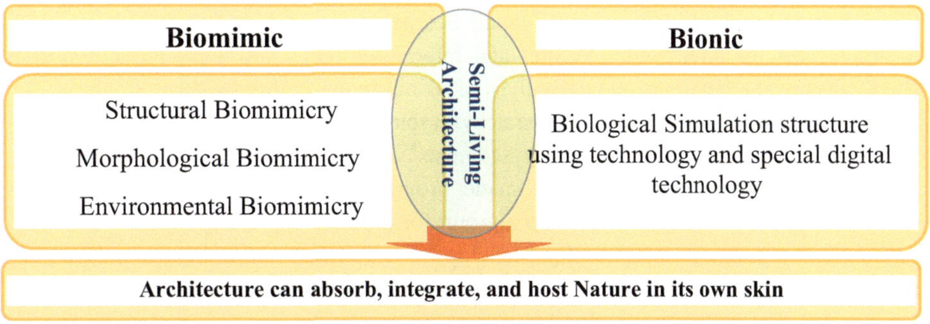

Fig. 1 Diagram shows that semi-living interior architecture is different from the traditional concept of the return to nature as a measure and guide, Author's own

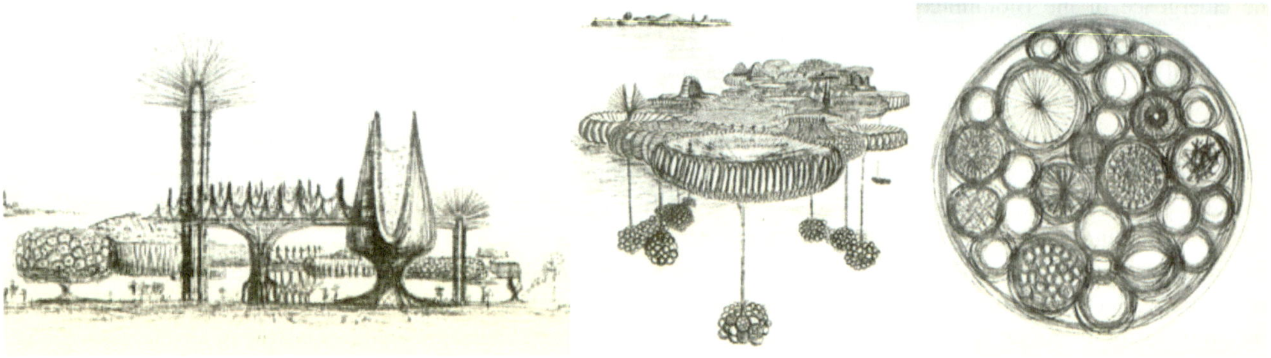

Fig. 2 William Katavolos experimental design of building architecture and interior space by seeds (Organics (n.d.))

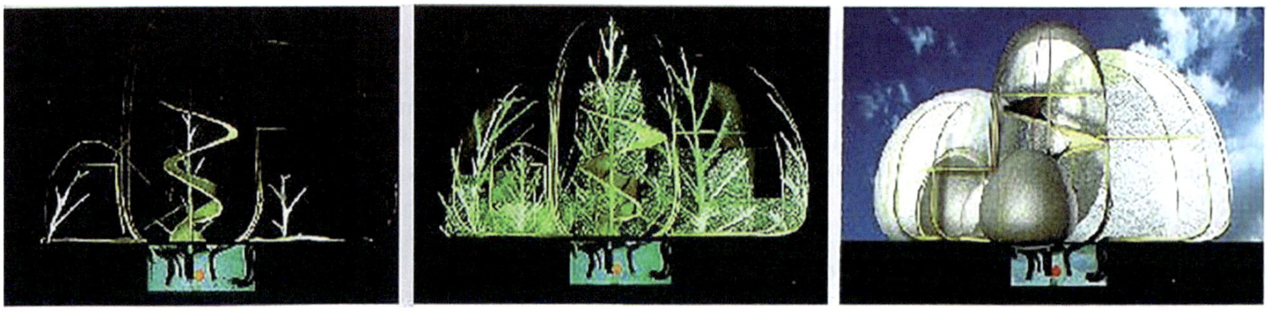

Fig. 3 It shows a sketch of the stages of "Molecular House" growth. It seems that it mimics the natural growth systems as in the forms of plants. Everything is integrated into one unit; the interior architecture and its cover show a continuous entity, Moclane (2002; Johansen, 2002)

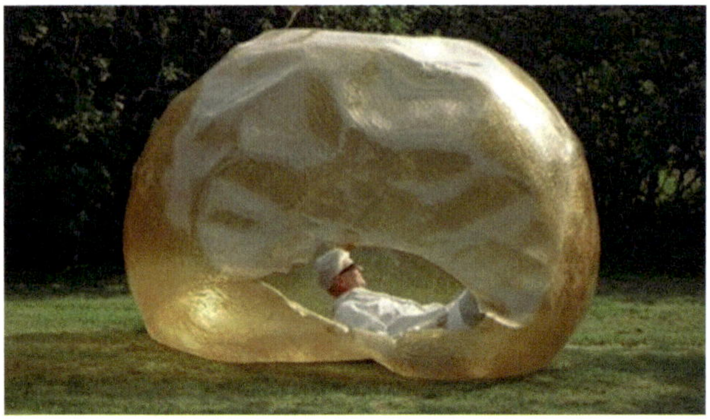

Fig. 4 The Spatium Gelatum, Biennale Venice, 2004 Sean (May–June 2009) (Rensselaer: Architecture. (n.d.))

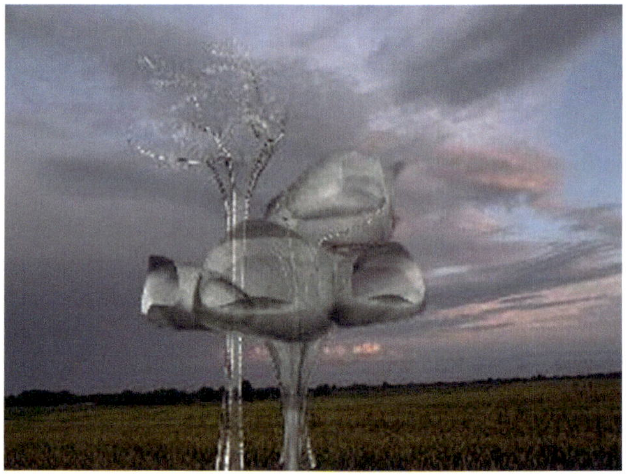

Fig. 5 The Multistory Apartment, 2001 (Johansen J., 2002)

4.1.2 The Bio-bricks

It was necessary to pay attention to small details of raw materials used for architectural design and interior design which concern the details of the daily lives of its occupants. This is what inspired designers to cooperate with scientists to create new a kind of brick, especially scientists who specialize in a particular area and have made progress. For example, BioMASON company set up a group for research over several years to organically develop a type of carbon-free brick, using certain bacteria that forms the sand into a crystalline environment to shape structural blocks of bricks (Fig. 7).

Bio-mason has focused on sustainable architecture and the use of biotechnology. Instead of taking nature as a guide or inspiration for ideas, the research team used it as the primary source of raw materials for manufacturing processes. The new bio-brick molds have the same strength and rigidity as traditional ones. Using materials available locally,

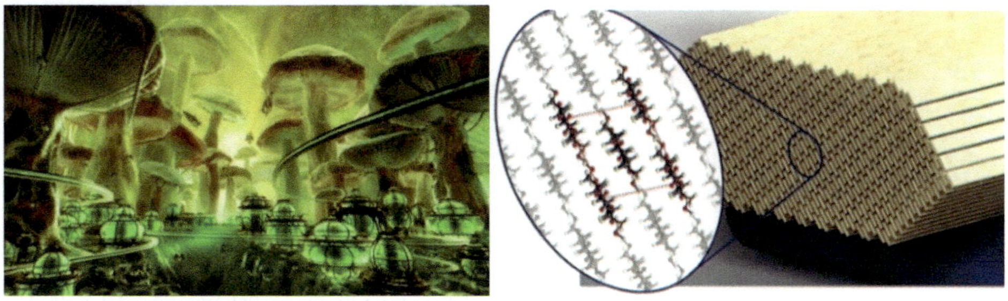

Fig. 6 The crystalline material interspersed with plant cellulose, where scientists used Quantum Mechanics to unlock the potential of super-cellulose, allowing the possibility of structural simulation and behaviour of plants. (Inhabitant—Green Design, Innovation, Architecture, Green Building. (n.d.))

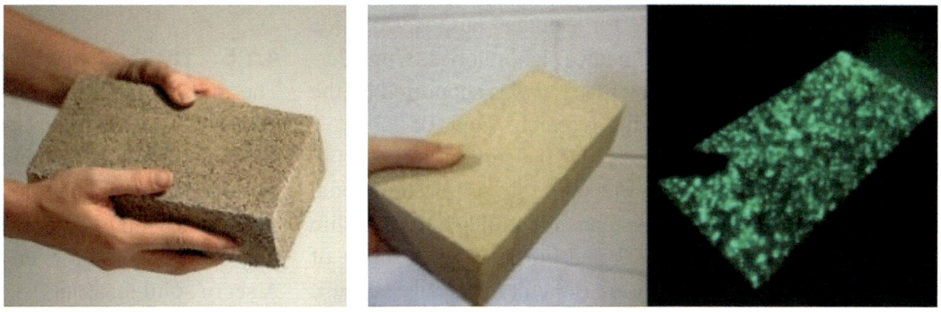

Fig. 7 Bio-bricks after the manufacturing showing the size and the density of bacteria developing in it. (Taylor-Foster, J. (2014, February 01))

it can be manufactured at the site in ambient temperatures without using fuel and energy to manufacture; saving cost and reducing carbon dioxide emissions (Winter, 2014).

4.1.3 The Bio-plastic

The bio-plastic morphologies model was presented by the designer Marilu Valente who used it in the design of some furniture units like (the Bio-morph Table) (Fig. 8). Bio-plastic is made of long chains of polymers, with a mixture of potato, starch, water, vinegar, and glycerin that make it more resistant. It has a multi-directional support system and shows good performance under pressure. (Bio-Plastic Morphologies—materiability. (n.d.))/(Studios, O. (n.d.)).

4.1.4 Chitosan

Chitosan is an organic chemical composite produced from the Chitin substance, found in shellfish and crustaceans. Nowadays, it is one of the most important scientifical achievements used by biomedical and biotechnology scientists. Being one of the sugars, it is one of the most abundant polymers available after cellulose, meaning it provides a compatibility with life and biodegradability. It has a large number of applications because of its physical and chemical properties, which allow modification by Hydroxyl Groups and Amino Acids. The biomedical applications became the architects and interior designers' dream for biomaterials (Adolfo & Fabio, 2017).

Interior space and its components can be composed of surfaces and formations that have the ability to self-repair, as well as living cells, such as chrysanthemums and oysters containing Chitosan. It can be a raw material formed and used because of its permeability, hardness, and anti-bacterial quality (Fig. 9; Winter, 2014).

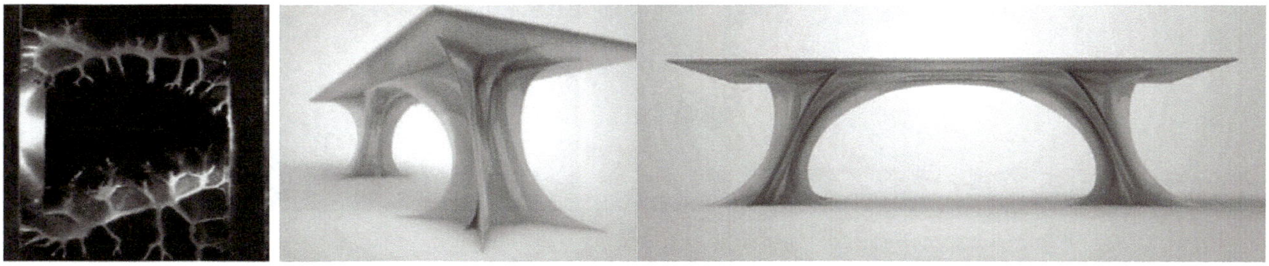

Fig. 8 3D modelling of the Bio-morph Table

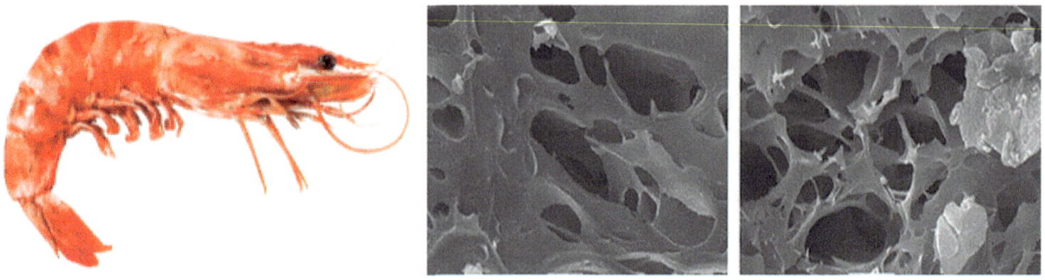

Fig. 9 A microchip image of Chitosan showing its soft organic structure, taken at de Polímeros laboratories, del Valle University, Colombia

4.1.5 The Vital Tissue

For more than a decade, artists and researchers, Oron Catts and Ionat Zurr, conducted research to find new and innovative ways to produce tissue for technical purposes outside of the ethical constraints of medical laboratories. Considering the living organism as raw material was supported by the progress of genetic engineering research (Fig. 10).

Then, one of its applications was the semi-living dolls in 2000 known as Patricia Piccinini. It is a functional hybridization of biological systems. It continued to carry life through stem cells until 2002 and emerged as a series of animal-like creatures which were difficult to classify because of their physical form of silicon and some biochemical mixtures and coating that had growing hair as if it were a living being. It was like a mole (Fig. 11; Cruz & Pike, 2008).

4.1.6 The Self-healing Glass

The self-healing glass research, conducted by NASA in 2009, shows that a surface can heal itself and return to its original state if it is damaged, broken, or even shot by a bullet. The polymer, as a built-in material, improves its properties using the exploitation of heat. It can rearrange the parts of the material even if the breakage remains apparent. A set of solid materials, including glass, plastic, and surfaces of ships and their components are known to ensure strength

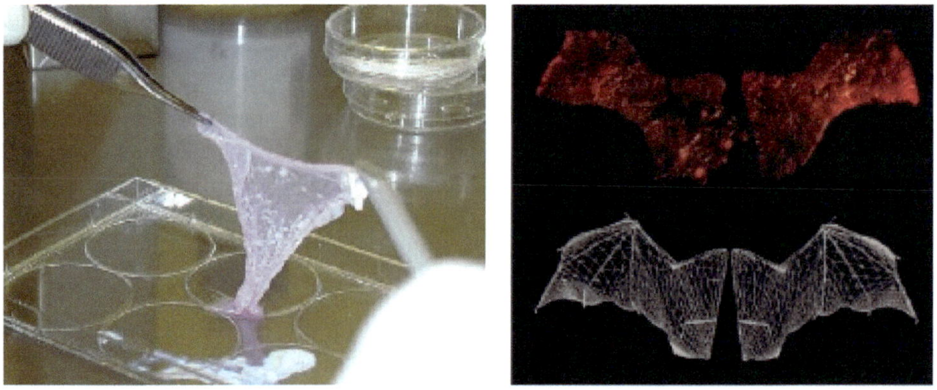

Fig. 10 A Vital tissue was one of the experiments by Oron Catts and Ionat Zurr

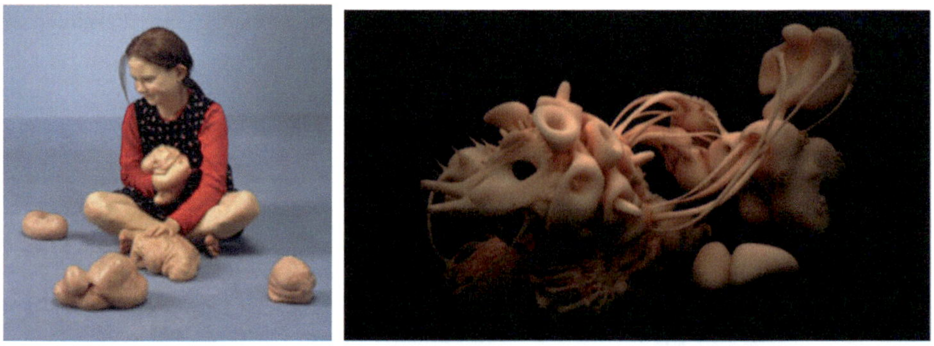

Fig. 11 Patricia Piccinini dolls 2000

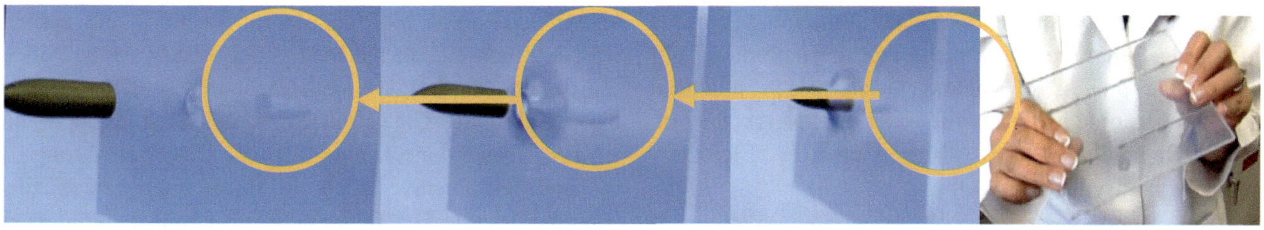

Fig. 12 A 3D shot shows the desired result of the healing self-material and how it works in the event of a breakthrough like a bullet (NASA researches)

and flexibility against pressure and potential hazards. Adding the self-healing quality predicts a safe future through the use of bioactive self-healing materials. (Fig. 12; Real World: Self-Healing Materials. (n.d.)).

4.2 Bio-treatments

4.2.1 Living Algae

This is a shift, from winning an architectural competition to developing the facade of a building in downtown Los Angeles in 2011 to a wider application in Germany, by HOK/Vanderweil in 2013 (Fig. 13). This construction of a complete building used double panels of glass containing algae producing energy. The façade is the world's first adaptive interface, transforming this primitive entity into a living machine. The adaptive quality of the facade takes advantage of the fact that algae grow faster in the sunshine, and use the unwanted heat in accelerating the rate of reproduction while preventing direct rays from entering the building, also stimulating photosynthesis. This also createaed a clean source of renewable energy.

Another another example of using living algaes to create a furnishing unit was the moss table (Fig. 14). It is an idea of

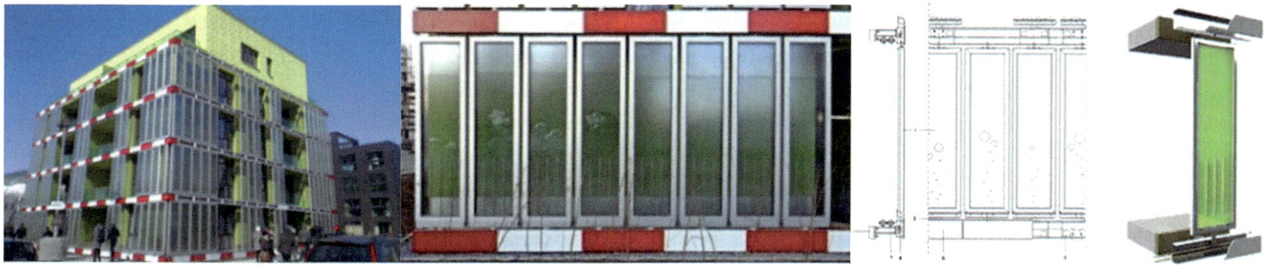

Fig. 13 The double glass panels, which contain living algae in the building facade, are an example of the use of the Passive House standard (Fytrou-Moschopoulou, A., 2015, June 01)/(Algae Powered House Biofacade Splitterwerk ARUP Colt International SCC Green Power Building (n.d.))

Fig. 14 A detailed sketch of algal photosynthesis work and how it produces low energy on the table top surface

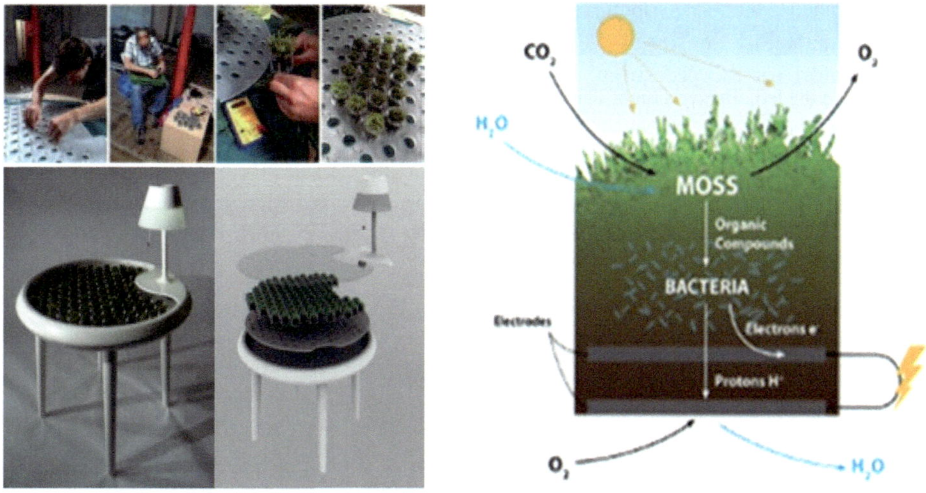

how one can apply Bio-Photo-Voltaic in the future, using a type of algae that produces energy that helps produce light through the process of photosynthesis, which produced energy to be stored in a battery for later use. (Moss table, (n. d.))/(Biophotovoltaic Moss Table Generates Electricity Through Photosynthesis (n.d.)).

4.2.2 Bio-lighting

This part shows two different examples of bio-lighting, one of them using two different types of creatures. But, both provide the same result of producing bio-illumination with low energy and cost. The first one was by the Glowees' team, and the other one was by the dutch electronics company, Philips.

Glowees' Team has been researching and experimenting with the creation of lighting units based on luminous marine organisms, as well as using the properties of microorganisms (Fig. 15) to produce an alternative to electricity. This alternative would reduce global warming which majorly comes from the excessive use of electrical energy. They discovered that their products can reduce 19% of the world's consumption of electricity, which is mainly used to run industrial lighting, which is responsible for 5% of gas emissions causing thermal containment. (*Glowee, enlightened by the sea.* (n.d.))

The Dutch electronics company known as Philips has attempted to adopt a natural approach to real-lighting. They extracted their new product from fireflies and deep-sea creatures to create a green light, not by electricity or sunlight, but by glowing bacteria that produce vital light, to be known as Philips bio-light. Other functions and activity wastes at home drive the bio-lighting system to feed off of the bacteria in the glass cells or crucibles to produce the biological light. The substance is delivered through thin silicon tubes connected to a tank. This was through a dynamic self-adaptive space in a researching project called (Philips Home). The proposal was submitted as an environmentally friendly microbial home (Fig. 16; *Philips Bio-light concept lights the home using bacteria*. New Atlas. (2015, May 2).

4.2.3 The Bio-concrete

A group of microbiologists at the University of Delft University in the Netherlands mixed concrete with a type of Calcite-precipitating bacteria, to fix the cracks quickly as soon as the water reached the surface of those cracks since with water the bacteria produces a kind of limestone, which gains strength and hardness that withstand stress and help it to stay as long as the conditions and the medium suitable for the work of bacteria exists (Fig. 17; Cuthbertson, A., 2015, May 15). The same idea was ment to be as some experiments and a trial section of bio-asphalt in Holand (Live Circular. (2015, October 13))/(Global Reach Internet Productions (n.d.)).

4.2.4 The Living Skin Wall Surface

The Living Skin Wall is a bacterial cellulose membrane that allows users to determine the level of privacy required, air quality within the interior, even the level of light and quality, by remote sensing the movements of the space's users and

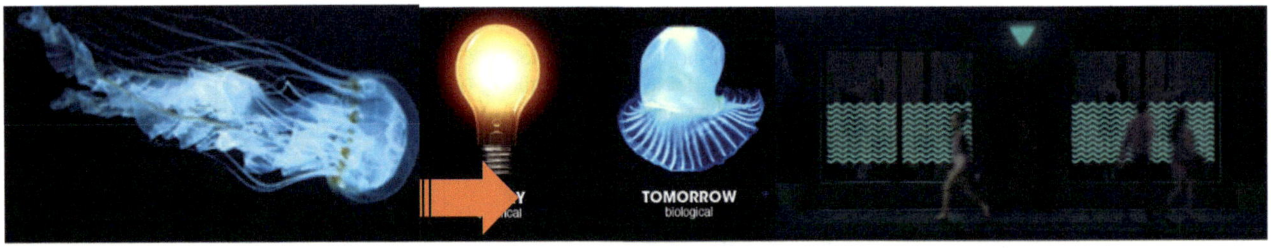

Fig. 15 On the left: The creation of lighting units based on luminous marine organisms (*Glowee, enlightened by the sea.* (n.d.)). On the right: A façade of a shop in France, which is covered by adhesive strips of vital lighting designed by the Glowee's Team

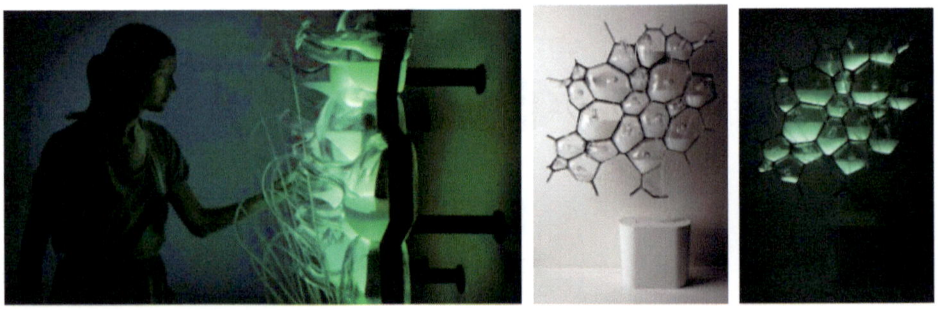

Fig. 16 The bio-light at "Microbial Home" by Philips company research

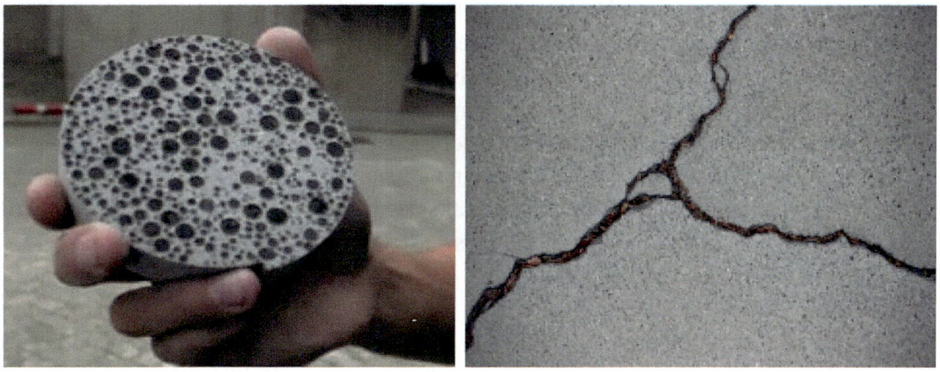

Fig. 17 The self-healing Bio-concrete

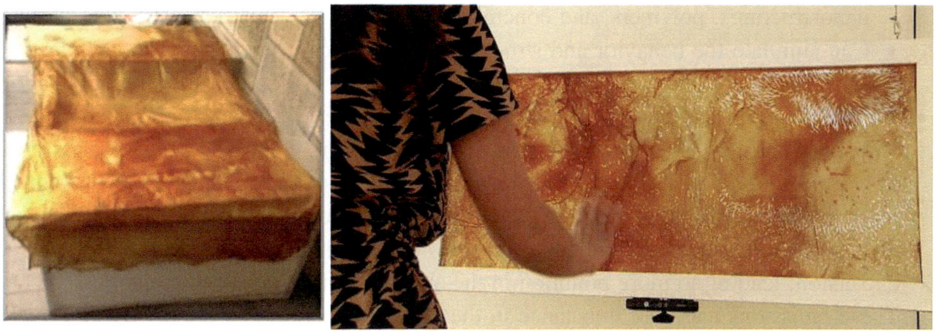

Fig. 18 A picture showing the bacterial cellulose wall surface with a visitor interacting with it, and the possibility of controlling the level of lighting, or the degree of privacy

their comfort levels (Fig. 18; sunny.chandel.50. (2013, November 16).

4.2.5 The Bio-detecting Surface

Drexel University researchers in Philadelphia conducted a series of research projects to design a model that uses synthetic biotechnology in interior architecture design. They created a new kind of smart and natural surfaces that are expected to be widely available. They suggested using surfaces with intercalation bacteria in the kitchens, which is known as Bio-Detecting Kitchen Surface. It detects the presence of pesticides, pathogens, salmonella or allergens. It has a feature that points out the presence of harmful components by changing the color around them on the countertop. For example, if someone is allergic to peanuts, then a specialized bacteria turn yellow and gather around anything that contains peanuts on the surface. The designer explained that this application can also have a great impact on health care, hospitals, surgical tools, and medical equipment, as they can detect surfaces' safety (Fig. 19; Alyn Griffiths|13 November 2013. (2015, September 14).

5 The Environmental Bio-sustainable Solutions

Through analyzing existing literature reviews on the use of bio and organic materials, we focus on designing solutions for the problems that did not receive sufficient attention. These problems such as resource, energy and fuel depletion increased emissions of carbon dioxide Co2 and heightened the global warming phenomenon. They provide solutions that would take the beauty of form and the depth of the idea and performance and use them to organize their environment while ensuring less damage and more renewal. These solutions are as follows (Author's own):

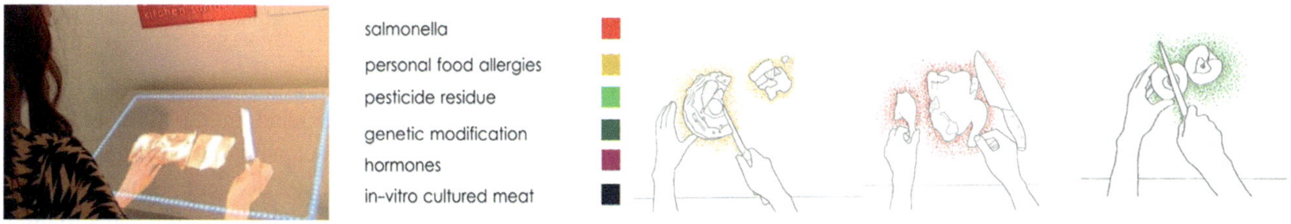

Fig. 19 Bio-Detecting Kitchen Surface detect the presence of harmful components by changing the colour of the countertop

- *Resource depletion*: Biotechnology and its cooperation with nanotechnology have enabled the creation of alternative raw biomaterials for the currently available materials. They may be cheaper and more energy-efficient as they are abundant and impermeable, considering nature as its raw source. They were found after paying attention to the smallest details of the existing biological systems. An example of this is the crystals of cellulose as they are distinguished by the following:
  - Widely found in plants, algae, and some marine organisms and bacteria.
  - May look brittle, but has a very high hardness.
  - Could be an alternative element to carbon tubes with nanostructures, polymers, and concrete
  - Can simulate the behavior and structural performance of plants
- *Carbon dioxide emissions*: They are one of the largest and most harmful climate changes in the environment. Since the middle of the eighteenth century, the rate of these emissions has been increasing rapidly, as the fields of raw materials production contributed to the urban expansion; leading to more pollution in the environment. These emissions cause an imbalance in the carbon dioxide quantities. The biomaterials like bio-bricks (Fig. 7) and bio-concrete (Fig. 17) found the solution to this problem which could reduce the massive CO_2 emissions.
- *Energy and fuel depletion:* Designers are looking for a way to take advantage of the most balanced energy sources, the availability of biomass fossilized in the ground, and the natural forces such as solar energy, wind, and others, which are the most widespread in the world, and then comes recycling. However, the research presents several design proposals and individual experimental projects that support this trend of using the available resources and reusing biological resources and waste and recycling them as a permanent natural resource. For example using algaes like in the proposal of the façade with the the double glass panels (Fig. 13) and the moss table (Fig. 14).
- *Global warming phenomenon:* It is caused by the industrial revolution which was one of the largest contributors to climate change, as a result of thermal emissions. It is a sign of disturbance of the ecosystem, which is followed by enormous environmental damage. Glowees' Team's achievements in bio-lighting support the reduction of the world's consumption of electricity and can lower the production temperature caused by it.

This analytical approach found that sustainable interior architecture offers a new vibrant view that allows the presence of new materials and treatments for space components, its parameters, and its protective cover. These treatments and materials organize their environment and ensure that they do not harm the natural environment. This analytical classification of research experiments on raw materials and living processing proves the why of using biomaterials.

6 From Digital Simulation to Bio-nature

After the descriptive-analytical study, it is necessary to discover how to convert to bio-design in the form of materials and design solutions. This involves the evolution of intellectual methodologies in biomedical and medical sciences and the development of architectural skin in both form and function. The new philosophy of design became based on vital tissues to include experiments that end with semi-living spaces through applications of a membrane and biotic systems as raw materials or a new way for design. That shows the necessity of classifying the most important methods of the integration of biotechnology with interior design to create biomaterial through a biological process (Table 1).

7 The Impact of Biotechnological Application on Interior Design Concept

According to the previous analysis and classification, it is necessary to discover the impact of utilizing biotechnology achievements and approaches on both function and form. Accordingly, it would affect the new role of the interior designer that will be clarified in the following part (Author's own).

7.1 The Impact of Biotechnology on the Functional Aspects

The most obvious impact on interior architecture design in the future, as a living nature and where the material has a new sense, is the emergence of new types of hybrid spaces combining several fields and expressing unprecedented forms of living and functions. Raw materials or design treatments for interior architecture were a result of biotechnology applications such as Growth, Bio-Illumination, Alternative energy production, Control the general atmosphere of the interior space, Cleaning and detoxification, Self-healing and Bio-Receiving Material (Author's own).

7.1.1 Growth

The use of technology to expand in the field of interior design is to adapt and interact with the surrounding environment and respond to changes to declare its vital entity and perform a function that makes it adaptive and sustaining

Table 1 The most important methods of interfacing biotechnology with the field of design of interior architecture for the creation of biomaterials and processors (Author's own)

Microbiology	
The self-healing bio-concrete The bio-bricks The living Skin Wall Surface The bio-lighting The living Skin Wall Surface The bio-Detecting Surface	Micro-Biology has been the main contributor to creating renewable and self-healing materials. Interactions and microorganisms as bacteria played a considerable role in the creation of new materials and biological treatments. Moreover, saving energy and costs that the traditional processors and raw materials used to produce them
Genetic Codes	
The bio-lighting The vital tissue	The manipulation of genetic codes is like the manipulation of the units of lego game, where the addition or exchange of one detail is dealt with very accurately, each of which expresses a new state of an organism, ie. reverse aging, resistance to viruses, probiotic prophylaxis, and infection. For example, dealing with the genetic codes of the ability to light and glow that characterize some marine organisms, and after extraction are added in other materials or even plants to give the glow feature and then used as one of the design processors (Fig. 14)
Cloning	
The vital tissue	Several studies have emerged to find new and innovative ways to produce growing vital tissues for technical purposes, removed from the ethical restrictions of medical laboratories. The question remains about the possibility of adapting the practical capabilities and the extent of exploitation of the living organism considering it as a raw material supported by genetic engineering
Nano-medical Purposes	
Chitosan The self-healing glass	The interior architecture and its determinants and components have become composed of surfaces and formations that have the ability to self-repair as in living cells. That is a part of a chemical decomposition process using nanotechnology to create a raw biomaterial. That newly treated biomaterial can be formed and used in the field of interior architecture design as it was previously prepared for applications in the medical field, like in the cloning of biological tissues and the idea of interior space control as "Glen Tomlin" suggested in his experimantal proposal using nano-surgical elements which are one of its medical purposes (Spiller, 2008)
Viscous Membrane	
The bio-plastic The living skin wall surface	Viscous Membranes are plastic materials that offer the possibility of using biomaterials as an alternative to traditional materials of constructions and space frames. They can be adapted to any form, transformation, mutation, expansion, and pressure because they are sticky. For example, the bio-plastic mentioned before (Fig. 8) and the living skin wall surface (Fig. 18)

space. Each project involve designers, scientists, environmentalists, and maybe urbanists to create a compatible state that allows materials to be responsive to environmental changes in the real-time of growth like the example of the bio-bricks through its production process. (Fig. 7) (Author's own), or like the species of light-averse protocells would form the reef, acting as a variety of bio-concrete to protect the Italian city from sinking into the delta's soft soil. (Words Jennifer Ferng, & Ferng, J.(n.d.))/(Pelfen, Winter 2014)/ (Armstrong, 2014).

7.1.2 Bio-illumination

The bio-illumination is the best solution for more eco-friendly, energy-saving, and everlasting lighting systems with compact fluorescent systems and LED systems. For non-glare illumination, bio-illumination can be utilized as warning signs, exit signs, entertainment venues, low-light corridors, road and highways signs at night. The two examples of bio-lighting are presented by Glowees'team and Philips Company (Figs. 14 and 15), and both can add value as interior designs.

7.1.3 Alternative Energy Production for Electricity

Some treatments have been developed to design the specifics of interior architecture or furniture units that can produce energy used to power devices as an alternative to electrical energy. They are natural, clean and safe. (Author's own).

For example, a table design with algae covering its surface for low energy production.

7.1.4 Control the General Atmosphere of the Interior Space

Biological treatments can help interior space cover-skin to change themselves as some organisms or to control their activity and change it as a kind of interaction with space users. Interior architecture can achieve self-control and functions as part of nature so that it can achieve self-regulating systems for ventilation, heating, hygiene, balance, and self-care. For example, the control of the general atmosphere of the interior space by using (living skin wall surface) or a bacterial cellulose membrane (Fig. 18).

7.1.5 Cleaning and Detoxification

The general idea of creating a biomaterial that can be self-cleaning and self-detoxifying is based on using bacteria to perform various functions and tasks using a kind of vital response, including cleaning, changing the properties of surfaces according to the needs of the interior space users, or even alerting the presence and capture of a toxic substance. It all depends on explaining how the bacteria can be penetrating, useful, and a natural source. It would richly present dynamic design treatments in the future and be available soon.

7.1.6 Self-healing

Biotechnology has given the privilege of the most biological treatments and raw materials for self-healing. It is the property of nature, as in self-healing bio-concrete, as well as the self-healing glass. (NASA. *Real World: Self-Healing Materials*(n.d.)). Both show that the material surface can return to its original state in case of damage through self-rearrangement to be sustained for a long time. (Author's own).

All this influenced the concept of formation and the merge of functions with an aesthetic look. It has emerged in individual experiments. The biomaterials emerged to form the future of the interior architecture, and make it alive, regenerative, growing, and balanced, as a kind of improvement within the ultimate system of nature, host and integrate with it.

8 The Impact of Biotechnology on the Formation Aspects (Author's own)

- Formulation of the function within the design of interior architecture.
- Formation using digital and manufacturing techniques.
- Design processors for modulation without considering functionality.

9 The Standard Model for Designing Semi-living Interior Architecture (Result Application 1) (Author's Own) (Fig. 24)

9.1 Criteria for using biomaterials.

The first part of the application study is a proposal for the determination of the new design thinking standards including ethical, moral, formal and functional criteria. They ensure the improvement of the quality of living (Table 2). The application of various systems of interior space needs coordination between form and function.

9.2 The Beauty of Function

Functional beauty refers to a reliable, healthy design for space users and the surrounding environment. Sometimes it may be renewing, self-healing or self-reforming. The use of biotechnology is fundamentally intended to achieve a new function or to improve an existing on. So, functional beauty remains in its performance ability and its positive effect on the usage of interior space. The space's harmony and compatibility with other functions perform as a part of the ultimate system of nature.

9.3 The Beauty of Form

There are several questions about the concept of form and structural beauty using these new tools of biological applications and protocell technique (Armstrong, 2014). Here, comes the question about how to determine if the biomaterials can look good or not. For example, there is a debate about whether the aesthetic form of the biomaterial may affect its function and the need to pay attention to it or that the function of this biomaterial is enough to overlook the beauty of the form to become "Design follows Technology".

Judging the aesthetical aspect of materials depends on the mental background about their origins and their visual and physical properties. Still, when it comes to biomaterials, they should be applied in an acceptable form and a compatible look with the background of existing materials and the needed knowledge of their original sources. For example, the comparison between "The alpaster"—one of the natural marble types—and the "living skin wall surface" in shape and function (Fig. 20) in terms of shape, color, configuration, layout, and texture, it is observed that they look very close and can be matched. Both can be described as beautiful and attractive, without differentiation of the raw material or the original source.

Table 2 Shows the main criteria of using biomaterials (Author's own)

Intellectual Criteria	Ethical Criteria
•The design should promote the human spirit as it works for its functional and visual benefit •Nature is the source of biomaterial and bio-treatments. That is by dependence on vital natural resources, considering the natural environment and its components as the raw material •The design should support the environmental balance, achieving what is known as the sustainable living interior architecture model •The designer should take into consideration the moral, literary, religious, and cultural aspects of society, according to where the interior space exists •Verifying the physical and psychological integrity of space users •Continuous and regenerative design: The design should be reproductive, creative, and continuous with time •The flexibility of use and durability	There are some religions that forbid the exploitation of vital tissues and deferent creatures' abuse. So, the designer should observe moral and religious ethics, and ensure that there is no harm or exploitation to human, animal, or plant, especially when it relates to medical applications or vital experiments **Functional and Technical Criteria** After classifying and analyzing the individual experiments of biotechnology usages in interior space design for futuristic sustainability achievements, the experimental interior bio-design announce that (Design follows Technology). Biotechnology is expressing both function and shape depending on the nature of interior space. It remains necessary to determine the criteria of the functional beauty and the formal (shape) beauty of Interior design

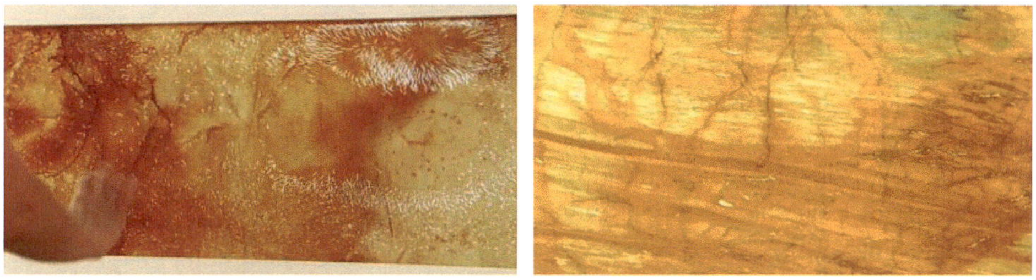

Fig. 20 (Bacterial Cellulose Wall Surface)—on the left—and (Alabaster Marble) —on the right—and the possibility of accepting both forms despite different origin and composition.

The natural marble is more acceptable when used as a wall cladding than (the surface of the living wall) because the first one is cut off from the original natural stone and used for use without reservation, while the second one (surface of the living wall) is a thin film of bacterial cellulose membrane, also called bacterial cellulose wall surface.

The presence of the bacterial element led to the material being rejected in the intellectual background of most. The bacteria and the aesthetic aspect have not linked in the past in any form. While the (living wall surface) can be formed and adapted more easily, less cost, and easy to clean, it can also change its look, density distribution, and surface pattern according to the bacteria's activity.

Another example is the comparison between biological leather and synthetic leather (Figs. 21 and 22). Leather and fabrics play a large role in the field of design in general, and the design of interior architecture, where the different colours and composition in different arrangements on them are either geometrical or printed patterns. Since the current fabrics, including the industrial ones, consume more energy for manufacture, the replacement of the artificial fabrics with the biological ones would provide the energy used for its formulation of these components, naturally, by using biotechnology.

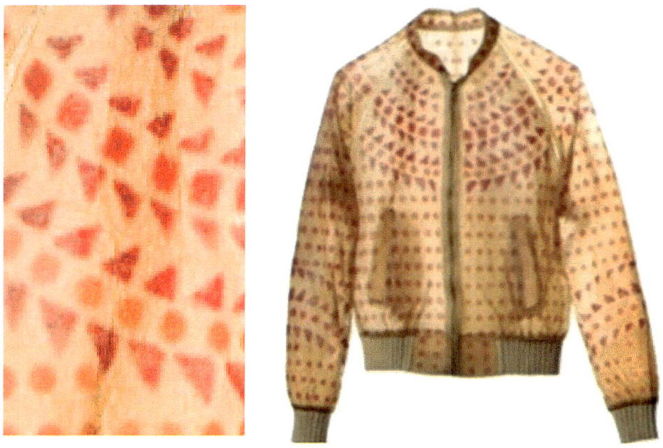

Fig. 21 Leather jacket, with a biological origin from yeast reaction, Bio-Bomber jacket, Designed by Bio-Couture. (Marcus Fairs | 12 February 2014. (2015, September 14)

Bio-couture, a London-based consulting firm that works in the field of design, is a pioneer in the use of biomaterials

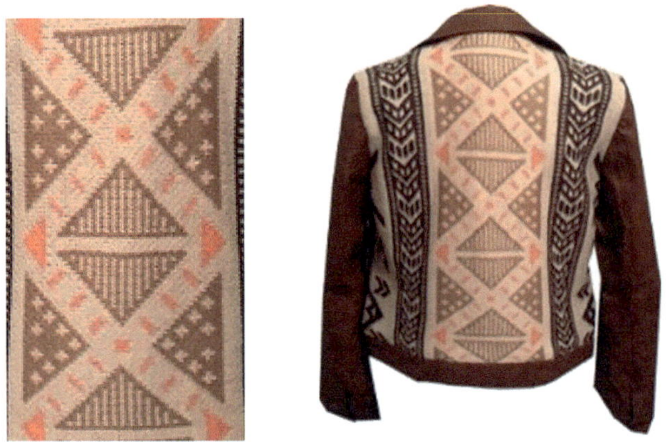

Fig. 22 Designed leather jackets made of synthetic and non-natural raw materials

for fashion and a supporter of how biotechnology can transform design thought into the fashion world.

The brand designed a set of jackets and shoes made of biomaterials produced by bacteria in a container that contains some liquid producing bacterial cellulose. It is a substance that has similar characteristics as skin. They called it (plant leather) as it is produced by cultivating a piece of clothing using a mixture of symbiotic yeast and bacteria.

Exploiting (plant leather) as a biomaterial in interior design will have the priority of using the processed leather, when considering the following:

- Using the microbe instead of the plant will allow the possibility of adapting its shape, pattern, and composition.
- Providing functional features that are non-degradable and can even regenerate and recycle.

On the other hand, the role of the interior designer should be determined as comprehensive and cooperative (Fig. 23). Using raw materials and design treatments with the development of the methodology needs solving complex problems, collecting and analyzing information and data, and experimenting with the proposed solutions until proving the appropriate idea. Then, the process of implementation, the time table, and the selection of the most appropriate biomaterials can be concluded in the following points:

- Collecting data that a designer can blend or integrate from science, technological knowledge, and art to formulate a design scenario.
- Analyzing information and creating a matrix or a network between data and information that was collected, where the design alternatives and the resulting possibilities increased as creative solutions, then identifying and arranging the action steps, and predicting the results. Hence the uniqueness of design and the success of the idea of living interior architecture (Vianna et al., 2012).
- Taking into account the functional and aesthetical design principles through visual and environmental balance, the design should take into account the foundations of the spatial boundaries, the area of each function, the complementarity of form with function. Then, considering the new role of each element of the interior space and its relation with the rest. Finally, not forgetting to meet the needs of the users of the new Bio-interior space.
- Collaboration with scientists and preparing the design environment.
- Design identity becomes of origin and returns to nature in the form of a sustainable living interior architecture as a kind of participation in the global and cosmic product of design (Dziersk and FIDSA, Spring 2014)/(Lawson, 2004)/(Lawson, 2005). This is due to the different tools used, including processes, biomaterials, and treatments, and each belonging to mother nature and everything that is vital as a source or application. The design identity of the future interior architecture no longer follows a specific design direction recognized as before.

9.4 Bio-materials

The biomaterials and design biotreatments have emerged as a kind of attempt to find alternatives to the existing traditional ones by putting them in a context that makes it easier for the designer to imagine how to achieve the most appropriate and most suitable design. (Hensel, Menges and Weinstock, March/April 2006). This can be considered a kind of prediction and expectation of the new material type development. Biomaterials emerged while seeking to create

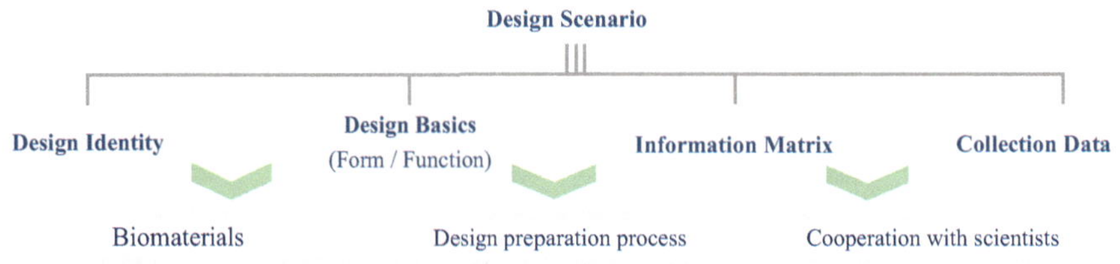

Fig. 23 Diagram shows what is included in the design scenario of the semi-living interior space and the designer new role (Author's own)

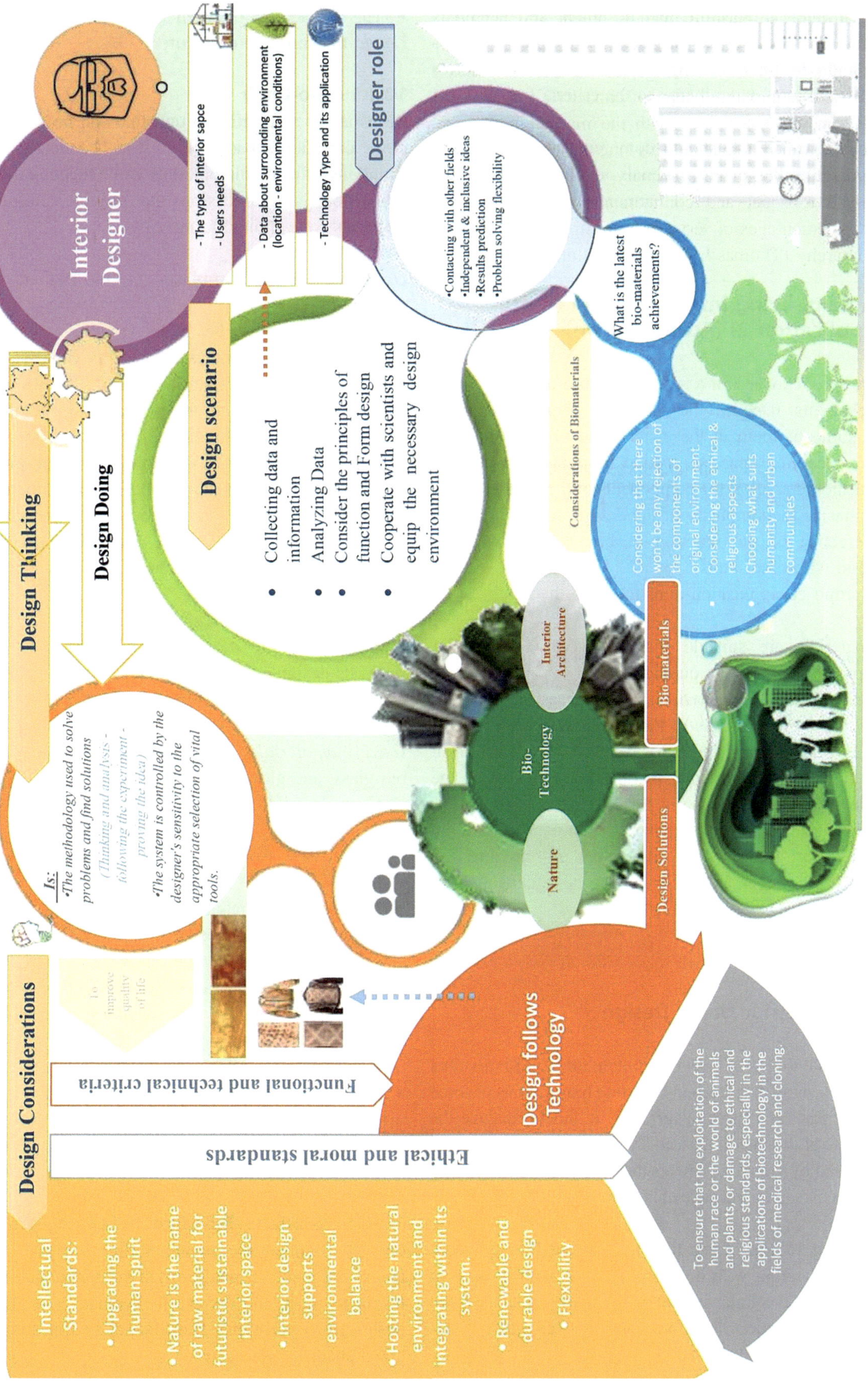

Fig. 24 Diagram shows the aspects of the standard model for semi-living interior architecture design. (Author's own).

a living interior architecture and designers and scientists collaborated to develop an organized list, then tried to classify and determine the impact on the design of interior space in the future, and adhered to the criteria for the most appropriate choice.

Selected biomaterials should communicate with tissues, cells, and molecules of the materials involved in the same design system to appear as a harmonious piece of music together.

10 A Semi-living interior Design (Result application 2) (Author's own)

It is a proposal to design a set of furniture units and individual elements. It shows the possibility of using biomaterials or some of them in the same piece of furniture in a complete reception space as a kind of public space. This proposal should analyze the components of the interior space and emphasize the simplicity of the design, form, lines, and the emergence of the function of biomaterial and their integration with other units.

The proposed application aims to prove that:

- Biomaterials are suitable for many types of interior design, either indoor or outdoor space.
- In case of using biomaterials, it was necessary to highlight that there is no need for complex forms, lines or shapes to keep them simple for whenever the function of raw material emerged and became integrated with the whole design.
- The proposal for using biomaterials will contribute to an actual saving of energy and add equivalent functions and additional services to the space without causing any damage to the environment or its users.

10.1 Bio-interior Units' Design

It is a proposal to design a set of furniture units and individual elements that show the possibility of using biomaterials or some of them in the same piece of furniture. It should analyze the components of the interior space. The design lines of the units were inspired by yoga movements that refer to soul integrating with nature and add a valuable character, as follows:

- Belonging to nature, hosting the surrounding environment, connecting with it, and integrating with its system help to achieve balance and consistency. It allows the ability of evolution, transformation, and growth of interior architectural elements.
- The Interactivity between space users, the space elements, and the surrounding nature.

10.1.1 Bio-Chair

Bio-Chair is a sustainable unit that can be used in different types of indoor spaces (Fig. 25). The design reflects the simplicity of using biomaterials while maintaining measures, proportions, and dimensions that provide comfort. The biomaterials were limited to only two.

Used biomaterials: Bio-plastic and bio-leather.

Form description: The main structure of the chair is bio-plastic, which is lightweight and flexible. This bioplastic is covered with bio-leather. (Fig. 21). It shows balance with the simple form, making the biomaterials the main hero and the catchy element while retaining the function.

It has some benefits as it is lightweight, non-biodegradable and can detect toxins and microbes. Besides, there is no additional cost required for maintenance. Finally, it is implemented at different arrangements to be allowed in a set that can be rearranged repeatedly and connected as a focal point.

10.1.2 Living-Counter

The Living Counter design comes in light and contemporary form. Its design can suit workspaces like information desks or front counters.

Used biomaterials: Bioplastic, the bio-detecting surface, bio-glass, and glowing plants.

Form description:

- The Living Counter structure is bio-plastic, which is flexible and easy to form. It is a great choice to replace the Corian material in furniture and countertops (Fig. 26).
- A thin layer known as the bio-detecting surface is covering the upper countertop surface to detect harmful bacteria; protecting users from infection and disease. It has a small bio-glass that separates the available part for customers. On the upper surface, there are some glowing plants, which are for lighting with a lively touch.
- The living counter has some added privileges as it is multifunctional, does not need electric power for lighting, and the biomaterials do not consume high energy or high cost during processing, formulation, or maintenance.

10.1.3 Garden's Bio-shaded Seating Unit

One of the dreams of sustainability was to increase green areas. Some designers resorted to horizontal and vertical levels to meet this dream. In keeping with this idea, the design of the Garden's Bio-Shaded seating unit included a green shading part.

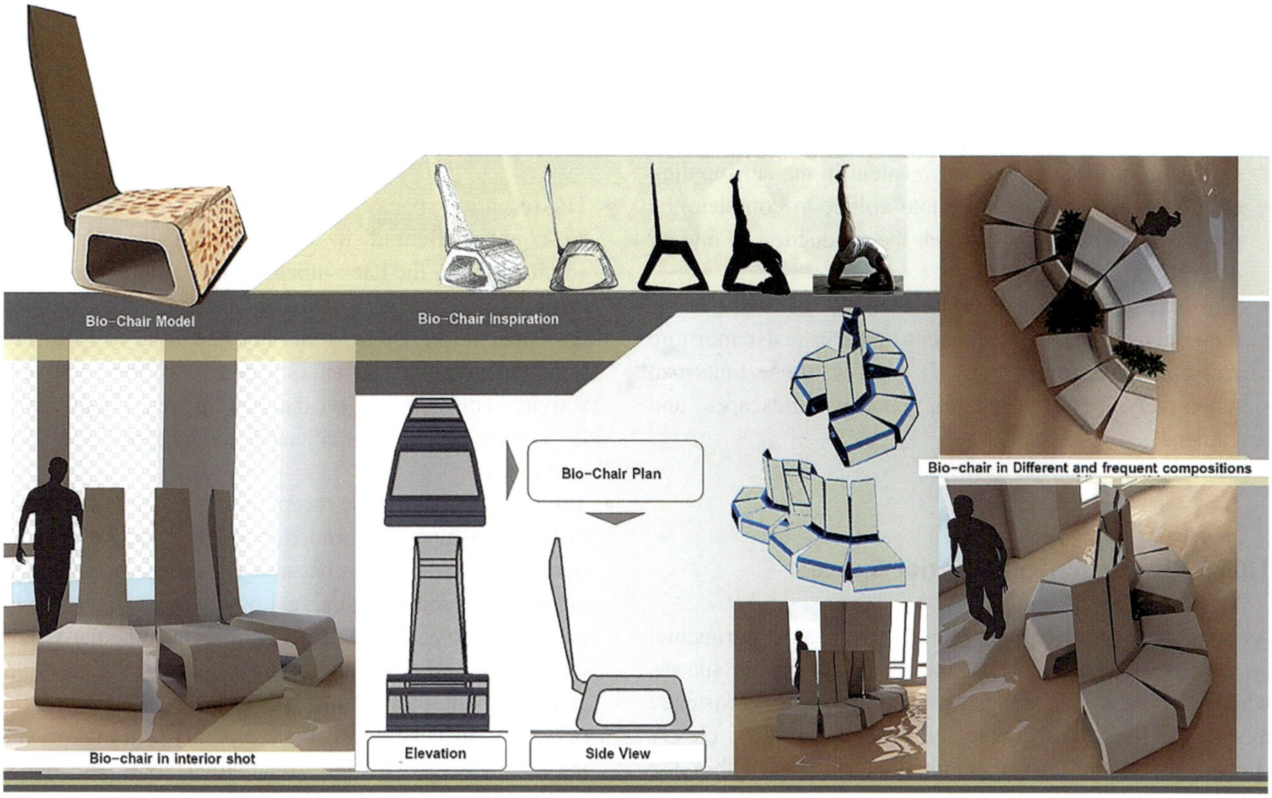

Fig. 25 Set of shots show the Bio-Chair as a separate unit or in a set using biomaterials with simple form (Author's own)

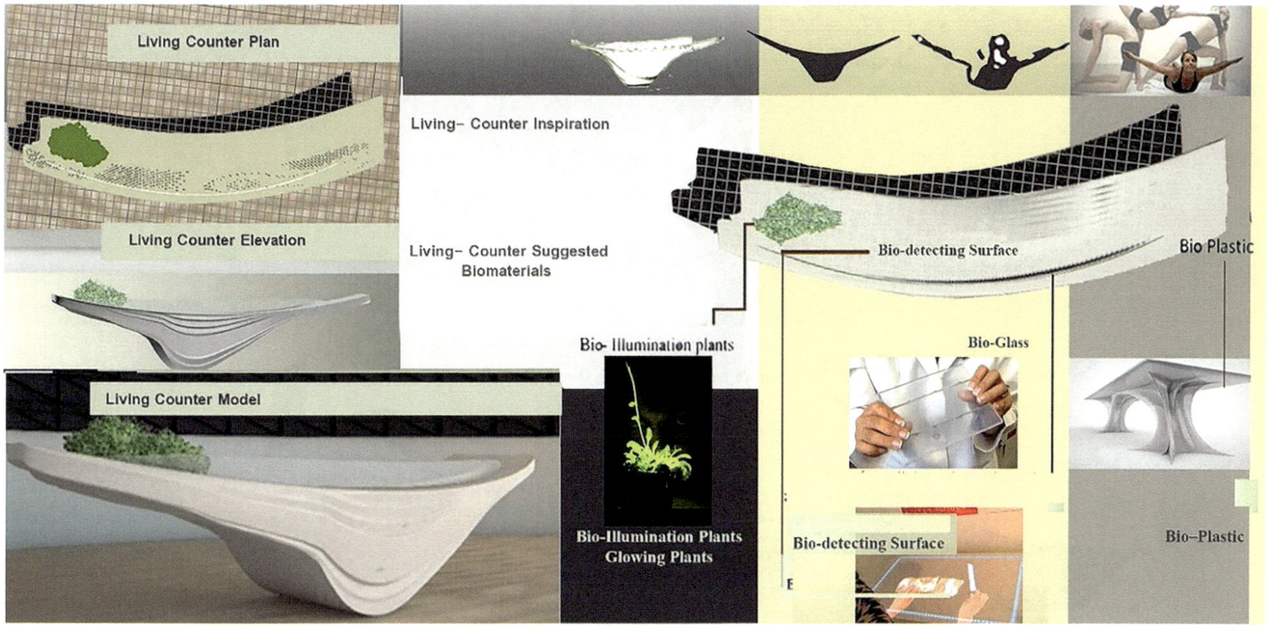

Fig. 26 Living Counter design details using biomaterials in a simple and contemporary form (Author's own)

Used biomaterials: self-healing bio-concrete, energy-producing algae.

Form description:

- Energy-producing algae covered the upper surface as an extension of the green areas prevalent in the surrounding environment, applying bio-photo-voltaic to complete the process of photosynthesis, then the production of energy needed for night lighting.
- The structural composition is made of self-healing bio-concrete. It is self-sufficient and can resist moisture and climate changes (Fig. 27). Finally, those units can repetitively be arranged for interior landscapes and gardens.

10.2 Bio-sustainable Interior Space

As an applied result of the research, here is an experimental proposal that presents a new model of sustainable spaces, where simplicity of design, modernity, and consistency meet. The biological treatment design harmony clearly shows the possibilities of synthetic formation within the interior space.

The proposal is for a reception space in a company that has many activities, offers many services, and hosts different types of users and visitors (Figure 29).

Space description:

- The reception space consists of a large area for the front bio-counter. Behind it, corridors are leading to the administration, the back offices, and the employees' work areas. Then, the central space of the reception has stairs to the upper floor. There is an allocated zone for presentations, advertising, and announcements of the company's activity. This zone is near the waiting area. Finally, there is a meeting room dedicated to meeting with VIP customers.
- The partition behind the reception bio-counter is designed to be multi-functional and raw and has integrated functions. It is self-lighting because it has basins containing green algae units producing low energy, which in turn contributes to adding indirect lighting in some parts. That lighting compliments the aesthetic form of the partitions as an element attraction as one of the basic vocabularies of space and increases the necessary lighting of the interior space. All the used biomaterials are either self-healing and repairing, or resistant to breakage, and

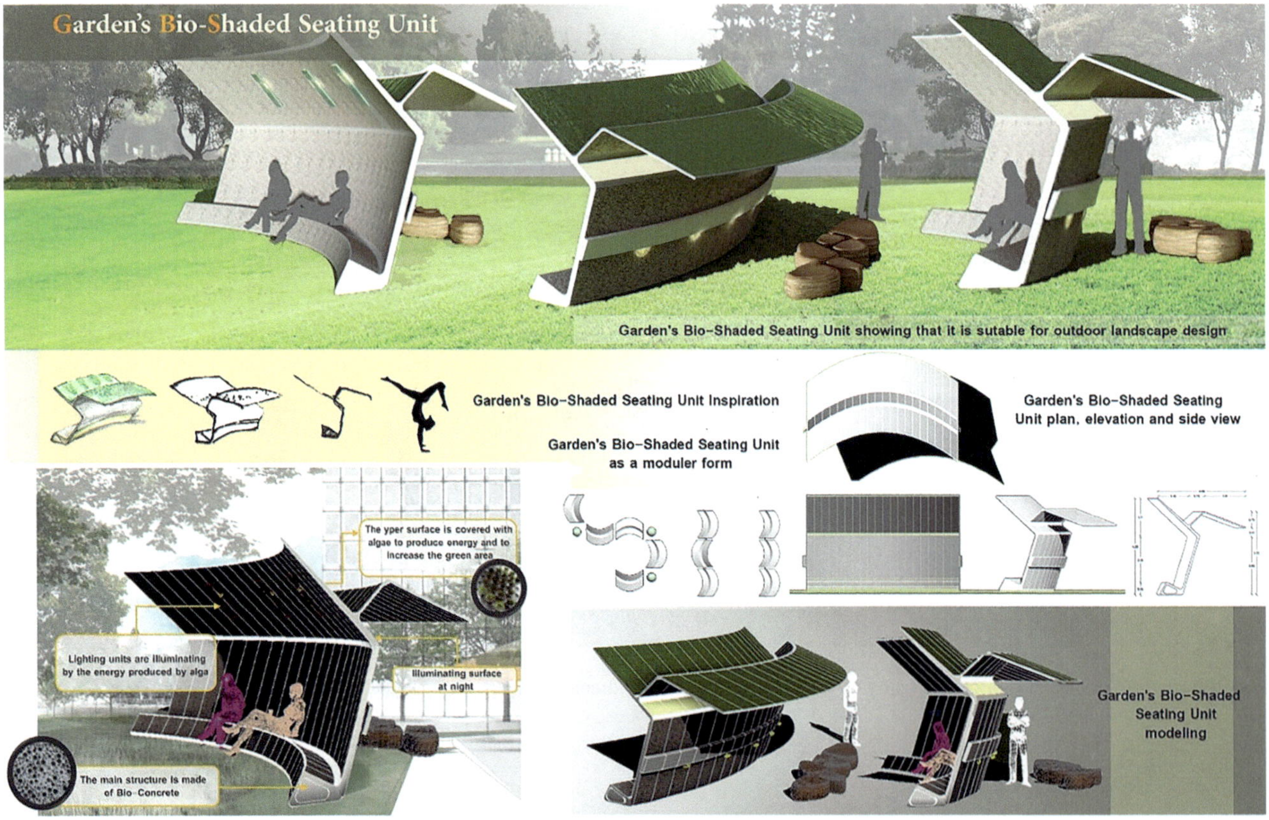

Fig. 27 The Garden's Bio-Shaded Seating Unit as self-sufficient form helps to increase the horizontal green area (Author's own)

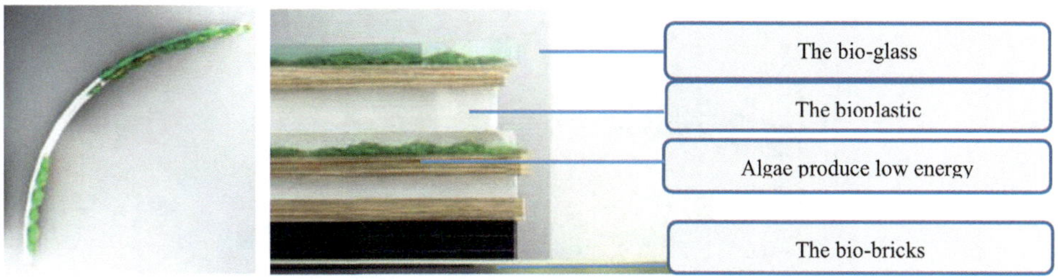

Fig. 28 The bio-materials used in the main partition behind the front counter. (Author's own)

did not consume much energy or cost during their manufacture (Fig. 28).

- The interior light mainly depends on natural lighting as a kind of integration with the external environment, and the preparation of the interior bio-lighting within the night.
- The lighting units are rounded ones formed using natural silk yarns using 3D printing technology. (Fig. 29).

- The interior landscape coordination in the stairs area using the biomaterials and plants is to achieve the integration of the functional and vital system in the space each of which plays a role in decreasing the proportion of contaminants and the provision of electrical energy used by the light. The facades are supported by bio-sticker which is a biomaterial extracted from illuminating sea creatures through photosynthesis and glowing. That biomaterial sticker is similar to one used in a retail facade design in France as one of Glowee's team designs (Fig. 15). Addiitonally, the floor is bio-detecting; meaning it can detect bacteria and control the general atmosphere of the space (Fig. 30).

- Self-illuminated advertisements, equipped with some of the walls using modified materials derived from glowing aquatic organisms, as well as signs on the paths, at doors and entrances, act as signs during daylight hours and provide a glow to interior space during the evening (Fig. 32).
- Some biomaterials cannot provide a specific service or function for a space; however, they remain harmless and vital. They can mostly self-repair or provide energy during manufacture, which saves maintenance costs and preserves the integrity of space elements and design elegance.

The current study is concerned with providing a futuristic solution for environmental sustainability, interior architecture, and their components as part of a new bio-sustainable space model. The analysis revealed that the use of bio-adaptive materials can serve as an innovative solution for environmental problems such as global warming, carbon dioxide emissions, and energy and resource depletion. Smart biomaterials were found to be used efficiently in the design artistically and biologically. The integration of biotechnology in interior design adds value as new functions and forms to create a new model of sustainable space. Therefore, we can conclude that synthetic biology is the future of adaptive interior spaces providing an ability to balance with the surrounding environment.

A highlight of using biomaterials in interior design is that they do not face rejection or cause any damage to the components of the original environment. Nevertheless, the

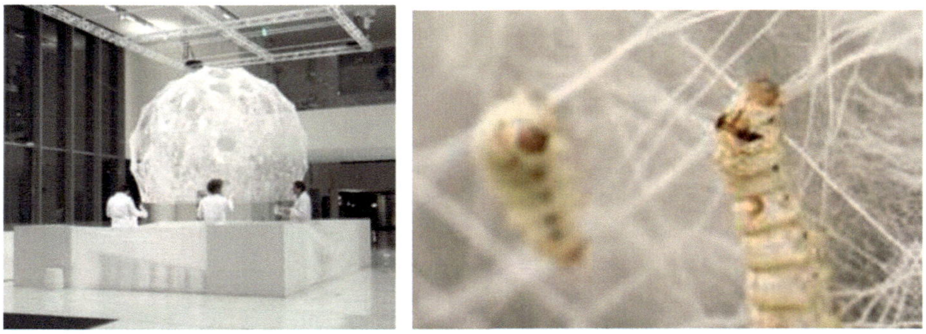

Fig. 29 Shots show that an MIT team, in 2017, designed a pavilion that is composed of silk fibres or a "3D Silk Wing". Its design manufacture is silk fibres from silkworm cocoons to create woven spaces and envelopes. Silk is considered a natural 3D printer, given how it built its cocoon, as no such technology or idea had been previously exploited in the interior architecture design process. (*MIT Researchers build silk pavilion using biological Silkworm 3D printer.* 3ders.org. (n.d.))

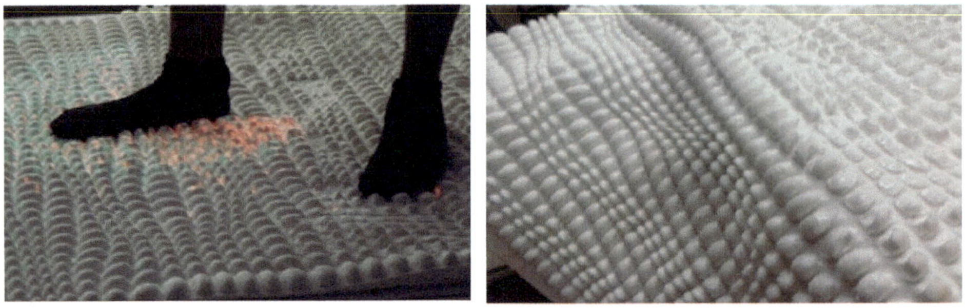

Fig. 30 The Bio detecting Floor prototype is a smart biomaterial designed by a student called Tashia Tucker in Philadelphia research, Drexel University. That Bio-detecting floor is embedded with synthetic bacteria that eat dirt and toxins to clean space users' feet (Alyn Griffiths| 13 November 2013 (2015, September 14))

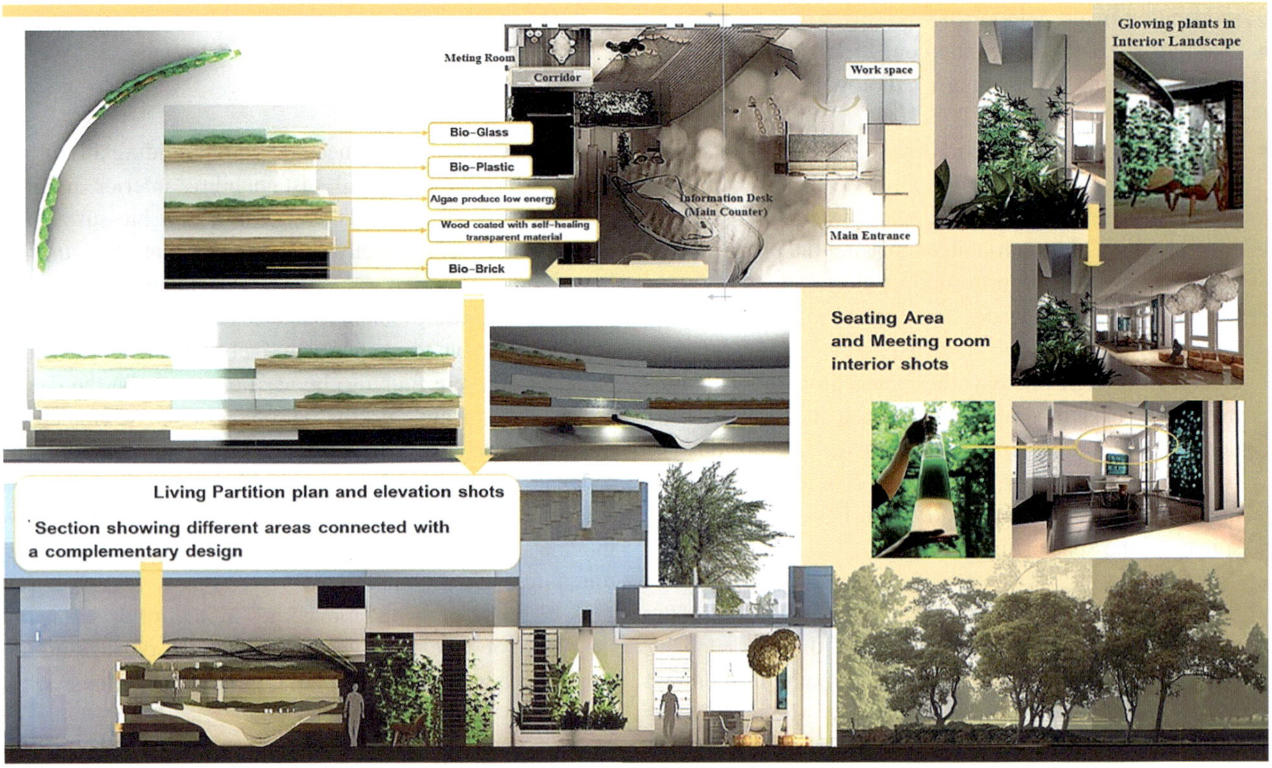

Fig. 31 The section and some details show how different areas are well connected with a complementary design using biomaterials (Author's own)

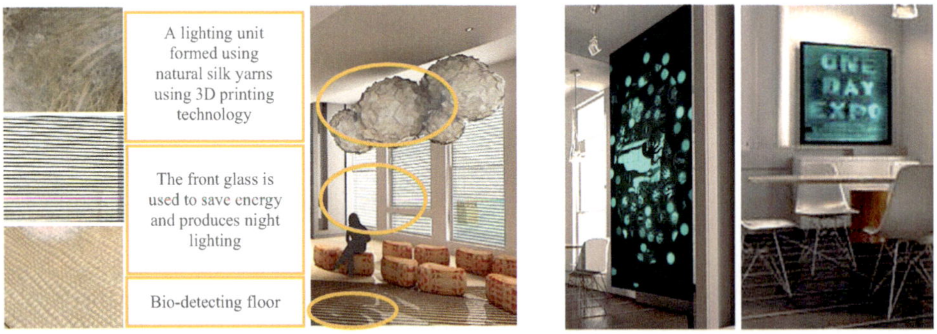

Fig. 31 Analytical shots show how to use different biomaterials in connected places with an integrated functional output (Right: waiting area using day light and bio-detecting floor (Author's own), Left: The illuminating signs and advertising boards)

use of some biomaterials is subjected to ethical and religious considerations, particularly in applications that include, for example, cloning and manipulation of genetic codes. Therefore, there is a necessity for increasing awareness of using biomaterials to facilitate their acceptance compared to traditional ones. Additionally, the new role of the designer's use of bio-sustainable space should be encouraged. Multidisciplinary research and further studies in the fields of bio-architecture and biomaterials are recommended.

Finally, the importance of this study lies in its attempt to suggestion for the possibility of integrating bio-architecture and biomaterials with one another to facilitate the design of a (semi-living) product or interior solution. Thus, the novelty of the study shows its importance in the future of sustainable design.

Acknowledgments I would like to express my gratitude to Prof. Dr. Mahmoud Hassan Ibrahim Al.Malah, The Vacant Professor of Interior Architecture, Decor department, Faculty of Fine Arts, Alexandria University, and Assistant Professor, Dr. Marwa Khaled Mahfouz, the Professor Assistant of Interior Architecture, Decor department, Faculty of Fine Arts, Alexandria University for the supervision of my Ph.D. Thesis which led to this research, their support, technical editing, and proofreading. And, for sharing their pearls of wisdom with me during the Ph.D. research. I am also immensely grateful to the examiners committee, Prof. Dr. Alia Mahmoud Abd El.Hady, The Vacant Professor of Interior Architecture, Decor department, Faculty of Fine Arts, Helwan University, and Prof. Dr. Ahmed Fouad Hassan Mahdy, The Vacant Professor of Interior Architecture, Decor department, Faculty of Fine Arts, Alexandria University, for their useful comments on the Ph. D. research.

References

Abo alkhier, A. H. E. (2008). Architecture Ideology's at the turn of century. Master degree of Science in Architecture, Faculty of Fine Arts, Alexandria University, Egypt.

Adolfo M. R. G., Fabio Z. C., Hector. (2017), "*Chitosan, Chitosan Derivatives and their Biomedical Applications*", University of Valle, Cali, Colombia, Chapter 5, P.92–94), Published by InTech, Rijeka, Croatia

Armstrong R. (2014). *Designing with Protocells: Applications of a Novel Technical Platform*. School of Architecture, Planning and Landscape, University of Newcastle, Newcastle-upon-Tyne, NE1 7RU, UK. Retrived from https://www.researchgate.net/publication/267874465_Designing_with_Protocells_Applications_of_a_Novel_Technical_Platform

Braham W. (January 2003). Biotechniques: Form Follows Flow?. Greenbuild 2003 Conference and Exposition, University of Pennsylvania, Department of Architecture, Pennsylvania. Retrieved from https://repository.upenn.edu/arch_papers/21/?utm_source=repository.upenn.edu%2Farch_papers%2F21&utm_medium=PDF&utm_campaign=PDFCoverPages

Connell R. (August 2009). *The Sentient Drape: Intelligent Building Layers*. Morgan State University, BTES Conference: Assembling Architecture.

Cruz M., Pike S., (December 2008), *AD - Neoplasmatic Design*, Architectural Design, Vol 78.

Dziersk M. and FIDSA, (Spring 2014), *Innovation : Design / Art / Craft*, Quarterly of The Industrial Design Society of America. Retrieved from https://issuu.com/idsa.innovation/docs/innovation_spring2014

Hensel M., Menges A and Weinstock M., (March/April 2006), AD "Techniques and Technologies in Morphogenetic Design, *Architectural Design*, Vol. 76.

HG Wells, (1975), H. G. Wells: Early Writings in Science and Science Fiction, *The limits of individual plasticity* (P. 36–39), in RM Philmus and DY Hughes (eds), ,University of California Press (Berkeley, CA), Retrieved from https://archive.org/details/bub_gb_TQ8eByH9QXsC/page/n2

Johansen, J. M. (2002). *Nanoarchitecture – A New Species of Architecture*. New York, Princenton Architectural Press. Retrieved from https://issuu.com/papress/docs/9781568983011

Lally S. (May- June 2009), AD - Energies, New Material Boundaries, *Architectural Design*, Vol. 79, Retrieved from https://onlinelibrary.wiley.com/toc/15542769/79/3

Lawson B. (2004). *What Designers Know*. Newgen Imaging Systems (P) Ltd., Chennai, India. Retrieved from https://issuu.com/hussien-mansour/docs/what_designers_know

Lawson B. (2005). *How Designers Think – The Design Process Demystified*. Great Britain by Biddles Ltd. Retrieved from https://glennisarndtlam.files.wordpress.com/2017/05/how-designers-think-by-bryan-lawson.pdf

Pazzaglini M. (2015). *Sustainable architecture and Complex Design*. LetteraVentidue Edizioni S.r.l. 96100 Siracusa, Italy. Retrieved from https://books.google.com.eg/books?id=8MFMCgAAQBAJ&printsec=frontcover&hl=ar#v=onepage&q&f=false

Pelfen S., (Winter 2014), *Architecture Alive*, Song Associates Oxford Brookes PLC, au.uk. Retrieved from https://issuu.com/pfsong/docs/dissertation_song

Spiller, N. (2006). *Visionary Architecture – Blueprints of the Modern Imagination*. Thames & Hudson Ltd.

Spiller, N. (2008). *Digital Architecture Now*. A global survey of emerging talent. Thames & Hudson Ltd.

Spiller N., Armstrong R., (March/April 2011), AD – Protocell Architecture, *Architectural Design*, Vol 81.

Vianna M., Vianna Y., K. Adler I., Lucena B. & Russo B. (2012). *Design Thinking* – Business Innovation. Rio de Janeiro RJ - Brazil: MJV Press. Retrieved from https://www.designthinkingbook.com/DT_MJV_book.pdf

Algae Powered House Biofacade Splitterwerk ARUP Colt International SCC Green Power Building. (n.d.). Retrieved December 30, 2020, from https://inhabitat.com/splitterwerk-architects-design-worlds-first-algae-powered-building-for-germany/algae-building/

Alyn Griffiths | 13 November 2013 (2015, September 14). *Surfaces that use "hacked" bacteria to detect dirt or clean themselves*. Dezeen. http://www.dezeen.com/2013/11/13/bio-surfaces-containing-hacked-bacteria-tashia-tucker/.

Biophotovoltaic Moss Table Generates Electricity Through Photosynthesis. Inhabitat Green Design Innovation Architecture Green Building. https://inhabitat.com/moss-table-by-biophotovoltaics-generates-electricity-through-photosynthesis/.

Bio-Plastic Morphologies – materiability. (n.d.). Retrieved December 30, 2020, from http://materiability.com/bio-plastic-morphologies/

Clara Rodríguez Fernández - 22/10/2019 7 mins - Tops. (2020, October 14). *Ten Ways Biotechnology Makes the World More Sustainable*. Retrieved December 28, 2020, from https://www.labiotech.eu/tops/sustainable-biotechnology/

Cuthbertson, A. (2015, May 15). *Self-healing bio-concrete uses bacteria to repair itself*. International Business Times UK. http://www.ibtimes.co.uk/self-healing-bio-concrete-uses-bacteria-repair-itself-1501475.

Fytrou-Moschopoulou, A. (2015, June 01). The BIQ House: First algae-powered building in the world. Retrieved December 30, 2020, from https://www.buildup.eu/en/practices/cases/biq-house-first-algae-powered-building-world

Global Reach Internet Productions, L. L. C.- A. *Bioasphalt binder*. Bioasphalt binder - Avello Bioenergy. http://www.avellobioenergy.com/en/products/bioasphalt_binder/.

Glowee, enlightened by the sea. Glowee. https://www.glowee.eu/.

Inhabitat - Green Design, Innovation, Architecture, Green Building. (n.d.). Retrieved December 30, 2020, from https://inhabitat.com/tag/cellulose-nanocrystals/

Johansen J. (n.d.). Retrieved December 28, 2020, from https://www.johnmjohansen.com/Future-of-Architecture.html

Live Circular. (2015, October 13). *Trial section of bio-asphalt hits the road in Holland.* Live Circular. https://livecircular.com/trial-section-of-bio-asphalt-hits-the-road-in-holland/.

Marcus Fairs | 12 February 2014 1 comment. (2015, September 14). *Microbes are "the factories of the future" says Suzanne Lee.* Dezeen. https://www.dezeen.com/2014/02/12/movie-biocouture-microbes-clothing-wearable-futures/.

MIT Researchers build silk pavilion using biological Silkworm 3D printer. 3ders.org. http://www.3ders.org/articles/20130601-mit-researchers-build-silk-pavilion-using-biological-silkworm-3d-printer.html

Moss table. https://www.mediamatic.net/en/page/37560/moss-table.

NASA. *Real World: Self-Healing Materials.* NASA. https://nasaeclips.arc.nasa.gov/playlists/realworld?v=real-world-self-healing-materials.

Organics (n.d.). Retrieved December 28, 2020, from http://www.joostrekveld.net/?page_id=522

Philips Bio-light concept lights the home using bacteria. New Atlas. (2015, May 2). http://www.gizmag.com/philips-bio-light-concept-taps-bioluminescence-for-home-use/20632/.

Rensselaer: Architecture. (n.d.). Retrieved December 28, 2020, from https://www.arch.rpi.edu/2011/10/oksiuta-zbigniew/

Real World: Self-Healing Materials. (n.d.). Retrieved December 30, 2020, from https://nasaeclips.arc.nasa.gov/playlists/realworld?v=real-world-self-healing-materials

Studios, O. (n.d.). *(bio)Plastic Morphologies Table by Marilu Valente.* Retrieved December 30, 2020, from https://competition.adesignaward.com/design.php?ID=37214

Song, P. F., & Pfsong. *Dissertation_Architecture Alive*! Issuu. https://issuu.com/pfsong/docs/dissertation_song.

(sunny.chandel.50. (2013, November 16). *Synthetic Bacteria will keep our homes clean in future.* HomeCrux. http://www.homecrux.com/2013/11/16/10367/synthetic-bacteria-will-keep-homes-clean-future.html.).

Taylor-Foster, J. (2014, February 01). *Bricks Grown From Bacteria.* Retrieved December 30, 2020, from https://www.archdaily.com/472905/bricks-grown-from-bacteria

Words Jennifer Ferng, & Ferng, J. *Reassembling biomaterials and modes of praxis.* Architecture Now. https://architecturenow.co.nz/articles/new-developments-in-architectural-materials-reassembling-biomaterials-and-modes-of-praxis/.

2011—Product and Energy Design / Design de produit et d'énergie—Carlos Peralta. *Alive: Biophotovoltaic Moss Table.* Alive Biophotovoltaic Moss Table Comments. http://thisisalive.com/biophotovoltaic-moss-table/.

Figures

Algae Powered House Biofacade Splitterwerk ARUP Colt International SCC Green Power Building. (n.d.). Retrieved December 30, 2020, from https://inhabitat.com/splitterwerk-architects-design-worlds-first-algae-powered-building-for-germany/algae-building/.

Alyn Griffiths | 13 November 2013 (2015, September 14). Surfaces that use 'hacked" bacteria to detect dirt or clean themselves. Dezeen. http://www.dezeen.com/2013/11/13/bio-surfaces-containing-hacked-bacteria-tashia-tucker/.

Biophotovoltaic Moss Table Generates Electricity Through Photosynthesis. Inhabitat Green Design Innovation Architecture Green Building. https://inhabitat.com/moss-table-by-biophotovoltaics-generates-electricity-through-photosynthesis/.

Bio-Plastic Morphologies – materiability. (n.d.). Retrieved December 30, 2020, from http://materiability.com/bio-plastic-morphologies/

Clara Rodríguez Fernández - 22/10/2019 7 mins - Tops. (2020, October 14). Ten Ways Biotechnology Makes the World More Sustainable. Retrieved December 28, 2020, from https://www.labiotech.eu/tops/sustainable-biotechnology/

Cuthbertson, A. (2015, May 15). Self-healing bio-concrete uses bacteria to repair itself. International Business Times UK. http://www.ibtimes.co.uk/self-healing-bio-concrete-uses-bacteria-repair-itself-1501475.

Fytrou-Moschopoulou, A. (2015, June 01). The BIQ House: First algae-powered building in the world. Retrieved December 30, 2020, from https://www.buildup.eu/en/practices/cases/biq-house-first-algae-powered-building-world

Global Reach Internet Productions, L. L. C.- A. Bioasphalt binder. Bioasphalt binder - Avello Bioenergy. http://www.avellobioenergy.com/en/products/bioasphalt_binder/

Glowee, enlightened by the sea. Glowee. https://www.glowee.eu/.

Inhabitat - Green Design, Innovation, Architecture, Green Building. (n.d.). Retrieved December 30, 2020, from https://inhabitat.com/tag/cellulose-nanocrystals/

Johansen J. (n.d.). Retrieved December 28, 2020, from https://www.johnmjohansen.com/Future-of-Architecture.html

Live Circular. (2015, October 13). Trial section of bio-asphalt hits the road in Holland. Live Circular. https://livecircular.com/trial-section-of-bio-asphalt-hits-the-road-in-holland/.

Marcus Fairs | 12 February 2014 1 comment. (2015, September 14). Microbes are 'the factories of the future" says Suzanne Lee. Dezeen. https://www.dezeen.com/2014/02/12/movie-biocouture-microbes-clothing-wearable-futures/.

MIT Researchers build silk pavilion using biological Silkworm 3D printer. 3ders.org. http://www.3ders.org/articles/20130601-mit-researchers-build-silk-pavilion-using-biological-silkworm-3d-printer.html

Moss table. https://www.mediamatic.net/en/page/37560/moss-table.

NASA. Real World: Self-Healing Materials. NASA. https://nasaeclips.arc.nasa.gov/playlists/realworld?v=real-world-self-healing-materials

NASA. Real World: Self-Healing Materials. NASA. https://nasaeclips.arc.nasa.gov/playlists/realworld?v=real-world-self-healing-materials

Organics (n.d.). Retrieved December 28, 2020, from http://www.joostrekveld.net/?page_id=522

Philips Bio-light concept lights the home using bacteria. New Atlas. (2015, May 2). http://www.gizmag.com/philips-bio-light-concept-taps-bioluminescence-for-home-use/20632/.

Rensselaer: Architecture. (n.d.). Retrieved December 28, 2020, from https://www.arch.rpi.edu/2011/10/oksiuta-zbigniew/

Real World: Self-Healing Materials. (n.d.). Retrieved December 30, 2020, from https://nasaeclips.arc.nasa.gov/playlists/realworld?v=real-world-self-healing-materials

Studios, O. (n.d.). (bio)Plastic Morphologies Table by Marilu Valente. Retrieved December 30, 2020, from https://competition.adesignaward.com/design.php?ID=37214

Song, P. F., & Pfsong. Dissertation_Architecture Alive! Issuu. https://issuu.com/pfsong/docs/dissertation_song

(sunny.chandel.50. (2013, November 16). Synthetic Bacteria will keep our homes clean in future. HomeCrux. http://www.homecrux.com/2013/11/16/10367/synthetic-bacteria-will-keep-homes-clean-future.html.).

Taylor-Foster, J. (2014, February 01). Bricks Grown From Bacteria. Retrieved December 30, 2020, from https://www.archdaily.com/472905/bricks-grown-from-bacteria

Water as an Element of Architectural Space Design Study the Psychological Impact of Water on the Occupants of the Space

Khawola F. Mahmoud, Hiyam Majeed Jaber Al-Bakry, and Assda Abdul-Hameed Al-Tuhafi

Abstract

In recent years, interest has increased in water and its preservation alongside environmental effects. Previously, the focus was on the geographical and health importance of water, but recent studies have discussed the use of water in architectural design and its effects on architectural spaces as well as its environmental effects on architecture. Some studies have examined the psychological impact of water properties on specific areas. The current research investigates the psychological impact of water on the visitors and determines the aspects of water and their effects on the users of architectural spaces to specify positive impact of water in order to utilize them in architectural design. Theoretical results of the current study were then applied in the form of a questionnaire to define the importance of an artificial waterfall in a park and its psychological effects on visitors.

Keywords

Psychological • Water • Design element

K. F. Mahmoud (✉) · H. M. J. Al-Bakry · A. A.-H. Al-Tuhafi
College of Engineering, Architecture Department, Mosul University, Mosul, Iraq
e-mail: khawola.mahmoud@uomosul.edu.iq

H. M. J. Al-Bakry
e-mail: hmj.arch@uomustansiriyah.edu.iq

A. A.-H. Al-Tuhafi
e-mail: assda.altuhafi@uomosul.edu.iq

A. A.-H. Al-Tuhafi
College of Engineering, Architecture Department, Mustansiriyah University, Baghdad, Iraq

1 Introduction

In different ways, "Water" can be noticed in various fields, including art, literature, planning, design, anthropology, geology, ecology, psychology, mythology, history, and religion. Sociologists emphasized the significance of water to humans' impression, estimate, and interpretation of spaces. The contemporary landscape planners and designers constantly acknowledge the importance of water in their works (Burmil et al., 1999).

There have been numerous studies that took into account water and its involvements with many fields of science, whether literary, geographical, health, or even psychological sciences, For example, Seçkin conducted studies that investigated the means in which man uses water in landscape design, basing Maslow prototype. Volume, shape, and sound of water as design factors that affect the way humans use water in exterior space design were also studied(SEÇKİN, 2010). Burmil argues that water has a great psychological impact on humans that must be closely observed besides the basic human needs for water in dry places (Burmil et al., 1999). Hawley, on the other hand, studies the aesthetic and psychological factors affecting humans when designing exterior spaces, with water being the most influential (Howley, 2011). Nasar and Li believe that people prefer reflections of objects on the water surface rather than reflections on other surfaces, i.e., mirrors (Nasar & Li, 2003). Lifang et al. researched the psychology of designing landscapes around river banks, using a hierarchic analysis method to lay a foundation for designing natural riverside scenes in urban areas (Lifang et al., 2008). By and large, Vernon and Tiwari investigate the effect of water on evoking feelings of belonging or nostalgia to places, as they point out that water gives a certain identity to places and increases feelings of belonging (Vernon & Tiwari, 2009). Yamashita, on her part, attempts to evaluate the perception of water in exterior spaces, using cameras. She claims that different age groups have a different perception of such designs, as adults can sense the medium dimension of water while children can only

F. Rosso et al. (eds.), *Advances in Architecture, Engineering and Technology*, Advances in Science, Technology & Innovation, https://doi.org/10.1007/978-3-030-86913-7_8

perceive shorter dimensions (Yamashita, 2002). Finally, in his study, Völker investigates the hygienic effect of water on the psychology of humans, as he presents water with the blue color, since blue is associated with comfort and serenity, which ultimately gives positive psychological impressions to humans, not to mention the direct hygienic effects of water on humans' well-being (Völker & Kistemann, 2011).

Accordingly, this study is an attempt to present water as an architectural element and its impact on the human psychology. Therefore, a necessary theoretical basis will be presented that can be used in the (Method).

1.1 Water In The Earlier Civilizations

The existence of water was a considerable factor to the emergence of many civilizations and their persistence for long periods. For example, the Roman and Greek civilizations, Egypt and Mesopotamia were founded along river deltas (Hall, 2001; Mumford, 1989). Ponds of water were found painted on the walls of Egyptian temples thousands of years ago, presenting the theme of Water. Here, water is used as an element in the gardening and landscape design. The use of water in Pompeii's villas, the old Roman city, was an important element, later it was used in gardens design. (Burmil et al., 1999). Also Water is an effective element in the Chinese gardens. When the Chinese were planning a garden, the first step was to check the source and flow of water available at the site. This is due to their belief that water serves the balance of other elements in nature. So pools, ponds, waterfalls, wells, and fountains were introduced into landscape that create fantasy and ambiguity (Burmil et al., 1999). Water in Islamic vision has a distinct location. Water is mentioned in many places in The Glorious Qur'an in addition to the ponds and rivers that flow in Paradise. Islam has organized water distribution systems, which is less similar in other civilizations. Islam orders people to use water in an economical form without extravagance and waste even when it is used in some worshipping rituals. Therefore, the garden design in Islamic civilization employed water as a central element, which integrated religious symbolism (Plumptre, 1993).

1.2 Water As a Design Element

1.2.1 Water in Landscape Design

From earlier civilizations till the present day, water has and does constitute a significant element of landscape design. Asakawa et al. believe that "to achieve highly preferred scenery in these stream corridors, there are three necessary components of natural scenery: water, vegetation, and sequential experience with variety"(Asakawa et al., 2004, p. 177). As a result of the formidable favorable response to water, its positive or substantial valuable general quality, and as a reason for its importance for ecosystems we perceive water as a major topic in design problems (Steinwender et al., 2008, pp..124–125). Karmanov and Hamel pinpoint the need of water for healthy urban environments. They integrated aspects regarding aesthetics, civilization, and environment in their definition (Karmanov & Hamel, 2008, p. 123). Through the history of garden design, water has been discussed. Repton finds a great significance of water in landscape design as a result of the contentment created by the glimmer of water and by the wonderful and joyful impact that can be felt by a small pond. Indeed, he used to include a pool in his designs even if it was not naturally appropriate (Burmil et al., 1999, p. 103). Water in design has oftentimes been linked to the emulation of water in nature. Natural water is also important as it can be considered as a muse when designing urban open spaces and fountains (Burmil et al., 1999).

It is important to note that designers have utilized a multitude of optical and non-optical attributes of water in landscape design water could be found in a stagnant state of motion at various places and it could also be shallow or deep or reflective of the surrounding elements and surroundings. It could also appear in several colors and produce many sounds, and when it's touched it can give a comforting sensation as well. Its color variations are associated with different attributes. Bluewater, for example, is linked to a kind of impressions of coldness, white water suggests strength, and thunderous sound (Litton, 1974).

1.2.2 Water in Interior Spaces

Water has been heavily used as a design element in interior spaces, for example, in fountains and small pools, alongside a multitude of uses. Arabs in Andalusia used to adorned their palaces, houses, gardens, and mosques with fountains of various shapes and sizes according to the spaces in which these fountains are placed, with the basic concept of squirting water upwards or in different directions. Water was mainly used in these fountains for drinking and ablution. For instance, water was also used aesthetically to reflect interior facades, add interesting motion in empty spaces that may seem dismal if left with no smart and attractive treatment, in addition to being a tool to mask noise coming from the surrounding environment. According to Feng Shui,[1] water fountains can be placed at one side of a room to add an aesthetic element to the room, as well as the bubbling of the water that gives feelings of relaxation and tranquility.

[1] Feng Shui is the Chinese art of design and placement. Chinese people thought that humans inhabit an energy filed. Feng Shui is concerned with the study of this fundamental energy. It is the basic comprehension of energy, in regard to its flow, manners and pattern, in a certain space.

1.3 Water and Human Perceptions

"Perception" is achieved through employing human senses to create emotions and stances toward anything. Water plays a vital part in the understanding of nature (Herzog et al., 2000, p. 341). Important aspects of the sensual perception of water are its sound, color, clarity, motion, and context (Völker & Kistemann, 2011). Designers tend to use water in design when they want to provide space that suggests joy, calmness, and comfort only (SEÇKİN 2010). Bachelard (1983) theorizes that water gains its aesthetic value as a result of its being a natural element which goes in harmony with the psychological theory that humans prefer natural elements (Nasar, 1990; Nasar & Li, 2003).

Waterscapes are designed for the enjoyment of humans alongside other reasons. As such, the designer has to provide important factors like explicitness and unity, in addition to creating a sense of participation through intricacy and ambiguity (Coeterier, J.F. 1996; Völker & Kistemann, 2011). Water has significant biophysical attributes which make the observer recognize blue spaces as an area that generates a feeling of pertinence and attachment towards water. Water is recognized as being nature's mirror, inspiring mystique by creating images that are not as clear as the images reflected by man-made mirrors (Burmil et al., 1999; Nasar & Li, 2003). Fascination, compatibility, attention, and calmness generated when looking at waterscapes are a translation of mental immersion (Laumann et al., 2001; White et al., 2010). Particularly, water could have a positive impact on the mental state in comparison to other environmental elements (Felsten, 2009). Also, water is not only used by individuals, as it could also be employed as a link between people to bring them together in leisure activities (Program & Fredrickson, 1999). Regan and Horn discovered that when people are in a stressed state of mind, they would feel relaxed when seeing waterscapes (Völker & Kistemann, 2011).

1.4 Water and Negative Ions

The atmosphere is filled with "negative ions" that present a plethora of several psychological and physical benefits. It is determined that the existence of water sprays in some places such as waterfalls results in waves of negative ions, which leads to high levels of serotonin. Serotonin plays an important role in human's temper, intellection, sleeping, and eating patterns. Serotonin levels in blood could treat some mild cases to moderate depression. Up to 10.000 negative ions can be found in the proximity of waterfalls, but their value does not come close to 100 at rush hour in major metropolises.[2]

[2] http://www.health-benefit-of-water.com.

1.5 Water and the Sense of Place

"Water sensitive urban design" was primarily a constituent of a policy strategy regarding water demand management. Nevertheless, "water physical urban design" has evolved, based on the realization that the "natural environment" is an important part of the sense of place (Hedgcock & Mouritz, 1993). "Water sensitive urban design" can also participate in creating a sense of place and establishing a local identity (Vernon & Tiwari, 2009).

"Sense of Place" is considered thematic for any practice of placemaking. It is substantial for the prosperity of a community through generating emotions of security, safety, a feeling of glory and belonging (Schultz, 1980). Placemaking does not only deal with the materialistic aspects of a place that serve the function but also with the manner those factors can create an environment by improving the "sense of place (psychological value)". Therefore, the integration of meeting the materialistic practical psychological requirements of users is considered crucial in making an appealing space. Furthermore, water sensitive urban design methods participate in the materialistic structure of space, and most significantly they participate in creating a suitable environment (Vernon & Tiwari, 2009). In addition, a sense of place in waterscapes is associated with feelings and "a symbolism difficult to achieve with any other natural element" (Whalley, 1988, p. 145).

1.6 Characteristics of Water

1.6.1 Sound of Water

Nature announces the presence of water in a joyous way that is by sound. The characteristics of music are found in water's sound, like diversity in volume, softness, rhythm, intensity, and harmony, which is the most important characteristic. Mass of individual water units falling from cascades that land on the surface create sound. The level of sound from water can vary from absolute silence to a roaring volume (architecture and water, 1994, p. 13). Sound can be considered a substantial element in the human experience of water in the landscape as a standalone element or combined with other visual elements. Sound is created as a result of the movement of water down cascades, through or above obstacles, and by lifeforms like fish and animals, making a movement through its surface. Water possesses a perpetual range of sounds. They range from the delicate sounds of single drops hitting the surface, the accelerated sounds of currents, or the thundering roar of a waterfall. Sound could indicate the presence of water even when it is nowhere in sight. (Burmil et al., 1999, pp. 100–104). On another hand, the sound of water performs several functions, water with vigorous energy and rapid flow falling from cascades or waterfalls invigorate the user. On the contrary, calm water with tranquil energy could perform the task of soothing and calming the user. Water, whether it comes in the forms of a trickle or a flow, can be tampered to achieve a certain set of design aims. Additionally, users can be informed of the existence of water through sound. Thus, they are provided with a sense of being in a safe environment. And from the design standpoint, this presents water as a heard element rather than seen. Another function that water could serve is disguising raucous detestable surrounding noise, though there is the possibility of the water becoming very loud in that process (SEÇKİN, 2010, p .7).

1.6.2 Reflection of Water

Water is a distinctive natural material. It can reflect light waves through its surface. Along with seeing the surface of water, images of the surrounding environment could also be reflected at its surface. n the condition of water having a calm surface, it can display quite vivid images of mountains, crags, plants, wildlife, and in some cases, the image of the observer himself is reflected. If the calm surface of the water is disturbed by a gust of wind or the movement of the water itself, the reflected images lose their acuity and high level of detail, producing images that resemble the works of an impressionist artist of the surrounding environment (Burmil et al., 1999, p. 100). Bachelard (1983) outlines that the images created by water are more natural and pure than the mirrors made by man. Seventy-two years earlier, Wright (1928) hypothesizes on the aesthetic value of reflective water and deemed it as "refreshing and beautifying in architecture, if architecturally used" (Nasar & Li, 2004). Water reflection is considered a significant feature. Hubbard and Hubbard claimed that harmony and enrichment could be achieved through the use of reflective water as it adds attractiveness. Small forms of water could catch the attention of the observer. These thoughts and notions are thought to have implications in design, because the reflectivity of water could be altered through varying the color of the containing surface. (Nasar & Li, 2003). Also, recent landscape theorists identified with the aesthetic value provided by reflective water (Burton & Litton, 1974). From a psychological viewpoint, Kaplan notes that "water itself provides a continuing, unifying theme to the landscape and one that calls attention to itself. It has a texture that easily sparkles, reflects images, or ripples with the wind" (Kaplan, 1977). It is thought that water itself or reflective water may possess restorative abilities (Nasar & Li, 2003, p. 237).

1.6.3 Movement of Water

A motionless small pool of water or a vast ocean can ingrain a feeling of serenity and peace. Undulations in water could be caused as a result of water flow in rivers or canals aided by gravity or caused by the tidal flow of seas. This mild stimulation of the water surface could cause changes in one's mood. The instilled feeling differs based on the water mass and velocity of the calm and soft or writhing, sensuous undulations. The term "black water" applies when the water surface is undisturbed regardless of how vigorous the movement is, whereas the term "White Water" is used to describe forms of water such as cascades, waterfalls, and tsunami. (Dreiseitl & Grau, 2005). The movement of water takes various forms, which could highly contradict with the tranquil elements on its borders (Burmil et al., 1999, P.100). Slow changes in the velocity of water often lead to a feeling of monotony, immobility, and boredom. (Burton & Litton, 1974). Few movements in nature around found attractive by humans; of which is the movement of water, with its variety and flow. Motion, by itself, or combined with sound has been viewed to have a substantial impact on people's understanding and evaluation of picturesque river landscapes (Burmil et al., 1999). What is more, Symmes explains that the water movement can be invitingly gentle or frighteningly destructive and there are no two water jets or waterfalls are exactly alike since moving water produces an infinite range of changing forms and sound designs (Symmes et al., 1998).

1.6.4 Colors of Water

Pure water has no odor, color nor taste. Although in nature, water oftentimes appears in various colors that range from a dark gray that seems impermeable to a marvelous blue or appears transparent. This multitude of appearance is a result of alterations in lighting, the daily and seasonal positioning of the sun, alongside cloud cover, atmospheric and aquatic particles (Burmil et al., 1999). Eroded materials that are found within the water could also play a part in its apparent color. The Colorado River was named the Red River due to the color of the mud it transfers. Colors could also indicate what sorts of aquatic species inhabit it (Leopold & Davis, 1966). The color of water has different notions to be associated with: blue could be linked to coolness while white with power and thunderous sound (Litton, 1974). In general, blue water is favored over yellow water due to high concentration of yellow substances. (Smith et al., 1995). Color and clarity are considered the determining factors to the public in perceiving water quality of a river (Pflüger et al., 2010). The majority of color we see in the water is a product of the surrounding environment which is reflected on its surface, or the seen objects underneath due to the transparency of water. There are various degrees in which one sees colors according to one's viewing angle. This variety could be explained by physics as a result of the differential refraction angle between air, water, and light. The passing light is absorbed by the transparent medium. That source of light is oftentimes the blue sky. The color of water gets deeper and darker when the light intensity is reduced (Dreiseitl & Grau, 2005).

1.6.5 The Form of Water

Water does not have a specific form. According to the laws of physics, water takes the shape of its container. Water could be naturally found in various forms, for example, it could fill in valleys and create lakes and take and follow a winding course through arroyos. These various forms can vary in size, shape, and depth. (Burmil et al., 1999). Alterations in the flow of water could result in altering the looks of streams, resulting in changing or the removal of particular forms of water. For instance, vertical forms (waterfalls), angular (cascades), and horizontal forms of water and other forms might cease to exist altogether. (Burmil et al., 1999). Seçkin discusses the formal and casual settings of water features, describing natural forms with organic shapes and edges that are on the softer side as casual while deeming angular shapes with geometric properties as formal. When it comes to designing water features, he believes that the water edge is a crucial factor that plays a part in creating the perspective that acts as a liaison between people and water. He describes it as the link between the space that surrounds the water feature and the water feature itself. He also states that the lack of the design's accuracy or the definition of the edge, would weaken the design and affect the clarity of the message it aims to send. (SEÇKİN, 2010).

2 Method

The psychological effect of water design characteristics in parks will be examined. The case study will be chosen by selecting a park that contains water in its design. After determining the possible psychological effects of the water characteristics, a questionnaire will be conducted for the visitors of the selected park. The questionnaire will include the design characteristics of water and possible psychological effects that have been studied in the theoretical study.

"Şelale Park' was chosen for this study to determine the psychological effects of water on humans and the most impactful characteristics of water. Şelale Park was built in 2009 (see Fig. 1) and it is located in the city of Eskişehir, Turkey. Şelale Park attracts thousands of visitors. The park consists of an artificial waterfall that ends in an artificial lake. The park also includes a cafeteria, children's playground, and gym. Moreover, the park sports a captivating panoramic scene over the city of Eskişehir.

A questionnaire was designed requiring visitors to "Şelale Park' in the city of Eskişehir in Turkey to determine the significance of the artificial waterfall for them and to determine the psychological effects of the existence of water in the form of waterfalls, rivers, seas, fountains, or lakes on humans in general and people being questioned in particular. Some aspects of water were excluded from the current study, including the strong movement of water [raging of water] as artificial waterfalls are calm by nature (as humans have different feelings towards calm and raging seas), and the aspect of time, i.e., summer and winter times (as water in the summer has an environmental function in refreshing the air). Values were given to questions listed in the questionnaire for further analysis, results, and conclusions (Table1).

3 Results

After utilizing the questionnaire and analysis of the values given for each question, the following results have been concluded:

- 60% of visitors primarily go to the park to see and enjoy the artificial waterfall, while 30% consider the waterfall their second reason to go to the park. Also, it is has found that the existence of water in the park was the primary element which grants joy to the visitors (see Fig. 2).
- percentages of the various characteristics of water were close, ranging between 20 and 27% for all characteristics except color of water as it has been found to have no significant effect [this may be because the waterfall is colorless as opposed to the blue water of seas and green water of rivers]. It has found that there is not a dominant characteristic among others, as different people have different preferences. Figure 3 illustrates the percentages for each Characteristic.
- It has been found that 50% of the questioned individuals feel calm and relaxed upon seeing the waterfall, 26% feel happy, 26% feel optimistic and hopeful, and 20% feel invigorated and enthusiastic. This means that water has various positive impacts on people (Fig. 4).

4 Conclusions

From the results above, it has been concluded that the waterfall at Şelale Park has added aesthetic value and identity for the park, as most people consider the waterfall to be a fundamental element of the park and the reason people go to the park. Moreover, sound, movement, and shape of water are found to be the distinctive elements of the waterfall that make visitors feel enjoyment and happiness.

It has also been concluded that people differ in their preferences of the various aspects of water; no aspect is dominant over others. Therefore, each aspect must be studied separately to create more appropriate spaces when designing aquatic planes, fountains, and waterfalls. Besides, water had various psychological effects on visitors, which

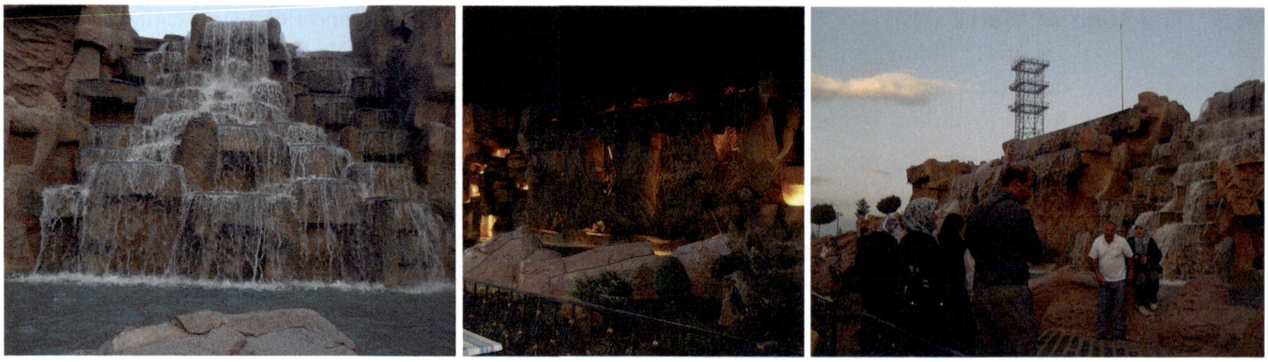

Fig. 1 "Şelale Park" (Authors)

Table 1 Show the possible values

Park contents [Design elements of Şelale Park] 1	Waterfall	1–1
	Cafeteria	2–1
	Green areas	3–1
	Games	4–1
	City panorama	5–1
	Other	6–1
Design Characteristics of water [waterfalls] 2	Sound of water	1–2
	Movement of water	2–2
	Reflections on water	3–2
	Shape of water	4–2
	Color of water	5–2
Type of Psychological effects of waterfalls design Characteristics 3	Joy, happiness, delight	1–3
	Love, excitement	2–3
	Hope, optimism	3–3
	Relaxation, calmness, comfort	4–3
	Activeness, vitality, enthusiasm	5–3
	Loneliness, pain, sadness	6–3
	Anger	7–3
	Anxiety, tension, stress	8–3
	Fatigue, boredom	9–3
	Fear	10–3
	No effect	11–3

ranged between feeling calm, happy, optimistic, invigorated, and active. It is safe to say that different aspects of water have different effects on people and that people differ in their preference for aspects of water.

Looking at previous studies and the results of the questionnaire, it has been concluded that the element of water is significant in parks and open areas due to their aesthetic, climatic, and psychological effects on humans. These effects must be taken into consideration when designing spaces.

Acknowledgements This research was supported by College of Engineering—Mosul University/ Mosul, Iraq and College of Engineering Mustansiriyah University/ Baghdad, Iraq.Also, thanks to Assistant Professor Dr. Ozlem Mumcu Ucar / Eskişehir Technical University, Faculty of Architecture and Design.

Fig. 2 A graph chart showcasing percentages of the most important elements in the park

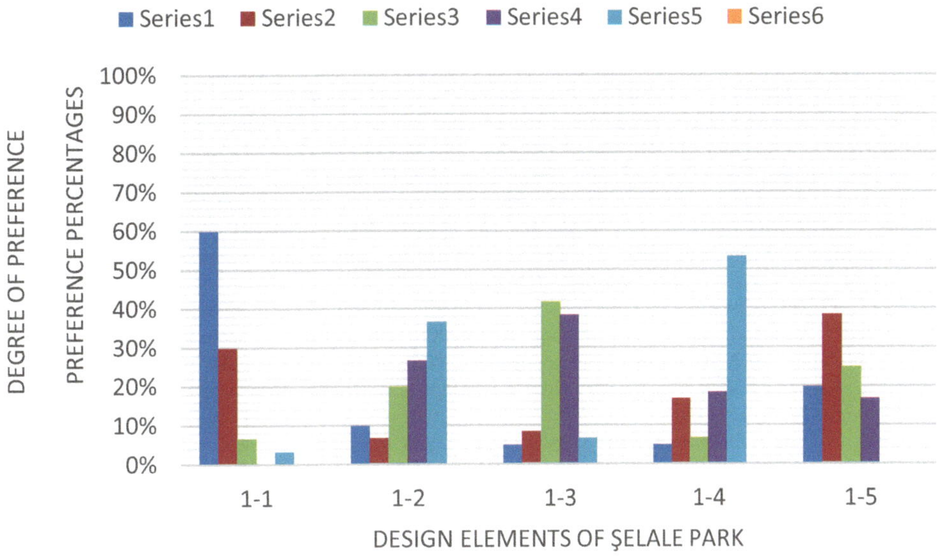

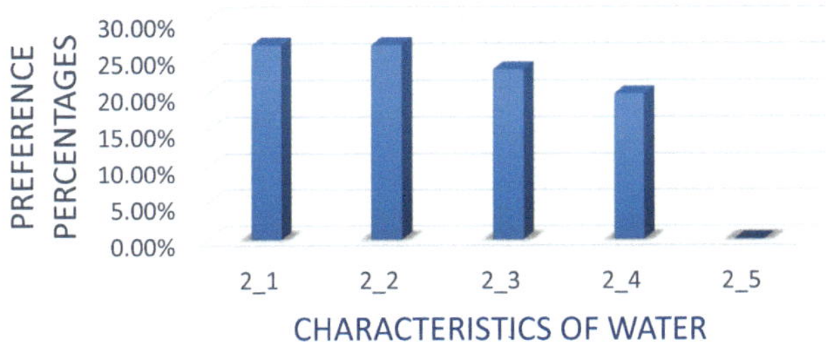

Fig. 3 Design characteristics of the water and the proportion of their preference according to the results of the questionnaire

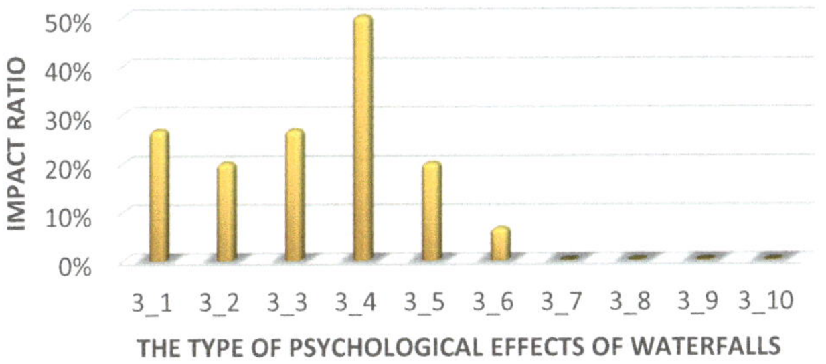

Fig. 4 Show cases The percentage of each type of psychological effects for waterfall design characteristics on the user according to the questionnaire results

References

Asakawa, S., Yoshida, K., & Yabe, K. (2004). Perceptions of Urban Stream Corridors within the Greenway System of Sapporo, Japan. *Landscape and Urban Planning, 68*(2–3), 167–182.

Burmil, S., Daniel, T. C., & Hetherington, J. D. (1999). Human Values and Perceptions of Water in Arid Landscapes. *Landscape and Urban Planning, 44*(2–3), 99–109.

Burton Litton. (1974). *Water and Landscape: An Aesthetic Overview of the Role of Water in the Landscape.*

Coeterier, J. F. (1996). Dominant Attributes in the Perception and Evaluation of the Dutch Landscape. *Landscape and Urban Planning, 34*, 27–44.

Dreiseitl, H, and D Grau. (2005). *New Waterscapes: Planning, Building and Designing with Water.* Birkhäuser. https://books.google.com.tr/books?id=U6uKAQAACAAJ.

Felsten, G. (2009). Where to Take a Study Break on the College Campus: An Attention Restoration Theory Perspective. *Journal of Environmental Psychology, 29*(1), 160–167. https://doi.org/10.1016/j.jenvp.2008.11.006

Hedgcock, D., & Mike, M. (1993). WATER SENSITIVE RESIDENTIAL DESIGN. *Australian Planner, 31*(2), 114–118. https://doi.org/10.1080/07293682.1993.9657618

Herzog, Thomas R., Eugene J. Herbert, Rachel Kaplan, and C. L. Crooks. (2000). "Cultural and Developmental Comparisons of Landscape Perceptions and Preferences." *Environment and Behavior* 32(3): 323–46. http://journals.sagepub.com/doi/https://doi.org/10.1177/0013916500323002.

Howley, P. (2011). Landscape Aesthetics: Assessing the General Publics' Preferences towards Rural Landscapes. *Ecological Economics, 72*, 161–169. https://doi.org/10.1016/j.ecolecon.2011.09.026

Kaplan, Rachel. (1977). "Down by the Riverside: Informational Factors in Waterscape Preference." *In: Proceedings on the Symposium on River Recreation Management and Research* USDA Fores: 285–89.

Karmanov, D., & Hamel, R. (2008). Assessing the Restorative Potential of Contemporary Urban Environment(s): Beyond the Nature versus Urban Dichotomy. *Landscape and Urban Planning, 86*(2), 115–125.

Laumann, K., Gärling, T., & Stormark, K. M. (2001). Rating Scale Measures of Restorative Components of Environments. *Journal of Environmental Psychology, 21*(1), 31–44.

Lifang, Qiao, Zhang Yichuan, and Cao Wei. (2008). "Evaluation of Urban River Landscape Design Rationality Based on AHP." *Water Science and Engineering* 1(4): 75–81. http://www.sciencedirect.com/science/article/pii/S1674237015300417.

Litton, R B. (1974). *Water and Landscape: An Aesthetic Overview of the Role of Water in the Landscape.* Water Information Center, Incorporated.

Nasar, J. L. (1990). The Evaluative Image of the City. *Journal of the American Planning Association, 56*(1), 41–53.

Nasar, J. L., & Li, M. (2003). Landscape Mirror: The Attractiveness of Reflecting Water. *Landscape and Urban Planning, 66*(4), 233–238.

Plumptre, G. (1993). *The Water Garden: Styles*. Thames & Hudson.
Program, Environmental Studies, and L Fredrickson. (1999). "A Qualitative Exploration of the Wilderness Experience As a Source of Spiritual Inspiration." *Journal of Environmental Psychology* 19(1): 21–39. http://linkinghub.elsevier.com/retrieve/pii/S0272494498901104.
SEÇKİN, Yasin Çağatay. (2010). "Understanding the Relationship between Human Needs and the Use of Water in Landscape Design." : 1–17.
Smith, D. G., Croker, G. F., & McFarlane, K. (1995). Human Perception of Water Appearance: 1. Clarity and Colour for Bathing and Aesthetics. *New Zealand Journal of Marine and Freshwater Research, 29*(1), 29–43.
Steinwender, A., Gundacker, C., & Wittmann, K. J. (2008). Objective versus Subjective Assessments of Environmental Quality of Standing and Running Waters in a Large City. *Landscape and Urban Planning, 84*(2), 116–126.
Symmes, M F, K A Breisch, S Astley, and Cooper-Hewitt Museum. (1998). *Fountains: Splash and Spectacle : Water and Design from the Renaissance to the Present*. Thames and Hudson. https://books.google.com.tr/books?id=O2VvQgAACAAJ.
Vernon, B., & Tiwari, R. (2009). Place-Making through Water Sensitive Urban Design. *Sustainability, 1*(4), 789–814.
Völker, S., & Kistemann, T. (2011). The Impact of Blue Space on Human Health and Well-Being - Salutogenetic Health Effects of Inland Surface Waters: A Review. *International Journal of Hygiene and Environmental Health, 214*(6), 449–460.
Whalley, J. (1988). Water in the Landscape. *Landscape and Urban Planning, 16*(1–2), 145–162.
White, M., et al. (2010). Blue Space: The Importance of Water for Preference, Affect, and Restorativeness Ratings of Natural and Built Scenes. *Journal of Environmental Psychology, 30*(4), 482–493. https://doi.org/10.1016/j.jenvp.2010.04.004
Yamashita, S. (2002). Perception and Evaluation of Water in Landscape: Use of Photo-Projective Method to Compare Child and Adult Residents' Perceptions of a Japanese River Environment. *Landscape and Urban Planning, 62*(1), 3–17.

Orientalism and Islamic Architecture

Lamiaa Abd-Elaziz, Khaled Mohamed Dewidar, and Mostafa Refat Ismail

Abstract

Orientalism' is a term that is commonly known as the fascination of occident scholars with studying eastern societies, cultures, and people; depicting them in their art and writings. Until Edward said introduced in his book "Orientalism" (1978) a new definition; He defined orientalism as the way the west views, understands, and assimilates Eastern cultures and how this affected their interpretation, judging, writings about this culture. Orientalism was accompanied by factors like culture totalizing, polarized view East/West and ethnology that contributed in shaping the understanding of eastern cultures including Islamic culture. Architecture is an aspect of the culture; therefore, it will be affected by orientalism the same way as the culture its part of it. In this paper will address how orientalism and architecture emerged as coupling and how Islamic architecture was affected by orientalism; through studying how western scholars and architects interpreted, understood, and described Islamic architecture in their writings and how this reflected on their buildings designs, which has Islamic architectural features. Reaching from this point to the main findings; that orientalism and architecture emerged as a coupling and both were strongly related; which had a great effect on Islamic architecture and contributed greatly in shaping ideas and limiting statements by western architects and historians about non-western architecture including Islamic architecture.

Keywords

Orientalism • Colonial architecture • Neo-Mamluk architecture • Islamic architecture • Heliopolis city • Postcolonial theories

1 Introduction

"Orientalism" is a descriptive term usually associated with the fascination of Western scholars, historians, etc. of the East. In 1992, Edward Said introduced a new definition to the term. In his book, "Orientalism", he explained it as a different sense of understanding, deeply linked to power. He explained that this was not a discourse that emerged by coincidence but rather a tool used to empower and justify the growing Western domination in many fields while at the same time, ensuring the superiority of the Western culture over the East. Orientalism affects all aspects of culture, including architecture.

This paper will address how architectural orientalism emerged and how it affected Islamic architecture. International expositions were platforms that deployed orientalism discourses. Writings by European scholars about Eastern culture and architecture also played a great role in the emergence of orientalism. Both international expositions and European writings offered similar statements about the orient culture and its architecture that was limited by the dominant racial rhetoric theories at that time. In general, these statements were abridged, reinterpreted, and taken away from its original context.

In this paper, we will explain the different phases of orientalism. First, the eighteenth-century orientalism with the little data it offered about Islamic architecture. Then, we will arrive at the nineteenth-century orientalism with the significant production of "Description de l'Egypte"—(Description of Egypt). Finally, we will reach to the new

L. Abd-Elaziz
Faculty of Engineering, Department of Architecture, German University in Cairo, Cairo, Egypt

K. M. Dewidar · M. R. Ismail
Faculty of Engineering, Department of Architecture, Ainshams University, Cairo, Egypt
e-mail: khaled.dewidar@bue.edu.eg

M. R. Ismail
e-mail: mostafa_ismail@eng.asu.edu.eg

F. Rosso et al. (eds.), *Advances in Architecture, Engineering and Technology*, Advances in Science, Technology & Innovation, https://doi.org/10.1007/978-3-030-86913-7_9

orientalism phase, in which an analytic approach toward Islamic architecture was introduced.

Orientalism greatly shaped European views about Islamic Architecture. It is becoming increasingly clear that architecture and orientalism are strongly related. Moreover, Edward Said's views and Post-Colonial theories are valid and should therefore be tested more through architecture.

2 Orientalism

"Orientalism" as a term is usually associated with the study and fascination of the western scholars and historians with the East that emerged with the enlightenment; it was a merely descriptive term. Until Edward Said in his book, "orientalism" introduced a new definition to it. This book was the foundation for post-colonial theories in humanities and was what created a significant transformation of ideas not only in his field: literary but also among other fields like architecture and visual arts. (Ackan, 2014; Prakash, 1995; Said, 1978).

In Said's work and statements "orientalism" implies a different sense of understanding that is deeply linked to power and spread across different activities. Based on that Orientalism is then a form of cultural paternalism, whereby abundant variety of societies and histories were abstracted, reduced, and restructured into a totalizing group of repetitious patterns which is later regarded as the "complementary opposite" to the orientalist's culture (Crinson, 1996; Said, 1978; Prakash, 1995).

In his book he explained how was the orient artistically and scholarly represented by the west and how these representations "Orientalized the orient" by describing it as the fanciful, exotic, barbarian and irrational "other" of the civilized and much progressive west; creating a polarized view and an imaginary border between the east and the west (Ackan, 2014; Burney, 2012; Massad, 2015). He also argued that "orientalism" was related to hegemonic dominance and that it wasn't a discourse of coincidence conducted by the means of modernity invisible apparatus and in that context we can find the most visible action for the imperialism of the west on the east and the great role "orientalism" had in empowering and justifying growing western domination. (Crinson, 1996; Prakash, 1995; Massad, 2015).

In addition to that, Edward Said divided orientalism into two main phases. The early orientalism phase in which the orient in the western literary was depicted as inferior and exotic, while the other phase was about contemporary orientalism. As previously mentioned these western depictions were what constructed the "orient" as the "other" while creating stereotypical images about it (Said, 1978). Resulting in; labeling the "occident cultures" as the rational, and the "orient cultures" as the emotional (Weidner, 2013).

Said's work has caused deep controversy and encouraged a variety of writings across different fields (Prakash, 1995). It started with the field of literature, human geography, and cultural studies until it reached the field of history of art. However, his debate and ideas barely found its way to the built environment and the role played by architectural orientalism in the larger cultural archive (Crinson, 1996).

3 International Expos and Orientalism

International expositions along with writings of European historians and scholars about eastern culture and architecture played a great role in the emergence and deploying of orientalism. In the book "colonizing Egypt", Timothy Mitchell argued that the displaying of the world as an exhibition at international expos was meant to promote a particular ordering principle that presented capitalist relations in the form of obvious relations of commodities (Crinson, 1996).

International expositions were recreational and educational places where Knowledge about Islamic culture was abstracted and displayed in a theatrical way. These expos offered a different way of consuming the Islamic culture other than the disciplinary one found in the professional and academic field. In many museums and international expos, Islamic architecture was presented through imitations and architectural pastiches. Especially in the British exhibitions where mock-up Islamic environments were created where Islamic products were displayed (Braga, 2017; Giese & Braga, 2018; Crinson, 1996).

The first international exhibition "the great exhibition" also known as the "crystal palace" (Fig. 1) was held in London in 1851. Later in 1854 it was transferred to Sydenham hill followed it the French exhibition "exposition universelle" that took place in Paris in 1855, which was intended to surpass the British great exhibition. During this time, orientalists like Owen Jones, Austen Layard, and James Fergusson were hired to design the pavilions and displays of these exhibitions. The first time nations were invited to design their own pavilions was in 1867 at the French exhibition, in which they built houses that were typical of their national architectural styles. Following that in 1878, these national houses were developed to become separate pavilions. Then at the "London colonial and Indian exhibition" held in 1886, artisans from India were brought to be seen by visitors whilst working on their crafts. Following that was the 1889 exhibition in Paris where Egyptian artisans were brought to construct buildings under French supervision (Crinson, 1996).

As previously mentioned these exhibitions were not for the mere purpose of entertaining on the contrary, the arrangement in which the abundant variety of objects was displayed showed overtly the difference between them. In

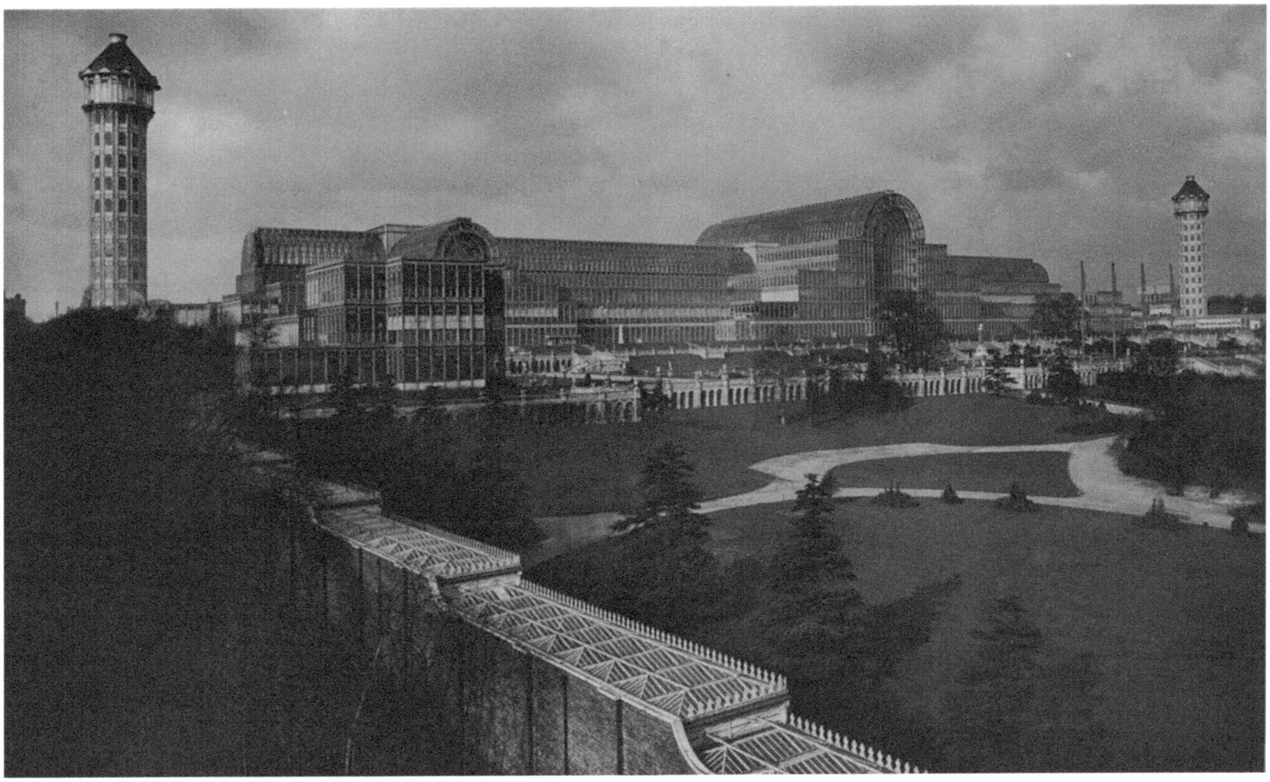

Fig. 1 Crystal palace, London. *Source* (Rotary Photographic Co., 1907–1914)

these exhibitions, there was no cover up of the rule of imperialism on the contrary; the obvious order of displaying made it more a matter of fact. Cultures products and their architecture were displayed as mere objects stripped of their background and the conditions under which they were created, becoming a coded possession of orientalism, subjecting both the cultural products and the people—visitors of the exhibition- to the mindset of consumption. (Crinson, 1996).

4 Orientalists and Islamic Architecture

European scholars' and architects' writings about the orient culture weren't much different from the way they were displayed at the international exhibitions they also were distant and limited by the racial discourses that ensured superiority of the west on the East (Ackan, 2014). Their writings about architectural history were always divided into two entities either East/West division, North/South division or Christian/Non-Christian division (Kilic, 2013).

There were also two generations of European scholars with different approaches toward Eastern Architecture particularly Islamic Architecture. A generation wrote about Eastern Architecture from a distance while following a descriptive approach. While the other generation visited some of the buildings, they wrote about and had an analytic approach toward its architecture. The former Generation included Edward Freeman, James Fergusson, and John Ruskin and the latter generation included Edward lane and Owen Jones (Crinson, 1996; Kilic, 2013).

For instance, John Ruskin in his book "the stones of Venice" showed great admiration for the hybridity of different cultures and styles he found in the Venetian arches including the influence of Eastern-Islamic architecture and culture. He later denounced this influence to follow the rhetoric racial discourses popular at that time; what he once considered an enriching influence has now become "the lower kind ornamentation" produced by distant, alien cruel persons (Ackan, 2014; Ruskin, 1925).

While in Owen Jones book "the grammar of ornament", he collected examples of ornament from different parts of the world and claimed that he found the universal law of ornament (Sloboda, 2008). In his book, he described the Islamic ornamentation as "rational, geometrical ordering of flat surfaces" and that the use of color was scientific. Moreover, he assimilates the Islamic ornamentation within his own frames of reference whilst using the "orient" as a base to prove the universality of his "law of ornament" claim (Ackan, 2014).

It might seem that Jones thoughts on Islamic culture are an exception during his time. Despite that, it was not an exempt of the picturesque elements like offering a rural appearance to the scene (Fig. 2), this exotic effect that nurtures the imagination of the Europeans encouraging them visit the near east. Owen's work challenged the common concept that the west culture was the only rational one. However, it was a tool that led to the development of the commodification of Islamic goods (Crinson, 1996; Lanford, 2001). Even though that was not Jones intention from abstracting Islamic patterns and reducing it to line and color (Fig. 3), his work made it suitable and easier to the modern means of mechanical production (Braga, 2016; Flores, 1996).

Both Ruskin's and jones writings are an example of the common approaches usually taken by western scholars toward non-western cultures that ensured presenting the (orient) as an entity separated from the self (west) sometimes as an alternative and other times as undeveloped, stagnant culture (Burney, 2012; Crinson, 1996) Moreover, the emulated Islamic goods and buildings displayed in the exhibition were just like most European scholars' writings; both of them were from a distance. These emulated displays were abridged, reinterpreted, and edited objects that were distanced from their reality and even in some cases; snatched from their original settings; viewed as elements of pedagogy, entertainment, or exchange; manifestation of subjugation or association (Crinson, 1996; Ackan, 2014). To the British public, these objects were representatives of a mediated Islamic world (Crinson, 1996).

Sir Banister Fletcher "tree of Architecture" diagram (Fig. 4) is an another example of this bilateral division in the

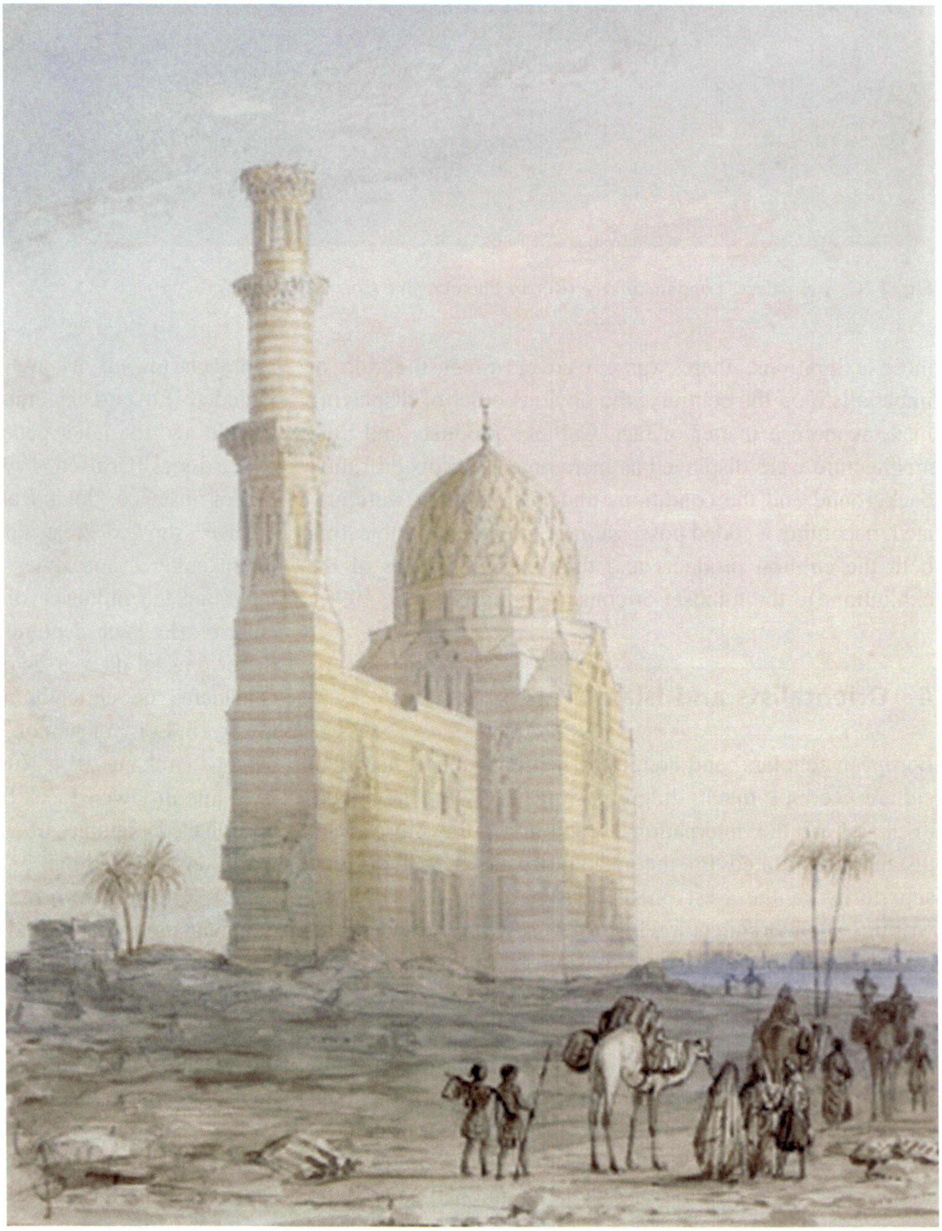

Fig. 2 "Tomb near Cairo" drawing illustration by Owen Jones. *Source* Braga (2016)

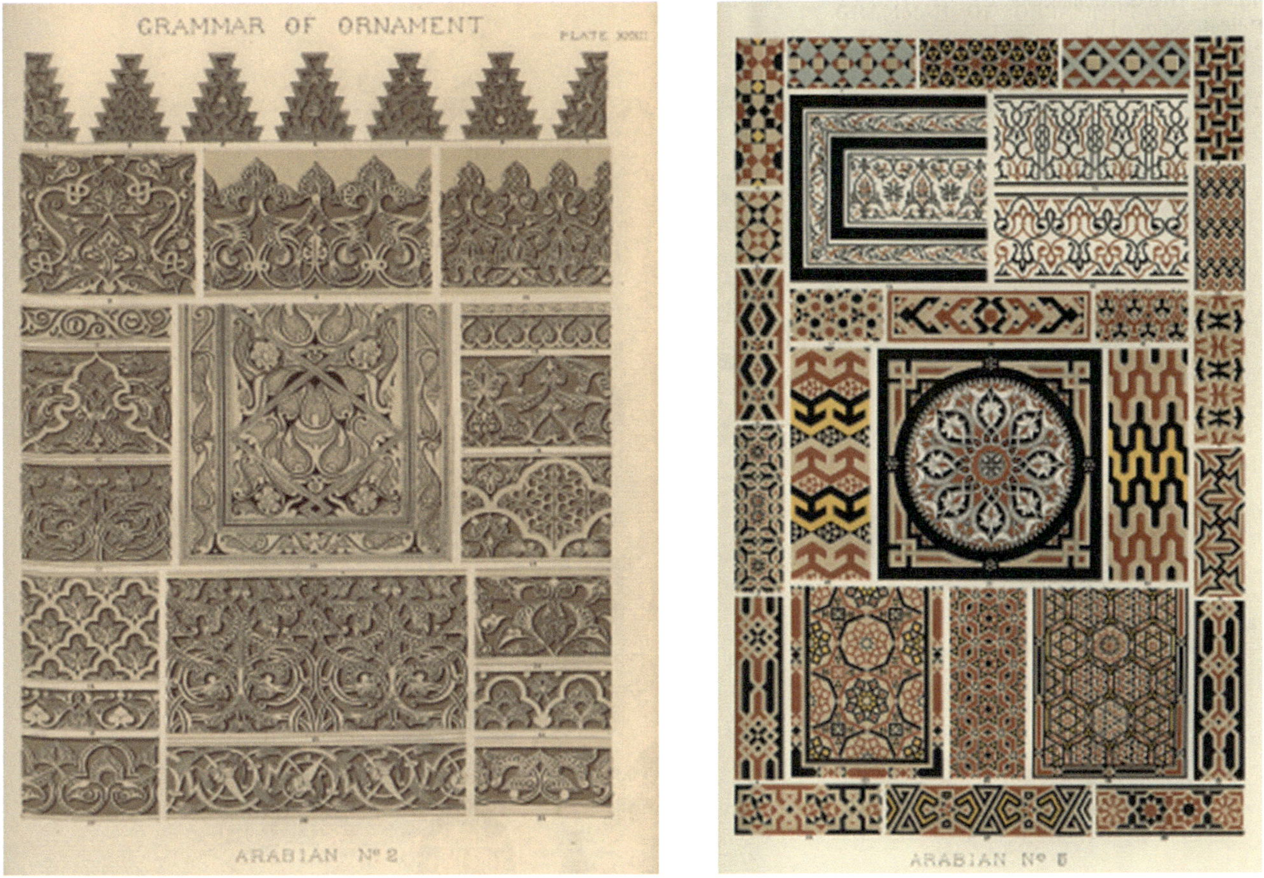

Fig. 3 Arabian ornament illustration, Owen Jones's book "Grammar of the ornament". *Source* Braga (2016), Jones, (1868)

representation of Architectural styles. In the diagram, he visualized the architectural styles and their evolution as a tree, where the main trunk represented the Greek, Roman, and Romanesque architectures. These styles evolve into the branches which in turn represented the American and European architectures, signifying the continuous succession and progress in these styles. While on the other hand, Eastern architectures are represented by side branches that neither evolve nor develop any further into any other architectural style. In this diagram, Eastern architectures including Islamic architecture are represented as styles that lack progress and change. Moreover, these features of progression and developing belong only to the Western architectures. This form of Orientalists knowledge that denies any progress or history to the Orient was one of Said's basic objections against Orientalism discourses (Ackan, 2014).

5 Phases of Orientalism

Orientalism and the way Islamic architecture was viewed, went through different phases (Fig. 5) starting with the "before eighteenth-century orientalism" phase which offered very little about Islamic architecture and passing by the "after the nineteenthcentury orientalism" period which one of its main and greatest product was the "Description de l'Egypte". Most of the illustrations included in it took a descriptive approach more than an analytic approach. It was until the 1820 that the "new orientalism" emerged and a new analytic approach toward Islamic architecture was taken.

5.1 Eighteenth-Century Orientalism

Before the nineteenth century, most travelers went to the Near East for trade and diplomacy purposes. Few went there for solely architectural purposes. Despite their interest in Islamic architecture, they mentioned it in their writing as a background to other aspects such as local customs. That is why their writings lacked important details, pictures, and drawings (Crinson, 1996).

Those who traveled there, whether, for diplomatic, commercial, or cultural purposes, their main reason was the desire to comprehend the geography and civilizations of the Bible while exploring Egypt for the traces of European origins that they thought it still held. Among that, aspects of Islamic culture might be mentioned for its exotica or as an

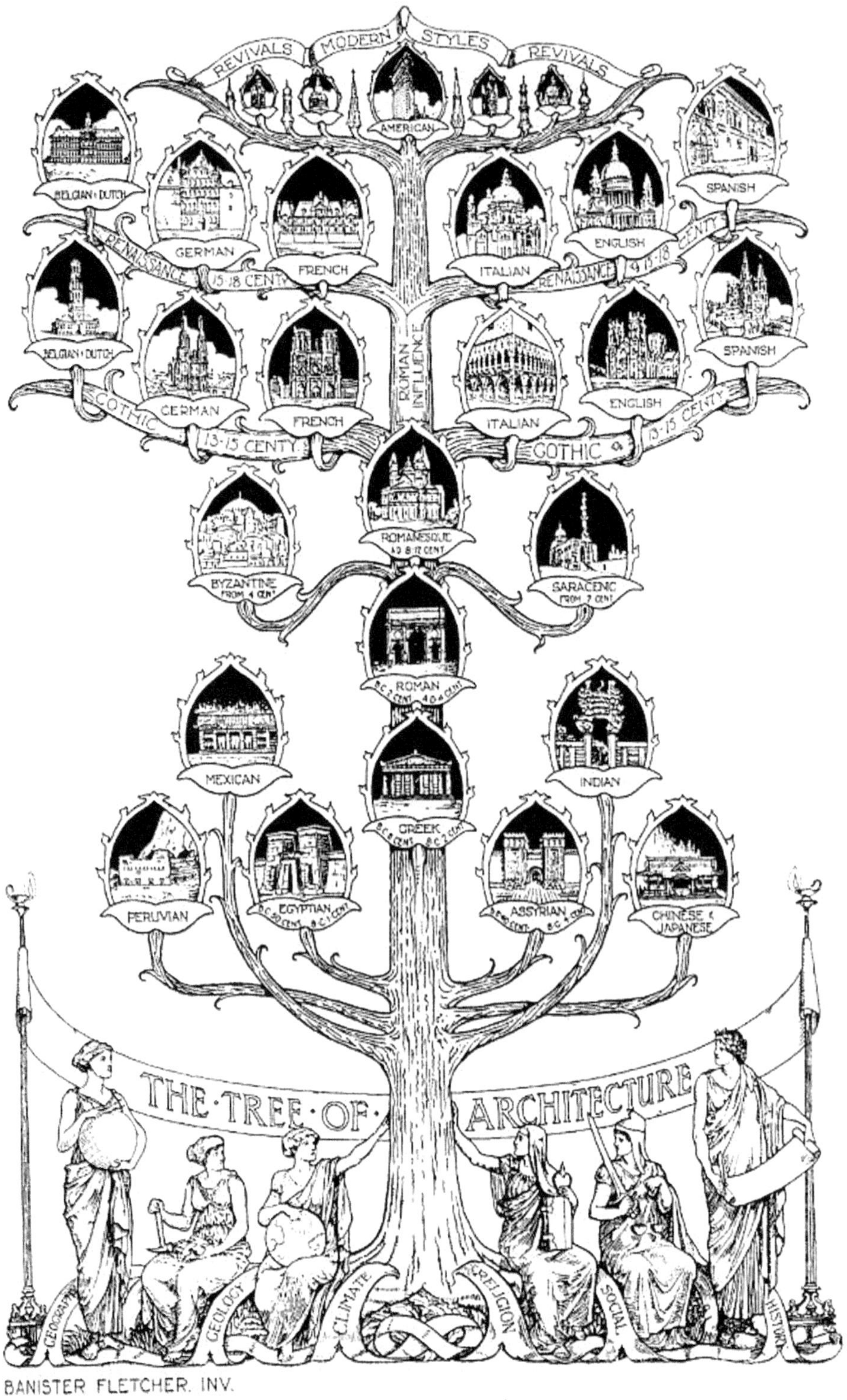

Fig. 4 "Tree of Architecture" diagram. *Source* Fletcher &and Cruickshank (1996)

undesirable overlay, which was commonly described as stagnant, undeveloped, and decaying (Crinson, 1996).

The eighteenth-century reports and writings about Islamic architecture were rare and if found was brief. In general, Islamic culture was deemed as a low matter and during that time, there was a lack of proper intellectual tools to analyze and assimilate it. That even in some cases writers or travelers used equivalent terms borrowed from classical architecture cannons and familiar terminologies to describe Islamic architecture buildings (Crinson, 1996).

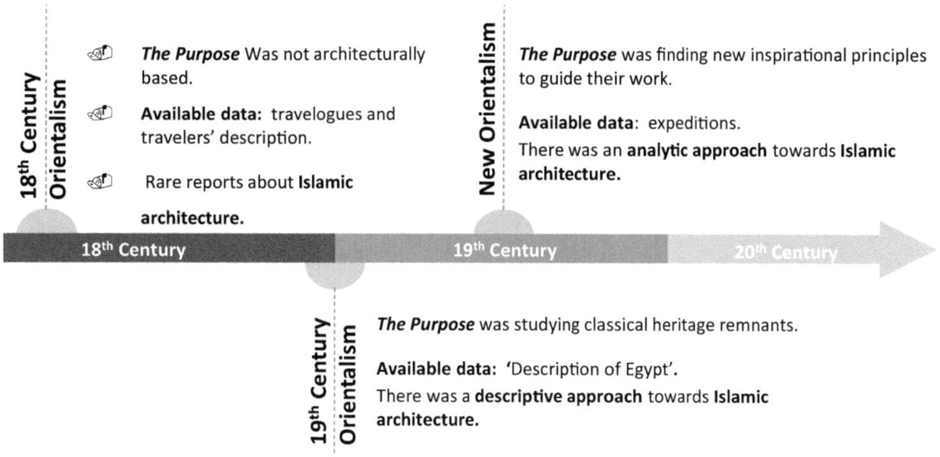

Fig. 5 Diagram showing phases of orientalism. *Source* Author

5.2 Nineteenth-Century Orientalism

Before the mid-nineteenth century, the majority of the British architects who visited the near east; shared almost the same purpose; studying classical heritage remnants. Despite them visiting Islamic lands, some of them regarded Islamic architecture as a non-classical one. While for others Islamic architecture simply did not exist. In both cases, it led to the same result (Crinson, 1996).

There was not much in the way of Islamic architecture historical and theoretical analysis up to the second quarter of the nineteenth century when Egypt was conquered by a modern European power that an expansive and resourceful account was made, since Egypt was one of the strategic cities for European interest and influence rivalries; which made both knowledge and power inseparable matters, in understanding Islamic architecture (Crinson, 1996).

In 1798 was Napoleon's invasion of Egypt and its significant production "le Description de l'Egypte"; which was a fourteen section dedicated to illustrations describing Egypt out of twenty–three massive volumes. Which Edward later described as "the orient as a body of knowledge in the west was modernized". The Islamic architecture illustrations in the "Description" were driven by the necessity of drawing maps (Mitchel, 1988). These illustrations were greatly descriptive than being analytic. While the ancient Egyptian structures were regarded as the ideal setting for the imperial dreams, Arab buildings were considered part of the "modern state" and there was only one article where Cairo's architecture was mentioned (Crinson, 1996).

"The Description" was an amateur attempt by napoleon's scholars to create an archive consisting of things Egyptian and Islamic. It was the first time that such many illustrations about Islamic architecture were assembled with great regard to their accuracy. In general, the information offered by the "Description" when it comes to an acknowledged interpretation of culture particularly architecture; is considered little. Besides, it confined the study of Islam to three images: ancient, medieval, and modern Egypt. Yet the modern Egyptians way of life was regarded as medieval and for the purpose of their modernization, practical information about their country needed to be gathered (Crinson, 1996).

5.3 New Orientalism

The difference between the nineteenth-century orientalism and the new orientalism could be clearly observed in the difference between the drawings illustrations produced in these two periods (Fig. 6). An example of new orientalism would be Lane's drawing of the Sultan Hassan mosque. Despite being presented in a poor and unskillful way, it offered a different kind of resource and reproducibility. He used a perspective that included most of the essential elements and features of the mosque such as the minaret, mihrab, minbar, dikkah, and courtyard. Also, it included two praying positions and text that described the orientation, materials, gendered arrangement of space, mosque opening times, and its decorum; which at the end makes the form of the mosque a natural and a logical result of the functions taking place inside of its space and its cultural environment. While on the other side when we look at the drawing illustration of Sultan Hassan mosque in the "Description de l'Egypte", we find that, the orientation compass was non-existent and a priority was given to the geometrical placement across the page. From such illustrations, one can deduct how to move through spaces or how to reconstruct and reproduce and not why it was formed this way; leading to stripping it off its meaning and presenting the mosque features as functionless decorative elements don't result from the cultural environment nor the function of the space (Crinson, 1996).

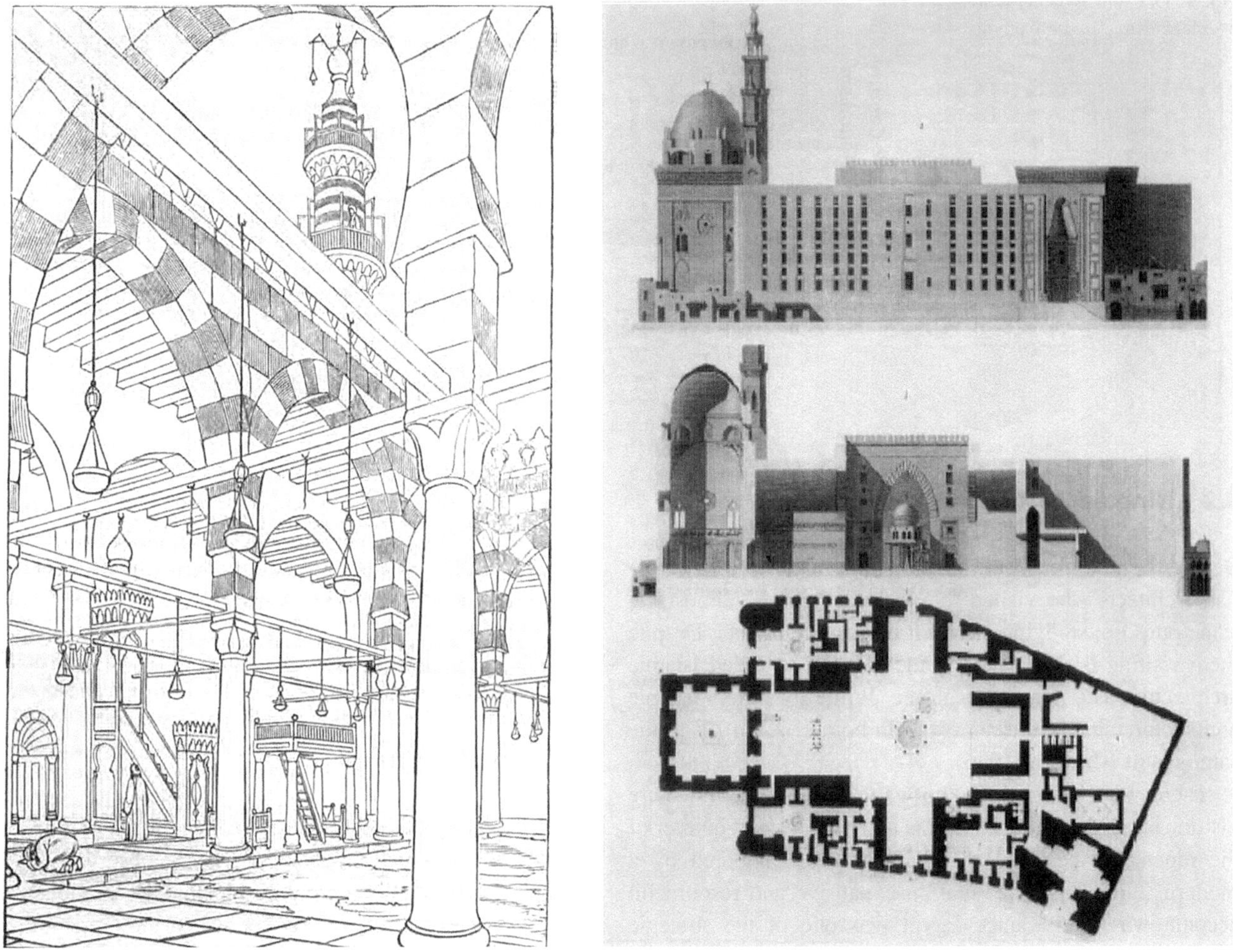

Fig. 6 Sultan Hassan mosque illustrations' comparison. *Source* Crinson (1996)

6 Orientalism and Architecture

For the Europeans to penetrate and control the Islamic culture it was fundamental to study Islamic architecture since it was a powerful sign of Islam and an embodiment of the ways of religion and life that was still going on into the present. Although Europeans reinterpreted, described, and even measured Islamic architecture, all of that was done in accordance with the idea that Islamic culture was powerless, stagnant, and voiceless. At first sight, Owen Jones's rational views about Islamic architecture might seem an exception to the predominant concept that the west was the only one known for its rationalism. However, his analysis for Islamic architecture was an important contributing tool toward further Islamic goods commodification. It should be noted that, although Ruskin and Jones had different views about Islamic architecture; these views were bounded by the unquestioned parameter that limited all statements about Islam. Which as Edward said argued before, the orient was considered an entity, in which its culture was confined by its religion, and race was a determining factor in the cultural work; while development and expansion were the exclusive possessions of the west; but even their duty (Said, 1978).

Architectural orientalism appeared as a cultural entity by the end of the second quarter of the nineteenth century. It emerged as a result to rejecting the classical paradigms and picturesque theory; however, the fascination with the new human sciences in particular ethnography and their association with modernization and rationalism had the greatest role and influence on the process of orientalism formation. This influence by the human sciences had its effect on the way other cultures and cultural products were judged, whether it was western, eastern, or historical. In general, Westerns scholars judged the cultural products in the same way they judged their creators. These judgements were politically and racially related. Moreover, when it came to Victorians judging Islamic architecture it was not an exception. Especially in the mid-nineteenth century judging Islamic architecture was often linked to political or racial

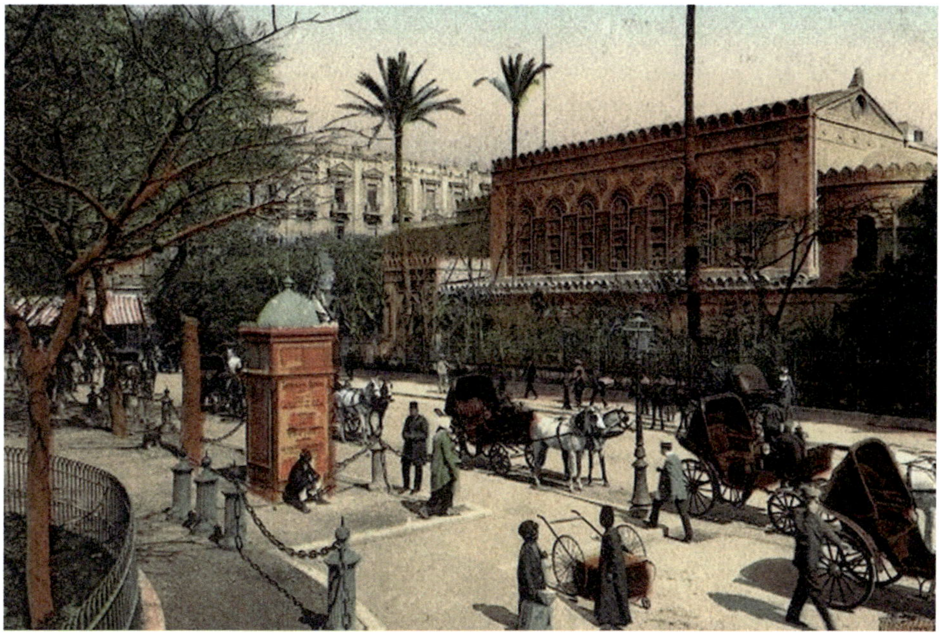

Fig. 7 St. Marks church. *Source* Humphreys (2019)

basis (Crinson, 1996). These writings influenced the way European Architects perceived and understood Islamic Architecture. Consequently, this affected their Architectural products that held Islamic features.

Therefore, architecture and orientalism arose as a pairing; architecture as a new profession emerged out of various range of training practices and buildings, orientalism as a formation built up discursively with its own specialists and its ability to spread among various activities and interests. Leading from this point, specific architectural concerns could be expressed clearly about discourses regarding oriental cultures aroused by travelers, artists, diplomats, and missionaries (Crinson, 1996). In Said's book "orientalism" he assumed that culture plays an important role in imperialism and that culture is more than a mere companion of imperialism, which could be tested through architecture particularly if imperialism could be confined to a more specific context of informal or economic imperialism (Said, 1978).

In Egypt, the eclectic architectural style of both St. Marks Church and Heliopolis city buildings is an example of how Orientalism influenced European architects' outcome of Islamic architecture. In general, the design of St. Mark's church (Figs. 7 and 8) is a hybrid between Revival gothic style and different Islamic styles. While the church plan was a typical early Christian church plan, the façade was greatly eclectic. In the design of the facade, James Wild borrowed architectural elements and Islamic Ornaments from Moorish

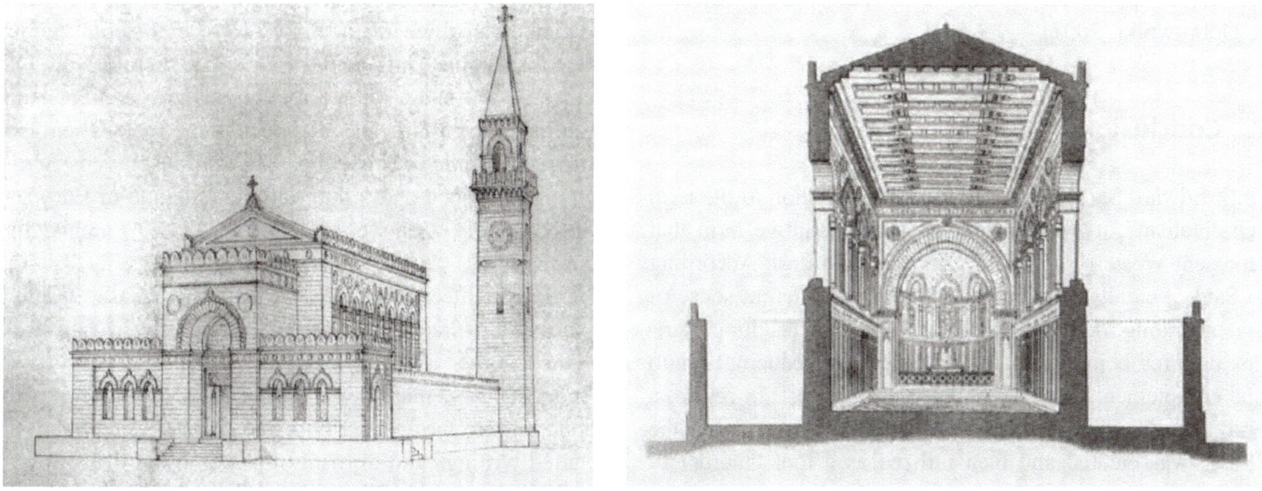

Fig. 8 St. Marks church perspective drawn by James Wild. *Source* Crinson (1996)

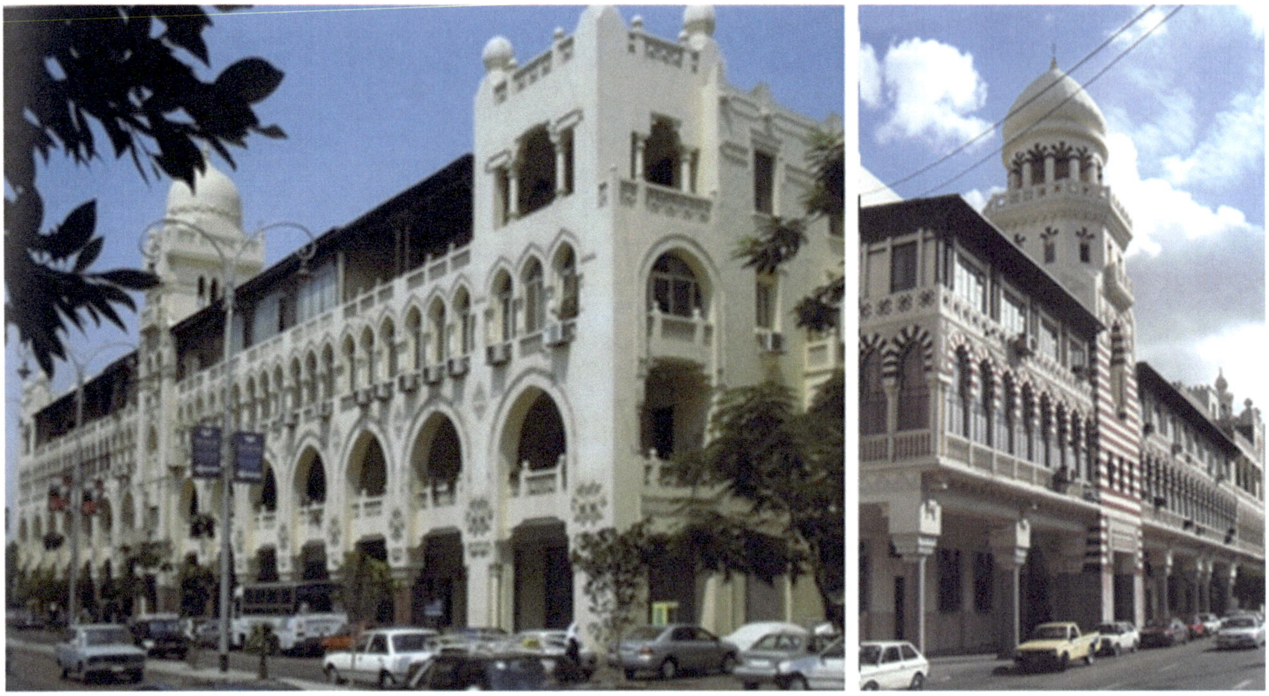

Fig. 9 Heliopolis city buildings. *Source* Dobrowolka and Dobrowolki (2007)

style, Ibn-Tulun mosque, and other buildings he was analyzing at Islamic Cairo (Crinson, 1996). On the other side Heliopolis city buildings (Fig. 9) were a hybrid between classical style and different Islamic styles, but mostly Moorish and Mamluk style. The plans of Heliopolis city buildings were European with Neo-Islamic facades. Even though most buildings were residential, minarets were utilized at the buildings' corners. The minarets were functionless decorative elements to grant the city an Oriental guise. Although in both cases the facades included elements from different Islamic styles, Heliopolis city buildings included more architectural elements from local Islamic architecture particularly Mamluk architecture (Dobrowolka & Dobrowolki, 2007).

7 Conclusion

Until Edward Said introduced the new definition to the term "Orientalism", it was regarded as the descriptive term that represent Western scholars fascination with East. According to Said, it entails a style of thought and a different sense of understanding, deeply linked to power. This results in various cultures being totalized in addition to reducing it into one complementary entity opposite to the west. In return, a body of knowledge influenced by Ethnology about the "East" was created and then utilized as a tool in order to empower and justify the growing western domination.

Colonial international exhibitions and European scholars' writings about Eastern Architecture Particularly Islamic architecture were the means by which Orientalism discourse disseminated. Orientalism shaped statements by western scholars about Eastern architecture especially Islamic architecture, which resulted in similarly limited views.

Orientalism greatly influenced the History of Islamic architecture narratives mentioned in European writings. Moreover, these European narratives had a bilateral division. This division varied between North/South, East/West, historic/ Non-Historic, and Christian/Non-Christian. These bilateral divisions were meant to ensure Western architecture superiority on its complementary other: Eastern architecture.

Orientalism three main phases were eighteenth century Orientalism, nineteenth century Orientalism, and New Orientalism. Each phase offered a different form of knowledge regarding the culture and architecture of the "Orient", in particular Islamic architecture. The various givens across each phase shaped and influenced the outcome of European architects of non-western or Islamic architecture accordingly.

In the eighteenth century Orientalism phase, writings about architecture were either rare or mentioned as a background to other aspects. The second phase of Orientalism —"nineteenth century Orientalism"—highlighted by the production of the "Description of Egypt", in which the included writings and drawing illustrations in it followed a descriptive approach. While the "New Orientalism" phase started by late nineteenth and early twentieth century and distinguished by a group of British scholars who decided to follow an analytic approach in studying Islamic architecture.

Orientalism and architecture jointly emerged and there were a strong correlation between them. Orientalism was a main factor in shaping the European architects' outcome of Islamic architecture. Since at each phase of Orientalism, Islamic architecture was analyzed and viewed using a different approach, European architects and scholars' understanding and interpretation of Islamic architecture varied based on the givens of each phase.

References

Ackan, E. (2014). Postcolonial Theories in Architecture. *A Critical History of Contemporary ArchiteCture, 1960–2010*, 119–140.

Braga, V. (2016). Owen Jones and the Oriental Perspective. *The Myth of the Orient. Architecture and Ornament in the Age of Orientalism, edited by Francine Giese and Ariane Varela Braga*, 149–164.

Braga, V. (2017). Owen Jones's Alhambra Court in the Crystal Palace of Sydenham. *A Fashionable Style. Carl von Diebitshc und das maurische revival. Edited by Francine Giese and Ariane Varela Braga*, 71–83.

Burney, S. (2012). Orientalism: The Making of the Other. *PEDAGOGY of the Other: Edward Said, PostcolonialTheory, and Strategies for Critique*, 23–39. Retrieved from https://www.jstor.org/stable/42981698

Crinson, M. (1996). *Empire building: Orientalism and Victorian architecture*. Routledge.

Dobrowolka, A., & Dobrowolki , J. (2007). *Heliopolis rebirth of the city of the sun.* Cairo: the American university in Cairo press.

Fletcher, B., & Cruickshank, D. (1996). *Sir Banister Fletcher's a history of Architecture* (20th ed.). Architectural press.

Flores, C. (1996). *OWEN JONES, ARCHITECT.* Georgia: Georgia Institute of Technology.

Giese, F., & Braga, A. (2018). The Alhambra in a global perspective. *The power of symbols*, 293–308.

Humphreys, A. (2019, FEBRUARY 7). *Church on the Square.* Retrieved from Egypt in the Golden Age of Travel: http://grandhotelsegypt.com/?tag=st-marks-church-alexandria

Jones, O. (1868). *Grammar of ornament*. Bernard Quaritch.

Kilic, S. (2013). *Byzantium between "East" and "west": perceptions and architectural historiography of the byzantine heritage.* Retrieved from http://etd.lib.metu.edu.tr/upload/12616697/index.pdf.

Lanford, C. (2001). Imperialism and the Parlor: Owen Jones's "The Grammar of Ornament". *ROMANTICISM AND INTERDISCIPLINARITY:"CENTERS AND PERIPHERIES"*, 38–43.

Massad, J. (2015). Orientalism as Occidentalism. *History of the Present*, 83–94. Retrieved from https://www.jstor.org/stable/https://doi.org/10.5406/historypresent.5.1.0083

Mitchel, T. (1988). *colonizing Egypt*. Cambridge University Press.

Prakash, P. (1995). Orientalism Now. *History and Theory, 34*(3), 199–212. Retrieved from https://www.jstor.org/stable/2505621

Rotary Photographic Co. (1907–1914). *10486–1. Crystal Palace, London. Rotary Photo, E.C.; verso: Rotary Photographic Series.* Retrieved September 10, 2019, from ARTSTOR: https://library.artstor.org/asset/SS35428_35428_22635670

Ruskin, J. (1925). *selections from the stones of Venice*. Macmillan.

Said, E. (1978). *Orientalism.* Routledge&Kegan Paul Ltd.

Sloboda, S. (2008). "The Grammar of Ornament": Cosmopolitanism and Reform in British Design. *Journal of Design History*, 223–236.

Weidner, S. (2013, december 9). *from the emotional orient to the distortion of Islam.* Retrieved from Qantara: https://en.qantara.de/content/translations-of-classical-islamic-literature-from-the-emotional-orient-to-the-distortion-of

Enhancing the Efficiency of Natural Ventilation Systems by Bio-mimicry Approach to Achieve Sustainability in Designing Office Buildings

Mona El Basyouni

Abstract

Bio-mimicry is a new growing area of research in the field of architecture. It offers the potential of enhancing natural ventilation systems. This happens by mimicking the adaptation methods found in flora and fauna to create sustainable designs. The problem is that mechanical air-condition systems consume the highest amount of energy among the building sections. This drives the need to apply Bio-mimicry to enhance natural ventilation systems. The main goal of the report is to explore the applications of Bio-mimicry in enhancing natural ventilation systems. Even though Bio-mimicry Science can dramatize the building's ventilation system performance and energy consumption, it is not popular in Egypt. The research provides the guidelines for applying Bio-mimicry concepts in enhancing the air ventilation system of the Egyptian office buildings. This is conducted by making a comparative analysis between the presented case studies that used Bio-mimicry and deriving the recommendations needed.

Keywords

Bio-mimicry • Bio-inspired design • Adaptive architecture • Office building • Sustainability • Enhancing natural ventilation

1 Introduction

Bio-mimicry is a new growing area of research in the field of architecture. It offers the potential of enhancing natural ventilation systems. This happens by mimicking the adaptation methods found in flora and fauna to create sustainable designs.

M. E. Basyouni (✉)
Modern Academy for Engineering and Technology, Cairo, Egypt

Bio-mimicry has been applied in architecture since early times by modeling forms based on proportions from nature to achieve aesthetic perfection. However, architects believe now that Bio-mimicry has a lot more to offer other than making a building look good. And according to Tim Ireland and Simon Garnier, the ventilation mechanism used by plants and animals can be applied to architecture buildings (2018). Nowadays, the problem is that mechanical air-condition systems consume the highest amount of energy among the building sections. This drives the need to apply Bio-mimicry to enhance natural ventilation systems.

The main goal of the research is to explore the applications of Bio-mimicry in enhancing natural ventilation systems. Bio-mimicry concepts will be applied to an office building in Egypt to transform it into a more humane and sustainable building.

Accordingly, the research has two parts: the first is an overview of Using Bio-mimicry to Enhance Natural Ventilation, while the second will be Analytical Case Studies. The research as well provides the guidelines for applying Bio-mimicry concepts in enhancing the air ventilation system of Egyptian office buildings. This is conducted by making a comparative analysis between the presented case studies that used Bio-mimicry and deriving the recommendations needed.

Finally, highlight the importance of examining and analyzing nature. Besides, always be up to date with all the recent developments in the Bio-mimicry field and try to benefit from them.

PART (A): Overview of Using Bio-mimicry to Enhance Natural Ventilation

1.1 Bio-mimicry and Natural Ventilation

The terms Bio-mimicry and Natural Ventilation will be defined to be able to examine the relation between them. It comes from the Greek word bio, which means life, and

F. Rosso et al. (eds.), *Advances in Architecture, Engineering and Technology*, Advances in Science, Technology & Innovation, https://doi.org/10.1007/978-3-030-86913-7_10

mimesis which means to imitate. Bio-mimicry seeks sustainable solutions to human problems by imitating nature (Benyus, 1997). It can be defined as the abstraction of good design from nature. The term Natural Ventilation refers to the process of supplying and removing air through an indoor space without using mechanical systems. Effective Natural Ventilation and Cooling systems can be inspired by the adaptation methods used by living organisms.

The application of Bio-mimicry is achieved by applying the underlying design principles of natural organisms to enhance natural ventilation systems. The main principles of the solutions produced by Bio-mimicry are listed as follows (Lasheen, 2009):

- Nature uses only the energy it needs.
- Nature fits form to function.
- Nature recycles everything.

1.2 Theories of Bio-mimicry

The approaches and levels of Bio-mimicry will be mentioned, as they will act as the design guidelines for innovating Bio-mimicry solutions.

The use of Bio-mimicry in the design process can be made by two approaches. The direct approach means defining a design problem and looking to the ways other organisms or ecosystems solve this. The indirect approach means identifying a particular characteristic or behavior of an organism and translating it into human designs (Biomimicry Guild, 2008).

Direct Approach to Bio-mimicry

The "Design Looking into Biology" involves four steps as shown in Fig. 1. An example is also added for each step for easier interpretation (Panchuk, 2006).

Indirect Approach to Bio-mimicry

The "Biology Influencing Design" involves four steps as shown in Fig. 2. An example is also added for each step for easier interpretation. (Panchuk, 2006).

Besides, both the direct and indirect approaches can be made in the form of three levels. The levels of Bio-mimicry vary according to the degree of involvement in the study of living organisms. The levels are listed in Table 1.

1.3 Techniques for Applying Bio-mimicry to Enhance Natural Ventilation

As well, Bio-mimicry has been applied in architecture since early times by modeling forms based on proportions from nature to achieve aesthetic perfection. However, architects believe now that Bio-mimicry has a lot more to offer other than making a building look good. The ventilation mechanism used by plants and animals can be applied to architecture buildings. The various mechanisms will be studied in the following subsections.

Bio-mimicry Adaptations Inspired from Plants

Plants offer numerous examples of adaptation methods to hot climates through physical characteristics, behavioral reactions or cooling processes. A brief description of these methods is summarized in Table 2.

Bio-mimicry Adaptations Inspired from Animals

Animals offer numerous examples of adaptation methods to hot climates through physical characteristics, behavioral reactions or cooling processes. A brief description of these methods is summarized in Table 3.

1.4 Conclusion and Deduced Criteria

Mimicking nature is formulated under the term "Bio-mimicry". Flora and fauna offer numerous examples of adaptation methods to hot climates. These adaptation methods can act as the basis for many architectural designs. In light of the above-mentioned points, the following criteria shown in Fig. 17 are designed to aid Bio-mimicry designers. It will be also used throughout the research to evaluate the analytical case studies and their application in Egypt.

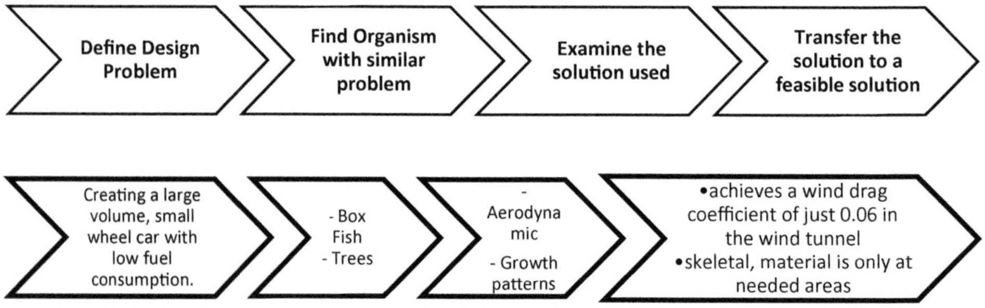

Fig. 1 Direct bio-mimicry approach. *Source* Researcher, based on data from Flghabawy (2006)

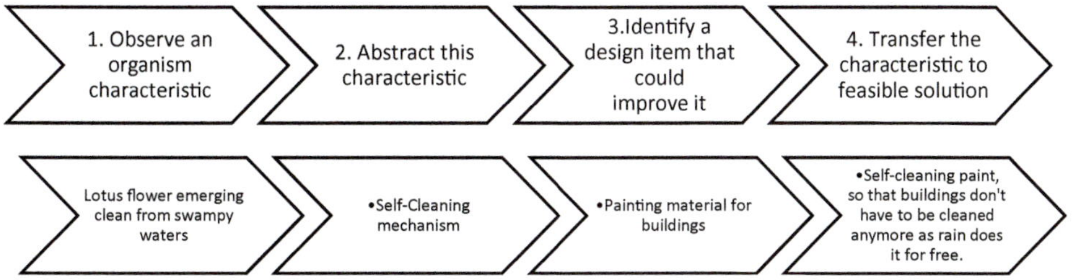

Fig. 2 Indirect bio-mimicry approach. *Source* Researcher, based on data from Sto-Sea-Brochures (2005)

Table 1 Levels of bio-mimicry

Point of Comparison	Organism Level	Behavior Level	Ecosystem Level
Definition	Mimicking the form of a certain living organism	Mimicking a process carried out by nature	Mimicking how natural ecosystems perform
Example: Organism Chosen	Beetles' ability to collect water in dry places like deserts	Termite mounds' ability to create a thermally stable environment	Proportions of the Ecosystem
Technique Used	Capturing moisture from the swift-moving fog that moves over the desert by tilting its body into the wind	Passive ventilation and temperature regulation techniques by using underground water for evaporative cooling	There is nothing called waste, and cycling of materials always occurs
Application	Fog-catcher design for the Hydrological Center for the University of Namibia	CH2 Building in Melbourne, Australia, that water is mined from the sewers beneath	La vasa Project which uses only recyclable materials, reforestation, rainwater harvesting
Illustration	See Fig. 3	See Fig. 4	See Fig. 5

Source Researcher, based on data collected from Elghawaby (2006, Berkebile and McLennan 2004)

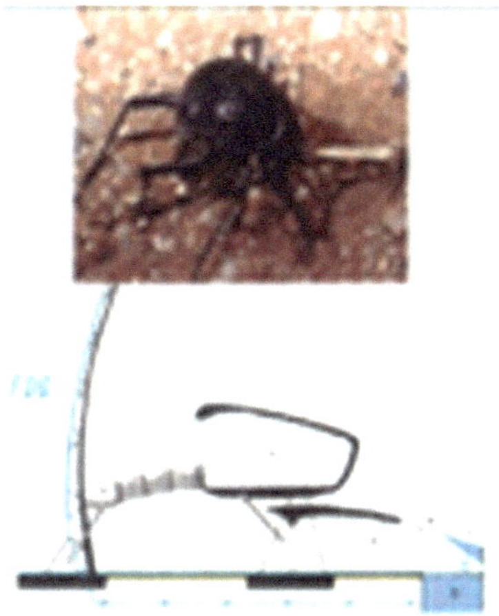

Fig. 3 Beetle-inspired fog catcher

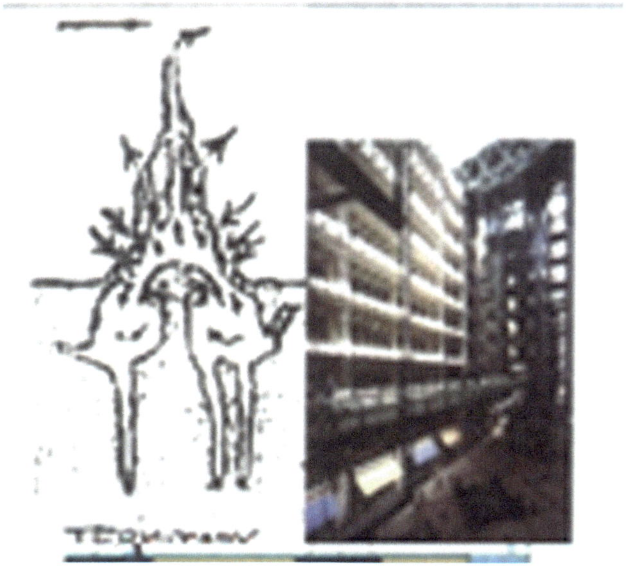

Fig. 4 Termite-inspired ventilation system

PART (B): Analytical Case Studies

The methodology of the analytical part is designed in order to deduce the preliminary guidelines, and buildings that enhanced their natural ventilation system by this science based on literature review. In this part, three international examples of using Bio-mimicry in office buildings will be studied and chosen as the first two examples are Bio-mimicry applications based on animals, while the last one is based on plants. A comparative analysis will be generated to deliver clear conclusions and recommendations serving the main research objective exploring the applications of Bio-mimicry in enhancing natural ventilation systems.

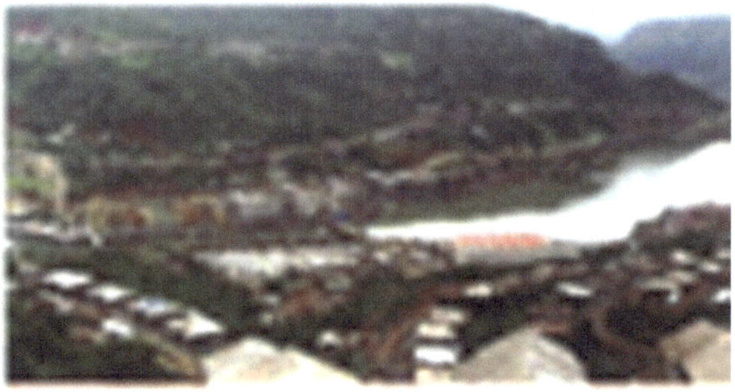

Fig. 5 Ecosystem proportions-inspired project

Table 2 Adaptation of Plants to hot weather

Organism	Technique Used	Level	Architecture Application and Adaptation Methods	Illustration
Raffle leaf	Shape-Resisting Wind	Organism level	Building Form that aids ventilation The form responds to the wind pattern by concave and convex walls. Largest flower in the world, with the shape that allows it to keep at a place if wind occurs	See Fig. 6
Plant leaf	Cooling systems	Behavior level	Ventilation System. Leaf surface having several stomata which evaporate water through it	See Fig. 7
Plant leaf	Ventilation System	Behavior level	Ventilation System Surface allows entry of air and wind and will be channeled into the building and filtered to produce clean and natural air conditioning Leaf surface having cellular openings involved in gaseous exchange	See Fig. 8
Flower	Ventilation System	Behavior level	Ventilation System Leaf-cluster systems for air filtration and ventilation. Airflow dynamics from flowers	See Fig. 9

Source Researcher, based on data collected from Elghawaby (2006), Lasheen (2009), The Biomimicry Institute (2013a, b, c, d, e, f, g)

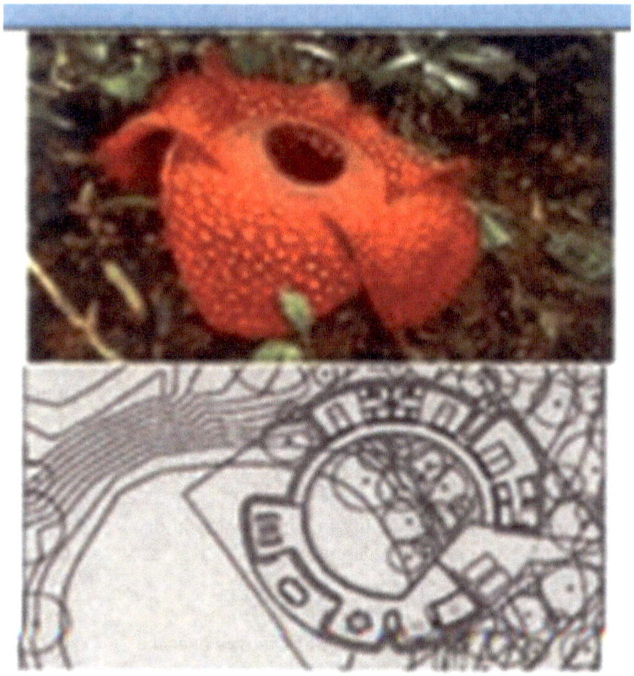

Fig. 6 Rafflesia's shape

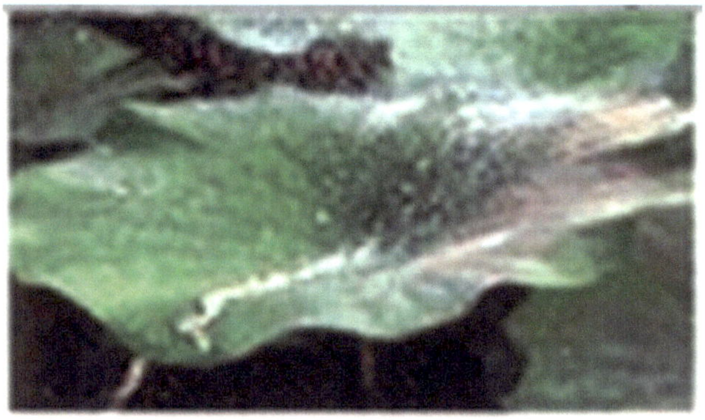

Fig. 7 Leaf shape

1.5 Eastgate Project, Zimbabwe

The Eastgate Project. A nine-story mixed-use office and retail building was completed in 1996. It was designed by architect Michael Pearce. It is located in the downtown business district of Harare, Zimbabwe, in South Africa as shown in Fig. 18.

Background information about the building's design and an evaluation of the ventilation mechanism of this building will be explained in the following sections.

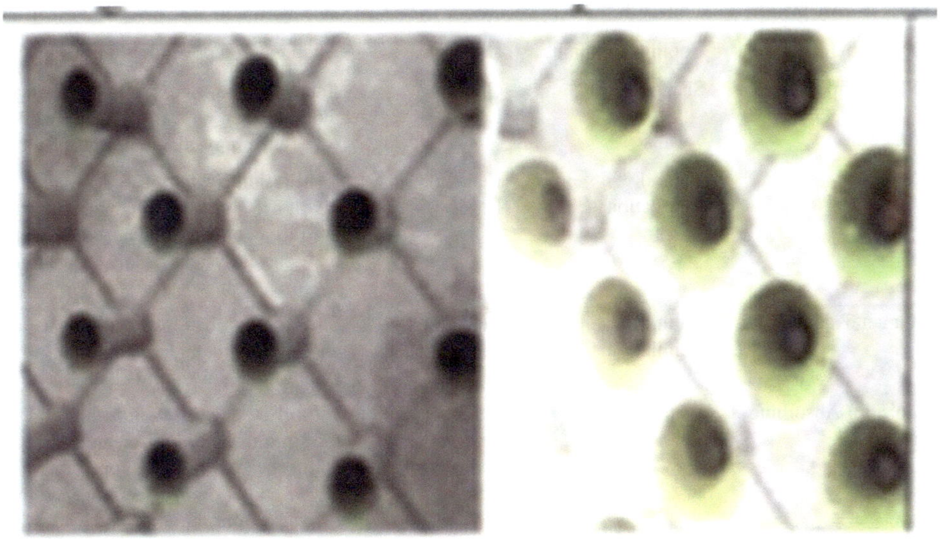

Fig. 8 Ventilation system inspired by plants

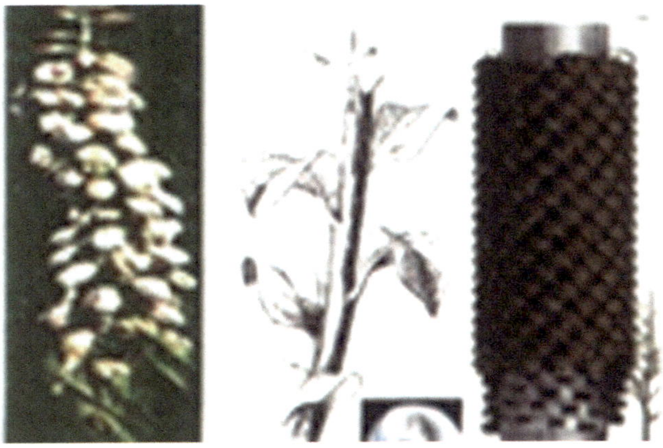

Fig. 9 Ventilation system inspired by leaves

Background information on the Building's Design

The project consists of two main complexes that surround a glass-covered atrium with retail shops on the first and second floors and offices above as shown in Fig. 19. It is largely made of concrete. The cooling and ventilating mechanism will be explained in the following section.

Evaluation of the Cooling and Ventilation System Inspired by Animals According to the Proposed Criteria

The Eastgate building is a Bio-mimicry-inspired design. The application of the criteria on the Eastgate Project is shown in Table 4.

Contribution to Sustainability

The goal of the Eastgate building in Zimbabwe was to reduce energy usage to be sustainable. In Zimbabwe, the temperature outside can vary from 3 °C up to 43 °C. The air condition plays a significant role. By using the ventilating system described before, the Eastgate building uses 90% less energy for ventilation than a conventional building its size (The Biomimicry Institute, 2009).

However, the building does not seem aesthetically appealing, therefore, newly constructed buildings that have the African termites' characteristic would be better if they were more structurally attractive.

Table 3 Adaptation of Animals to hot weather

Organism	Cooling technique	Level	Architecture application and adaptation methods	Illustration
Whale Flippers	Shape	Organism Level	**Fan** Produces a greater lift allowing it to move *25* percent more air than conventional fan while using 20 percent less energy. Bumpy-Shape design of the humpback	See Fig. 10
Termites	Natural Ventilation	Behavior Level	**Ventilation System** The ventilation system is sustainable as it will use about 10 percent of the energy that will have been used in a conventional air-conditioning system. Staying cool by a clever system of air pocket to drive natural ventilation by convection	See Fig. 11
Human Trachea	Natural Ventilation	Behavior Level	**Fan** This fan has lower turbulence and higher efficiency for cooling than conventional ones. Airflow Pattern (The logarithmic spiral shape found in such phenomena)	See Fig. 12
Moisture-Absorbing Insect	Water Retaining	Behavior Level	**Liquid Desiccant De-humidifier** Device absorbs the moisture in humid air creating drier air. Humans will feel comfortable at higher temperatures which reduce the energy for air conditioning. Secreting hydrophilic solution to absorb moisture from the air	See Fig. 13
Glass Sponge	Food Retaining	Echo-System Level	**Ventilation System** Air is ventilated by the same mechanism, as it is sucked from the base and expelled through holes at the top. A group of sponges filters nutrients from the water by sucking water from their base and expelling it through the holes at their top	See Fig. 14
Human Skin	Thermoregulation Process	Echo-System Level	**Breathing walls** Defined as allowing airflow to pass through walls made of natural materials, in addition to their capability to absorb the moisture from the air (Fathy, 1986) It has *3* layers: A . Exterior: minimizes direct sunlight and allows airflow. B. Middle: thermal insulation layer and cools airflow by evaporative cooling C. Interior layer: controls airflow Multiple layers of skin that perform interrelated tasks	See Figs. 15, 16

Source Researcher, based on data collected from Straube and Acahrya (2013), Lasheen (2009) and The Biomimicry Institute (2013a, b, c, d, e, f, g)

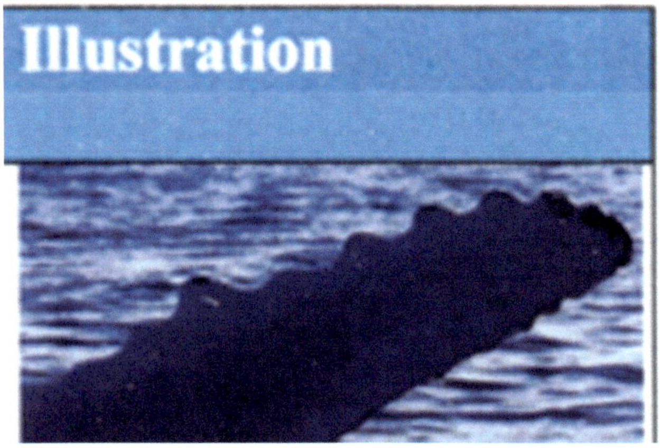

Fig. 10 Whale design

1.6 Council House 2 (CH2 Building), Australia

The CH2 Project is probably Australia's first building that is directly based on Bio-mimicry. It is designed by architect Mick Pearce. It is located in Melbourne in Australia as shown in Fig. 23.

Background Information on the Building's Design

The project consists of a public square with a retail shop. Above are 9 floors with open office spaces as shown in Fig. 24. Nature has inspired various aspects of the design:

- The epidermis or skin (which influenced the design of the facade).

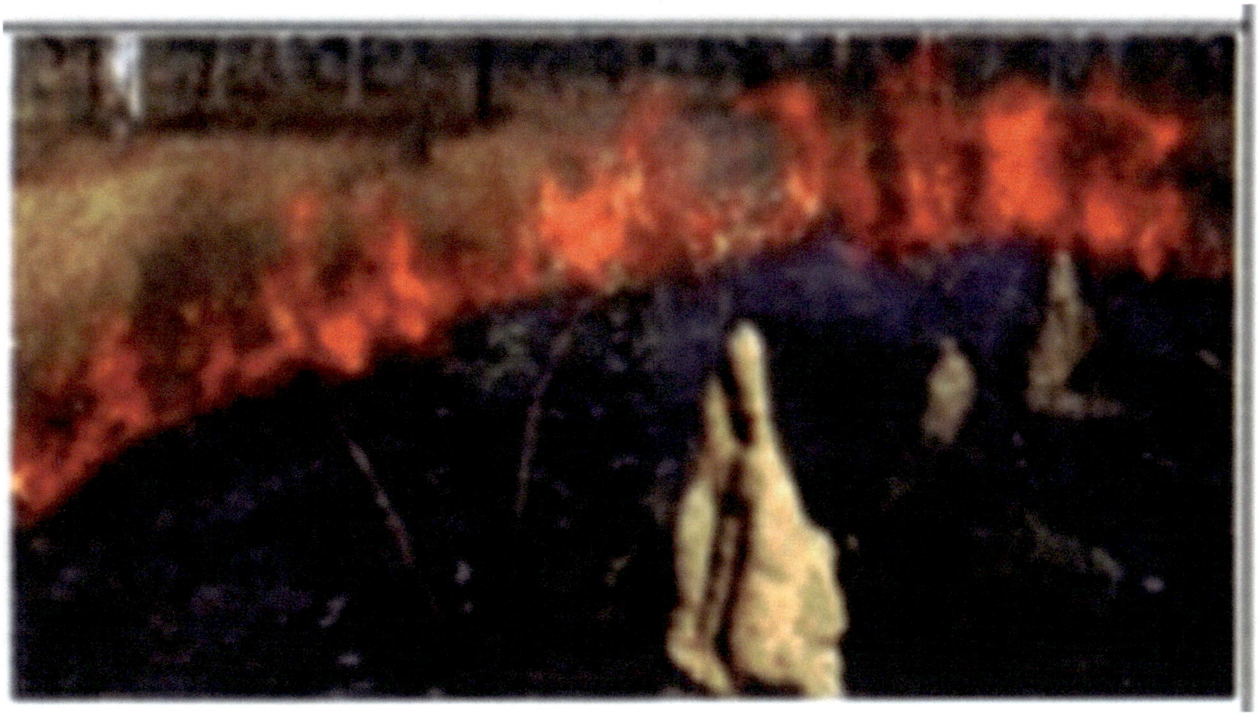

Fig. 11 Termite effect

Fig. 12 Trachea's effect

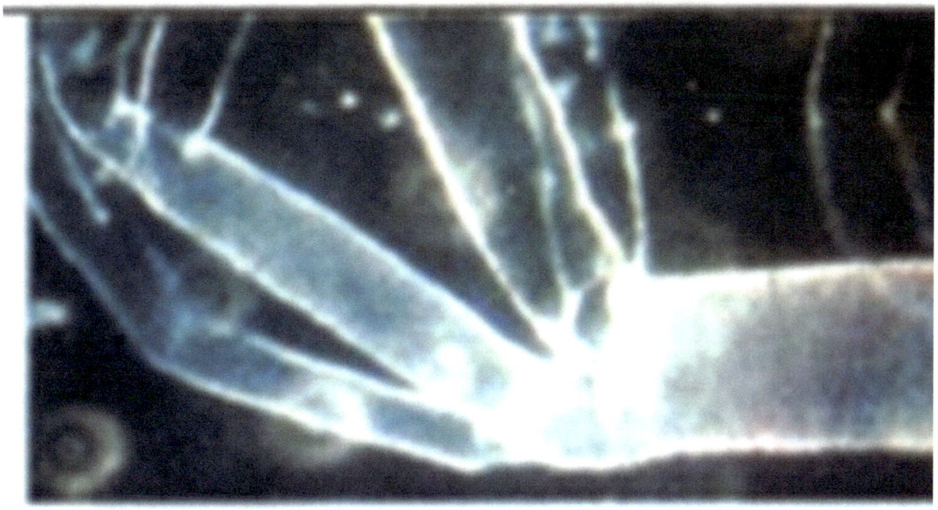

- Bark-inspired eastern facade.
- The termite mound using thermal mass and natural air movement.

Evaluation of the Cooling and Ventilation System Inspired by Animals According to the Proposed Criteria.

The CH2 building is a Bio-mimicry-inspired design. The application of the criteria on the CH2 Project is shown in Table 5.

The ventilation process can be described as follows:

- For all office's spaces:

1. Air present at the roof is treated by filtering, heating or cooling and drying (if needed) in the plant rooms.
2. Treated air at approximately 20 °C is pumped from the roof's plant room down the vertical supply duct that delivers air to the horizontal duct on each floor.

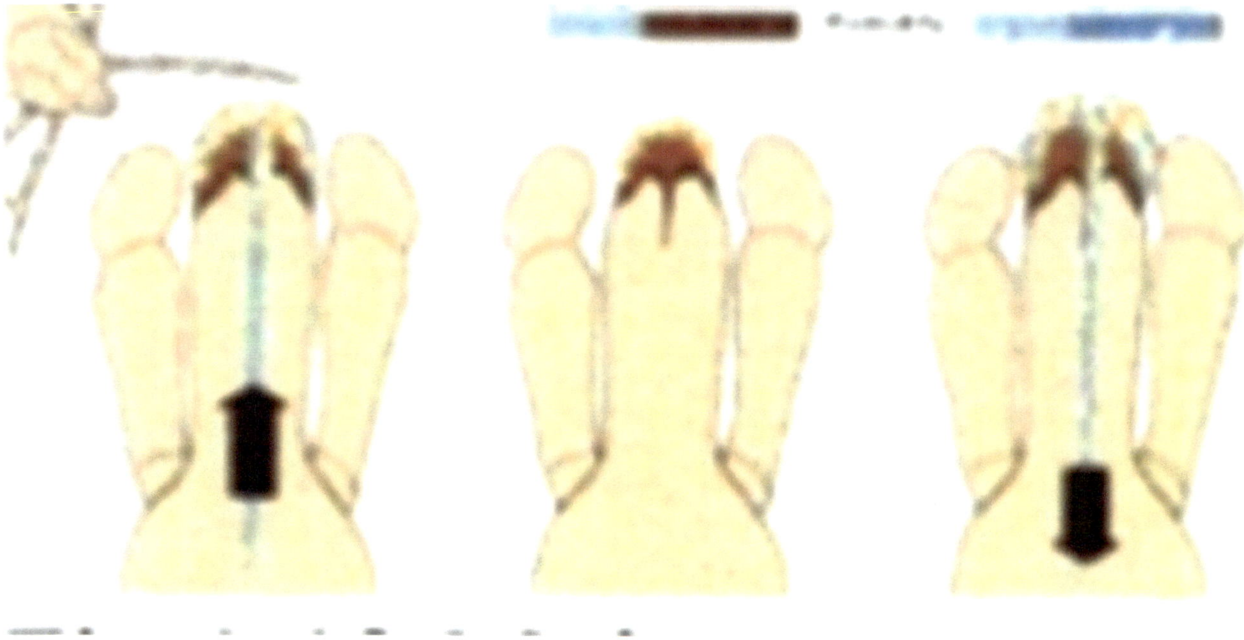

Fig. 13 Moisture absorption

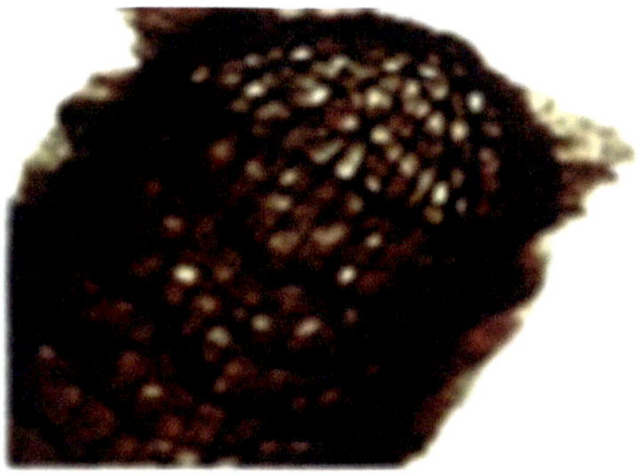

Fig. 14 Whale flipper design

3. Fresh air is released from the floor, with swirl diffusers located on movable floor tiles that allow individual placement to suit workstations.
4. By natural convection, warmth will rise and move out of the space via vents in the ceiling formed by the "belly" of the curved concrete ceiling as shown in Fig. 26.
5. The vents are connected to the exhaust-air shafts (located on the north side) by a short metal duct. Stack ventilation occurs.
6. Six wind turbines will extract air from the offices through ducts on the north façade. Air is drawn as the wind turbines induce a slightly negative pressure.

Fig. 15 Skin layers

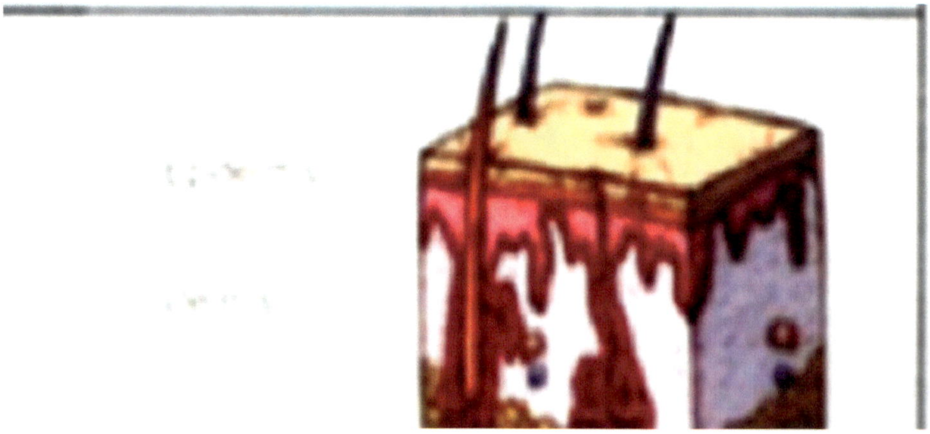

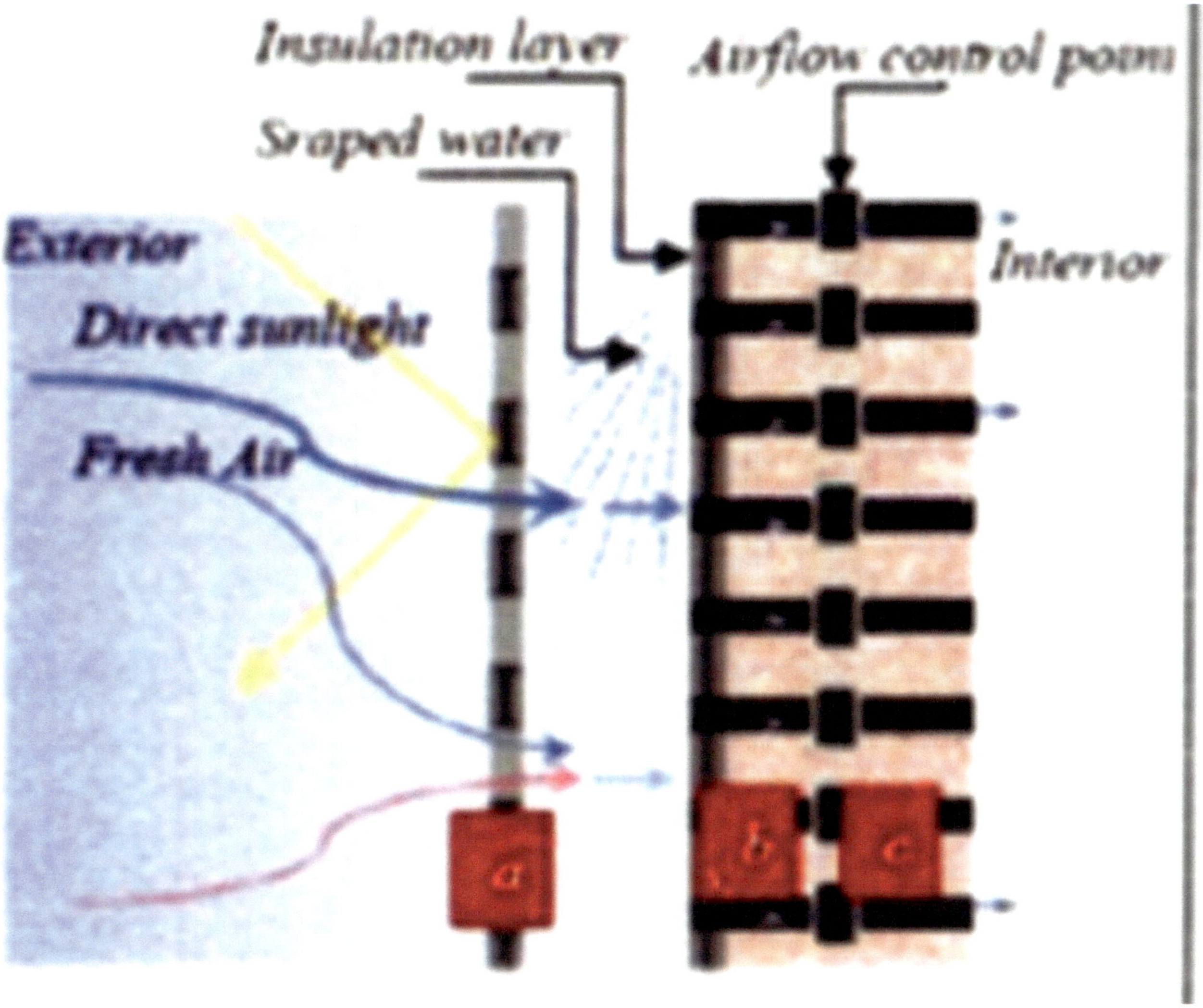

Fig. 16 Breathing wall

- For Retail space:

1. A 13 m tall shower tower is located on the building's south façade as shown in Fig. 27.

Inside a water shower, air movement is induced and cooling occurs. Water falls through the three-story tube, pulling air in from openings at the top.

2. The cool air assists the cooling of the ground floor lobby, shops and arcade.
3. The water then enters a tank in the basement.

The process used to ventilate the retail space is illustrated in Fig. 28.

Contribution to Sustainability

All air in CH2 is 100% fresh air and is drawn in from outside. Air is only used once and is exhausted by natural convection to the atmosphere. This air system also means that the output from coughs and sneezes is being taken straight out of the building rather than being spread around the floor and the building. This creates a healthy workplace. CH2's emissions, when compared to a Five Green Star building, will be 64% lower, thus it's a very sustainable approach (Lasheen, 2009).

1.7 Breathing Building—Habitat 2020 China

Habitat 2020 is designed by Philips Architects. It is 2020-proposed mixed-use with an office building design that applies Bio-mimicry principles that fuse high-tech ideas with basic cellular functions to create a "living structure like natural organisms". It will be located in Shanghai City, China, as it's the most populated city in the world as shown in Fig. 29.

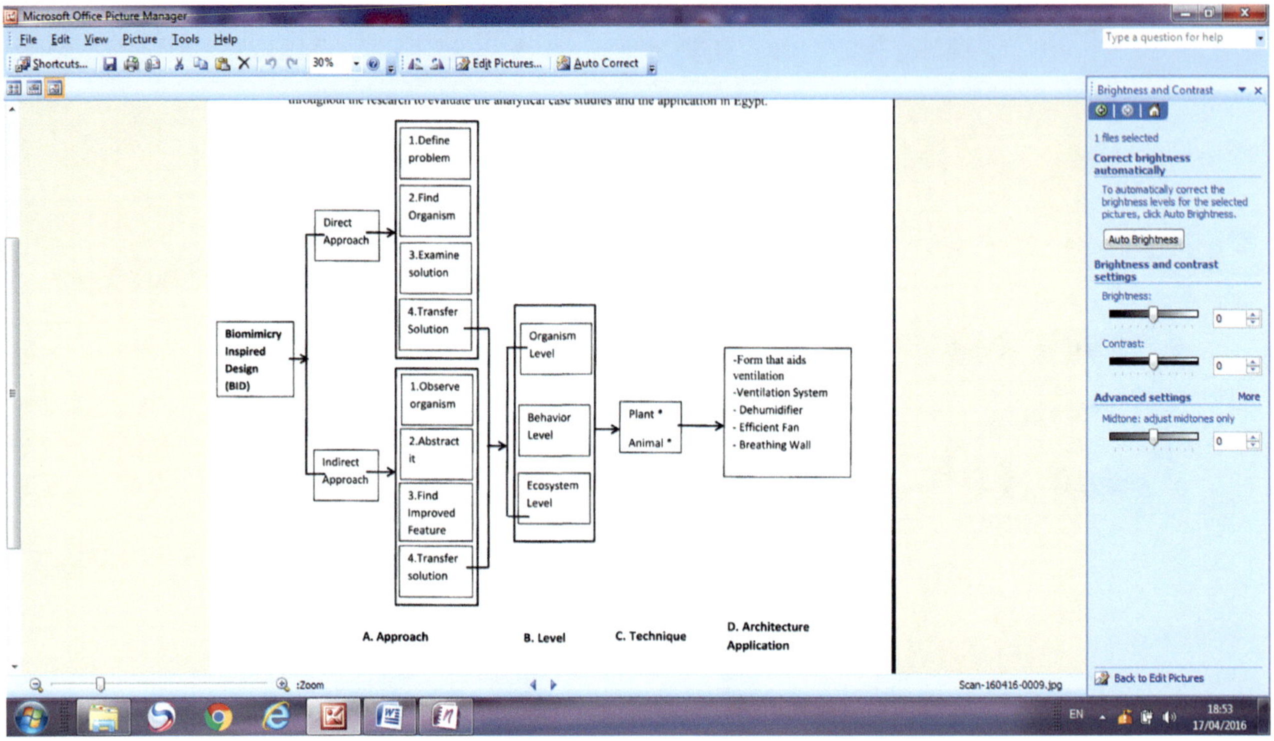

Fig. 17 Flowchart showing the deduced criteria. *Source* The Biomimicry Institute (2013a)

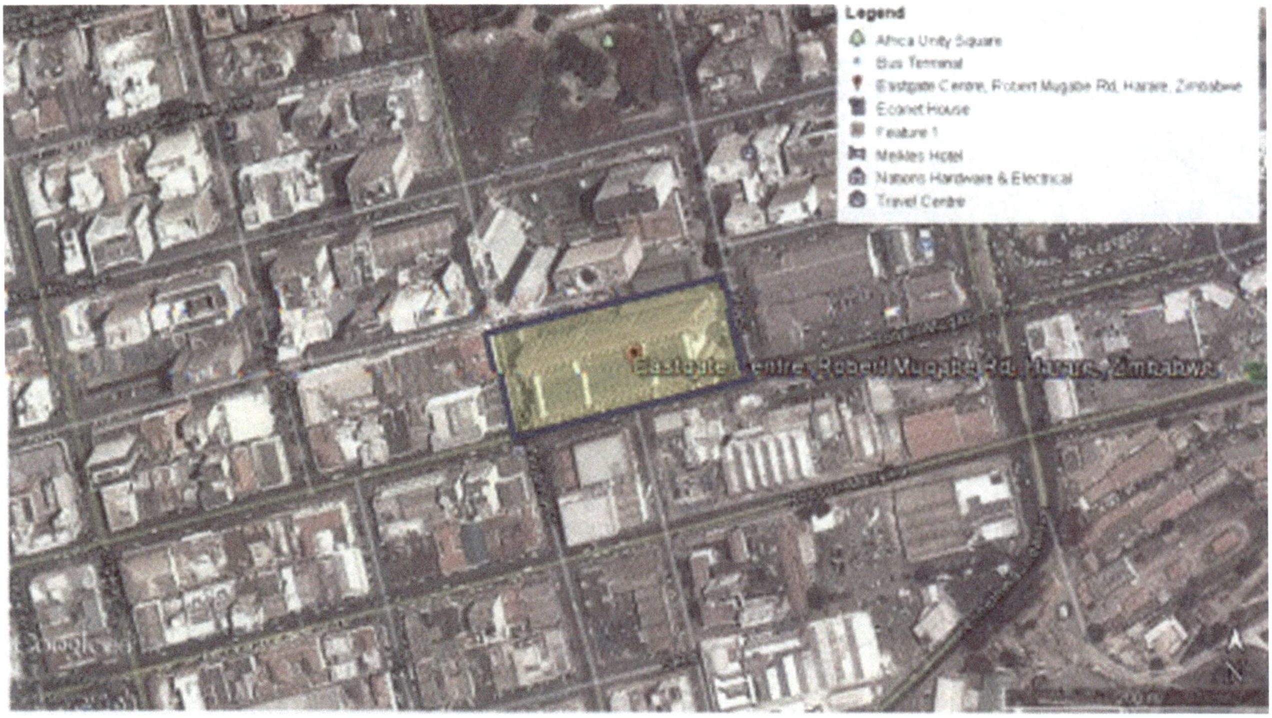

Fig. 18 Eastagate Project Location in Zimbabwe. *Source* Google Maps (2009)

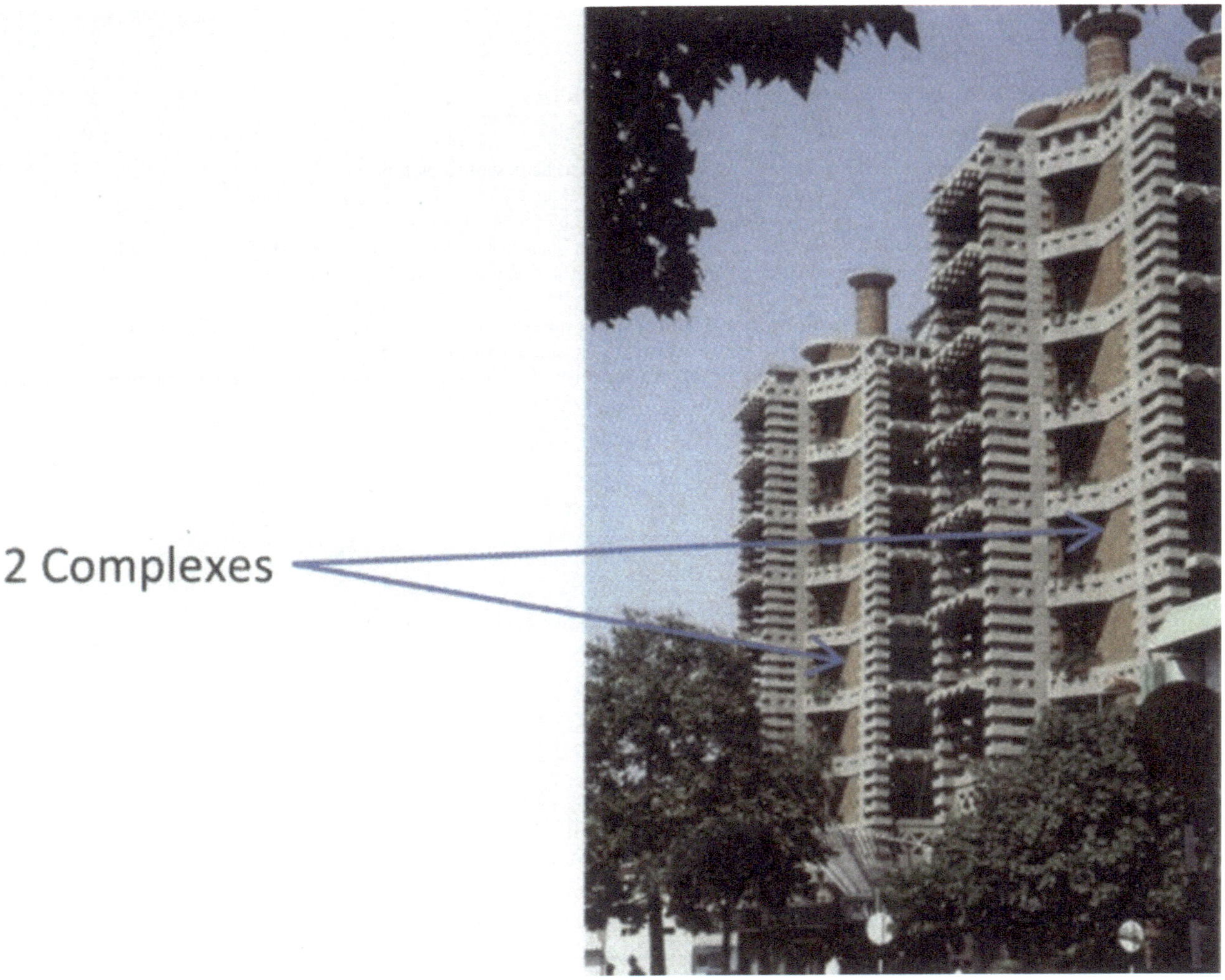

Fig. 19 Eastgate Project Building. *Source* Lasheen (2009)

Table 4 Application of Criteria for Eastgate Building. *Source* Researcher, based on data collected from Lasheen (2009), Zari (2007) and Klein (2009a, b)

Assessment Point	Application	
A. Approach	The Eastgate building is a Direct Approach for a BID	
B. Procedure	1. Define Problem	In Zimbabwe, the outside temperature varies from 3 °C up to 43 °C, thus the air condition plays a significant role. The challenge was to create a sustainable ventilation system
	2. Find Organism	Animal-Termite's passive ventilation and temperature regulation technique. Termites build gigantic mounds in which they farm a fungus that is their primary food source. The fungus must be kept constant at exactly 87 degrees, in a climate that fluctuates greatly between *35* degrees at night and 104 degrees during the day
	3. Examine Solution	Termites achieve this constant temperature by air exchange 1. They build breeze catchers at the base of the structure which draws in air 2. The air is cooled by pulling it through chambers made of the wet mud at the base 3. The cold air replaces the hot air which is expelled through flues at the top of the mound as shown in Fig. 20
	4. Transfer Solution	Ventilation system

(continued)

Table 4 (continued)

Assessment Point	Application
A. Approach	The Eastgate building is a Direct Approach for a BID
C. Level	Behavior Level as it mimicked a process
D. Architecture Application	Ventilation System which has a mechanism that can be described as follows: (The process is illustrated in Fig. 21.) 1. Air is continuously drawn from this open space by fans on the first floor 2. It is then pushed up the vertical supply sections of ducts that are located in the central spine of each of the two buildings 3. At each floor level, it enters a horizontal duct that lies between the ceiling concrete slab of the office and the floor slab of the office above 4. When the air leaves the horizontal ducts, it enters the office above through low-level grilles. The fresh air replaces stale air that rises and exits through exhaust ports in the ceilings of each floor 5. Finally, it enters the exhaust section of the vertical ducts before it is flushed out of the building through chimneys Fig. 22

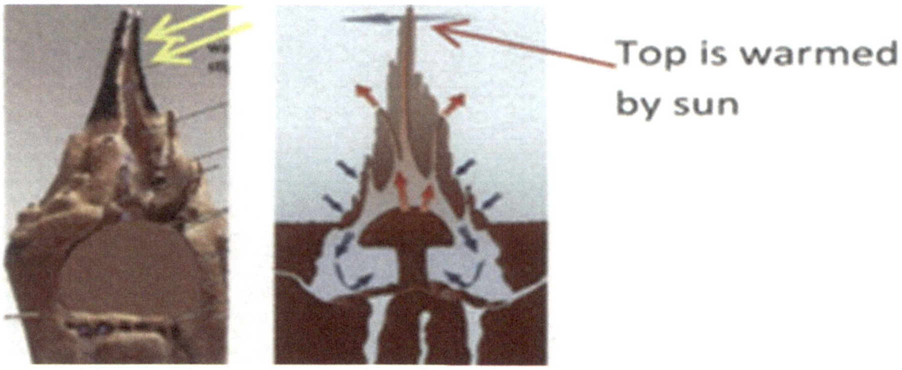

Fig. 20 Termites mechanism. Zari (2007)

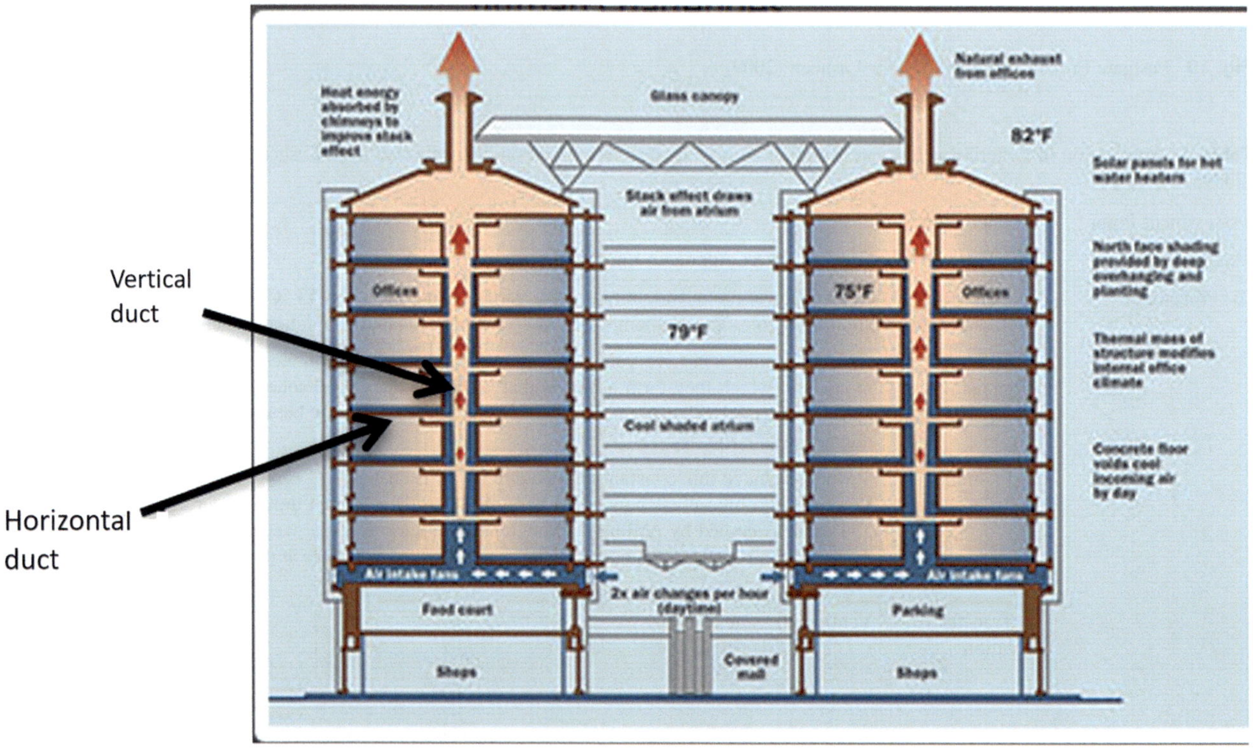

Fig. 21 Eastgate project ventilation system. *Source* Lasheen (2009)

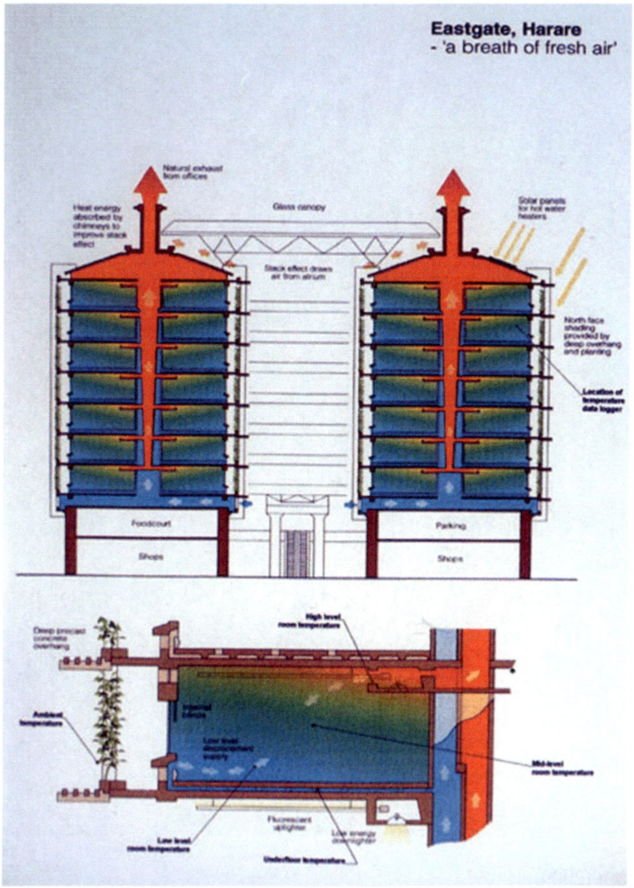

Fig. 22 Horizontal andvertical duct details *Source* Lasheen (2009)

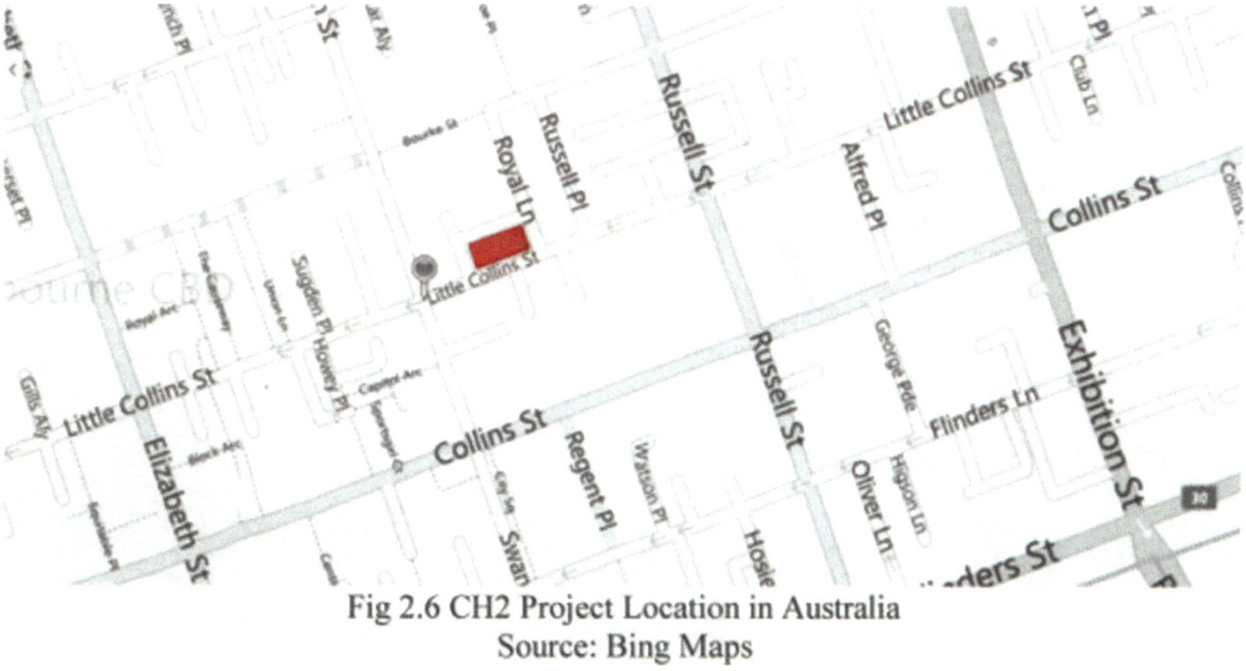

Fig. 23 CH2 project location in Australia. *Source* Bing maps 8 Aug. (2008)

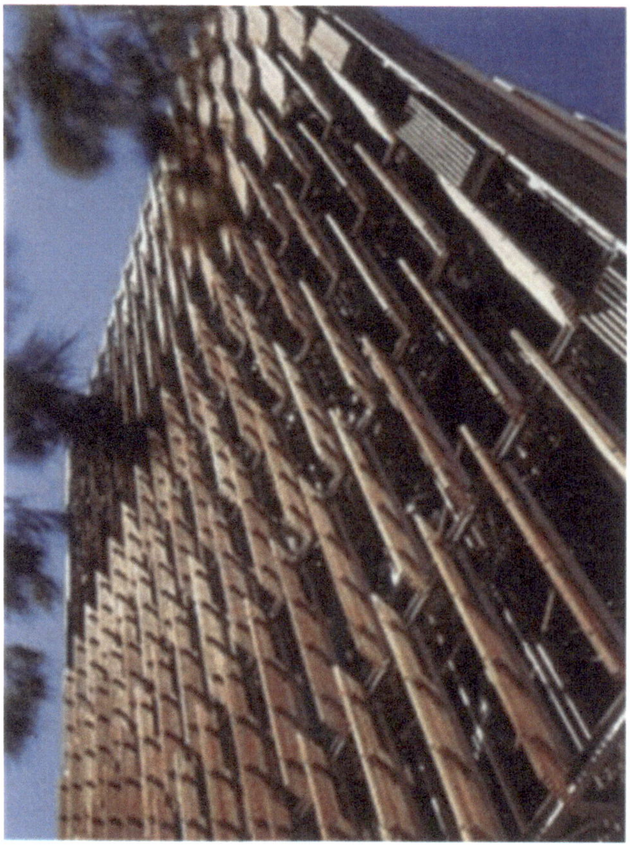

Fig. 24 CH2 building. *Source* Lasheen (2009)

Table 5 Application of Criteria for CH2 Building. *Source* The Biomimicry Institute (2013b)

Assessment Point	Application	
A. Approach	The CH2 building is a Direct Approach for the BID	
B. Procedure	1. Define Problem	Creating a sustainable air ventilation system in Australia
	2. Find Organism	Animal-Termite's passive ventilation and temperature regulation technique
	3. Examine Solution	Termites achieve this constant temperature by an air exchange process which is described in Sect. 2.1.2
	4. Transfer Solution	Ventilation System. Concepts from the termite mound were taken as the use of natural convection, thermal mass, ventilation stacks and water for cooling
C. Level	Behavior Level as it mimicked a process	
D. architecture Application	See Fig. 25	

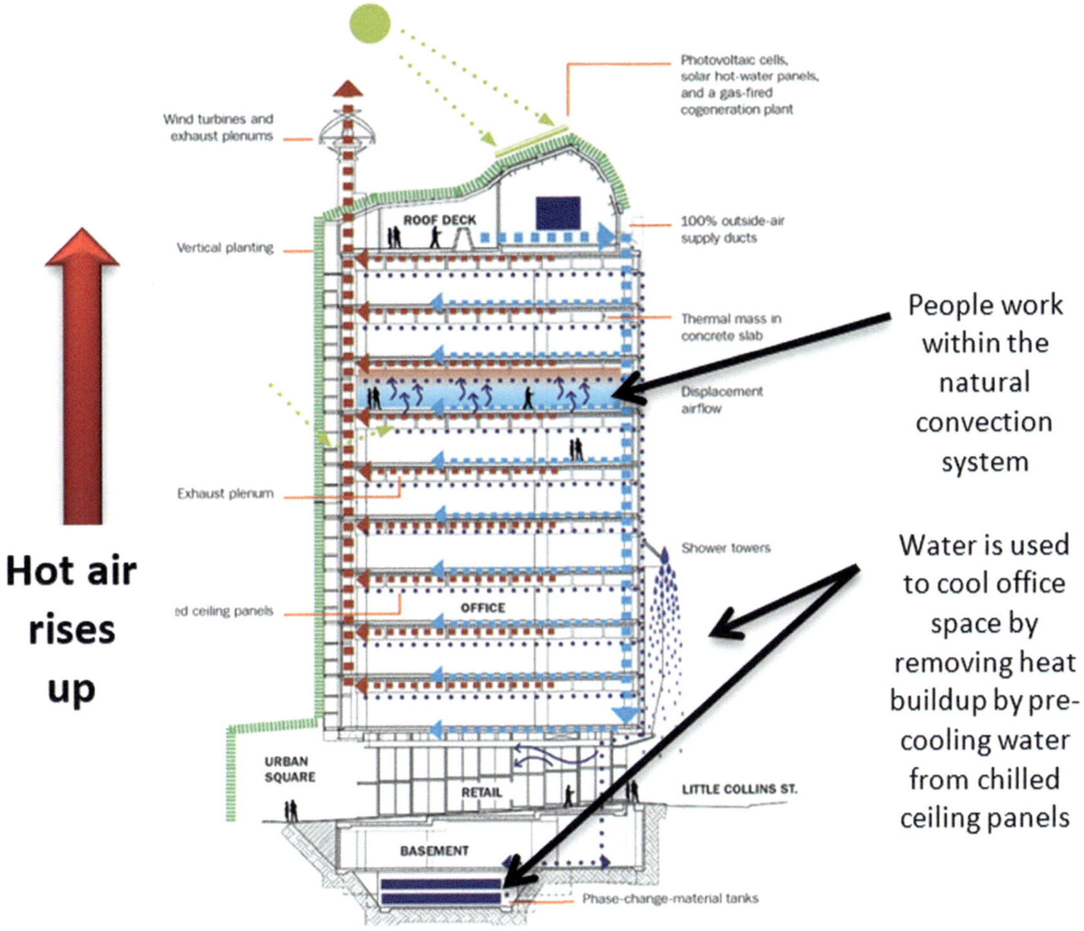

Fig. 25 CH2 project ventilation system. *Source* Lasheen (2009)

– Background Information on the Building's Design

The active skin of the building shown in Fig. 30 is designed to react to the surrounding environmental conditions listed follows:

- Sunlight automatically moves its skin to the most efficient position to channel light and generate energy.
- Air for natural ventilation will be explained in the next subsection.
- Rain harvesting and water undergoes purification, filtration and recycling processes.

– Evaluation of the Cooling and Ventilation System inspired by Plants According to the Proposed Criteria.

The Breathing Building is a Bio-mimicry-inspired design. The application of the criteria on it is shown in Table 6.

Contribution to Sustainability

It is a sustainable design as it depends on a natural ventilation system. It will be built in the most populated city in the world which will help reduce drastic effects on the environment. By collecting and channeling the natural light, no electricity will be needed during the day. Bringing natural light into our homes will not only save energy but will be healthy too. The building acts as a filter, reducing the amount of carbon dioxide in the air (Lasheen, 2009).

1.8 Comparison Between the Case Studies and the Deduced Preliminary Guidelines

To deduce the preliminary guidelines for local application, a comparison between the presented case studies will be conducted as shown in Table 7.

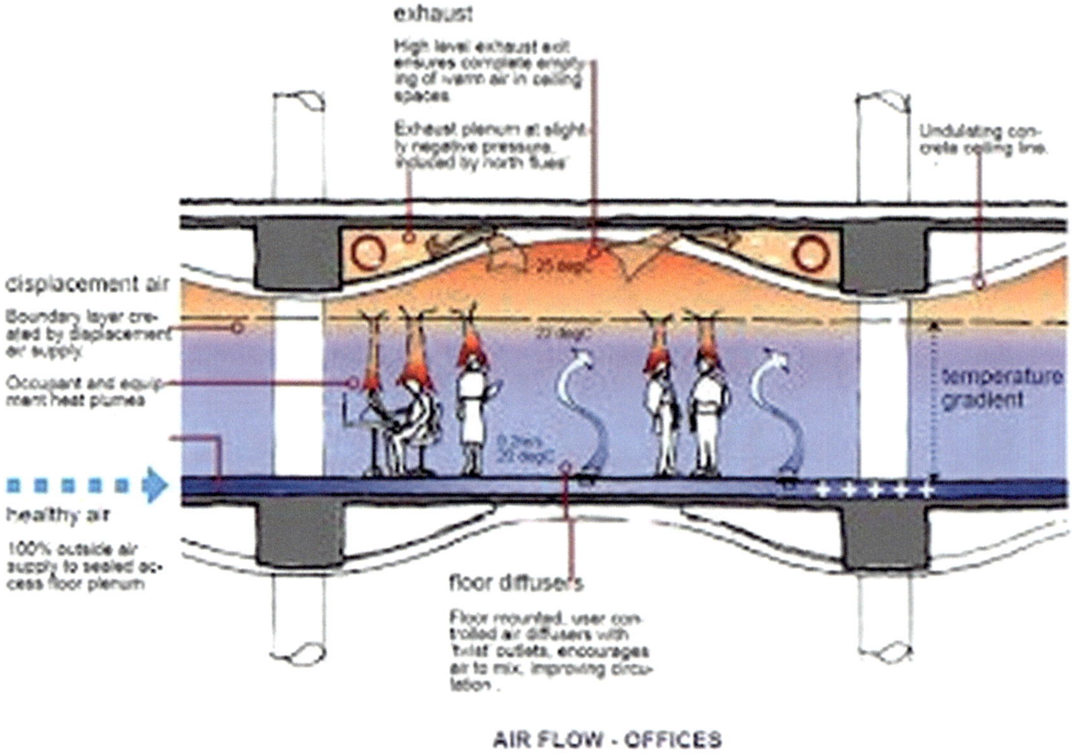

Fig. 26 Exhaust vents in CH2 ceiling mechanism. *Source* Lasheen (2009)

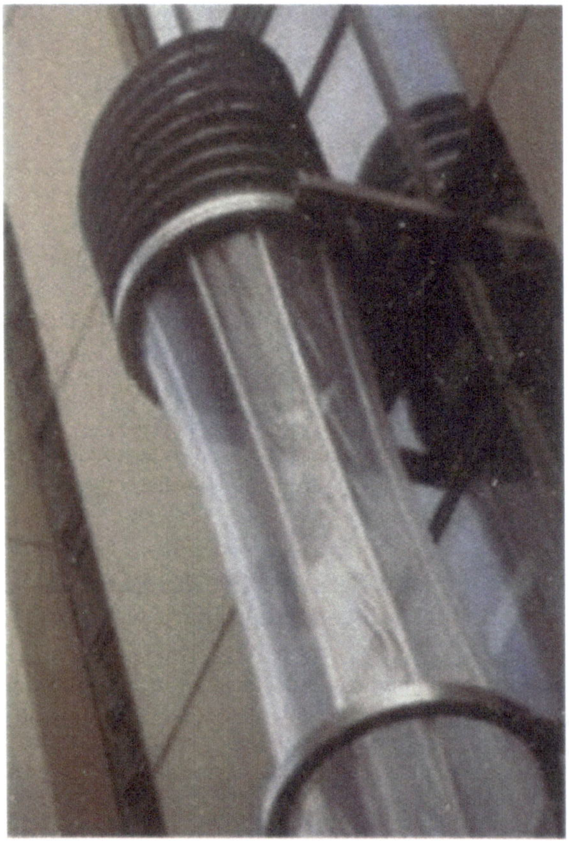

Fig. 27 Shower Tower in CH2 Building. *Source* Lasheen (2009)

After conducting the above comparative analysis, it was important to examine the feasibility of applying each project in Egypt. The results are presented in Table 8.

From the comparative analysis, the researchers deduced the following preliminary guidelines which will help to enhance the natural ventilation system by Bio-mimicry and at the same time is suitable for local application. The points presented in Table 10.9 should be followed when applying Bio-mimicry on any Egyptian office building.

Conclusion

Bio-mimicry is a great tool to enhance natural ventilation systems and make them sustainable. Since the research focused on introducing Bio-mimicry science, its main theories, levels and approaches. Numerous adaptation techniques are used by plants and animals to breathe. These adaptation methods can be mimicked and transferred to architecture applications. Applying such concepts will transform office buildings into more humane and sustainable designs which will reduce their harmful effects on the environment. Through the comparative analysis of the presented three case studies, the guidelines needed for the application of Bio-mimicry in Egypt were deduced because of the advantages of applying Bio-mimicry; it is recommended that architects should extensively use it in the design process. This requires that they

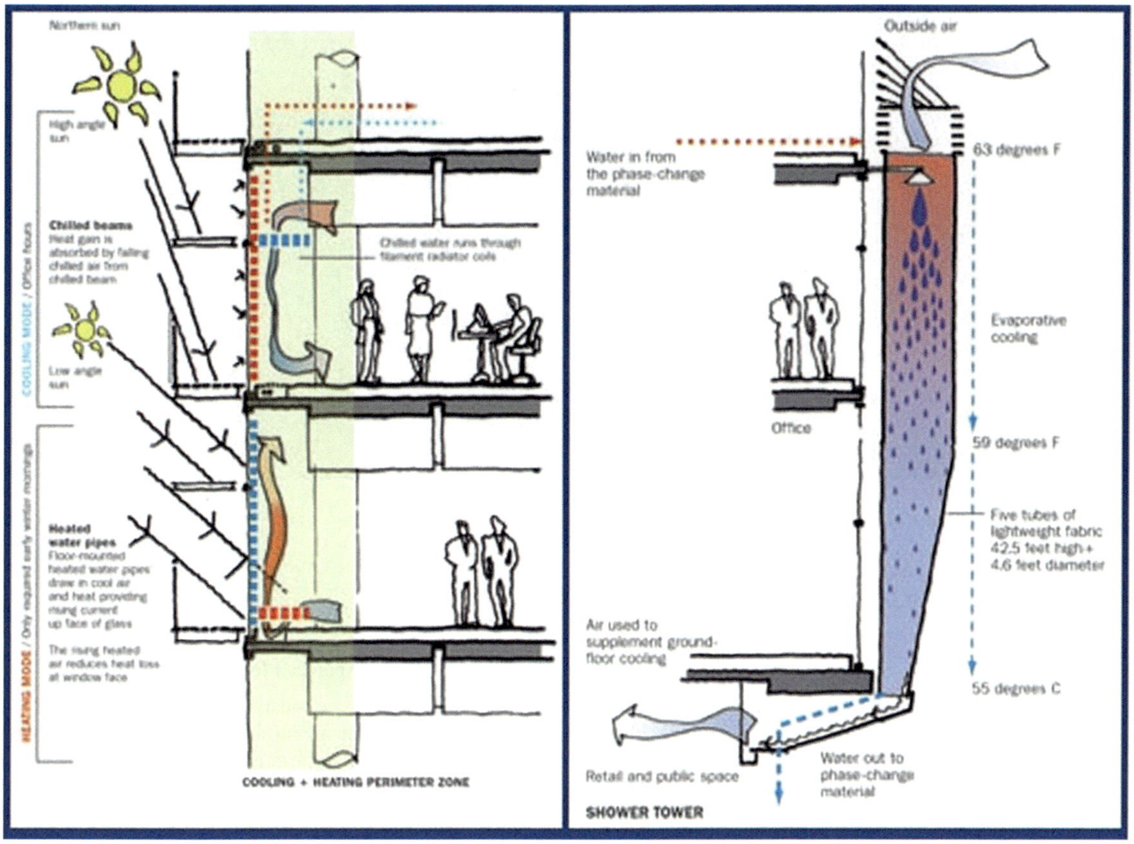

Fig. 28 Shower tower ventilation mechanism. *Source* www.solaripedia.com, 2018

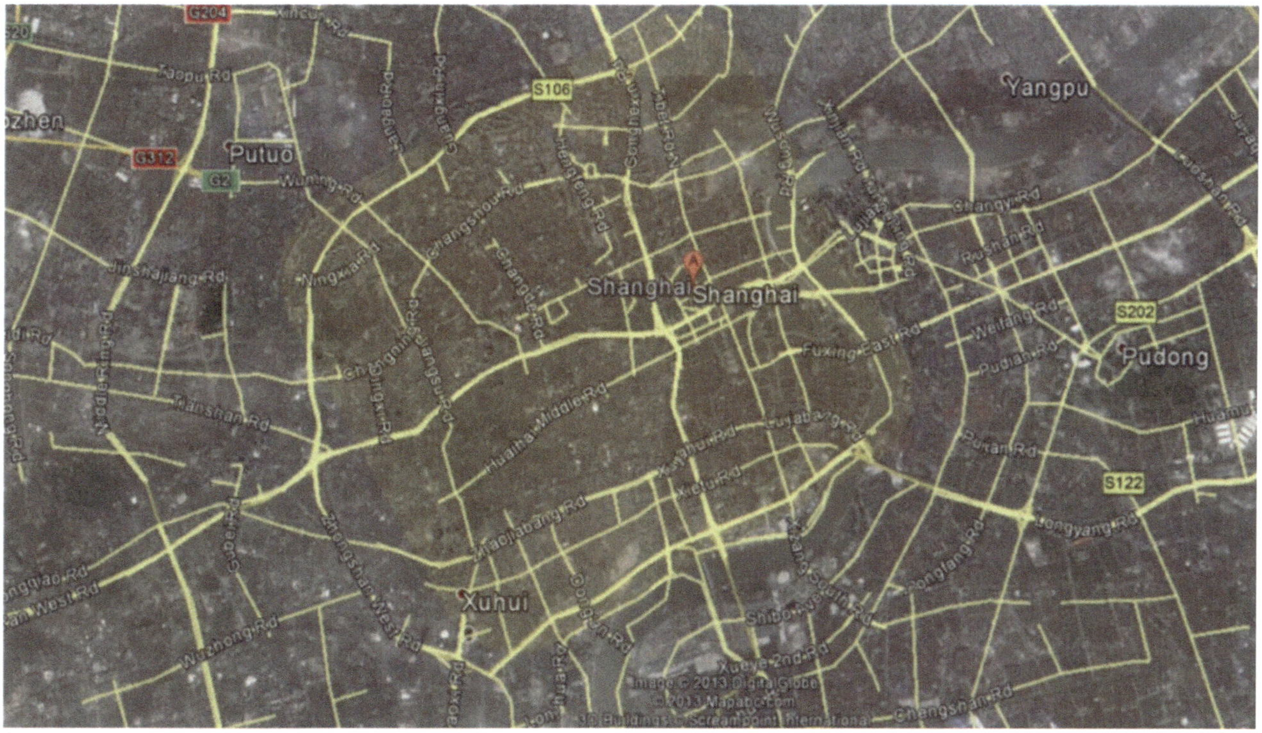

Fig. 29 Shanghai City Boarder. *Source* Google Maps

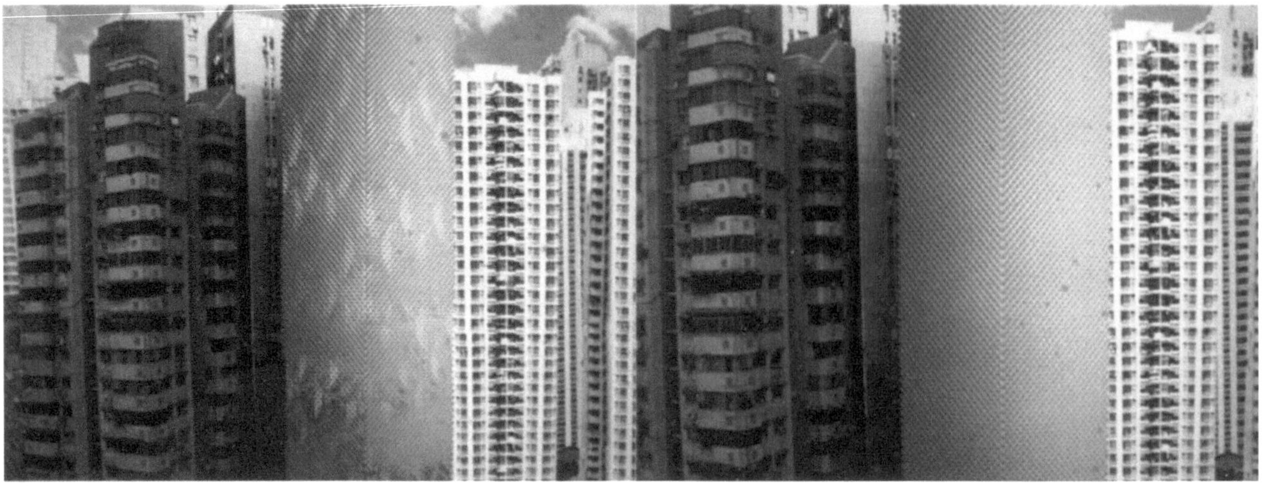

Fig. 30 Breathing Building in China's Context. *Source* Lasheen (2009)

Table 6 Application of Criteria for Breathing Building. *Source* Researcher, based on data collected from Lasheen (2009)

Assessment Point	Application	
A. Approach	The Breathing Building is a Direct Approach for BID	
B. Procedure	1. Define Problem	China is the world's secondary carbon dioxide producer. Thus, building a ventilation system shouldn't consume energy and help in filtering the air. The problem is in the dumb construction materials that don't respond to the surrounding conditions
	2. Find Organism	Leaf's surface has several stomata (cellular openings) involved in gaseous exchange as shown in Fig. 31
	3. Examine Solution	The stomata provide an essential connection between the interior and the exterior. Opening, closing and changing in size regulate the gaseous exchange process. When the stomata open, production and consumption of oxygen and carbon dioxide in the leaf are sufficient to maintain a concentration gradient steep enough to facilitate gas exchange
	4. Transfer Solution	Ventilation System depending on sensitive functional skin with stomata-like openings that act as a membrane used for transporting, collecting and channeling the elements of air, water, and light as shown in Fig. 32
C. Level	It is considered an Ecosystem Level. The architect did not only mimic the behavior of the stomata but also incorporated it to perform other natural functions. The building even interacts with the environment by purifying the air	
D. Architecture Application	Ventilation System, in which the building skin has a mechanism that can be described as fol1ows: 1. The active skin of the building reacts to the wind 2. The direction of the funnel moves to face and capture on-coming breezes and airflow increases as the tubes transform into trumpets as air pressure is increased 3. Openings in the active skin would allow the entry of air, light and water into the building as shown in Fig. 33 4. Through the osmosis process by tubes, air and wind will be channeled through the skin of the building. Energy will be generated which will assist the active skin as shown in Fig. 34 5. Air will be filtered to provide clean air inside the building. Also, moisture will be removed 6. Air will be compressed and dissipated through funnels and cooled 7. providing natural air-conditioning as shown in Fig. 35	

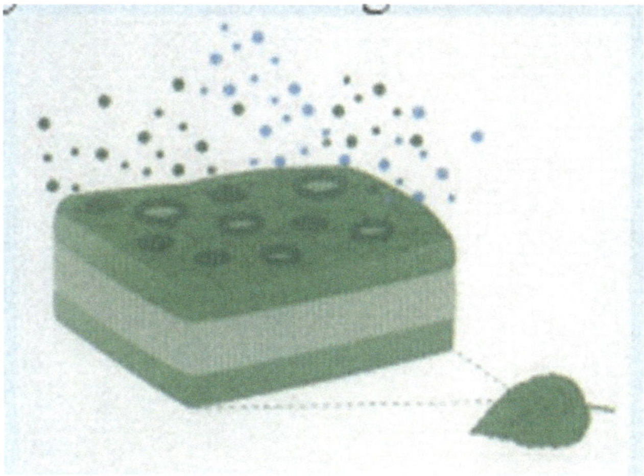

Fig. 31 Stomata openings in leaf

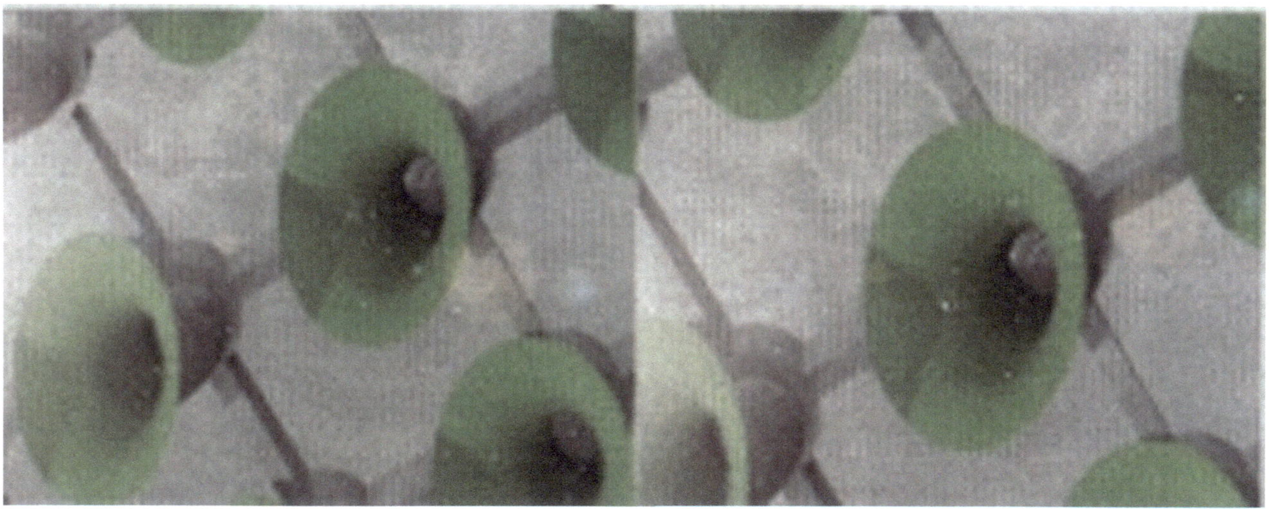

Fig. 32 Building skin with stomata-like openings

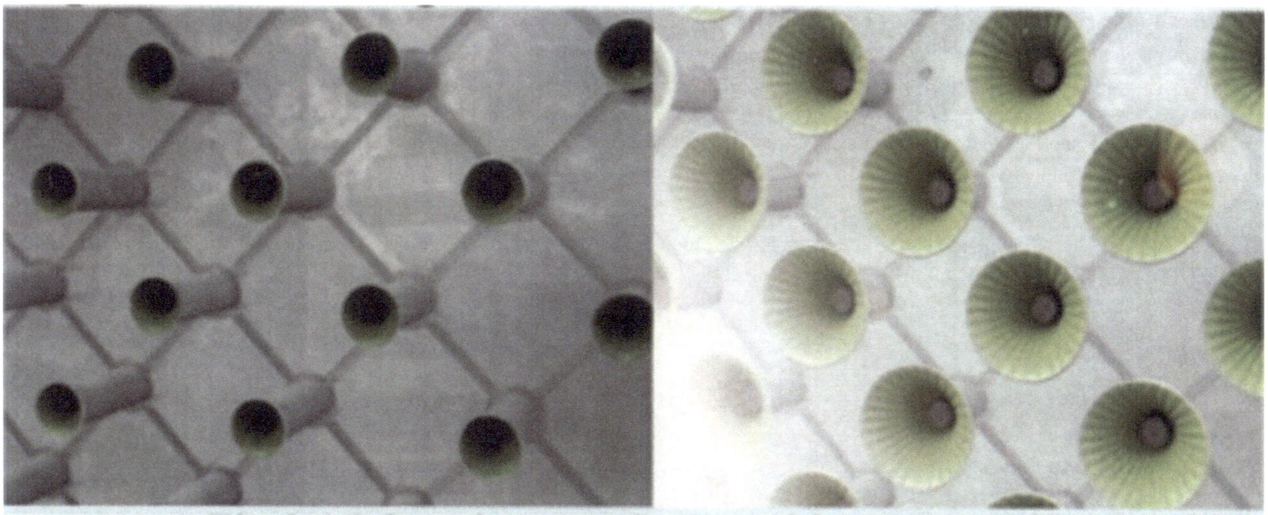

Fig. 33 Openings on the Breathing Wall Skin

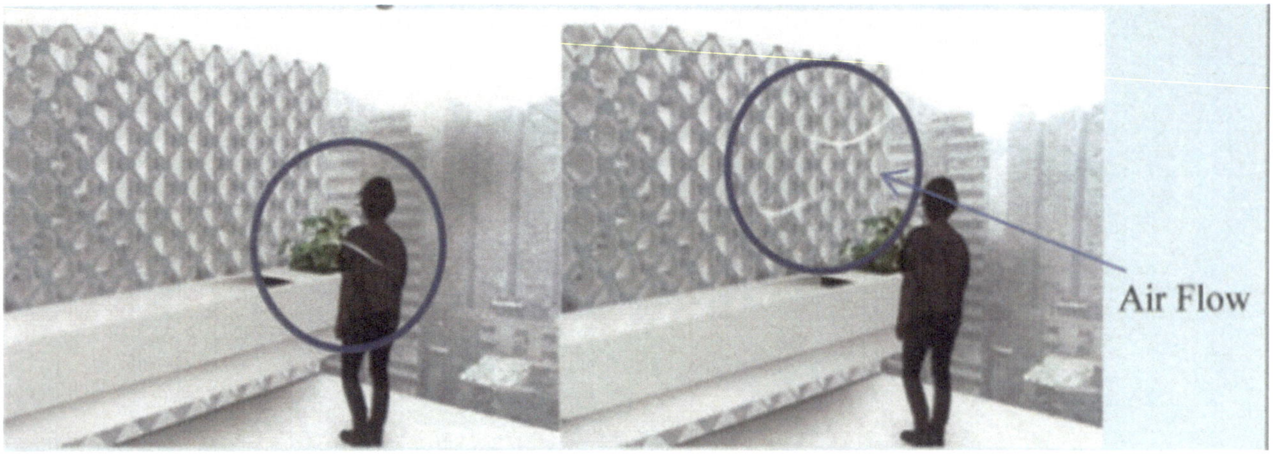

Fig. 34 Airflow through breathing wall

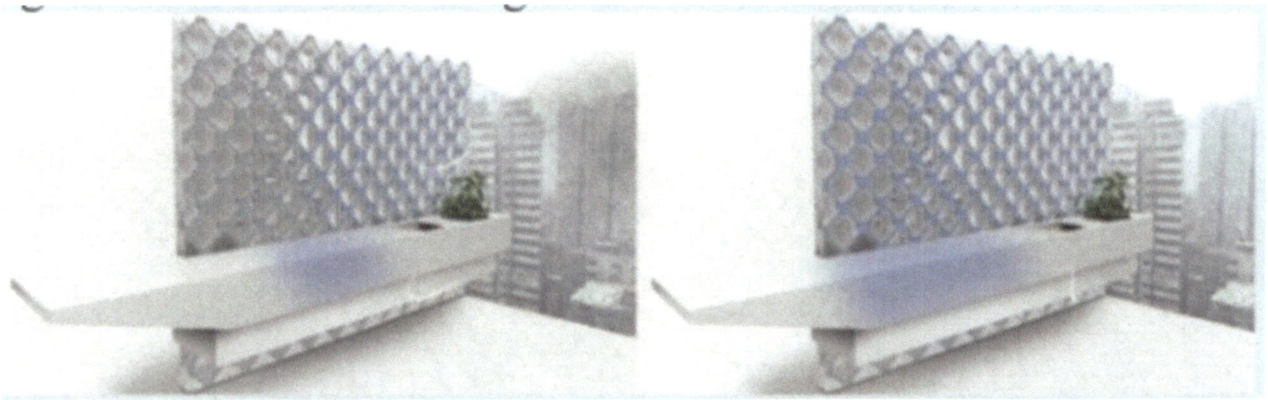

Fig. 35 Cooling or Air through Breathing walls 7. Before air is exhausted from the building, it is cleaned and stripped of carbon dioxide

Table 7 Comparison between case studies. *Source* Researcher based on The Biomimicry Institute (2013a, b, c, d, e, f, g)

Project Description			
Point of comparison Case study	Eastgate Project	Council House 2 Building (CH2)	Breathing Building (Habitat 2020)
Location	Zimbabwe	Australia	China
Description	The project consists of two main complexes that surround a glass-covered atrium	The project consists of a public square with a retail shop. Above are 9 floors with open office spaces	The project consists of a sensitive functional skin that encloses all office spaces
Climatic Constraint			
Climatic Constraint	The outside temperature varies from 3 °C up to 43 °C during the year	Summer is warm to hot (below 30 °C, however, it can reach 4o's in some days). Winter is cold (temperature doesn't reach negative values)	Summer is hot and humid exceeding 35 °C. Winter is cold but the temperature doesn't reach negative values
Bio-mimicry-Inspired Design			
BID Approach	Direct	Direct	Direct
BID Level	Behavioral Level	Behavioral Level	Ecosystem Level
BID Technique	Animal (Termites)	Animal (Termites)	Plant (Levels)
Application			
Architecture Application	Ventilation System	Ventilation System	Skin responsible for many functions including ventilation
Method of supplying air	–Air is drawn from the atrium by fans on the first floor	By 2 Techniques • Office Spaces:	Openings on the skin transform into trumpets so air pressure is increased

(continued)

Table 7 (continued)

Project Description			
Point of comparison Case study	Eastgate Project	Council House 2 Building (CH2)	Breathing Building (Habitat 2020)
	–It is then pushed up vertical ducts –Horizontal ducts supply air at each floor level	–Treated air is pumped from the roof through vertical ducts –Horizontal ducts supply air at each floor level • For Retail Space: Shower tower that induces air movement	– Air is drawn by openings on the skin – It is then cooled and dissipates into the spaces
Method of removing air	Stack Ventilation – By natural convection, warmth will rise – It moves out via ports in the ceiling –Vents are connected to the exhaust section and are flushed out through the chimney	Stack Ventilation –By natural convection, warmth will rise –It moves out via vents in the ceiling – Vents are connected to the exhaust–air shafts	Air Pressure Difference (Osmosis) – Hot air is vented through openings in the skin
Opening	Rectangular openings	Exposed roof with vertical	Stomata-like openings
Shape, operation location	Facing the atrium sides	ducts to transfer air to floors	Forming the skin of the building from all orientations
Capability and Efficiency			
Apparatus that assists ventilation (chimneys, court, fans)	Cool shaded atrium to supply air –Chimney to exhaust air –Fans to draw air to the interior	– Wind turbines to draw air from the interior to exhaust it – Shower tower to supply cool air to retail offices	– Active Skin – Energy converted from sunlight or wind assists in cooing of air
Capability with daylighting	The ventilation system does not affect daylighting	The ventilation system does not affect daylighting	A ventilation system is integrated with day lighting mechanism as both occur through the same openings
Effect on Thermal control	Air heating/cooling is integrated with ventilation system	Air heating/cooling is integrated with ventilation system	–Air heating/cooling is integrated with the ventilation system –Moisture is removed which improves and feels like temperature
Occupant comfort	– Air ventilation can't be controlled manually by users – The healthy environment is due to the presence of air exhaust	– Swirl diffusers located on floor tiles to allow individual placement to suit occupants – All air in CH2 is 100% recycled air – Healthy space as the output from coughs is being taken straight out of the building	– Air ventilation can be controlled manually as each group of openings can be modified separately – The healthy environment due to the presence of air exhaust – Air is cleaned and filtered from smoke
Sustainability and Energy Efficiency	Sustainable as air is exhausted by passive means. The building uses 90% less energy for ventilation than a conventional building its size	Sustainable as air is exhausted through passive means. CH2's emissions are 64% lower than a five green star building	Sustainable as air is naturally ventilated. The energy needed by the active skin is provided by PVC cells present. By collecting the natural light, no electricity will be needed during the day
Aesthetics	Ventilation system doesn't affect the aesthetics of the building; however, presence of an atrium is necessary	Ventilation doesn't affect the aesthetics of the building. However, the presence of wind turbines is necessary	The ventilation system depends on the building of active skin form. However, it matches the technological advances and looks good
Cost	Cost–Effective	Cost–Effective	Very Expensive as extensive research is needed

Table 8 Applicability of Case Studies in Egypt. *Source* Researcher based on The Biomimicry Institute (2013a, b, c, d, e, f, g)

Point of comparison Case study	Eastgate Project	Council House 2 Building (CH2)	Breathing Building (Habitat 2020)
Applicability in Egypt, Cairo	Applicable in Cairo	Applicable in Cairo but some concerns are present like – Wind Turbines: Winds in Cairo come mainly from the south-west; however, their speed is not high enough to ensure continuous natural ventilation (Suitable in Zafarana) – Shower Tower: Should be in areas where continuous water supply is provided. Mainly areas next to the River Nile. However, this water should be carefully recycled or it will be considered a waste of resource	Not applicable in Cairo The system used is very expensive for a building with office use Also, the research needed is expensive and hard to be conducted in Egypt

Table 9 Preliminary Guidelines for Application in Egypt. *Source* Researcher based on The Biomimicry Institute (2013a, b, c, d, e, f, g)

Bio-mimicry-Inspired Design Steps	A. Approach	Direct Approach to solve an existing problem in Egypt
	B. Procedure	1. Define Problem: Enhancing the natural ventilation system of an office building using Bio-mimicry techniques The climate in Cairo can be considered warm to hot summers (temperature is below 40 °C) and cold winters (temperature doesn't reach negative values) with about 70% humidity)
		2. Find Organism: Animal (Termites)
		3. Examine Solution: The termite mechanism is explained in Sect. 2.1.2
		4. Transfer Solution: Ventilation System
	C. Level	Behavioral Level
	D. Architecture Application	Ventilation System
Applied methods	1.Method of supplying air	Air is drawn from Atrium. The atrium is considered the best solution for office building designs - Fans can help the process of drawing air from the atrium to the first floor -Presence of vertical ducts through which *air* will be pushed up -Presence of horizontal ducts supplies air at each floor level
	2. Method of removing air	The main concept is stack ventilation through the present exhaust section. In the office space, by natural convection, warmth will rise - Presence of vents in the ceiling through which air can move out. Vents should be connected to the exhaust section - The presence of a chimney can simulate the process of flushing air out through the exhaust section
Architecture Application Guidelines	3. Opening typology, operation, location, shape	Effective openings for natural ventilation are needed on sides facing the atrium. Other openings will contribute to daylighting only Generally, for sustainable design of openings: - Openings on the south should be limited to avoid heat gain - Shading techniques are needed for openings in the south and west - No recommendations for openings on the north
	4. Apparatus that assists ventilation	-Atrium: its presence is a necessity and is suitable in Egypt. A court is not preferable; as the atrium is shaded by a canopy, it will let cool air enter the building -Fans: Suitable for use in Egypt. Although fans consume a small amount of energy, to be 100% sustainable, they should be integrated with a building system that produces energy. This energy produced will be used by the fans -Chimneys: Suitable in Egypt as there is a lot of sunlight which will make the chimney hot and enhance the stack effect
	5. Capability with daylighting	The ventilation system should not affect daylighting at least. However, for a better sustainable design, they should integrate like what happened in the Breathing Building in China
	6. Effect on Thermal control	-Air heating/cooling should be integrated with the ventilation system -Moisture should be removed which improves and feels like temperature as humidity is high in Cairo
	7 Occupant comfort	- Air ventilation should have an option to be controlled manually by users (Swirl Diffusers is one technique that can be used.) - Air Exhaust should be provided to ensure a healthy environment - Outside air should be cleaned and filtered from the smoke

(continued)

Table 9 (continued)

Bio-mimicry-Inspired Design Steps	A. Approach	Direct Approach to solve an existing problem in Egypt
	B. Procedure	1. Define Problem: Enhancing the natural ventilation system of an office building using Bio-mimicry techniques The climate in Cairo can be considered warm to hot summers (temperature is below 40 °C) and cold winters (temperature doesn't reach negative values) with about 70% humidity)
		2. Find Organism: Animal (Termites)
		3. Examine Solution: The termite mechanism is explained in Sect. 2.1.2
		4. Transfer Solution: Ventilation System
	C. Level	Behavioral Level
	D. Architecture Application	Ventilation System
		-All air in the office spaces should be 100% un-recycled fresh air
	8. Sustainability and Energy Efficiency	The design should be sustainable and reduce energy consumption by a big percentage when compared to a conventional building of the same size and type
	9. Aesthetics	The ventilation system shouldn't affect the aesthetics of the building as it is one of the goals of the architecture profession. However, an essential requirement of an office building design is the presence of an atrium
	10. Cost	The design should be cost-effective

should examine and analyze nature, and always be up to date with all the recent developments in the Bio-mimicry field and try to benefit from them.

References

Benyus, J. (1997). *Biomimicry - Innovation Inspired by Nature*. Harper Collins Publishers.

Berkebile, B. & McLennan, J. (2004). The Living Building Biomimicry in Architecture, *Integrating Technology with Nature*, Biolnspire 18

Biomimicry Guild. (2008). Innovation Inspired by Nature Work Book. *Biomimicry Guild*, April

Elghawaby, M. (2006). *The future of smart architecture in Egypt; a way to design environmental building,* M.Sc. thesis, Ain Shams Un., Cairo. Egypt.

Fathy, H. (1986). *Natural energy and vernacular architecture: Principles and Examples with Reference to hot arid climates*, Edited by Chicago, U.States.

Ireland, T., and Garnier, S. (2018). *Architecture, space and information in constructions built by humans and social insects: a conceptual review*, the royal society.

Klein, L. (2009a). *A Phenomenological Interpretation of Biomimicry and its Potential Value for Sustainable Design*, M.Sc. thesis, Kansas State University, Kansas, United States.

Klein. L. (2009b). *A Phenomenological Interpretation of Biomimicry in Two Sustainable Designs*, United States: Edra.

Lasheen, H., L. (2009). *Biornimetic Architecture*, M.Sc. thesis, Alexandria University, Alexandria, Egypt.

Panchuk, D., & Vickers, J. N. (2006). Gaze behaviors of goaltenders under spatial temporal constraints. *Human Movement Science, 25*, 733–752.

Solaripedia, http://www.solaripedia.com/images/large/1735.jpg, 2018

Sto-Sea-Brochures, (2005).

Straube, J. F., & Acahrya, V. (2013). *Indoor air quality, healthy buildings, and breathing walls*. Iris communications Inc.

The Biomimicry Institute (2013a). *Large cars used to cool off: Jackrabbit*. http://www.asknature.org/strategy/a250478ba7f69e68c71405d931c91

The Biomimicry Institute (2013b). *Air scoops provide cooling: Ants*. http://www.edra.org/sites/default/tiles/publications/EDRA4O-Klein

The Biomimicry institute (2013c). *Blood vessels regulate temperature vertebrates*. http://www.asknature.org/strategy/3e94c41

The Biomimicry Institute (2013d). *Gular fluttering dissipates heat: nightjars*. http://www.asknature.org/strategy/df479edb

The Biomimicry Institute (2013e). *Leaves optimize internal state: Mangrove*. http://www.asknature.org/strategy/7368abb7b463b9c81e8aa721

The Biomimicry Institute (2013f). *Sweating aids thermoregulation: mammals*. http://www.asknature.org/stratcgy/e427ca4215ca4bc83ab647869c2643c9

The Biomimicry Institute (2013g). *What is Biomimicry*.

Zari, M. P. (2007). *Biomimetic Approaches to Architectural Design for Increased Sustainability*. New Zealand Government: Ministry for the Environment. New Zealand.

Applied Systems and Data Analysis

Integration of BIM as a Process in the Architectural Education Curriculum and Its Impact on the Egyptian AEC Industry

Ahmed Gad El Karim and Gehan Nagy

Abstract

Building Information Modeling (BIM) is the process of involving all project stakeholders at the right time to deliver the project in its best way. BIM as a program was introduced to the architectural engineering and construction industry since the 90s. Since then, the Integration of BIM as a program has increased greatly worldwide; however, there is a conflict between the use of BIM as a program and the complete implementation and understanding of BIM as a process. A review of the previous literature indicated that the majority of BIM users in the Middle East were self-taught users, and only 14% of users learned BIM through college-taught courses. Even more, within the Egyptian context, there is a lack of understanding of BIM as a process in the Egyptian AEC industry. Thus, this research aims to reduce the gap between architecture education and the AEC industry in Egypt, through the implementation of BIM as a process in the Egyptian architecture curriculum. To achieve this aim, a methodology comprised of one main part will be adopted. This will entail a review of the current literature about BIM and the current architecture education system. Then the research will conclude by proposing a framework for the integration of BIM as a process in the architectural engineering curriculum in Egypt.

Keyword

Building information modeling (BIM) • Architectural engineering and construction industry (AEC) • architectural education • Egyptian architecture curriculum

1 Literature Review

This section of the paper will be going through three main parts that the research aims to investigate in: first is the building information modeling (BIM), then we will go through the implementation environment, which will be divided into the education facility and the AEC industry; lastly, we will discuss the implication strategies globally and the risk that faces this implementation.

1.1 Main Aim

The main aim of this research is to propose/conclude a set of guidelines for the implementation of BIM as a process within the Egyptian architectural education curriculum.

1.2 Methodology

A review to introduce key terms related to BIM, its history, and importance, and identify current BIM adoption methods within the architectural education system.

A comparative analysis of two case studies: One international case that adopts BIM within its educational system and a local university. To highlight how and why, BIM should be implemented.

A. G. E. Karim (✉) · G. Nagy
Department of Architecture, Faculty of Engineering, British University in Egypt Cairo, Cairo, Egypt
e-mail: Ahmed.Gad@bue.edu.eg

G. Nagy
e-mail: Gehan.Nagy@bue.edu.eg

F. Rosso et al. (eds.), *Advances in Architecture, Engineering and Technology*, Advances in Science, Technology & Innovation, https://doi.org/10.1007/978-3-030-86913-7_11

Nomenclature

A BIM (Building Information Molding)
B Implementation Environment
C Implementation Strategies.

2 Building Information Modeling

The building information modeling has been in the market since Prof. Charles Eastman began practicing BIM at the Georgia University in the 1980s (El-Shazly, 2018), thus the first part of this chapter will discuss the BIM development since the early begging till its condition nowadays, reach the result of a timeline for BIM developments and important milestone, the following part of this chapter will be a studying the different definitions of the BIM and will categorize team according to understanding reaching the conclusion of table dividing the definition according to category, then the flowing two sections will be taking about two different ideas of BIM understanding as software and as a process and the conclusion of every section of them will be a diagram explaining the integration of every idea in the construction industry, the following section will be discussing different applications used in the BIM answering the who, why and when questions about it and this chapter will be finished with a chart explains when this application is used in the construction industry and the finial section will be discussing the importance of BIM for every stakeholder in the project and when the stakeholders will be involved in the projects.

2.1 History of BIM

The building information modeling roots go back to 1970 when the USA started researching parametric modeling; the first practice was by Prof. Charles Eastman at the Georgia tech school (Azhar et al., 2012). Then later in 1982, 2D CAD was made followed by radar CH in 1984, and Graphisoft company invited ArchiCAD and tried to construct a 3D visual model. In 1992, the building information modeling term was officially known to the public in an article; then in 2000, Revit was made followed by naves works in 2001; in 2002, Autodesk bought Revit to add to the BIM program. Four years later Autodesk updated the Revit program than in 2007 bus Navisworks and add it the BIM (Fig. 1).

2.2 BIM Definitions

The building information modeling has been in the field for a long time and along its life, the organizations have been developing different understanding of the BIM, Therefore, there are many definitions of BIM, for example, the National Institute of Building Science (NIBS) is defining the BIM as the use of cutting and edging digital tools to make a digital representation for the physical and functional properties of the building and it also carries the building life cycle information (Mandhar & Mandhar, 2013), however, Autodesk defines it as a process of involving a creation of 3D digital model to communicate with the design, decision-making, and simulation team and to make the information more clear for all of the project stakeholders and facilitate the project managing along its life cycle (Mordue, 2015); moreover, Royal Institute of British Architects (RIBA), Construction Project Information Committee (CPIC), and Building Smart have jointly defined BIM as a "digital representation of physical and functional characteristics of a facility creating a shared knowledge resource for information about it forming a reliable basis for decisions during its life cycle, from earliest conception to demolition". One more definition by Building Smart is that BIM is a digital model of a building with information arranged and shared in a form of 3D/4D or even 5 dimensions integrating components of time and cost. Lastly, Eastman in 2008 defined BIM as a collection of different tools and processes to produce or analyze a building.

2.3 BIM as a Technology

As discussed before, BIM has a long history, thus there are many definitions and understandings for it. Building Smart defined it in 2010 as a product or a model that contains shredded information in terms of 3D, 4D, 5D, and 6D. Thus, BIM is considered as a 3D visualization program, but there is a principle difference between it and conventional 3D CAD such as the 3D CAD describes the building through independent views as plans, sections, and elevations. Moreover, editing any of the views will require the updating and checking of the other views, which allows the error to happen, in addition to 3D CAD being considered a dumb program, because it deals with the building as entities such as circles, arches, and lines; on the other hand, BIM intelligently deals with the model as it defines the building as components and elements, spaces walls, beams, and columns

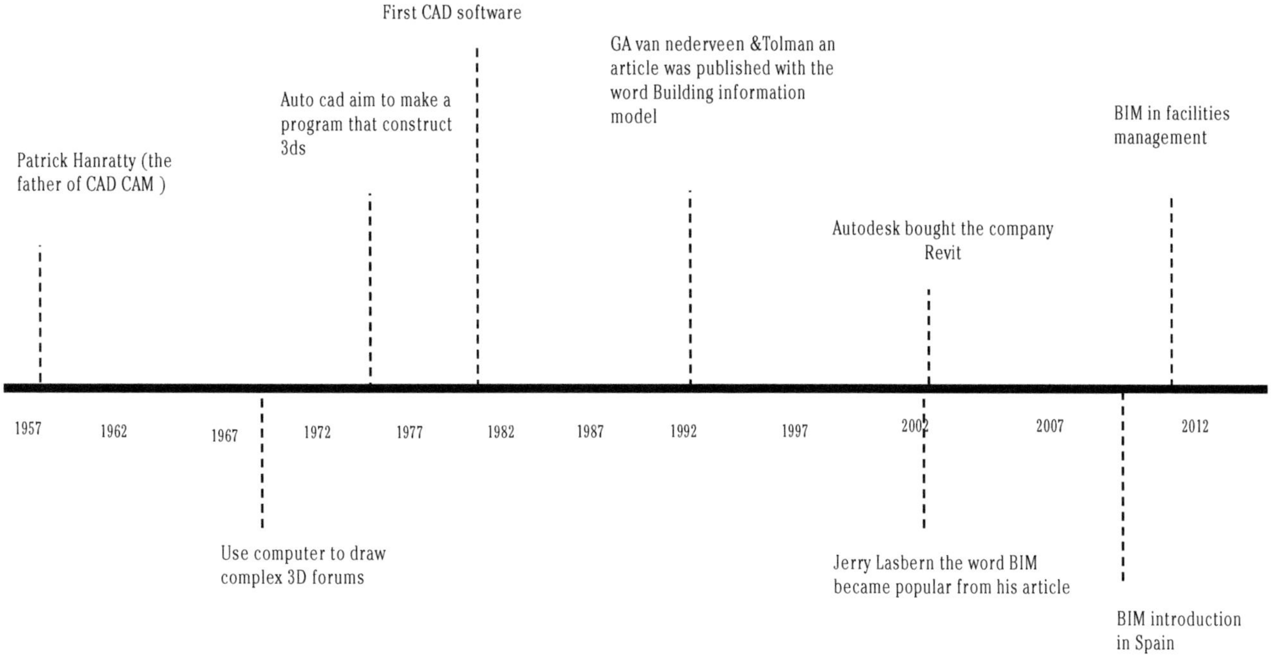

Fig. 1 BIM timeline (Author, 2019)

(Azhar et al., 2012). The BIM model carries all of the project data from the design phase until the end of the project lifecycle (Charef et al., 2018); the data is carried in the form of dimensions.

2.3.1 BIM Dimensions

This section of the research will be discussing the different parts and levels of BIM, breaking down each level and explaining its main parts.

First Dimension

This dimension is defined as the baby steps stage of the project where the design team makes the sketches and the ideas, after understanding the main scope that the owner is trying to reach (Carpenter-Beck, 2017).

Second Dimension

(X, Y).

The second dimensions are mainly the plans and sections and elevations of the project which contain the information in the shape of the working drawing (McPartland, 2017).

Third Dimension

(X, Y, Z).

It's the addition of the third dimension Z to the drawing process; thus, it is defined as the creation of graphical and non-graphical data for the model in a common data environment (CDE); the information becomes richer as the project life cycle develops (McPartland, 2017). At this level, the user uses the tools to create the graphical form of the building checking on the codes and standards required for the building and the clash detection, which is the prevention of any geometric conflict ("BIM dimensions—3D, 4D, 5D, 6D, 7D BIM explained—BibLus", 2018; Carpenter-Beck, 2017).

Fourth Dimension

(x,y,z) + Time, (scheduling).

This dimension adds the time factor to the BIM model which is represented as scheduling. This allows knowing how every component of the project will develop through time and how the project will run sequentially. The time information could be how much time it will take to get the product, how it will be installed or constructed in the project, and how much time it will need to harden or be ready to use (Charef et al., 2018). The knowledge of such data allows the planners to make an accurate program for the project time, which saves a lot of time during the project construction and allows more inventions to take place by minimizing the amount of risk (McPartland, 2017; "BIM dimensions—3D, 4D, 5D, 6D, 7D BIM explained—BibLus", 2018).

Fifth Dimension

(x,y,z) + cost, (estimating).

It's simply the 3D geometry by adding the cost factor to it; at this dimension, the planner should be able to specify every component cost during the project. The main components of the cost estimating for every element are the initial cost of the element, maintenance, and renovation (Carpenter-Beck, 2017). The elements of any project are divided into two quantities based on the actual model (as door and windows), quantities derived from the model components (as door and windows frames), and the non-modeled quantities (as the framework and joints). The calculation of the elements minimizes the waste in cost and provides an accurate way of estimating (McPartland, 2017).

Sixth Dimension

(x, y, z) + cost + time, (sustainability).
The 6D BIM comes to use after the construction phase of the project is done; it is mainly used by facility managers. They use the information gathered from the design and construction process to facilitate the decision-making and managing of the facility along with its life (Carpenter-Beck, 2017). The gathered data help the manager to reduce the amount of energy consumption along with the project life.

Seventh Dimension

(x, y, z) + cost + time, (Lifecycle management).
The seventh dimension uses the same data as the sixth dimension, where the data isn't used for the reduction of energy consumption, but in the facility management and maintenance of the project because it helps track down the relevant assets data such as operation manuals and elements renewal time, which reduce the building run cost ("BIM 3D,4D, 5D, 6D & 7D", 2018).

2.4 BIM as a Process

BIM is a Process, not a Software as published in 2015 by the contractor magazine; the misunderstanding of BIM as only software that is used to coordinate between different fields of a project is spread in the Middle East, however, BIM is mainly the process of involving all project stakeholders in the right time in the project to deliver the project is the best way, so there is some organization that has defined BIM as a collaborative design, procurement and building operation and is defined by EASTMAN as "A disruptive technology" as it will transform many aspects of the AEC industry (Eastman et al., cited in Sabongi, J. F); thus, the researcher has come to a conclusion that BIM implantation is needed to mainly support the combination and collaboration, in addition to the involvement of the important stakeholders in the early stage of the project, which conflicts with the national way of workflow (Design-Bid-Build), so BIM as a process changes the total workflow of the AEC industry as it involves the manufactures, managers, owner, design team, and suppliers to share experts in the early life of the project (Charef et al., 2018).

2.5 The Relation Between Different Stakeholders

The output of this part is summarized in two diagrams: the first one shows the relation between the different stakeholders of BIM and Gantt chart representing their duration of work and relating this work to BIM as a process and as a technology (Fig. 2); the second diagram is dividing BIM as a technology and as a process to components also showing how much information every BIM user needs to reach every position (Fig. 3).

3 Implementation Environment

This section will discuss the environment, where the implementation will take place. The environment plays a big role in the implementation process as the strategy varies from one environment to another; thus, the section will be divided into two main parts, the first part will be discussing the BIM implementation in the AEC industry on their major levels internationally, the Middle East and Egypt getting out with the conclusion of what is the current state of BIM and guidelines about the BIM in the education studying the graduate skills of every education system.

3.1 Brief About the AEC Industry

The term AEC stands for the architecture engineering industry, which contains three different players that work together to achieve the project conditions. The ingratiation of such different players such as the design team, engineers (mechanical, electrical, and civil engineers), and the contractors gives the industry the value of efficiency during the trip of reaching a common goal, Fig. 4 ("IMSCAD—Serving the Architecture, Engineering & Construction (AEC) Industry", 2018). The AEC industry ingratiates the information from the main player to make the project stronger and have no error and minimize the risk taken in every step, because in every step all the three players are there checking and analyzing, thus throughout the industry life the participants and the firms have been adding new different ways to make the integration easier and facilitate the connection process (Brian Bass, 2017).

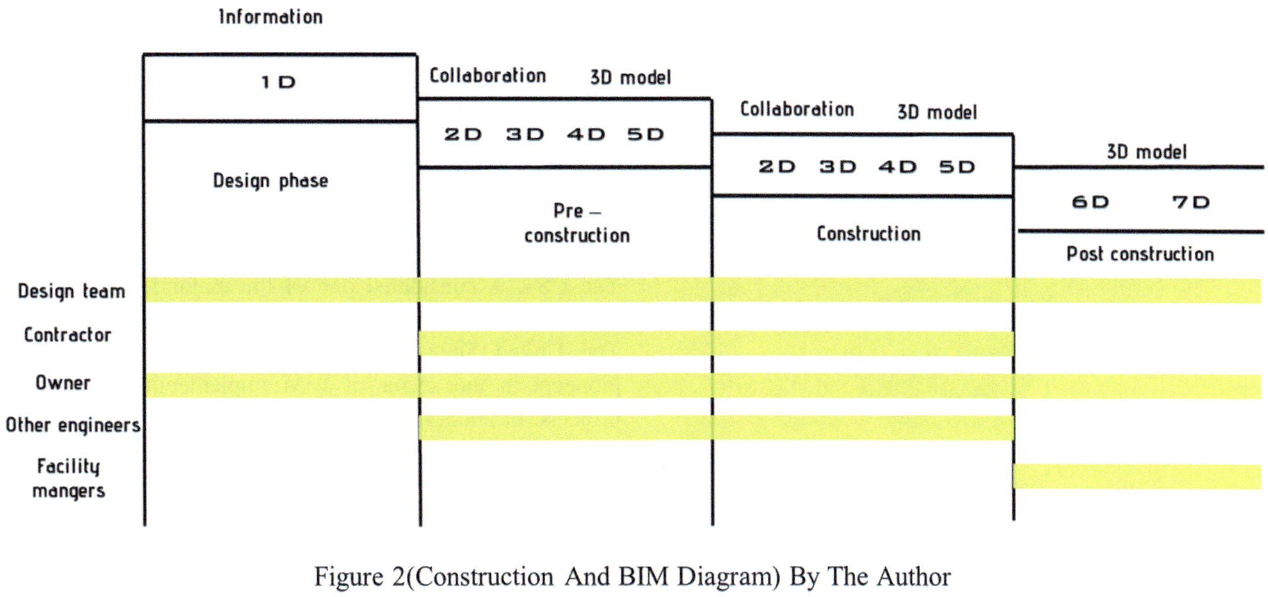

Fig. 2 Construction and BIM diagram (Author, 2019)

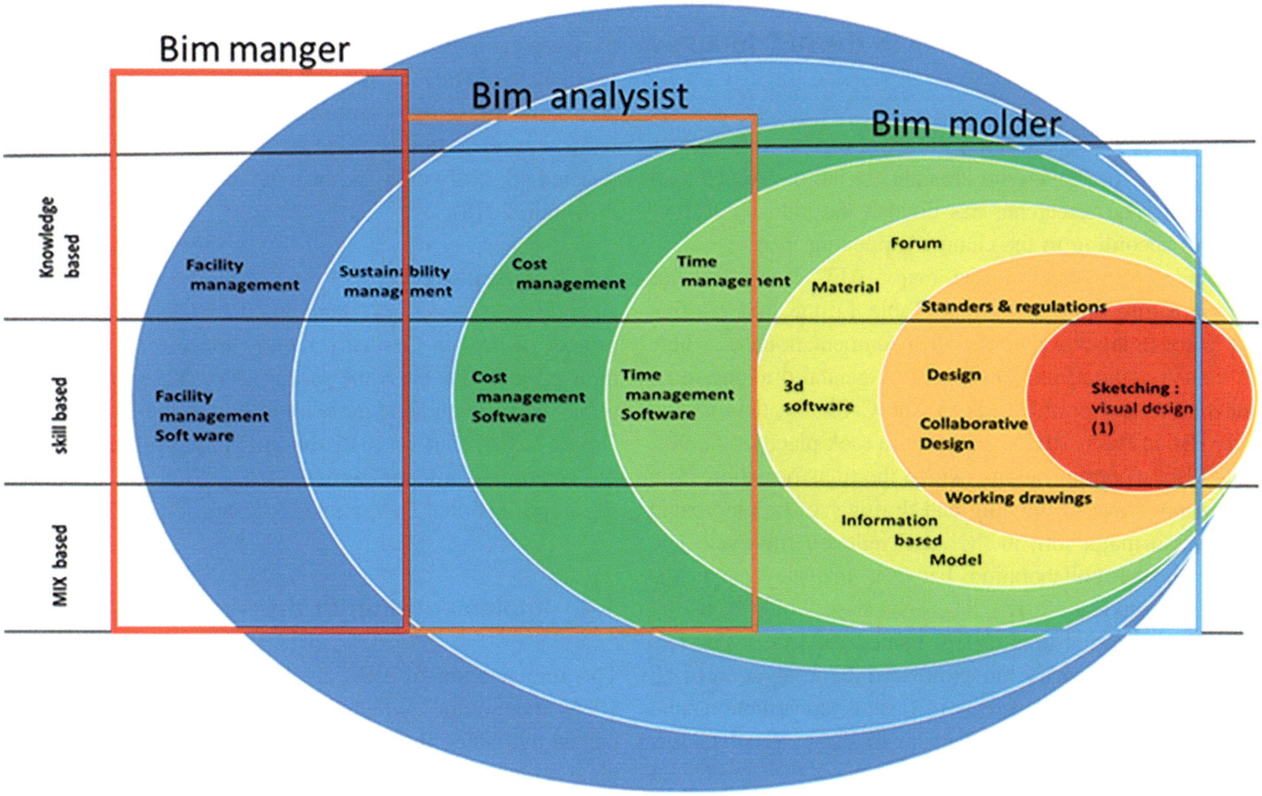

Fig. 3 BIM implementation components (Author, 2019)

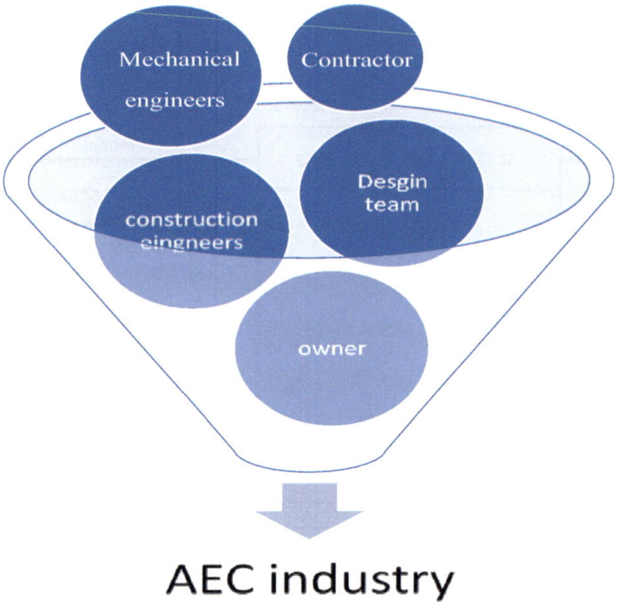

Fig. 4 AEC industry players (Author, 2019)

3.2 Implementation of BIM in the AEC Industry Eras

The industry has been around since the requirement of a human being for shelter, but the industry has developed a lot since then. The researcher has divided the industry history into eras according to the change happening in it (

Autodesk, 2018), thus the major AEC researcher has divided it into three main eras (Table 1): the first era that took place in the 80s is named the documentation era, where the industry started moving from the manual drawing, documentation, and graft boards to AutoCAD and data format study (Brian Bass, 2017); the next era took place in the 90s, named Era of optimizations, where the transformation from the manual ways of drawing and drafting to the automated ones was a major turn in the AEC industry life cycle and helped in the collaboration between the players in the industry (Autodesk, 2018). The last phase which is taking place in the current time (2018) named the Era of connection, where the industry moved toward the integration of all the stakeholders in all phases of design, construction, and managing. also of the involvement of the external factors such as environmental and economic factors by using Cloud uploading, which helps in improving the project delivery (Autodesk, 2018; Brian Bass, 2017).

3.3 Implementation of BIM Global States

As discussed before in the BIM history, it has started back in the 1980s and the implementation of such a technology was quick in many countries which made them a leader in BIM and the AEC industry nowadays. The below section will discuss the development of BIM in various countries at the international level; it also will be summarized as a chart that shows the difference between implementation percentages from 2013 to 2015 (Fig. 5).

3.3.1 United States of America (USA)

The USA is considered one of the major leaders of BIM implementation in the AEC industry (Wong et al., 2009). The United States general administration is one of the pioneers in the sector of BIM implementation in public projects. In 2003, they established a common 3D–4D for all offices to use; then in 2007, they maintained the use of BIM as an official tool in the public projects; then they established common standers and guidelines for BIM users all over the USA (Seglias & Greenhall, 2012).

3.3.2 United Kingdom (UK)

The United Kingdom strategy of BIM implementation is considered by many the most ambitious and advanced implementation program (Cable et al., 2013); these strategies have a great impact on the AEC firms as they start to update their technologies to meet the new requirement (Smith, 2014). In 2011, the government had mandated the use of BIM in publicly funded projects, then in 2014 BIM adoption reached 48; lastly in 2016, they started aiming for BIM level 2 (Singh, 2018).

3.3.3 Germany

According to a survey done in 2014, 90 percent of the project owners in Germany require the use of the BIM in their project; however, the same survey has shown that the public sector is the one more interested in the BIM adoption (Jones, 2014), but in 2015 the government has announced the creation of a team from an industry-led organization to develop a national BIM strategy (Singh, 2018).

3.4 Implementation In the Middle East

The implantation of BIM in the AEC industry started in 1987 under many different names and by different companies and since then countries have made many different plans to implement BIM in the AEC Industry; on the other hand, in the Middle East there are a few countries that have implemented the BIM. According to a survey made in 2016 on 130 of the top-ranked contracting and consulting firms, Egypt has shown that most of the construction and design firms have a high requirement for engineers with the skill of BIM. The researcher in the same paper has pointed out that BIM is expanding in the industry and the requirement for engineers and architects with such skill is

Table 1 AEC development eras (by the author)

Era name	Date	Transformed from	Transformed To	Description
Era of documentation	Took place back in the 80s	Manual drawing and documentation and draft board	AutoCAD (computer-aided design) and data format study	The transformation from the manual ways of drawing and drafting to the automated ones was a major turn in the AEC industry life cycle and helped in the collaboration between the players in the industry
Era of optimizations	Started in the 90s	The coordination between players At AEC was through a common manger	The building information modeling (BIM) implementation in the AEC industry	The industry moved to use 3D models for simulation and analysis for the building also the workflow changed from design-build to a more integrated way
Era of connection	Nowadays	Connection through meeting and feedbacks	Use of Cloud and mobile social networks	To integrate all the stakeholders in all phases of design, construction, and managing. Also of the involvement of the external factors as environmental and economic factors by using Cloud uploading, which helps in improving the project delivery

Sources Brian Bass (2017) (Autodesk)

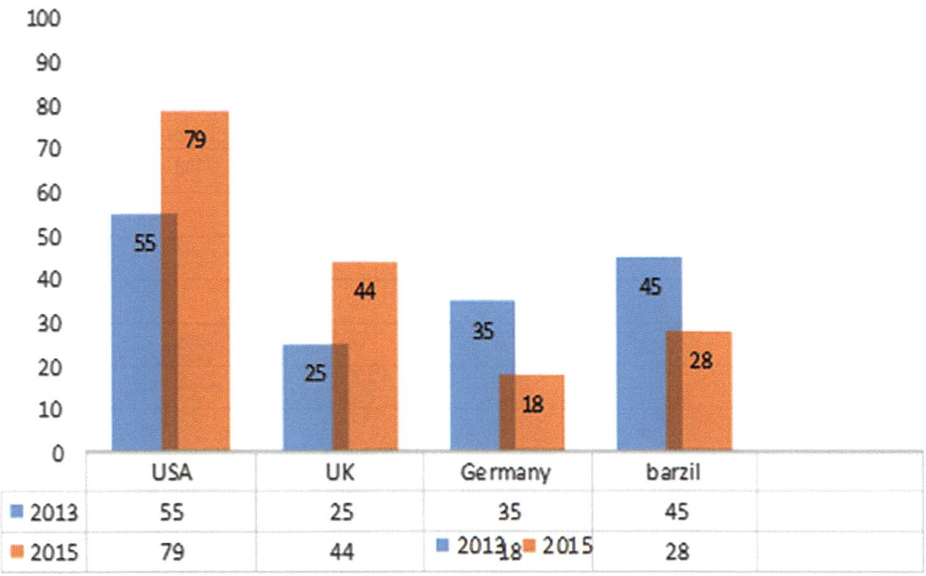

Fig. 5 Countries implementing BIM diagram (Author, 2019)

only increasing, which is creating a lack in the architecture and engineering and construction industry (AEC) (Elyamany, 2017).

In the same context, the Donya mahram research in 2016, the BIM implementation in the Middle East industry is increasing. although it's not mandatory. According to a survey made in 2016 by Michel Gerges, it was showed that the top three countries implementing BIM are the United Arab Emirates, Egypt, and Qatar, on the other hand, Lebanon and Jordon came with the lowest rate of BIM use.

3.5 Project Delivery Methods

Collaborative project delivery methods such as design-build, integrated project delivery (IPD), and public-private partnerships (PPP) are gaining ground on traditional approaches. The execution of projects is becoming more and more interactive and needs innovative practices to work together. Design and development companies seek alternate teaming arrangements, such as joint venture alliances, end customers test "big space" methods involving co-location and Cloud sharing, and multinational national teams look to more effectively collaborate.

3.5.1 Design-Build

A design-build project consists of two or more teams working together, which allows for one of the teams to concrete all the efforts on the design process, and the other team or teams concentrate on other project tasks such as the concretion process, which is different from the conventional design-bid-build project in which the design and construction team bid independently on the project, the design-build project demands that both the design and construction teams bid collaboratively on the project (Mehran, 2016) (Fig. 6).

3.5.2 Design-Bid-Build Versus Design-Build

The distinction between design-build and design-bid-build is that the former is a design-building team with one point for the owner and the latter includes the owners seeking different structures for design and development (Marco & Karzouna, 2018).

3.5.3 Contractor-Led Design-Build

Design-build designs typically have a contract between an investor and a builder under which an architectural firm is contracted for the completion of the building. This style of the design team is managed by the contractor rather than by the architect (Mehran, 2016).

3.5.4 Architect-Led Design-Build

An architect-led design-build approach is defined as an agreement between the project owner and the architect, in which the Architecture engineer is contracted to finish the project as Design and as a construction (Gazder et al., 2018).

3.5.5 2Integrated Project Delivery

Integrated Project Delivery, or IPD, is a project delivery method distinguished by early Cross-functional team coordination through all design, planning, and construction processes. Teams will collaboratively leverage the skills and insights of all participants by entering into this form of business agreement to improve project performance, increase value to the owner, decrease waste, and maximize productivity across all design, manufacturing, and construction phases. The entire process is characterized by early substantive involvement by all main stakeholders, from definition to construction (Pöyhönen et al., 2017).

3.6 New Jobs for Architects in the AEC Industry

The implementation of BIM into the AEC industry has changed the industry workflow, thus some new jobs, opportunities, and newly made positions to adapt to the change happing in the workflow as BIM manager which came as result of the need for the one responsible for guidelines, standers, and workflow (Pell_Frischmann, 2017), also BIM coordinator one responsible for coordinating the

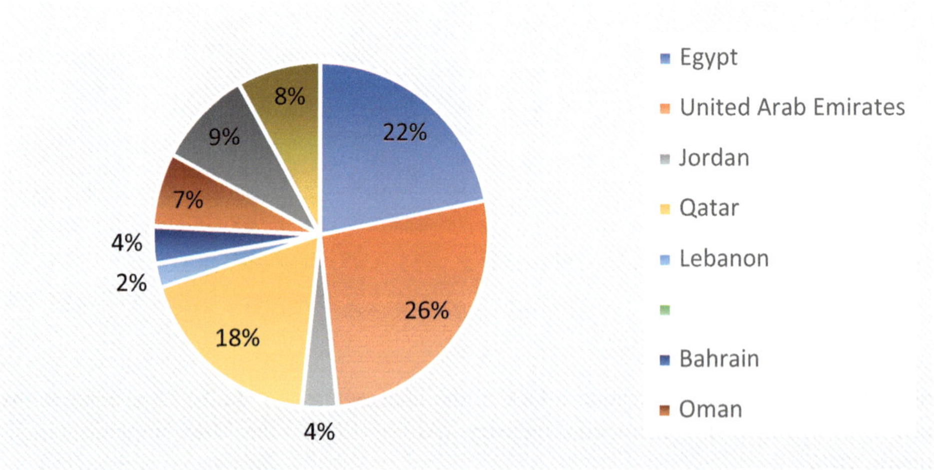

Fig. 6 Implementation in the middle east (Author, 2019)

process of generation models and managing data, the BIM coordinator must be well educated in term of architecture (Barison, 2010), Bim consultant is an expert that aids the firm in the baby steps of implementation (Uhm et al., 2017).

3.7 The Architecture Education System in Egypt

In 2016, a study concerning Middle East BIM users showed that the percentage of people having any training on BIM is that 34.89% are self-taught, while 26.25% undergo in-house training in addition to 25% that was undergo industry-led training, and only 14% undergo college courses. That justifies the source of lack created in the AEC industry firms, as 86% of the trained people are only familiar with BIM as a program and not as a process, which is considered as a major conflict for practitioners (Gerges et al., 2013). Such a study concluded that the implementation of BIM in the current education system is necessary, thus the flowing section will discuss various examples using the BIM in the education system and others that are not implementing BIM going through them in the approach of knowing the skill and practice-based modules and what is the exact graduate skill for every university (Fig. 7).

3.8 AEC Industry Needs for BIM Practitioners

This section ends with the conclusion of knowing what the current AEC industry requirements in terms of the fresh graduates are; in addition to knowing that the required changes such as jobs, for example, the requirements of the BIM manager is different from BIM molder, thus the following matrix shows different jobs and their different requirements, Fig. 4. In addition to knowing the difference between ordinary workflow and integrated project delivery method which is used in the BIM process (Figs. 8 and 9).

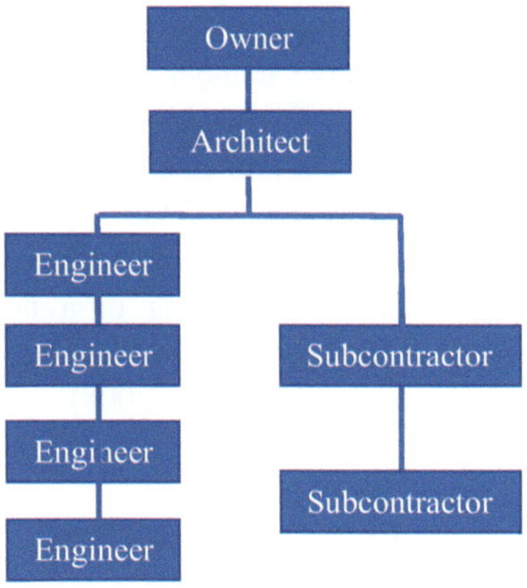

Fig. 8 Typical design-build team (Author, 2019)

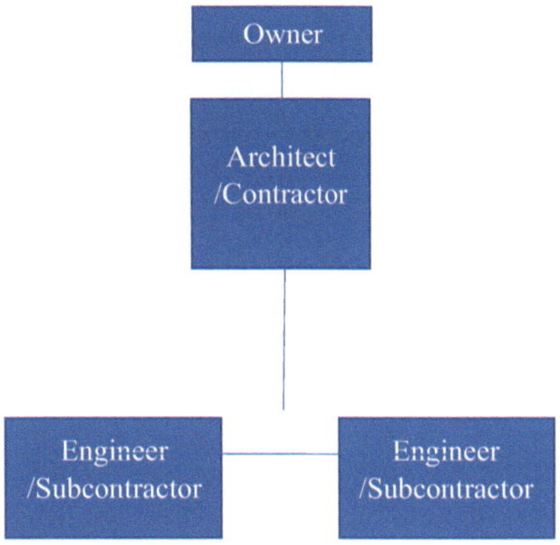

Fig. 9 Integrated team (Author, 2019)

4 Implementation of BIM in Education

Throughout the previous studies, it has been clear that there is a large number of firms and companies that are used or starting to use the BIM in their projects, thus the requirement

Fig. 7 BIM education in the middle east (Author, 2019)

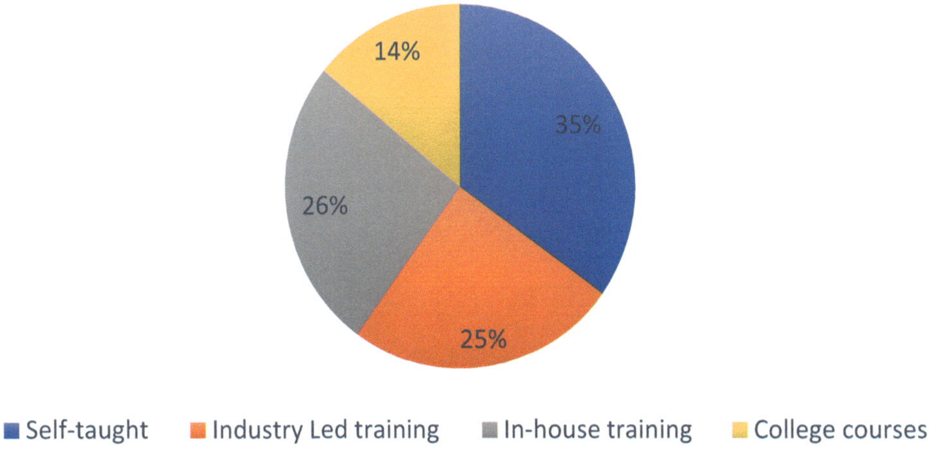

for engineers who are familiar with such a skill is increasing and as a response, a large number of universities started implementing BIM into their education curriculum (Holland & Messner, 2010), but the implementation of such a new trend in the education system will face a large number of obstacles, which are divided into three main categories: the difficulties of learning and teaching in a new way, the misunderstanding of BIM as a process, and the issues of the surrounding environment (Kymmell, 2008).

4.1 History of the Implementation:

Since the early 90s, BIM researchers have started implementing BIM in education and the very first one was the *college of architecture at Texas A&M University;* then in 2005, many other universities began to combine the construction management and integrated design studios; then in 2008, the interdisciplinary studios began to start mixing six different engineers in the AEC industry, followed by many other university plans for implementation (Fig. 10) (Barison, 2010).

4.2 Implementation of Collaborative Methods

The researcher has concluded after studying different ways of collaborative working methods and that was to divide collaborative methods into three main types a study was made in the same paper showing the percent of every category used in the international school (Fig. 11).

4.2.1 Single-Courses

This way of teaching is only giving the student the information through only one discipline as to how to create, develop, and analyze BIM models or real collaboration, but with the student in the same department there are two examples of such a method, first, the construction management students at Tongji University, second is the Queensland University of Technology that simulates a real BIM collaboration (Nielsen et al., 2009) (Fig. 12) (Fig. 13).

4.2.2 Interdisciplinary

This category comes from a unique experience in Australia at the University of New South Wales, where the students from different departments worked together on a common project, then some other college started following the same method, such as Penn State, Oklahoma University, Auburn University, Cal Poly, and the University of Maryland (Barison, 2010; Holland & Messner, 2010). To Simplify, the researcher has come to a definition for this method, which is a design team or an individual designer is integrating different departments knowledge to reach a common goal (Kocaturk & Kiviniemi, 2016) (Fig. 14).

4.2.3 Distance Collaboration

In this category, the students from different departments and different schools are working together to reach a common goal (Kocaturk & Kiviniemi, 2016). Examples for such a method is existing in different countries such as the University of Nebraska Lincoln, Montana State University, and the University of Wyoming which have developed two different programs for applying such a method (Fig. 14).

4.3 Levels of the Implementations

The implementing method also varies depending on the level of the students' knowledge and the level of the staff teaching the courses, thus there are three main levels for integration: first is introductory, second is intermediary, and last is the advanced level (Barison, 2010).

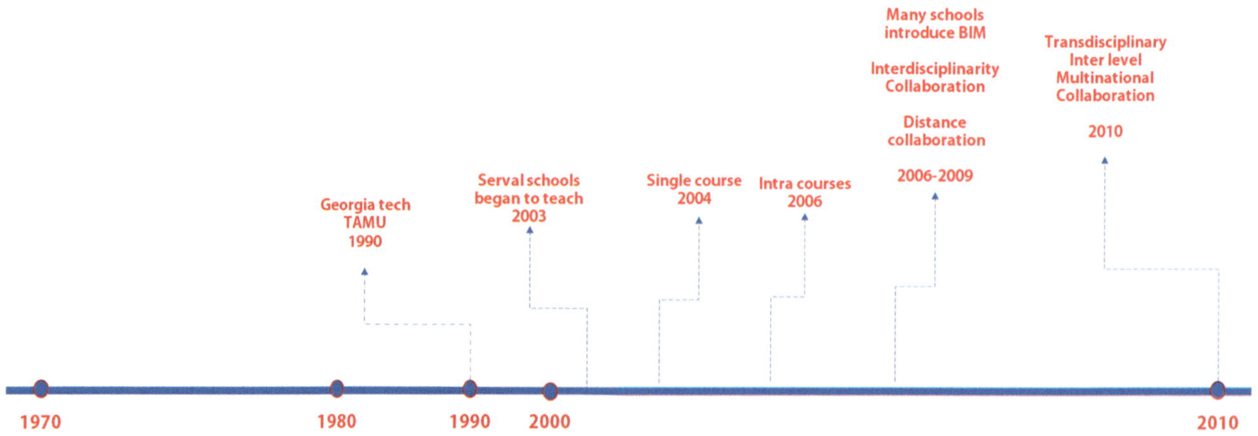

Fig. 10 Universities implementing the BIM timeline (Author, 2019)

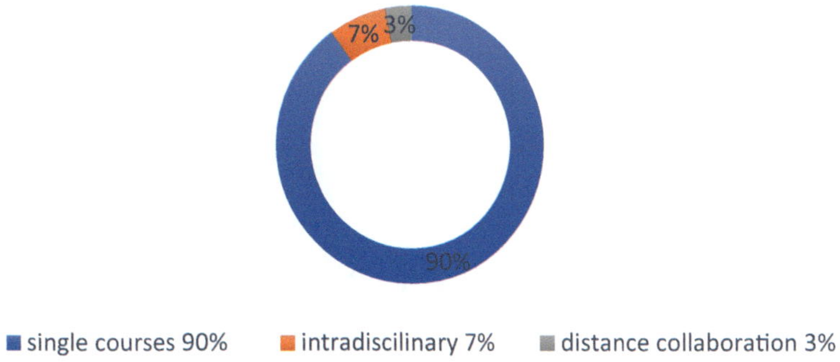

Fig. 11 BIM education (Author, 2019)

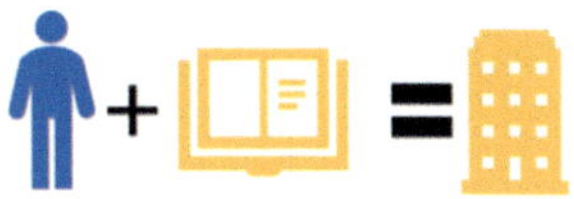

Fig. 12 Single course (Author, 2019)

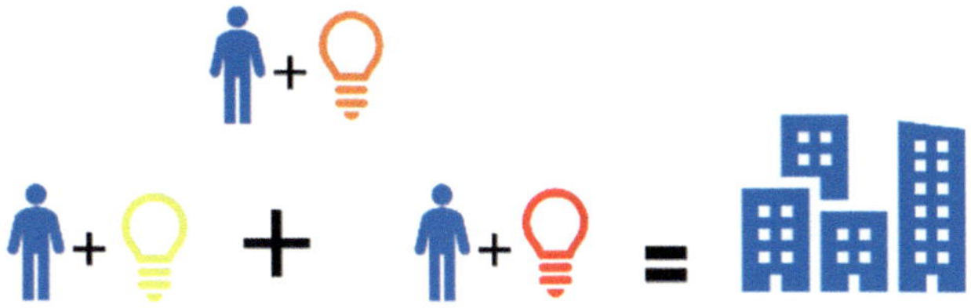

Fig. 13 Interdisciplinary (Author, 2019)

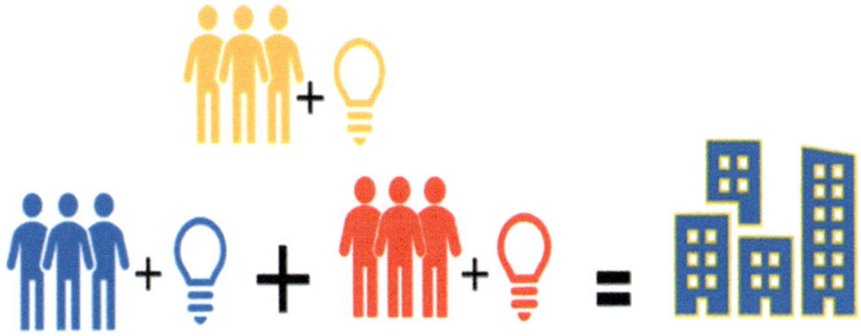

Fig. 14 Distance collaboration (Author, 2019)

4.3.1 Introductory

At this level, the students start from scratch, thus there are no requirements for this level of education. The student's main course at this phase is a digital graphics representative course, then it's required of him to submit a simple project at the ending of the project. The graduate from this course can work as a BIM modeler.

4.3.2 Intermediary

At this level, the student is required to BIM concepts, construction materials, and BIM tools. The student's main course at this phase is an integrated design studio and building technology, then the student is required to submit two projects: a common project and a simple rectangle project. The graduate from this course can work as a BIM analyzer.

4.3.3 Advanced

At this level, the student is required to know four main points to enter the course construction methods, BIM and application tools, building technology, and professional practice. There are two main courses at this level, a collaborative design studio and a construction management studio, then the student is required to submit a real-time comprehensive building project. The graduate of such a course is ready to work as a BIM manager (Fig. 15).

4.4 Obstacles and Risks

Many obstacles are surrounding the implementation of BIM in the architecture education system, first, the lack of BIM knowledge among the staff and also a lack of BIM-educated tutors, second is the conflict of teaching it as a technology or a process theses where according to a paper published in 2013 (Mandhar & Mandhar, 2013). In addition to that,

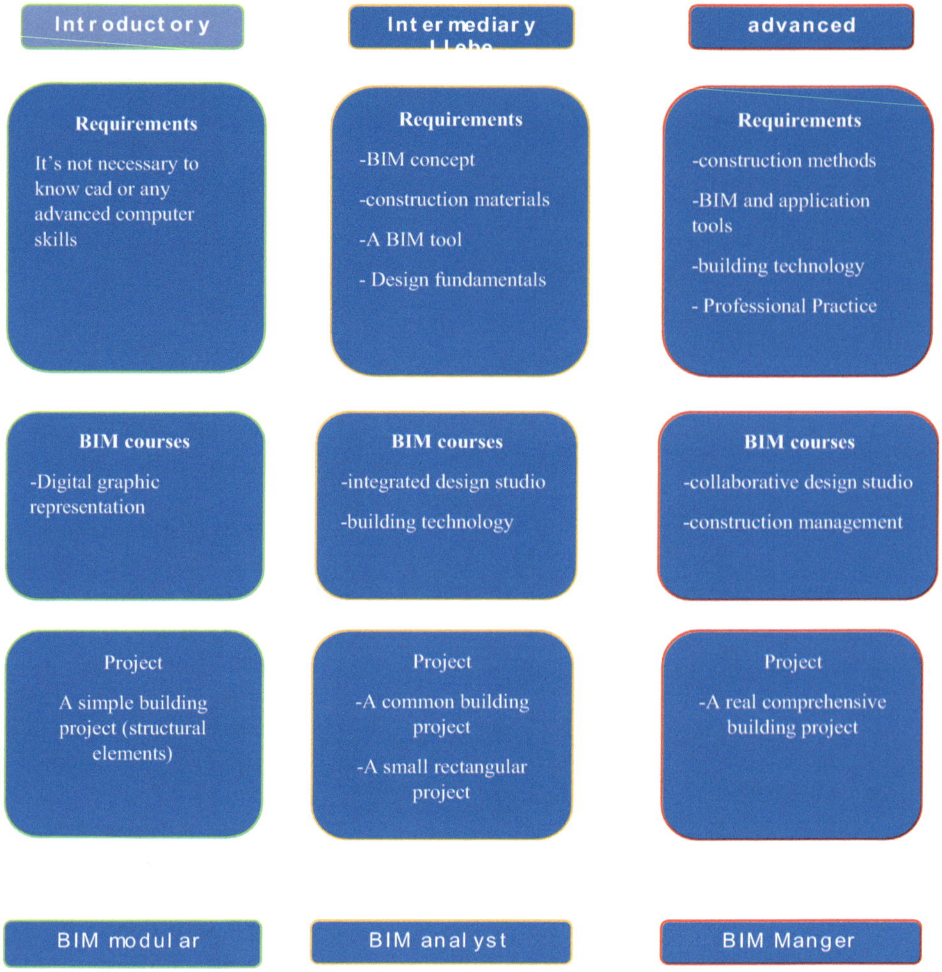

Fig. 15 BIM levels of implementation summarized (Author, 2019)

Fig. 16 Obstacles checklist

Obstacles check list:

- ☐ Staff needs training
- ☐ Misunderstanding of BIM as a process
- ☐ Insufficient resources
- ☐ Student have difficulty in learning
- ☐ Resistance to change
- ☐ No room in curriculum
- ☐ No reference materials
- ☐ lack of teamwork am
- ☐ knowledge
- ☐ Staff availability
- ☐ student's resistance

Kymmell divided them into three main categories: first is the misunderstanding of BIM as a process, the difficulty of student learning it, and lastly, the training of the academic staff and their resistance to change (Kymmell, 2008). A survey made in 2011 among year 4 undergraduate architecture students has shown that one of the obstacles for such an implementation is that there is no space in the current curriculum and no resources (Sabongi, 2011); thus, to summarize all the obstacles and risks, the following chart was made based on 2011 (Becerik-Gerber et al., 2011), lastly, a checklist was designed based on the review to previous case studies of implementation (Fig. 16) (Fig. 17).

Fig. 17 Guideline diagram

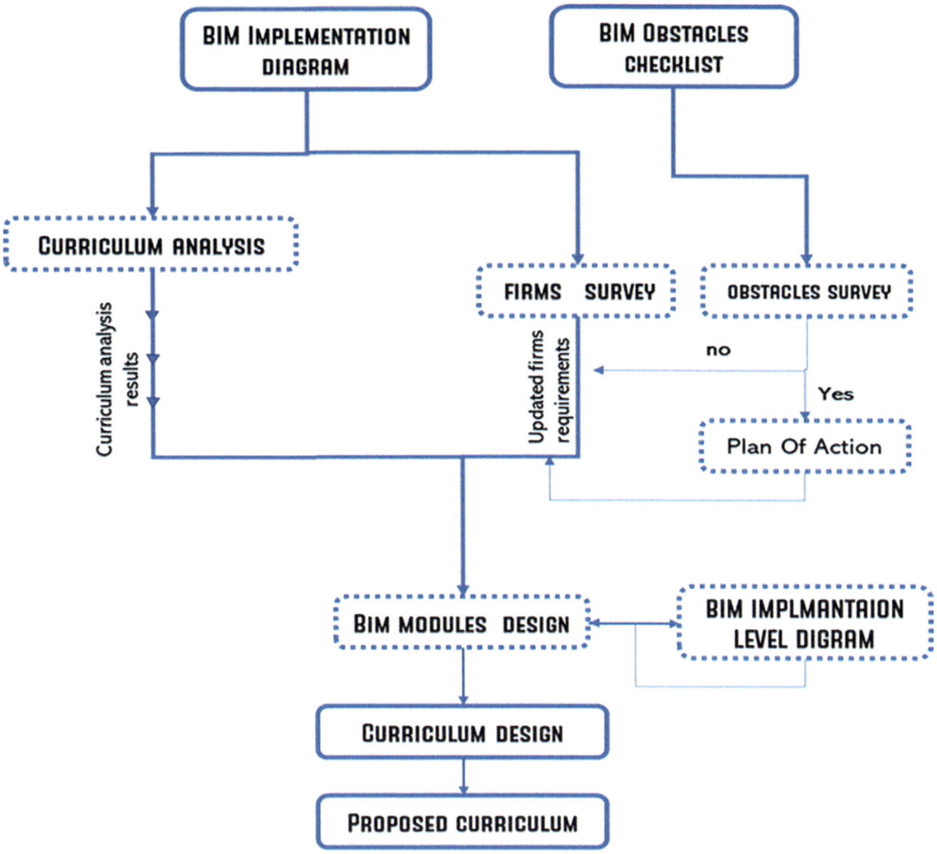

5 Conclusion

As conclusion, implementation Guidelines are designed; these implantation guidelines are designed to be applied to any architecture education system in Egypt: first, the one who is working on this curriculum needs to fully understand the BIM diagram that is discussed before, then there are three main tasks to be done; all of these steps can be derived from the BIM diagram, first, analyzing the current curriculum of the university, second, a survey among the AEC firms is made to know the priorities of the current state of the market, third is a survey to understand the obstacle in the implementation environment and develop a plan of action to overcome these obstacles, last, the curriculum is modified based on the current requirements of the AEC industry in Egypt. All the steps are summarized in the guideline diagram (Fig. 17).

References

Autodesk. *How technology will disrupt and transform the AEC industry*. youtube. https://www.youtube.com/watch?v=HIM_RUYX3wk.

Azhar, S., Khalfan, M., & Maqsood, T. (2012). building information modeling (BIM): Now and beyond. *Australasian Journal of Construction Economics and Building, 12(4), 15–28*. https://www.researchgate.net/publication/272491129_Building_information_modeling_BIM_Now_and_beyond.

Azhar, S. (2011, 11). Building Information modeling: Trends, Benefits, Risks, and challenges. *ascelibrary.org*. Retrieved from ascelibrary.org.

Barison, M. B. (2010). An overview of BIM specialists roles and responsibilities of BIM specialists. https://www.researchgate.net/publication/289437352_An_overview_of_BIM_specialists.

Bass, B. (2017). What is the AEC industry. *bizfluent*. https://bizfluent.com/info-8477602-aec-industry.html. Accessed 10 Nov 2018.

Becerik-Gerber, B., Gerber, D. J., & Kum K. (2011). The pace of technological innovation in architecture, engineering, and construction education: Integrating recent trends into the curricula. *Electronic Journal Of Information Technology In Construction, 16*, pp. 411–432. https://www.researchgate.net/publication/238316836_The_pace_of_technological_innovation_in_architecture_engineering_and_construction_education_Integrating_recent_trends_into_the_curricula.

Cable, V. F., Michael, and David Higgins. (2013). Construction in 2025. http://www.bis.gov.uk/assets/biscore/innovation/docs/b/12-1327-building-information-modelling.pdf%5Cnhttps://www.gov.uk/government/publications/construction-2025-strategy%5Cnhttps://www.gov.uk/government/uploads/system/uploads/attachment_data/file/210099/.

Carpenter-Beck, D. (2017). Defining the dimensions of BIM | Sage Advice US. SAGE ADVICE US. https://www.sage.com/en-us/blog/defining-dimensions-bim/%0A.

Charef, R., Alaka, H., & Emmitt, S. (2018). Beyond the third dimension of BIM: A systematic review of literature and assessment of professional views. *Journal of Building Engineering, 19*(April), 242–57. https://www.researchgate.net/publication/324947498_Beyond_the_Third_Dimension_of_BIM_A_Systematic_Review_of_Literature_and_Assessment_of_Professional_Views.

Don, Y., et al. (2014). Outcomes and impact assessment on skill courses program in community college Malaysia. *European Scientific Journal,* 1857–7881. https://www.researchgate.net/publication/267331399_OUTCOMES_AND_IMPACT_ASSESSMENT_ON_SKILL_COURSES_PROGRAM_IN_COMMUNITY_COLLEGE_MALAYSIA.

Elyamany, A. H. (2017). Current practices of building information modelling in Egypt current practices of building information modelling in Egypt. *Inderscience Enterprises Ltd.* https://www.researchgate.net/publication/309277267.

El-Shazly Amira (2018) "The implementation of BIM towards sustainable construction industry in egypt." The American University in Cairo. http://dar.aucegypt.edu/bitstream/handle/10526/5454/THE IMPLEMENTATION OF BIM TOWARDS SUSTAINABLE CONSTRUCTION INDUSTRY IN EGYPT.pdf? sequence=1

Gazder, U., Shakshuki, E., Adnan, M., & Yasar, A. U. H. (2018). Artificial neural network model to relate organization characteristics and construction project delivery methods. *Procedia Computer Science, 134*, 59–66. https://doi.org/10.1016/j.procs.2018.07.144

Gerges, M., Abbasnejad, B., & Moud, H. I. (2013). BIM and basic challenges associated with its definitions, interpretations, and expectations. *International Journal of Engineering Research and Applications (IJERA), 3*(2), 287–294. http://www.ijera.com/papers/Vol3_issue2/AQ32287294.pdf.

Holland, R., & Messner, J. (2010). Integrated design courses using BIM as the technology platform. In *Proceedings of the BIM ….* http://php.scripts.psu.edu/users/r/l/rls5008/CV/LinkedPapers/Holland et al 2010 Integrated Design Courses Using BIM as the Technology Platform.pdf.

IMSCAD—Serving the Architecture, Engineering & Construction (AEC) Industry. (2018). Retrieved from http://www.imscadglobal.com/industries-AEC.php.

Jones, S. (2014). Recent smart market BIM research by *McGraw Hill Construction.* https://cdn.ymaws.com/www.nibs.org/resource/resmgr/BSA/20140108_moa_jones.pdf.

Joseph, J., BIM Cad, and Technologies Saic. (2010). BIM titles and job descriptions: how do they fit in your organizational structure? http://aucache.autodesk.com/au2011/sessions/4436/class_handouts/v1_DL4436_Joseph_BIM_Titles_Job_Descriptions_JJ.pdf.

Kocaturk, T., & Kiviniemi, A. (2016). Challenges of integrating BIM in architectural education challenges of integrating BIM in architectural education. https://www.researchgate.net/publication/296704398_Challenges_of_Integrating_BIM_in_Architectural_Education.

Kymmell, W. (2008). Building information modeling: Planning and managing construction projects with 4D CAD and simulations (McGraw-Hill Construction Series). New York, Chicago, San Francisco, Lisbon, London, Madrid, Mexico City, Milan, New Delhi, San Juan, Seoul, Singapore, Sydney, Toronto. https://www.accessengineeringlibrary.com/browse/building-information-modeling-planning-and-managing-construction-projects-with-4d-cad-and-simulations-mcgraw-hill-construction-series#fullDetails.

Mandhar, M., & Mandhar, M. (2013). © I A E M E BIMING the architectural curricula—Integrating. *International Journal of Architecture (ija), 1*(1), 1–20.

Marco, A. D., & Karzouna A. (2018). Assessing the benefits of the integrated project delivery method: A survey of expert opinions. *Procedia Computer Science, 138*, 823–828. https://linkinghub.elsevier.com/retrieve/pii/S187705091831754X.

McPartland, R. (2017). BIM dimensions—3D, 4D, 5D, 6D BIM explained. https://www.thenbs.com/knowledge/bim-dimensions-3d-4d-5d-6d-bim-explained%0A. Accessed 9 Nov 2018.

Mehran, D. (2016). Exploring the adoption of BIM in the UAE construction industry for AEC firms. *Procedia Engineering, 145*, 1110–18. https://doi.org/10.1016/j.proeng.2016.04.144.

Mordue, S. (2015). BIM definitions, pp. 6–7. https://www.ace-cae.eu/fileadmin/New_Upload/3._Area_2_Practice/BIM/Other_Docs/1_S.Mordue_Definition_of_BIM_01.pdf.

Nielsen, D., Fleming, M., & Kumarasuriyar, A. C. (2009). Queensland University of technology "An Innovative Learing Model for Teaching Architectural Technology Using Building Information Modeling: A Queensland University of Technology Prespective." Queensland University of Technology.

Pell_Frischmann. (2017). BIM manager. https://www.indeed.co.uk/cmp/Pell-Frischmann/jobs/Bim-Manager-4d750fbff2515796?sjdu=QwrRXKrqZ3CNX5W-O9jEvRHL52HFQyECqJqzhtRwd6dQk4gshM5ufBx2ZOFuHbkRu7VWnQgt5yol422-xJD-0Q&tk=1c3h4d92h944sfrv

Pöyhönen, P., Sivunen, M., & Kajander, J. K. (2017). Developing a project delivery system for construction project—A case study. *Procedia Engineering, 196*(June), 520–526. https://doi.org/10.1016/j.proeng.2017.07.233

Sabongi, F. J. (2011). *Proceedings of the 45th annual conference associated schools of construction "The Integration of BIM in the Undergraduate Curriculum: An Analysis of Undergraduate Courses."* https://www.researchgate.net/publication/267703625_The_Integration_of_BIM_in_the_Undergraduate_Curriculum_an_analysis_of_undergraduate_courses.

Seglias, C., & Greenhall P. (2012). Tasmanian building and construction industry training board building information modelling (BIM): An introduction and international perspectives. https://www.researchgate.net/publication/316375059_Building_Information_Modelling_BIM_an_Introduction_and_International_Perspectives.

Smith, P. (2014). BIM Implementation—Global strategies. *Procedia Engineering, 85*, pp. 482–492. https://www.sciencedirect.com/science/article/pii/S1877705814019419.

Technology Disruptors to AEC industry | Autodesk. (2018). Retrieved from https://www.autodesk.com/solutions/bim/hub/three-technology-disruptors-to-aec-industry#ad-technology-disruptor.

Uhm, M., Lee, G, Jeon, B. (2017). Automation in construction an analysis of bim jobs and competencies based on the use of terms in the industry. *Automation in Construction 81*(October 2016), 67–98. https://doi.org/10.1016/j.autcon.2017.06.002.

Wong, A., F. K. W. Wong, & Nadeem, A. (2009). Comparative roles of major stakeholders for the implementation of BIM in various countries. In *Proceedings of the international conference on changing roles:* New Roles, New Challenges. https://www.researchgate.net/publication/228743459_Comparative_Roles_of_Major_Stakeholders_for_the_Implementation_of_BIM_in_Various_Countries.

Application of 3D Printing in Architecture: Taking Jiangning Sewage Treatment Equipment House in Nanjing as an Example

Tianran Hua

Abstract

Construction industry is gradually affected by the progress of 3D printing technology in recent years, and this technology is already widely used in architecture design. This article analyzes the application situation of architectural 3D printing technology to expound the advantages of architectural 3D printing technology applied in building construction and to point out the difficulties of using this technology in construction. Finally, the article analyzes the application and development of 3D printing technology in the construction with the case of the sewage disposal system house in Nanjing city, hopes to provide some reference for the 3D printing technology of architecture, and aims to put forward some instructive ideas for architectural 3D printing.

Keywords

3D printing technology • Status of application • Standard component split

1 Introduction

For a long time, construction methods of architectural engineering have been limited by traditional construction tools and techniques. On the one hand, architects' rich imagination and creativity make it difficult to put their works into practice. On the other hand, the rough construction technology also damages the environment seriously, which is resulting in huge resource consumption and waste. With the development of global economy and the progress of social science, the new architectural form of light quality, high strength, and green environmental protection intellectualization has become the development trend of future architecture. As a result the traditional construction methods with high consumption, high pollution, and low efficiency are bound to be eliminated, and the construction industry needs to find new methods. The 3D printing digital construction technology will be an effective solution, and its industrialized construction mode will bring earth-shaking changes to the construction industry.

This article summarizes the current status of the application of 3D printing technology in the construction field, discusses the advantages and disadvantages of various 3D printing technologies used in the construction field, and proposes some 3D printing construction methods that conform to the current technology status. At the same time, the design and construction of the actual case of "Nanjing Jiangning Wastewater Treatment Equipment House" will be summarized to analyze the advantages and disadvantages of the existing 3D printing technology in the construction field. Finally, a solution was proposed: the construction was completed by printing the components at the factory and transporting them to the site for assembly. This method can solve the current problems such as the limited volume of printed buildings.

2 The Basic Principle of 3D Printing

3D printing is an emerging technology that is developing rapidly in the manufacturing industry. It is called "a manufacturing technology with industrial revolution significance". The technology is to design data model by using computer software, and then print the product layer by layer with liquefied, powdered, and silky solid materials for specific forming equipment.

3D printing digital construction technology is to combine 3D printing technology with the design and construction of buildings. It integrates computer technology, numerical

T. Hua (✉)
Department of Building Science and Technology, School of Architecture, Southeast University, Nanjing, 210096, China
e-mail: 512418882@qq.com

F. Rosso et al. (eds.), *Advances in Architecture, Engineering and Technology*, Advances in Science, Technology & Innovation, https://doi.org/10.1007/978-3-030-86913-7_12

control technology, material forming technology, etc. and adopts the basic principle of material layering and superposition, the basic information of the 3d model of the building is processed and output by computer, the model is decomposed into laminated files, then the laminated files are corrected, and the correct numerical control program is generated. Finally, the CNC system controls the mechanical device to move according to the specified path to realize the automatic construction of the structure. In addition, construction completed by mechanical arm or robot is also a digital construction technology of building 3D printing.

3 Current Status of 3D Printing Applications

3D printing in architecture was first proposed by American scholar Joseph Pegna in 1997. It is a construction method in which cement materials are added layer by layer and selectively solidified into components. The relatively mature 3D printing methods can be divided into four categories: D-shape printing, Contour Crafting, Concrete Printing, and the construction method of three-dimensional structure of materials driven by large mechanical arms.

3.1 D-shape Printing

The D-shape printing was invented by Italian inventor Enrique Dini. The D-shape printer has hundreds of nozzles at the bottom, which can spray magnesium adhesives. Sand spraying on the adhesives can be gradually cast into stone solid. In the working state, the 3D printer moves back and forth along the horizontal axis beam and four vertical columns and form 5 to 10 mm thick layer. The printer can be operated by AUTOCAD software, the complete structure similar with marble material, which is stronger than concrete and does not require a built-in iron pipe for reinforcement. In fact, this method is similar to selective powder deposition, and the material used for printing is magnesium oxchloride cement. At present, the printer has successfully constructed the internal curve, splitter, catheter, and hollow column structures (Fig. 1) (Ding et al., 2015).

3.2 Contour Crafting

"Contour Crafting" was proposed by Biloc Horshnevis, professor of industrial and systems engineering at the University of Southern California. Different with the D-shape printing, the materials of Contour Crafting are extruded from the nozzle. The nozzle will spray concrete materials at the designated direction, and then the spatula attached to both sides of the nozzle will automatically stretch out, and the plasticity of concrete materials is required by the computer. The characteristic of Contour Crafting is that it does not need to use molds, and the outline of the building printed by the printer will become part of the building, which greatly improves the building efficiency (Fig. 2) (Yin et al., 2018).

3.3 Concrete Printing

Concrete printing was proposed by the faculty of Architectural Engineering, University of Loughbourg, UK. The technology is similar to the contour technology which is based on the way of concrete spraying and piling forming. The concrete components are constructed by stacking method using nozzles according to the designed path according to the requirements of computer. The studio developed the polypropylene fiber concrete suitable for 3D printing, passed the stress test, and confirmed that the concrete can be used for concrete printing technology and also used concrete printing technology to produce concrete components. Now many studios have used the self-developed concrete composite materials to print out the satisfactory concrete components with the method of concrete 3D printing (Fig. 3) (Cui, 2017).

3.4 Construction Method of Large Mechanical Arm Drive

Since 2006, the teams of Fabio and Matthias from Switzerland have been working on the digital design and construction of large robotic arms, including brick stacking technology. Bricks are stacked with bricks as material units, and the stacked bricks are grasped by NC robots in dislocation form. The upper and lower bricks are reinforced with epoxy resin binder. By this way, they construct a "dynamic brick wall" which has a more than 300 square meters external facade. In the past two years, researchers have developed a new technology of brick grabbing and stacking with small robotic aircraft, which make the work more efficiency and freedom. This method takes bricks as building materials, combines the original building materials with digital construction technology, and highlights the beauty of non-linear building. However its construction size is still limited by machinery.

Now many studios focus on the research and development of on-site construction robots for 3D printing buildings. The main idea of this method is to develop a multiple and combinable on-site intelligent robot, including on-site wall printer robots, grouting and plastering robots, steel bar assembly robots, etc. These different types of robots cooperate to complete the building (Ding et al., 2015).

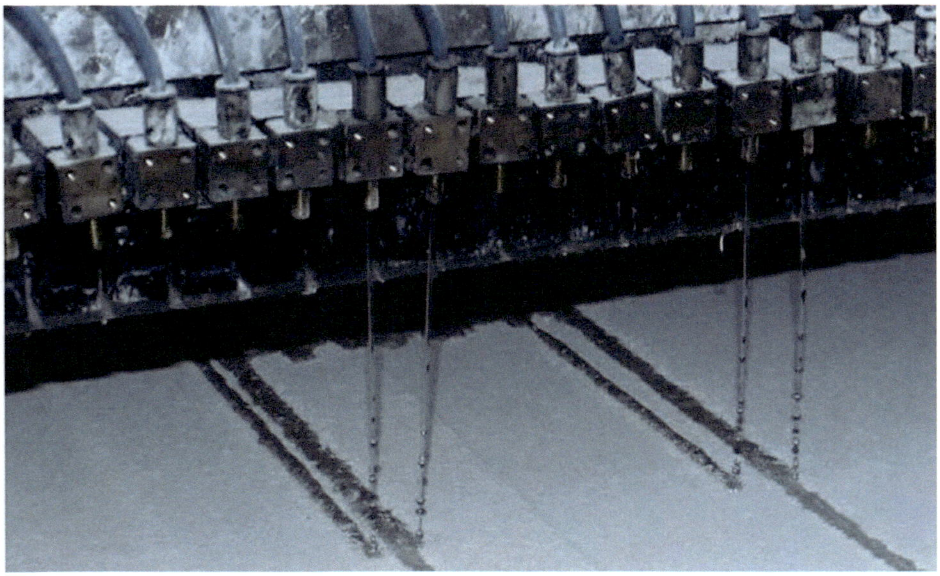

Fig. 1 D-Shape printing

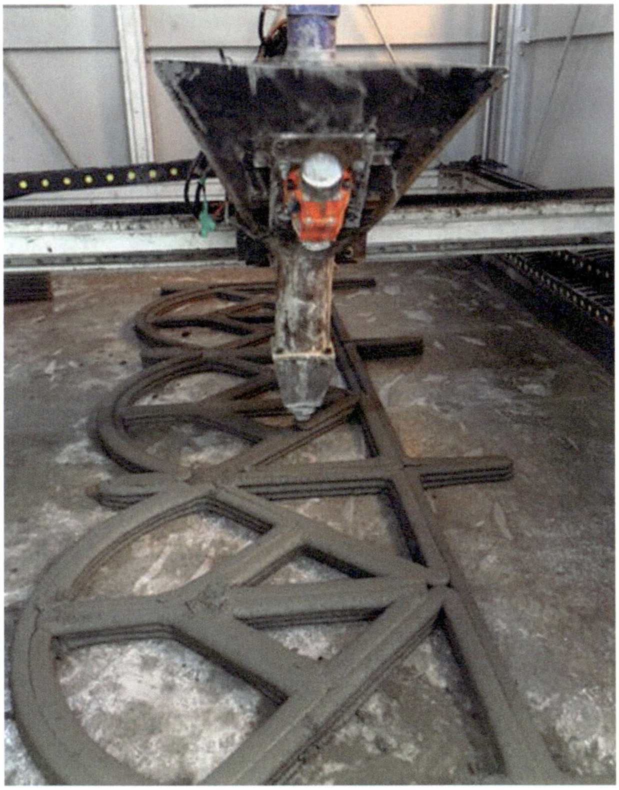

Fig. 2 Contour crafting

4 Advantages and Difficulties of 3D Printing Application in Architecture

4.1 Advantages of Building 3D Printing

4.1.1 Simplified Construction Process

Traditional construction requires a large number of workers to build from the foundation and gradually complete the complicated building construction. Traditional construction mode costs high and complicates the construction process, which reduce construction efficiency while consuming more time. Obviously, traditional construction mode has begun to fall behind the needs of the developing society. However, if the construction 3D printing technology is applied, most of the work will be completed by 3D printing equipment, which can effectively save human resources, reduce construction cost, simplify construction process, and improve construction efficiency.

4.2 Construction is More Environmentally Friendly

Abundant construction waste will be generated in the traditional construction process, building resources are not rationally utilized, and the phenomenon is relatively serious.

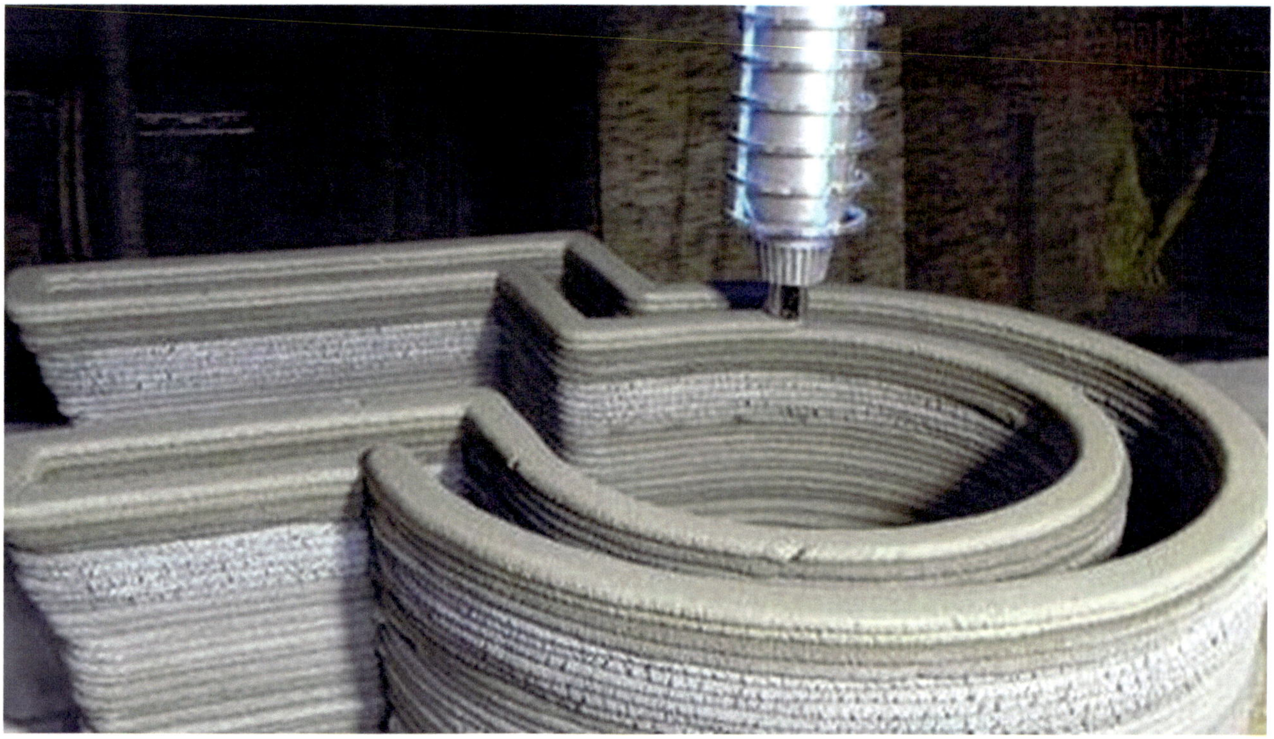

Fig. 3 Concrete printing

Otherwise, the materials of 3D printing buildings can be derived from the waste materials, industrial waste, and mine tailing, which not only reduces the production cost of buildings but also recycling the waste and realizes the protection of environment and natural resources as well.

In addition, the traditional construction mode will cause noise pollution, dust pollution, and garbage pollution, which not only affect the natural ecological environment, but also reduce the quality of neighbor residents' life. As 3D printing avoid extensive construction activities, fundamentally reduce the noise pollution, dust pollution, and garbage pollution during construction, which is environmentally friendly.

4.2.1 Reduce Construction Difficulty

With the rapid development of social economy, personalized requirements are getting higher. Architectural style is developing diversified. According to the architects' imaginative and creative works, the difficulty of building construction has raise, and the traditional technology is difficult to reach the requirements. However, with 3D printing technology, many construction problems can be solved easily, which lower the construction difficulty, also improves the construction level.

4.2.2 Ensuring the Personnel Safety in Construction

In traditional construction, the workers have to working at heights. It is a high-risk work, which amid a massive security operation. What is even worse is that the phenomenon of reduce security facilities in order to save the production cost is exist, this leads to the frequent accidents in construction, threatening the safety of the workers. However with the application of robot construction technology used on site, the manual operations are replaced by robot operations, which reduce the human resources while reducing security risks.

4.3 Difficulties of building 3D printing

4.3.1 Limitations of Printing Materials

At present, the application of 3D printing technology in construction is still in development. Different materials are needed for different components, which are expensive and not suitable for all printers. At present, bearing construction materials have not been developed to replace the traditional structural in high-rise buildings.

4.3.2 Limitations of Printing Equipment

The 3D printing equipment is huge, also difficult in operation. It is hard for many 3D printing manufacturers to ensure the stability of 3D printing equipment in the process of printing. At the same time, there is a direct relationship between the equipment and the building volume. In large construction, the volume of printing equipment needs increase. But not all the 3D printing equipment can match the size of buildings, which is the main reasons why it is difficult to print large buildings.

4.3.3 Irreversibility of Printing Process

The irreversibility of printing process is another problem of 3D printing technology at present. All data and models must be set up correctly, and the workflow should be worked out before print the buildings. If there are mistakes in the printing process or the design scheme has been changed, the operator cannot solve the problem immediately by adjusting the printing equipment. Once the situation happened, the completed components have to be discarded, which causes unnecessary waste.

4.3.4 Lack of Relevant Design Standards

It is difficult for 3D printing building to be surveyed because standards for 3D printing have not been formulated in the existing building codes, so it is hard to popularization, which limits the development of 3D printing technology fundamentally.

5 Jiangning Sewage Treatment Equipment House in Nanjing

The 3D–printed concrete sewage treatment equipment room (Fig. 4) besides on the Qinhuai river located in Jiangning district, Nanjing city. It is a regular rectangular building for placing sewage treatment equipment (6 m * 3.2 m * 3.2 m). The equipment is built by concrete printing technology. The 3D printing factory is Nanjing Jiayi Precision Machinery Manufacturing Co., Ltd.

At the beginning of the design, we tried to print the entire building at one time by in-situ printing, but due to the size of the 3D printer, it was impossible to complete the printing construction at one time, so we considered printing the components at the factory and then transporting them to the site for assembly. To complete this construction in a way that requires the design team to consider the detailed division of the entire building into various structural components when designing, and these components must meet the maximum print size of the 3D printer, while considering the Connection method, so as to facilitate subsequent construction and reduce errors.

So this time the 3D printing construction is different from the overall printing method of most 3D printing construction sites. In consideration of the printer's volume and the shape of the project, during the research, we split the structures into separate components and redesigned in order to simplify the variety of components (Figs. 5 and 6). Instead of the common method, Contour Crafting, we focused on individual components. By using standardized measurement and a reserved cavity in components, we successfully reduced the variety of components. At the same time, each cavity component to be connected by reinforcement verticality, and the horizontal component to be connected by mortar. Finally, the project printed five kinds of enclosure components, a total of 44 pieces (Table 1).

In the process of on-site assembly, we found that this construction method has high requirements on the accuracy of the printed components because the connection nodes between the components are designed in advance. If the printing is not accurate enough, the embedded parts and the reserved holes cannot be accurate, eventually it will be difficult to complete

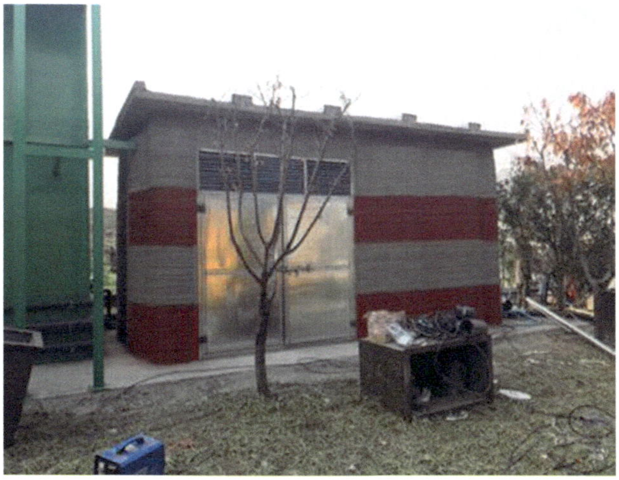

Fig. 4 3D printed concrete sewage treatment equipment room

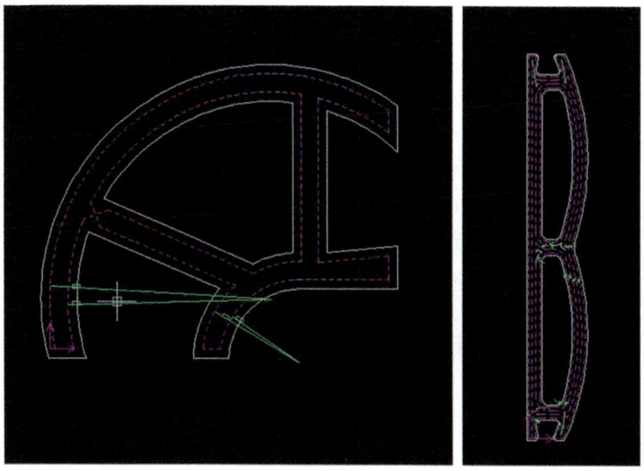

Fig. 5 3D printing track

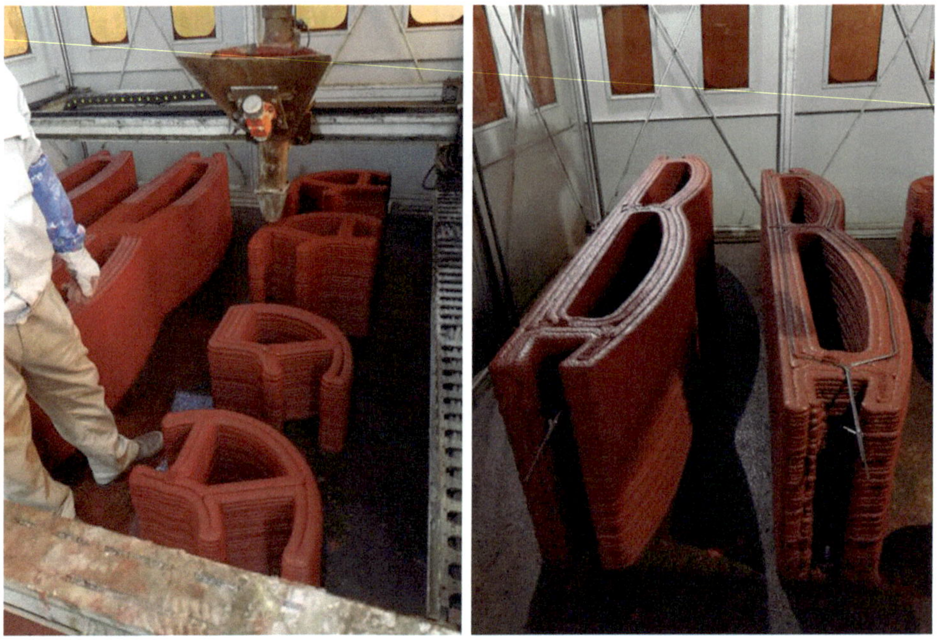

Fig. 6 Printed components

Table 1 The details of the component type and quantity

The type of the component	Block number
South and north side enclosure components	12
East side enclosure components	8
West side enclosure components	4
Corner connecting components	16
Roof components	4
Total	44

the assembly of the components. At the same time, the components are stacked in both the horizontal and vertical directions, errors will continue to accumulate, resulting in failure to meet the design and use requirements. In addition to the requirements for printing accuracy, the protection of components during transportation and on-site assembly is also necessary because damage to components often occurs during this process, which also requires attention.

This method resolves the problem that "How to print large buildings with small 3D printers", and it confirms to the stress standards. However, this method has disadvantages as well. Firstly, the 3D printing for separated components is not accurate as integral printing because of the construction error. Secondly, the transportation and installation of separated components is not efficient and cost friendly. Thirdly, the instrumental error and algorithmic error will cause increasing material waste and construction cost (Fig. 7).

6 Conclusion and Future Work

3D printing technology promotes the industrialization of the construction industry. This technology simplifies the construction steps and makes the traditional building materials garbage recyclable, which saves costs, increases the environmental protection level of construction, guarantees the personal safety of construction personnel, and improves the construction quality of buildings. However, because the 3D printing technology of buildings is still in the research and development stage, in the actual use process, the performance of printing materials and the volume of printing equipment have limitations, there are still some difficulties in the printing of the subject structure, and the most important thing is that there is no relevant standard to regulate this field. Therefore, 3D printing technology is still difficult to be used in large-scale construction. However, in the highly

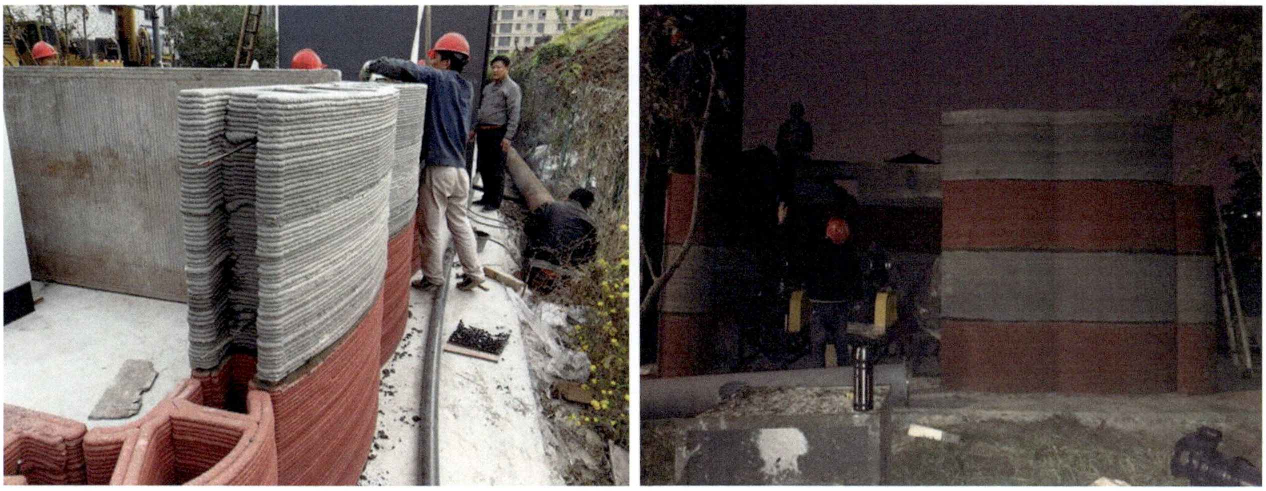

Fig. 7 Construction site

developed future of 3D printing technology, these problems will be solved one by one, and the digital construction of 3D printing will have a broader application prospect.

References

Biswas, K., Rose, J., Eikevik, L., Guerguis, M., Enquist, P., Brian Lee, R., & Jackson R. (2017). Additive manufacturing integrated energy—Enabling innovative solutions for buildings of the future. *Journal of Solar Energy Engineering, 139*, 015001–015010.

Bos, F., Wolfs, R., Ahmed, Z., & Salet, T. (2016). Additive manufacturing of concrete in construction: Potentials and challenges of 3D concrete printing. *Virtual and Physical Prototyping, 11*(3), 209–225.

Bosscher, P., Williams, R., Bryson, L., & Castro-Lacouture, D. (2007). Cable-suspended robotic contour crafting system. *Automation in Construction, 17*(1), 45–55.

Calì, J., Calian, D., Amati, C., Kleinberger, R., Steed, A., Kautz, J., & Weyrich, T. (2012). 3D-printing of non-assembly, articulated models. In *ACM transactions on graphics (TOG)—Proceedings of ACM SIGGRAPH Asia 2012* (Vol. 31, p. 10). Singapore.

Castelli, G., Ottaviano, E., & Rea, P. (2014a). A Cartesian cable-suspended robot for improving end-users' mobility in an Urban environment. *Journal of Robotics and Computer-Integrated Manufacturing, 30*(3), 335–343.

Cesaretti, G., Dini, E., De Kestelier, X., Colla, V., & Pambaguian, L. (2014b). Building components for an outpost on the Lunar soil by means of a novel 3D printing technology. *Acta Astronautica, 93*, 430–450.

Cui, X. (2018). Analysis of application trend of 3D printing technology in construction. *Development of Architectural Technology,* (2), (11) (2017).

Ding, L., Jie, Xu., & Tan, Y. (2015). Overview of research and application of 3D digital construction technology in architecture. *Journal of Civil Engineering and Management, 3*, 1–10.

D-shape. (2017). D-shape the large-scale 3D printer. https://d-shape.com/what-is-it/.

Gibson, I., Rosen, D. W., & Stucker, B. (2010). *Additive manufacturing technologies rapid prototyping to direct digital manufacturing* (pp. 19–243). Springer.

Gosselin, C., Duballet, R., Roux, P., Gausilliere, N., Dirrenberger, J., & Morel, P. (2016). Large-scale 3D printing of ultra-high performance concrete—A new processing route for architects and builders. *Materials & Design, 100*, 102–109.

Gramazio, F., Kohler, M., & Willmann, J. (2014). *The robotic touch: How robots change architecture* (pp. 20–235). Park Books.

Hack, N., Lauer, W., Gramazio, F., & Kohler, M. (2015). Mesh mould: Robotically fabricated metal meshes as concrete formwork and reinforcement. *Ferro, 11*, 347–359.

Hack, N., Lauer, W., Langenberg, S., Gramazio, F., & Kohler, M. (2013). Overcoming repetition: Robotic fabrication processes at a large scale. *International Journal of Architectural Computing, 11* (3), 285–299.

Hunt, G., Mitzalis, F., & Alhinai, T. (2014). 3D printing with flying robots. In *2014 IEEE international conference on robotics and automation (ICRA)* (pp. 4493–4499).

Keating, S., & Oxman, N. (2013). Compound fabrication: A multi-functional robotic platform for digital design and fabrication. *Robotics and Computer-Integrated Manufacturing, 29*, 439–448.

Keating, S., Leland, J., Cai, L., & Oxman, N. (2017). Toward site-specific and self-sufficient robotic fabrication on architectural scales. *Science Robotics, 2*, 1–15.

Khoshnevis, B. (2004). Automated construction by contour crafting—Related robotics and information technologies. *Automation in Construction, 13*(1), 5–19.

Perez, A., Roberson, D., & Wicker, R. (2014). Fracture surface analysis of 3D-printed tensile specimens of novel ABS-based materials. *Journal of Failure Analysis and Prevention, 14*(3), 343–353.

Sachs, E., Cima, M., & Cornie, J. (1990). Three-dimensional printing: Rapid tooling and prototypes directly from a CAD model. *CIRP Annals—Manufacturing Technology, 39*(1), 201–204.

Schwartz, M., & Park, J. (2017). robotic control strategies applicable to architectural design. *Technology | Architecture + Design, 1*(1), 73–82.

Yin, H., Qu, M., Zhang, H., & Lim, Y. (2018). 3D printing and buildings: A technology review and future outlook. *Technology| Architecture + Design*, 94–111.

Framework for Integrating Buildings' Energy Simulation Tools (BESTs) with Intelligent Decision Support Systems (IDSS)

Omar O. Elrawy, Tamer S. Hamza, and Samir S. Hosny

Abstract

Along with the uprising need for sustainable buildings, the use of Building Energy Simulation Tools (BESTs) became widely common among building engineers and architects. Challenges within the energy simulation process are also increasing; challenges like uncertainty, sensitivity, and multivariate analysis within the energy simulation process are widely discussed nowadays. The power of today's computational methods within the field of Artificial Intelligence (AI), in fusion with the well-established science of Decision Support Systems (DSS), forms Intelligent Decision Support Systems (IDSS). IDSS is expected by the researchers to provide BESTs with the needed intelligence and expertise to overcome the energy simulation and optimization challenges. This article points-out major challenges in dealing with Building Energy Simulation Tools BESTs; the article also points out recently developed AI capabilities and significant applications. The article finally concludes a framework for integrating BESTs with IDSS.

Keywords

Building performance simulation (BPS) • Energy modeling • Energy optimization • Machine learning (ML) • Artificial intelligence (AI)

O. O. Elrawy (✉)
American University in Cairo (AUC), Cairo, Egypt
e-mail: omarelrawy@aucegypt.edu

T. S. Hamza
Ain Shams University, Cairo, Egypt

S. S. Hosny
Future University in Egypt (FUE), Cairo, Egypt

1 Introduction

Buildings account for 40% of all energy use in the United States (EIA, 2019). For achieving an environmentally sustainable building, optimizing the building's direct energy consumption is the most challenging aspect, as energy optimization involves all building disciplines. A sound energy model should be established at early design phases, develops as the building goes through design development, and does not stop developing until the building goes into operation, where the energy model is expected to be in its most detailed, sophisticated, and realistic form, and it nearly simulates the actual constructed building's performance (AIA, 2012).

As the energy model develops it becomes more detailed; therefore, the energy optimization process gets more complex. The building becomes more determined by design and construction constraints, and all disciplines tend to evolve and develop. This yields to more constraints on the energy optimization measures, and accordingly on the building's project team, therefore, building performance simulation needs an interdisciplinary problem oriented, dynamic tool using numerical methods that approximate a solution of a realistic model (Hopfe, 2009).

The occurring variation in constraints along the building process requires an energy simulation tool that is capable of coping like an expert with the change in different building parameters—shown in Fig. 1—all along the building process timeline. This tool must exist as an aiding tool to the simulation engineer from the moment the energy model is built, all through the construction phase, to the moment that the building comes to operation, this tool must be also capable of producing justifications, provide certainty and sensitivity assistance to the engineer, learn the effect of changing a certain parameter before attempting to change another, and must be flexible enough to fit from one building to another, providing rigorous energy related decisions during the energy model's lifecycle.

F. Rosso et al. (eds.), *Advances in Architecture, Engineering and Technology*, Advances in Science, Technology & Innovation, https://doi.org/10.1007/978-3-030-86913-7_13

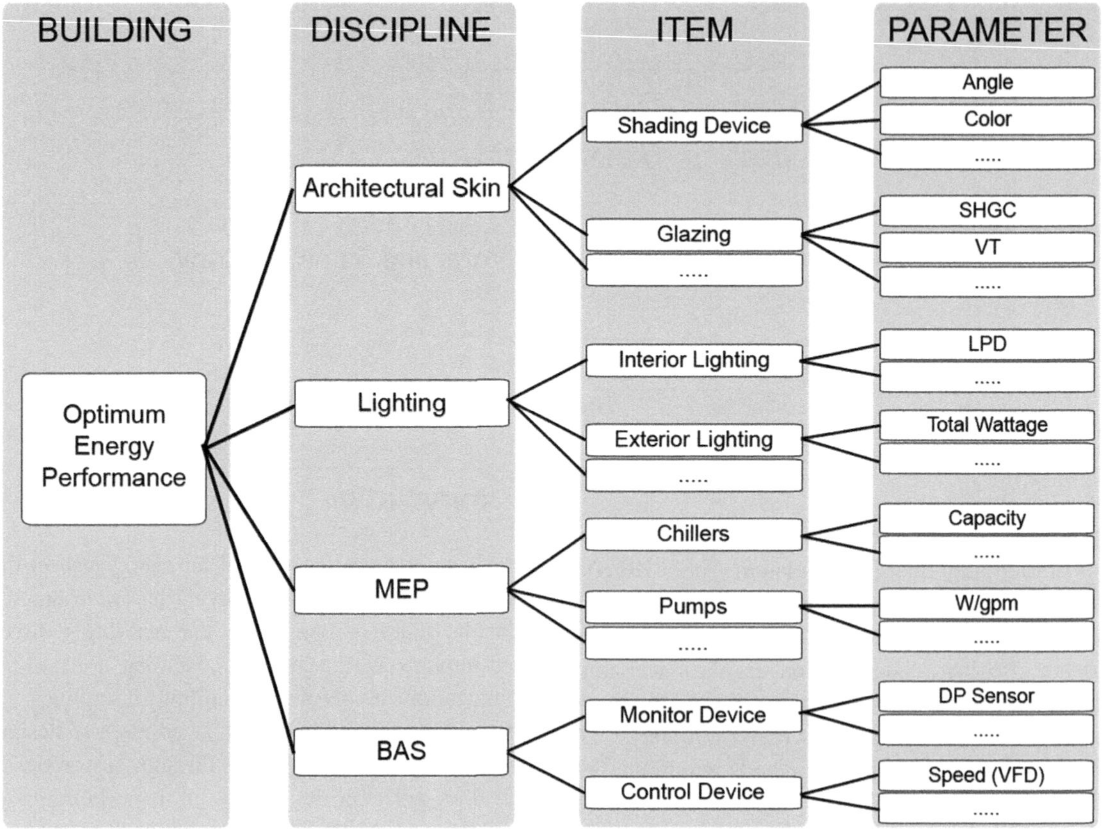

Fig. 1 Building energy simulation (BES) parameters. *Note* BAS: Building Automation System, SHGC: Solar Heat Gain Coefficient, VT: Visible Transmittance, LPD: Lighting Power Density, W/gpm: Watt/Gallon Per Minute, DP: Differential Pressure, VFD: Variable Frequency Drive

This interdisciplinary parametric property of the energy optimization process hence needs an intelligent Decision Support System (DSS). "Fundamentally, decision support systems are systems designed to support the decisions of managers and some operating personnel. They can be defined as an approach to effective decision making involving an interface between the individual and the computer, from problem formulation to solution." (Robert, 1988). One "very important characteristic of DSS is that they provide decision makers with a set of capabilities to apply in a sequence and form that fits each person's cognitive style." (Robert, 1988); hence, integrating Intelligent Decision Support Systems (IDSS) within BESTs is expected to be valid and to provide the needed intelligence and expertise which are able to solve the existing building's energy simulation challenges.

2 Literature Review

The importance of uncertainty challenge among energy modelling professionals is proven through a survey done by Christina Hopfe (2009). The survey included interviews with fifteen professionals and an online questionnaire by seven services professionals (Hopfe, 2009).

The questionnaire was concerned with the final design stage and was assessing three main issues: How is software currently used in the final design stage, what are tool requirements during the final design, and what should/could be improved in currently available simulation tools. Figures 2 and 3 summarize the survey's results, confirming that uncertainty analysis is a major need among practitioners.

Uncertainty and sensitivity analysis is discussed in depth by Eisenhower et al. (2011), emphasizing the need to trace the uncertainty and sensitivity flow inside the simulation process, instead of just analyzing the results' uncertainty. Eisenhower et al. (2011) applied this decomposition in their research entitled "Uncertainty and Sensitivity Decomposition of Building Energy Models", as shown in Figs. 4 and 5.

The decomposition conducted by Eisenhower et al. (2011) "For each node, a circle is drawn around it which represents the coefficient of variation. There is no appropriate scale for these circles, and they are intended to be viewed relative to other circles in the figure. The thickness of the wires corresponds to the magnitude of the sensitivity index. Where there is no wire, the sensitivity index is

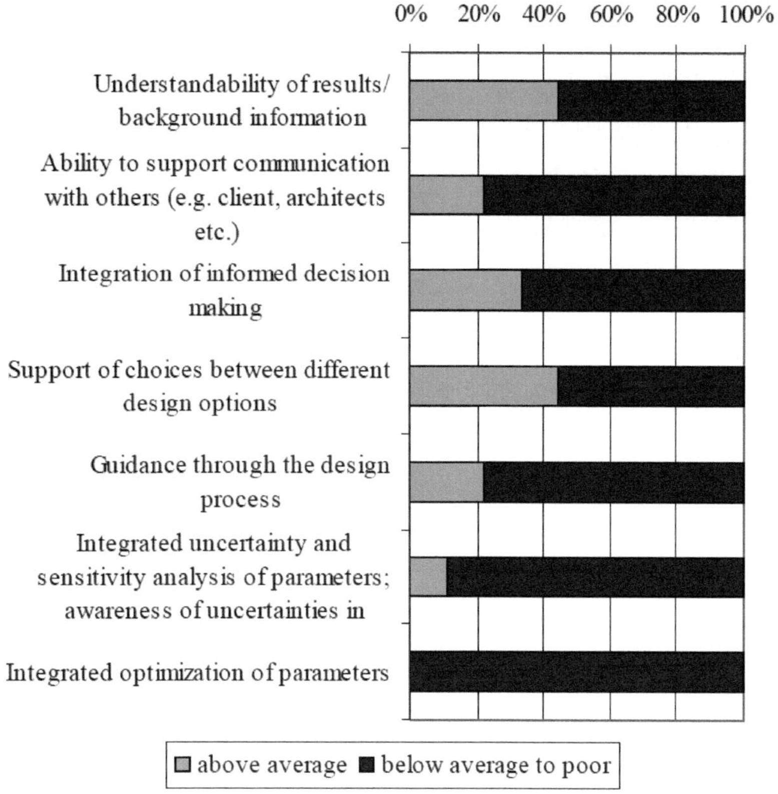

Fig. 2 Summary of the current satisfaction level in BPS according to professionals' perception. Adapted from "Uncertainty and sensitivity analysis in building performance simulation for decision support and design optimization", by Hopfe (2009, p. 25)

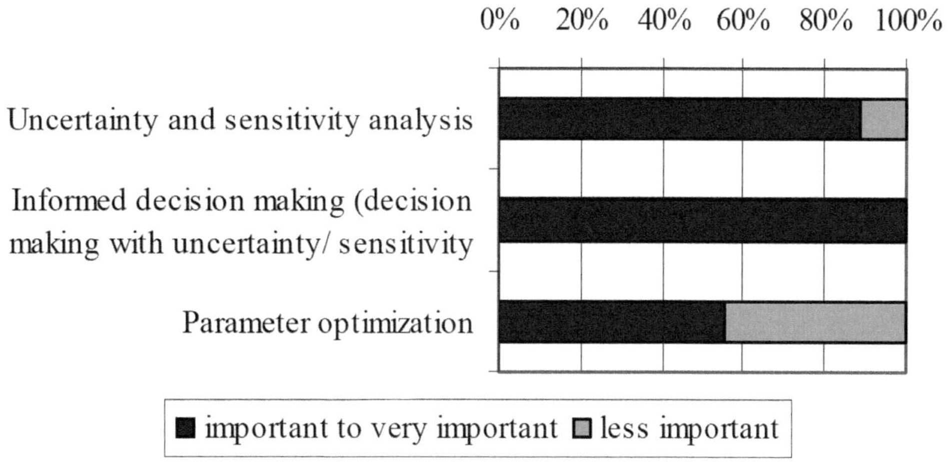

Fig. 3 Wish list of techniques for the integration in BPS for detailed design use according to professionals. Retrieved from "Uncertainty and sensitivity analysis in building performance simulation for decision support and design optimization", by Hopfe (2009, p. 26)

negligible, and the thickest wires represent the strongest influence between the variables" (Eisenhower et al., 2011).

3 Methodology

The current paper points out major energy simulation challenges using conducted surveys on building professionals and using experimental approach by running an energy optimization process on an existing building's energy model. The paper then proposes a new framework for buildings' energy simulation tools, which is expected to overcome the discussed energy simulation and optimization challenges.

3.1 The Multivariate Challenge in Building Energy Simulation (BES)

Building energy optimization encompass many possible objective functions and main design variables. The multivariate challenge is specific to the building's main design

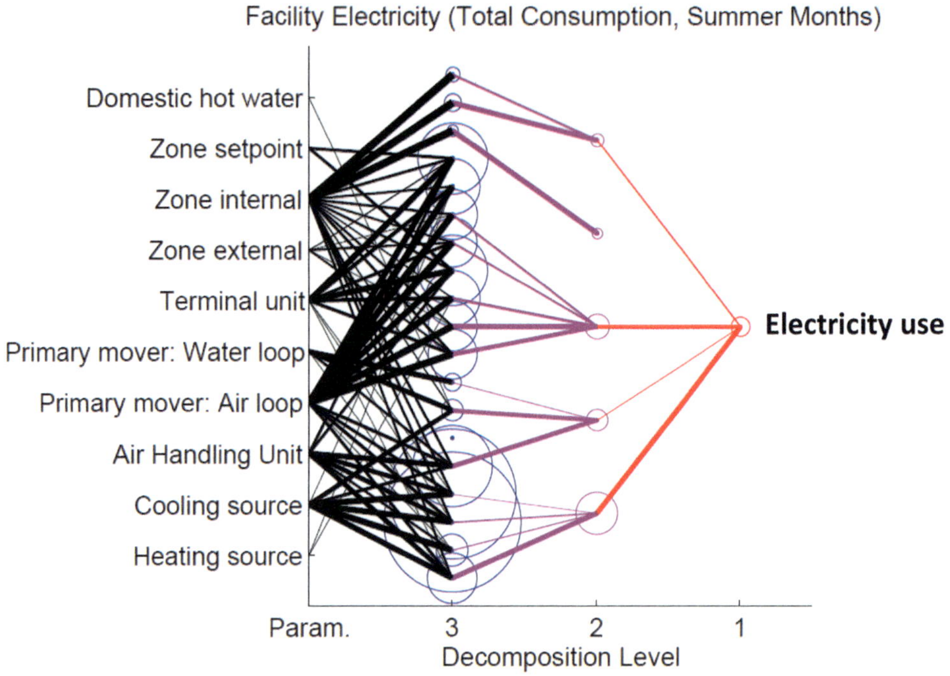

Fig. 4 Sensitivity decomposition of the electricity consumed by the facility (total consumption over the summer months). Retrieved from "Uncertainty and sensitivity analysis in building energy models", *Journal of Building Performance Simulation*, by Eisenhower et al. (2011, p. 14)

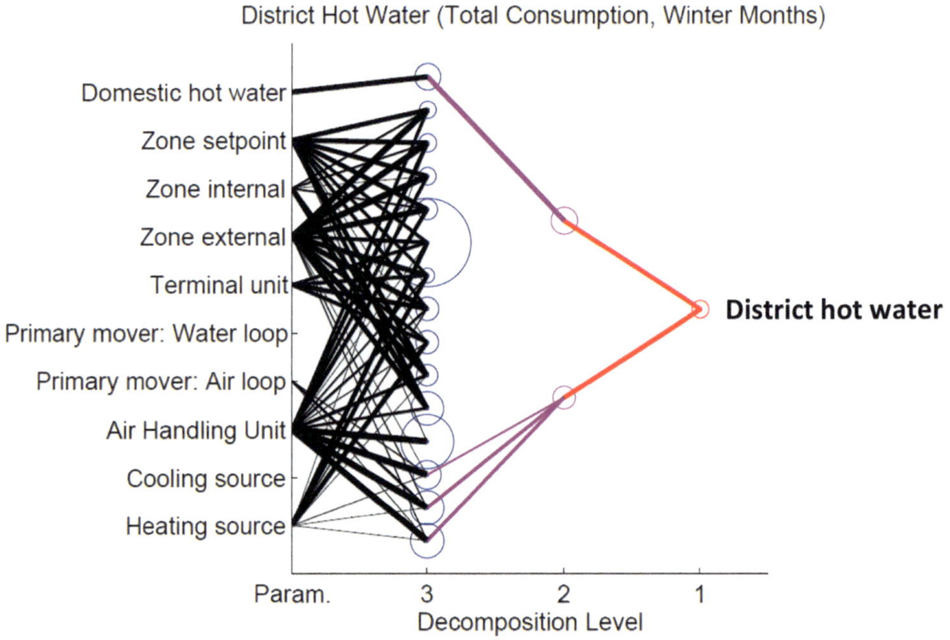

Fig. 5 Sensitivity decomposition of the district hot water consumed by the facility (total consumption over the winter months). Retrieved from "Uncertainty and sensitivity analysis in building energy models", Journal of Building Performance Simulation, by Eisenhower et al. (2011, p. 14)

variables (Fig. 1) assuming one objective, which is to reach the minimum energy consumption. The multivariate challenge during BES optimization occurs when one parameter's optimum value causes a change in another parameter's optimum value. The following example is from a new construction mid-rise office building. During design phase, a decision was to be made concerning the building's façade curtain-wall, and the building was designed with Lighting Power Density (LPD) of (0.6 W/ft2), giving an optimum glazing U-value of 0.18 (BTU/h °F ft2), during design development, LED lighting system was modified, and daylight harvest system was considered, as well as occupancy

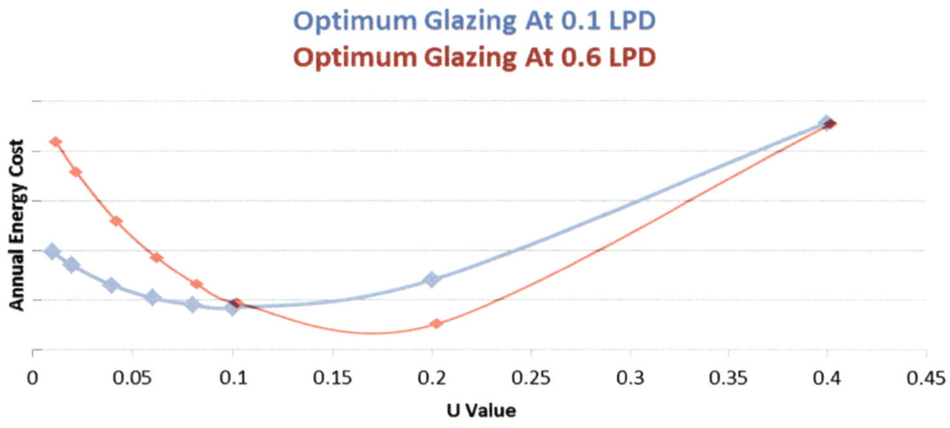

Fig. 6 Optimum glazing at 0.1 & 0.6 w/ft2 lighting power density (LPD)

sensors for all private office spaces, leading to LPD of 0.1. This decision did cut down energy costs but did as-well shift the optimum U-value from 0.18 (BTU/h °F ft2) to 0.10 (BTU/h °F ft2) (Fig. 6).

This example, modeled by the researchers, is a repetitive case during design development stage among the energy model's parameters shown in Fig. 1 and becomes more challenging with the increase in building's complexity.

3.2 Uncertainty and Sensitivity in Building Energy Simulation (BES)

BES model's accuracy and reliability are two major parameters that determine the model's avail; hence, uncertainty and sensitivity analysis of BES model became an uprising concern among building energy modellers. Although uncertainty and sensitivity are two distinguishable expressions, they are used in conjunction with each other; sensitivity analysis studies the impact of the change in input parameter's value on the simulation output, while certainty is the measure of belief in a certain output; hence, for a single BES parameter, sensitivity analysis do determine the importance of parameters' uncertainty during the Decision Making (DM) process (Loucks et al., 2005).

According to Eisenhower et al. (2011), it is essential to trace the uncertainty and sensitivity flow inside the simulation process instead of just analyzing the results' uncertainty, so they developed a study with a presentation technique for showing how each parameter's uncertainty and sensitivity changes along the simulation process.

Although Eisenhower's study is essential, it revealed a complex side of the uncertainty and sensitivity analysis. Conducting Eisenhower's decomposition is very significant, yet it is not practical due to the need for third party statistical tool(s), and due to the needed time and effort, which are hardly available in practice during the energy model's lifecycle.

4 The Developed IDSS Framework

IDSS is the fusion between the well-established DSS structure and expert systems.[1] According to Sprague and Carlson (1982) the DSS framework involves a database, a model base, a Database Management System (DBMS), a Model Base Management System (MBMS), and a Dialogue Generation Management System (DGMS), all within the task environment[2]; This DSS is transformed to an IDSS as shown in Fig. 7.

Changes and additions were made to the framework of Sprague and Calrson to add intelligence, user support, and the needed expertise. The main change was adding the expert system components within the intelligent agent, placing the intelligent agent inside the DGMS. Another major change was in replacing the database with a knowledge base.

Components of the expert system added to the intelligent agent includes explanation engine, inference engine, and knowledge base editor to include the expert systems' features in the software agent which is expected to enhance the dialogue between the IDSS and the user; hence, the intelligent agent becomes a significant aid to the DGMS.

Involving a software agent provides the framework with several characteristics; such as being autonomous, adaptive, proactive, reactive, communicative, cooperative, mobile, goal-Directed, and persistent (Adam & Humphreys, 2008).

[1] "Expert systems are a sub field of AI that is focused on using expert knowledge to encode the decisions that an expert would make in the field" (Tracy, 2015).

[2] "The term task environment refers to an environment coupled with a goal, problem, or task-the one for which the motivation of the subject is assumed". (Newell & Simon, 1972).

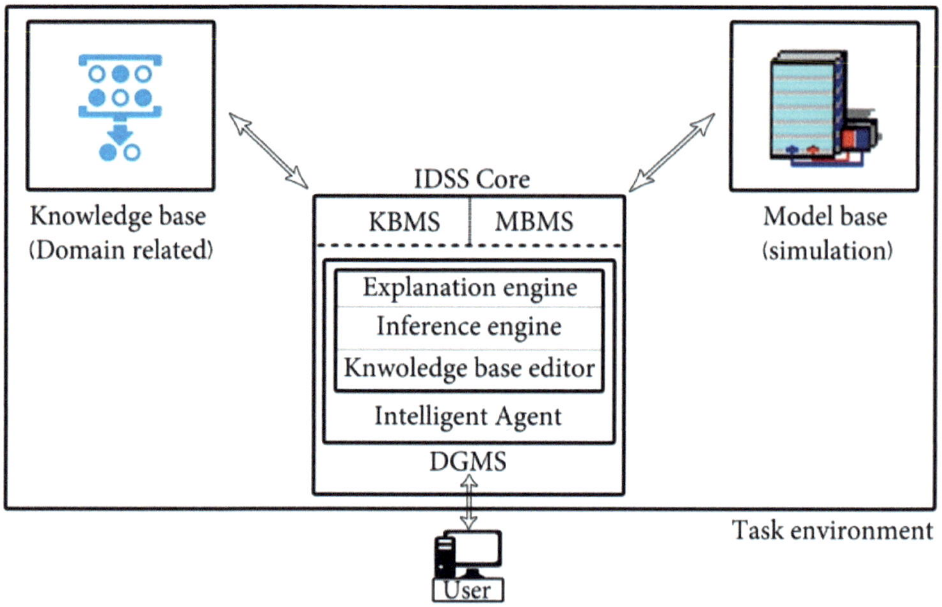

Fig. 7 The IDSS framework

5 The Enhanced BEST Workflow

BESTs are in continuous development, and the researchers-practitioners mutual interaction showed effective impact on this development, while interaction needs to include a key player in this loop, which is the manufacturer. Manufacturers are responsible for the actual performance of the building; hence, they must be involved in the tools' development process so to close the loop. BESTs development must be also technology-driven to cope with today's intelligence and connectivity advances. Figure 8 illustrates this idea in the form of a mechanism for BESTs development.

BESTs already provide a simulation process which is quick, detailed, and interoperable, and the energy optimization process is already being widely discussed and developed, yet, it is still thought to be a "black box". "One of the main issues of BES is that it is being a black box, opaque to outside users with little knowledge of what is running underneath" (Galanos, 2017). This is considered one of the most accurate expressions which emphasized the need for better communication between what is happening during the BES process and the energy modeler/project team. This is also essential to show what is excluded/assumed during the simulation process.

Also two of the significant BESTs needs were to be "more up-to-date with current computer science technology", and the need for "data sharing over networks, based on model servers", as stated by Maile et al. (2007). Hence, the development within the BEST workflow is intended to enhance the communication with the user and emphasize the role of computational power within the tool's framework.

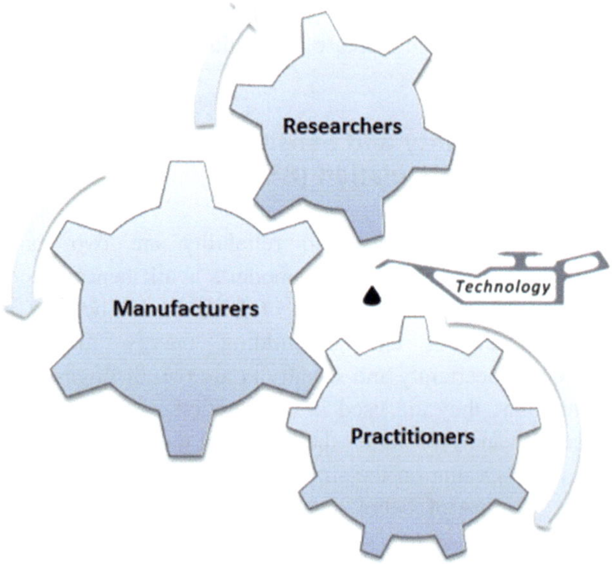

Fig. 8 BESTs development mechanism

The BEST workflow is developed with a modular approach to allow for integration with IDSS. The modular approach also allows for adding ready module add-ons during simulation, like the manufacturer's modules; for example, HVAC suppliers can prepare a module of their own products within the added Manufacturer Module. The modular approach also allowed for adding the Model Formulation and the Simulation Controller modules and communicating all modules via an added software agent, as shown in Fig. 9.

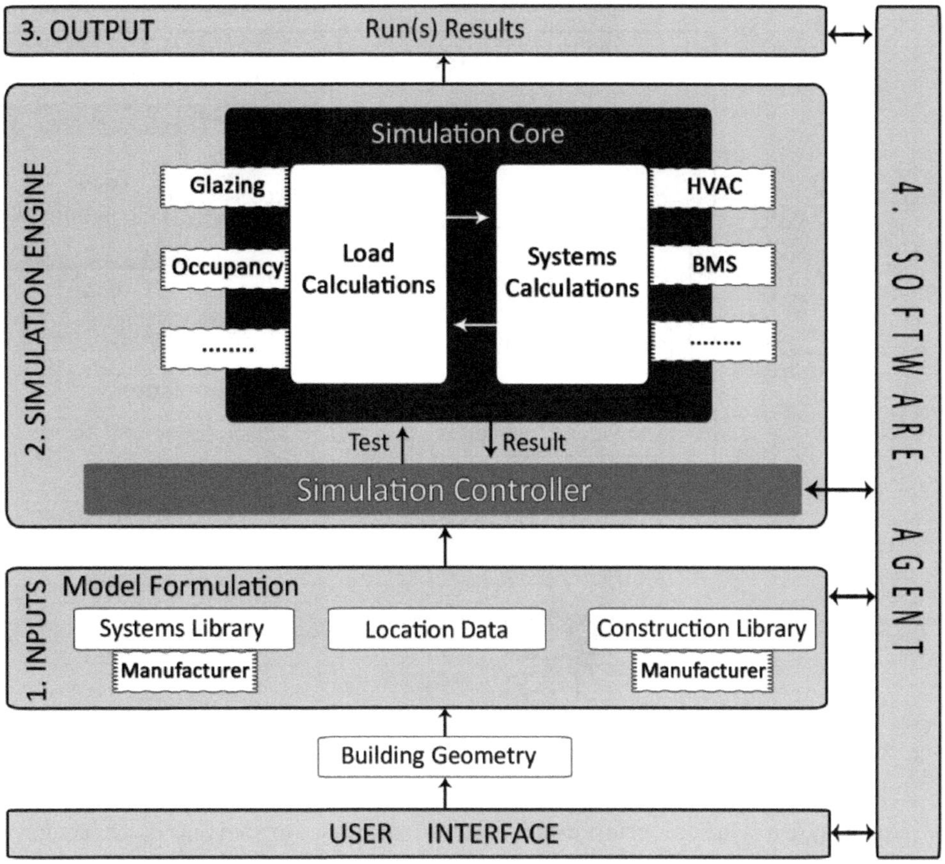

Fig. 9 The enhanced BEST workflow

6 The Outcome: BEST-IDSS Framework

The proposed BEST-IDSS framework is a result of the integration between the developed IDSS (Fig. 3), and the enhanced BEST workflow is shown in Fig. 9. In the formulated BEST-IDSS framework (Fig. 10), the enhanced BEST framework replaced the traditional MBMS. The only activity happening outside the task environment is defining the building geometry; since the building geometry is the component that must be initiated and edited directly by the user, while the information within this geometry is processed inside the task environment.

6.1 The Knowledge Base

According to Newell and Simon (1972), "problem solving is achieved by accumulation of knowledge". Within the formulated BEST-IDSS, the knowledge base is the main player in the machine learning process, since each run by the BEST is inferred in the IDSS core, and new knowledge is developed and stored within the knowledge base.

For example, after successive simulation runs in the model base, it may be observed that changing the chiller capacity results in a change in the glazing's optimum U-value, while apparently both parameters seem to be independent. After successive runs the inference engine will detect this change by forward chaining, as it experienced a case that showed new data, and therefore this knowledge will be stored within the knowledge base, stating that the optimum glazing is a function of the chiller capacity.

Additionally, and since the expert system has the ability "to explain the reasoning process that it used to reach a recommendation." (Jain, 2009), the explanation provided by the expert system will tell that it is based upon the user's request, the expert system may provide an explanation for its results or decision at a certain step, in that case the energy modeling and optimization process is no longer a black box.

6.2 The IDSS Core

The IDSS core is the component which encompasses the DSS components (KBMS and MBMS), in addition to the expert system's components that make this DSS an IDSS; those components are the explanation engine, the inference engine, and the knowledge base editor. The three components are placed inside an intelligent agent that replaces the DGMS of the known DSS.

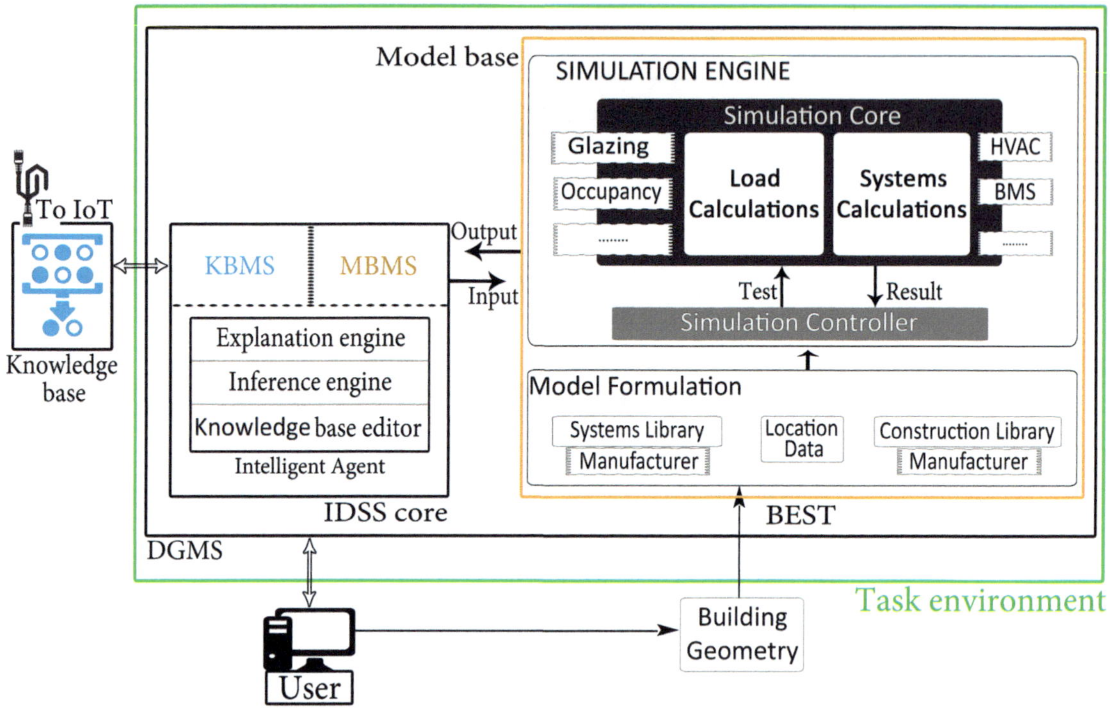

Fig. 10 The formulated BEST-IDSS Framework

Expert systems include many expert features, such as goal driven reasoning, data driven reasoning, providing explanations, coping with uncertainty, and conducting reasoning with uncertainty (Jain, 2009). Those are mostly needed to provide the user with the explanation behind the provided results and edit the knowledge base based on these results too.

For example, during a glazing's U-value optimization process, the IDSS might select an optimum U-value that is higher than the user's expectations, in that case, and within the intelligent agent, the inference engine will infer this by running all applicable U-values as per the pre-defined rules within the knowledge base, then it will show the trend of the U-value against total annual energy consumption, additionally, it will trace the simulation results, and based on some predefined rules (such as the months (winter or summer) where the increased energy consumption occurred, and also the component where energy consumption increased (Chiller for example)), it will state that this is the optimum value since the building is internal load dominant[3] building.

A most suitable interaction between the user and a BEST is a major challenge, and the reason lies in the fact that users' needs from a BEST vary according to many factors related to their needs. According to Djunaedy et al. (2006), "In various fields, including building and system designers, people can be classified according to their innovativeness. In the "interviews conducted" by Hopfe (2009) it was found that "Obviously it will be very hard to create a single software which would satisfy all." Since the Intelligent Agent possesses autonomous, adaptive, and proactive characteristics (Adam & Humphreys, 2008), therefore it can overcome this challenge.

7 BEST-IDSS Significance and Capabilities

7.1 The Knowledge Base-Internet of Things (IOT) Connection

According to Maile et al. (2007) BEST needs include "data sharing over networks, based on model servers "; the knowledge base in the formulated BEST-IDSS is connected to the internet, and as it is controlled via the intelligent agent, infinite set of possibilities is potential through this connection.

For example, and as today's AI can use reinforcement learning to self-learn through the web and fill-in the knowledge gaps (Dormehl, 2016), the intelligent agent will be able to get the suitable knowledge that makes it able to utilize the up-to-date energy optimization strategies and algorithms, store them within the knowledge base, and

[3] Internal load dominant is the case when the thermal loads inside the building exceeds the thermal loads due to conductance through building envelope. In case of using eQUEST this can be calculated from LS-C report.

implement them in the task environment; furthermore, it can test their impact, feedback to the internet after inferring the impact. This connection can be established within one BES cloud server.

This connection allows the establishment of a BES cloud where building's model, its optimization results, and the actual operating building all can be shared, allowing for more autonomous knowledge base development process and allows connecting the simulated building to a similar simulated or operating building, which is a significant step toward involving building's energy performance in today's IOT.

7.2 Uncertainty and Sensitivity Analysis

Since uncertainty is one of the major BEST challenges, therefore the intelligent agent can be considered as the most significant component within the IDSS; since that intelligent agents in IDSSs can "quantify uncertainty", as stated by Adam & Humphreys, 2008. This formulation is not only capable of quantifying uncertainty but also capable of coping with it, since that the expert system is introduced within the intelligent agent, and "Coping with uncertainty" is one of its features (Jain, 2009). Intelligent agent can also create explanations and justify them (Adam & Humphreys, 2008).

As the "simulation controller" is directly connected to the software agent; the uncertainty/sensitivity module is a controller module that can receive the inputs' level of detail from the software agent, controls the simulation to be running while testing the target parameter, then reporting this parameter's level of uncertainty and its sensitivity, which can be displayed to the user with the desired statistical/graphical representation.

For example, taking glazing's U-value with the uncertainty/sensitivity module installed, the user interface will be as shown in Fig. 11. The user will input the U-value along with its uncertainty measure based on the available data; assume U-value = 0.45 Btu/h-ft2-0F with Certainty Factor (CF) = 0.5 for a building at its design phase.

The expected simulation mechanism is that the software agent will store these values and will communicate with the model formulation step to formulate a model with all parameters unchanged except for the glazing U-value, which will be changing on each iteration, then the software agent will store the run results, alternate the U-value, and then perform the next run, and the simulation controller can limit the simulation process to show only results affected by the variations in U-value. This process shall stop once the change in results is considered negligible. The result of this process is a *sensitivity* value that will be shown on the user interface instantaneously so that the user is able to assess the inputs in real-time, as shown in Fig. 11.

Additionally, when the user performs a whole run and requires the building's total annual energy consumption, the result is to be shown along with its CF, as the total annual energy consumption's uncertainty is affected by the input parameters' level of uncertainty, and the model's sensitivity to these parameters; hence, the annual energy report is expected to be shown as in Fig. 12.

7.3 Multivariate Analysis

The simulation controller can be directed by the IDSS core to customize the simulation process such that each parameter is studied separately, and instead of the user studying each parameter independently using an external computational model, the tool module will report each run to the knowledge base, and the IDSS core will calculate and display a dynamic chart to the user, showing each parameter's optimum value.

Additionally the IDSS core, with an added intelligence level and by learning from previous runs, can show the expected changes based on one parameter being changed, and it can suggest a change as well. This can be done using reinforcement learning (Dormehl, 2016).

For this mode to represent real-time variation in the optimum value of the input parameters, it must direct the simulation controller to formulate successive runs—aided by the connection to the IDSS core—and to report the results internally; however, the IDSS core will have to show the most affected parameter which is detected based on the parameter's sensitivity, as suggested in Fig. 13.

8 Going Further with Algorithmic Building Operation

In this section one further application of the IDSS is being outlined, which is the IDSS for building operation. The IDSS of building operation is approached by formulating a simplified model which shows major BAS components of a typical medium rise office building. The model shows a sample of each item within the BAS, and how it can be **perceived** as a model, and then this model is to be integrated within the IDSS in its model base.

For the purpose of the framework, all the commonly used BAS components are categorized under sensors, actuators, energy meters, and inverters/variable frequency drives (VFDs). Each is communicated within the building's Local Operating Network (LON) and works as either input (I) device, output (O) device or both (I/O), and each communicates either in a digital node (on/off data) or in an analogue node (varying data). All these devices on the LON are interfaced with the BAS computer via LON to Local Area

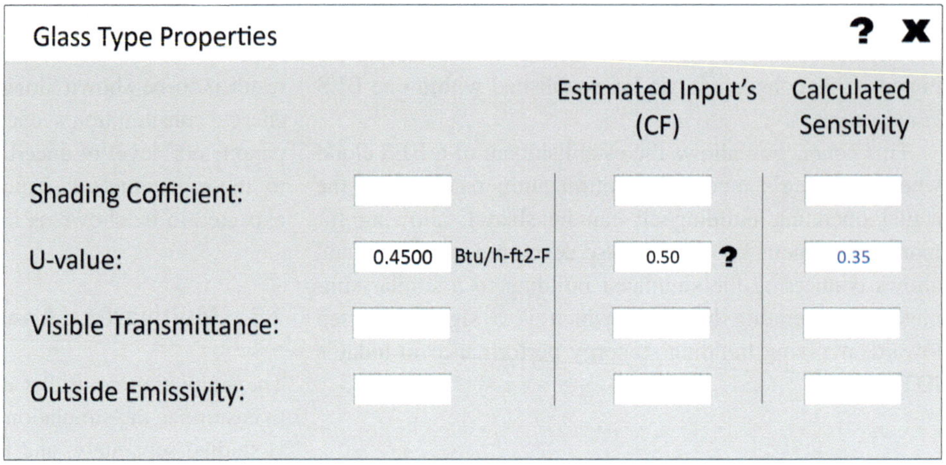

Fig. 11 Part of the user interface in the uncertainty/sensitivity mode

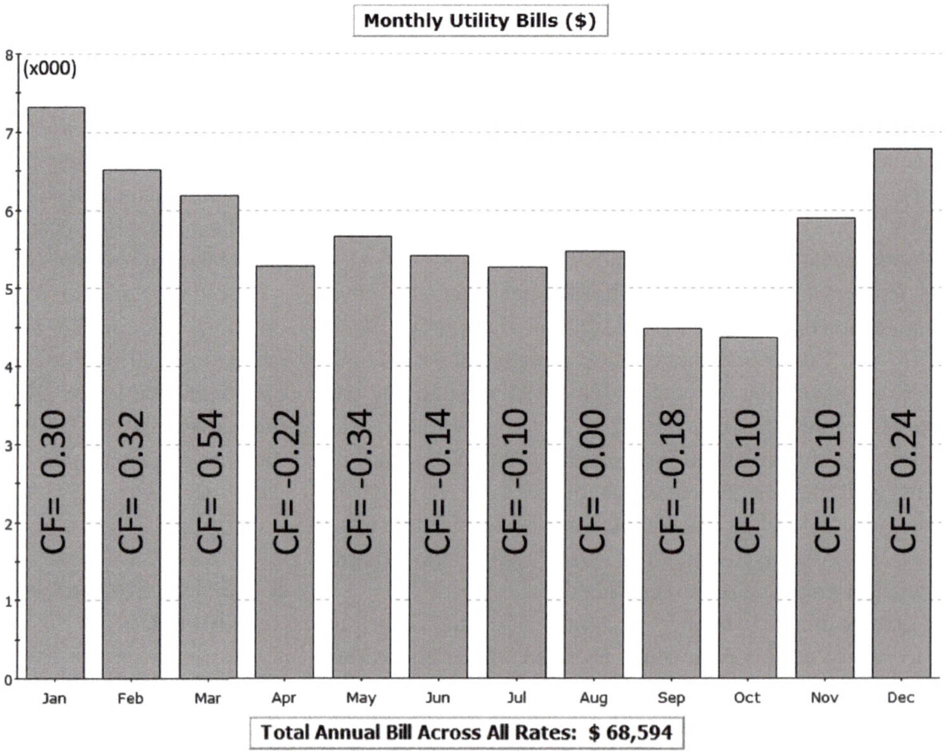

Fig. 12 Expected results in the uncertainty/sensitivity mode

Network (LAN) gateway to be read by the control server, which is a server supplied by the BAS manufacturer, this server is capable of monitoring and controlling the network interfaced devices, it is also capable of reading from and writing to the building's automation database.

Adding the SQL is the key-step for converting the operation process to a real-time model, where all data within the automation database are readable, including both input and output data, so that the model can show the input parameter along with its effect on a determined output value. For the shown framework the SQL is to show and export the data in the form of a CSV/excel model, and the IDSS will perceive this framework as its model base, as shown in Fig. 14.

The formulated IDSS of building operation is the same as the BEST-IDSS framework except for the model base; instead of relying on a simulation model the model here is an actual operating building; hence, the operational model is considered as a real-time model base, as it shows the instantaneous impact of changing a certain parameter by reading the impacted output. And as the intelligent agent utilizes a reinforcement learning algorithm, this IDSS is to operate heuristically until the intelligent agent is determined to be eligible for taking control of the operation process.

Fig. 13 Part of the user interface in multivariate analysis mode

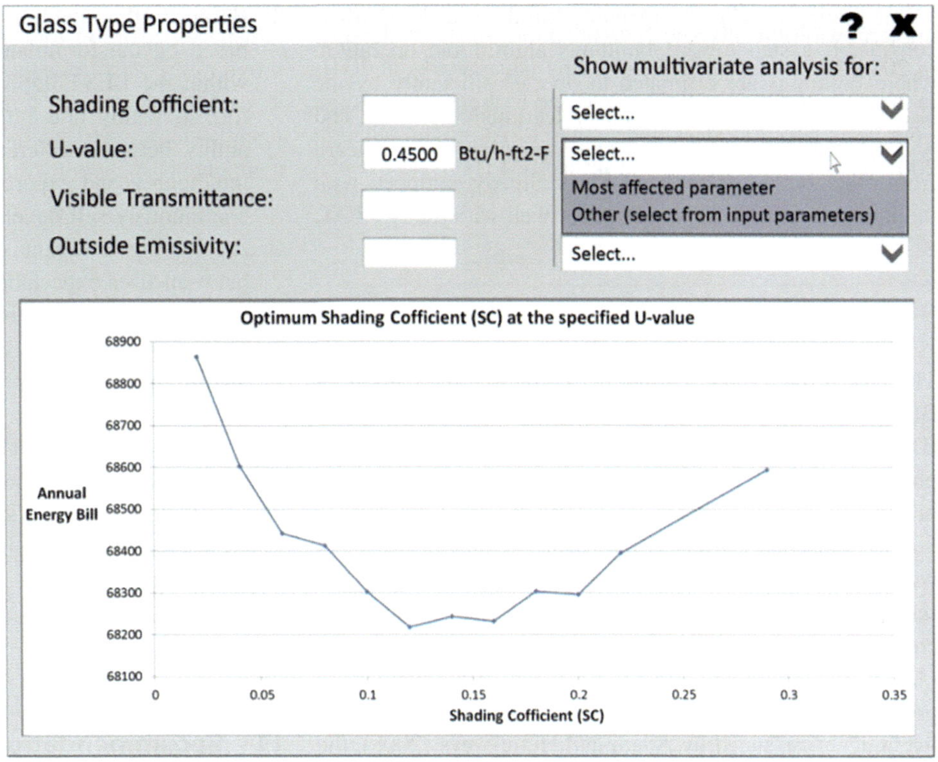

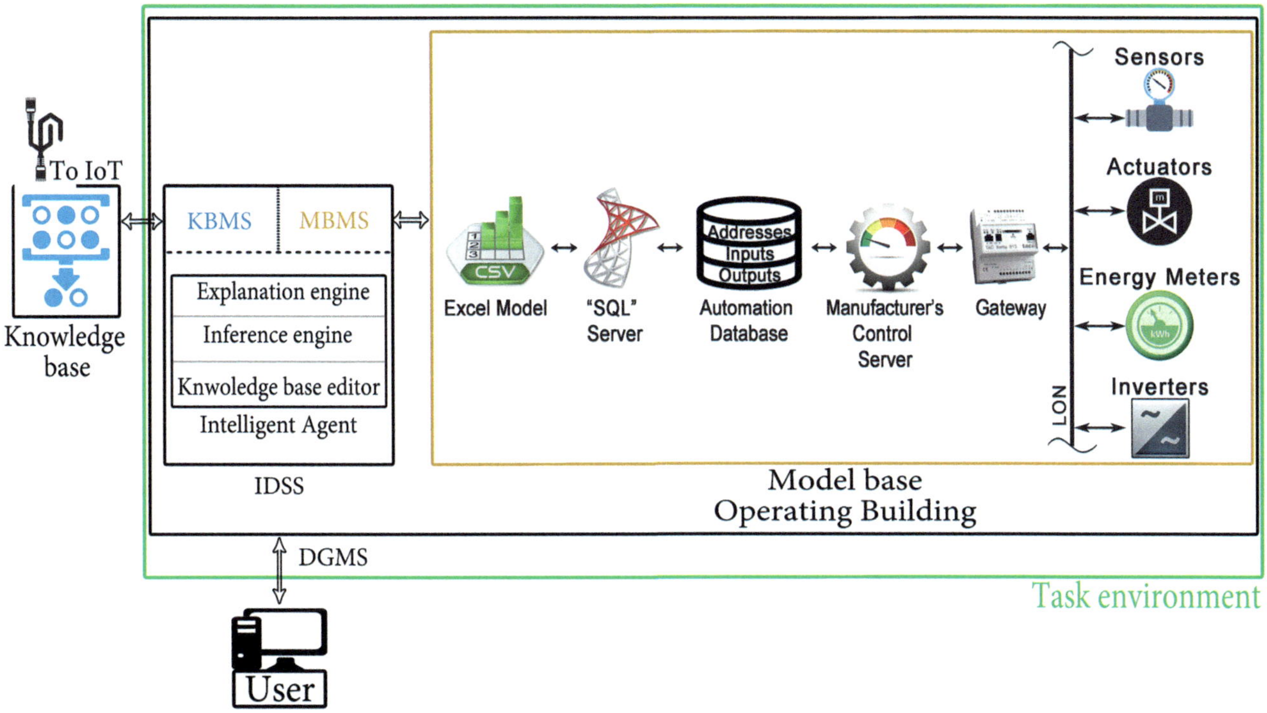

Fig. 14 The IDSS of building operation

Integrating an already operating building to the IDSS is considered a step toward buildings algorithmic operation, where buildings are estimated to operate efficiently by the aid of the most recent reinforcement learning algorithms, and will not only learn from their own operation but will learn from other buildings as the IDSS is already equipped with the internet connection, which copes well with today's IOT.

9 Discussion

This paper presented a framework for the integration of BESTs and IDSS and revealed that the techniques within IDSS encompass the solution for common BEST challenges; pointing out that IDSS can revolutionize today's known BES process.

The paper also revealed that using an intelligent agent would be a major part of the solution to uncertainty and sensitivity analysis. Intelligent agents also help and facilitate creating "forecasts, recommendations, and explanations" (Adam & Humphreys, 2008). As those recommendations produced by the IDSS are justified to the user by the intelligent agent, as stated by Adam and Humphreys (2008), the user can get optimum, justified, and consequently more reliable results.

The discussed challenges within the BES process are also surpassed as the formulated BEST-IDSS is connected online; and once big data gets involved, each building will be a supporting case to develop the knowledge base within the current case as well as future cases.

The approach taken during this research, of cross-case analysis for the latest AI applications and integrating them to the workflow of BESTs unlocked some ideas; and much more ideas are expected to ascend based on this research as the AI science ascends. All these ideas contribute to a smarter, more autonomous, and more reliable BES process and consequently less buildings' energy consumption.

10 Conclusion

There is a lot of potential in today's AI technology and in connectivity technologies as well. All those witnessed and all the expected developments will have great benefit in enhancing buildings' energy efficiency; either by utilizing them in hypothetical buildings' simulation, or in existing buildings' operation. The paper's conclusion can be summarized as follows:

1. Based on the integrative nature of the BES process and based on the formulated BEST-IDSS, it is proven that within the BEST-IDSS the needed analysis can be performed within one entity, with less need for interoperability between different software tools as the level of intelligence and expertise involved is able to produce a continuously self-developing tool.
2. An intelligent agent can overcome the "discrepancy between user expectations and simulation tool capabilities and features" (Djunaedy et al., 2006, para. 19), as the agent comprises a user-based reinforcement learning algorithm, it can adapt to the user's needs, taking into consideration the user's innovativeness level and the model's level of detail/complexity.
3. Linking building's model, either simulation or operational model, to the knowledge base, and linking the knowledge base to the internet is a significant step toward involving building's performance in today's IOT networks.

11 Recommendations

It is strongly recommended for researchers to minimize the attempt of different BEST tools with different capabilities and inter-operating different simulation programs; alternatively, it is strongly recommended that researchers maximize the research within IDSS for better BESTs.

Taking the path of independent AI algorithms do not cope with the interdisciplinary property of the BES process and do not match with building's integrity as well. Also high error occurs and is expected to occur when non-computational professionals attempt to apply Machine Learning algorithms within one or more step during energy simulation and/or optimization; hence, this research strongly recommends collaboration toward one framework which encompasses all the needed and possible IDSS components and which addresses all BES needs as well, this approach is more applicable during the construction process as it saves time and effort and is more convenient to building engineers and architects.

The proposed framework within the current research is limited to one optimization objective, which is to minimize buildings' energy consumption. It is recommended for future research to investigate more possible objective functions, following the multi-objective framework methodology developed by Ascione et al. (2019).

Acknowledgements Sincere gratitude to Professor Dr. Samir Sadek for his kind supervision, for the intense knowledge, and above all for all the inspiration he provides. Deep respect and gratitude to the supervision of Dr. Tamer Samir for all the scientific expertise which made this research possible. This research paper is based on Elrawy's MSc thesis which was accomplished under both supervisions.

References

Adam, F., & Humphreys, P. (2008). Encyclopedia of decision making and decision support technologies. In *Encyclopedia of decision making and decision support technologies*. Hershey, New York: InformatIon Science reference.

AIA. (2012). An architect's guide to integrating energy modeling in the design process. The American Institute of Architects (AIA). http://www.aia.org/aiaucmp/groups/aia/documents/pdf/aiab096060.pdf.

Ascione, F., Bianco, N., Mauro, G. M., & Vanoli, G. P. (2019). A new comprehensive framework for the multi-objective optimization of building energy design: Harlequin. *Applied Energy, 241*, 331–361. https://doi.org/10.1016/j.apenergy.2019.03.028

Djunaedy, E., Hensen, J., Hopfe, C., Struck, C., Trcka, M., & Yahiaoui, A. (2006). Notes on building performance simulation and the role of IBPSA. In *11th Brazilian congress of thermal sciences and engineering* (Vol. 11). Braz. Soc. of Mechanical Sciences and Engineering. Retrieved from https://www.academia.edu/3059084/Notes_on_building_performance_simulation_and_the_role_of_IBPSA.

Dormehl, L. (2016, November 11). Machine-learning system aggregates knowledge by surfing web for information. Retrieved December 11, 2016, from http://www.digitaltrends.com/cool-tech/ai-surfs-the-web-learning/.

EIA. (2019, May). U.S. Energy Information Administration (EIA). https://www.eia.gov/tools/faqs/faq.php.

Eisenhower, B., O'Neill, Z., Fonoberov, V. A., & Mezić, I. (2011). Uncertainty and sensitivity decomposition of building energy models. *Journal of Building Performance Simulation*. https://doi.org/10.1080/1940149.

Galanos, T. (2017, March 27). Populating vast design spaces: looking back for a way forward. Retrieved April 4, 2017, from https://www.linkedin.com/pulse/populating-vast-design-spaces-looking-back-way-forward-galanos.

Hopfe, C. J. (2009). *Uncertainty and sensitivity analysis in building performance simulation for decision support and design optimization* (Ph.D. Thesis). Geboren te Jugenheim, Duitsland, Dutchland.

Jain, V. K. (2009). *INFORMATION technology issues & challenges*. Excel Books India.

Loucks, D. P., Beek, E. van, Stedinger, J. R., Dijkman, J. P. M., & Villars, M. T. (2005). *Water resources systems planning and management: An introduction to methods, models and applications*. France: Paris, UNESCO. Retrieved January 10, 2018, from http://unesdoc.unesco.org/images/0014/001434/143430e.pdf.

Maile, T., Fischer, M., & Bazjanac, V. (2007). *Building energy performance simulation tools—A life-cycle and interoperable perspective* (Working Paper No. #WP107) (p. 43). Stanford, CA, USA: Stanford university. Retrieved July 10, 2017, from http://web.stanford.edu/group/narratives/classes/08-09/CEE215/ReferenceLibrary/BIM%20and%20Building%20Simulation%20Research/Building%20Energy%20Performance%20Simulation%20Tools-%20A%20Lifecycle%20and%20Interoperable%20Perspective.pdf.

Newell, A., & Simon, H. A. (1972). *Human problem solving*. Prentice-Hall.

Robert, T. J. (1988). *User-oriented decision support systems*. Prentice-Hall.

Sprague, R. H., & Carlson, E. D. (1982). *Building effective decision support systems*. Prentice-Hall.

Tracy, K. (2015, October 31). What is the relation between expert systems, knowledge based systems and artificial intelligence, where do they overlap, and how can they be differentiated?—Quora. Retrieved March 26, 2017, from https://www.quora.com/What-is-the-relation-between-expert-systems-knowledge-based-systems-and-artificial-intelligence-where-do-they-overlap-and-how-can-they-be-differentiated.

Feasibility Study of Nano-Technology-Based Insulation Materials' Usage to Decrease the Cooling Loads in High-Income Housing in KSA

Abdulhafeez Ahmad M. Alwafi

Abstract

Analyzing the economic status in KSA in recent years, especially the investment sector, a major increase in the residential buildings sector is noted. This growth led to a massive energy consumption that was disproportionate to the Saudi's growth in power generation. It is also worth noting that the residential sector accounts for 42.3% of the total energy consumed in KSA where the Heating, Ventilation, and Air Conditioning (HVAC) is considered one of the key factors in energy waste. Meanwhile, the architects who design those buildings usually don't pay enough attention to the energy in the design process. This paper will tackle a strategy for the environmental control of building designs which is the insulation especially with the high technology in material manufacturing that led to many new materials that are developed with nano-technology, it is vital to take into consideration buildings' energy efficiency and the building compatibility with the environment by improving the design of the building envelope elements such as building shell. This research sheds light on the feasibility and profitability of nano-technology insulation materials' implementation and associating it with the thermal and environmental quality of the envelope of the building. The current building energy and HVAC performance in Al Riyadh City, one of the largest cities in KSA, was measured using a computer-based simulation tool (Design-Builder). Toward the end of the research, a conclusion about some of the aspects of the building envelope will be reached like thermal insulators' implementation into the construction layers of any building shell element seeking to minimize the energy waste in the case study and improve the indoor environmental quality, in addition to the various criteria of residential buildings' designing process in KSA in the near future.

Keywords

Low-energy buildings • Energy and thermal performance • Simulation • KSA

1 Introduction

Worldwide energy use is predicted to be enlarged with 53% of the current consumption within the following 10 years based on the prediction of the International Energy Agency (IEA). This is caused by the major increase in urban and industrial activities as a result of the country's remarkable development and intensively increasing population in recent years (Ong et al., 2011). The growing demand for energy is predicted to escalate, particularly in developed countries like KSA—Kingdom of Saudi Arabia—because of the dramatic growth in new constructions, whereas the implementation of the techniques of energy efficiency in constructions is usually got lower attention by the constructors and design professionals (Hui & others, 2000). As a significant demand for energy raises, environmental affairs are becoming more manifested. An extensively known example of the pollution causative is carbon dioxide (CO_2) which has been considered as a damaging component to human well-being (Oh & Chua, 2010). Moreover, carbon dioxide contributes strongly to the greenhouse gas effect, which causes the median worldwide temperature to rise (Lau et al., 2009). If no essential steps are taken toward reducing greenhouse gases and CO_2 emissions, the world's surface average temperature is expected to surpass the current temperatures by 1.1–6.4 °C by 2100. However, an increase of 2 °C in the worldwide average temperature will lead to irreversible effects on the

A. A. M. Alwafi (✉)
Assistant Professor, Department of Islamic Architecture, Faculty of Engineering and Islamic Architecture, Umm Al-Qura University, Mecca, Saudi Arabia

F. Rosso et al. (eds.), *Advances in Architecture, Engineering and Technology*, Advances in Science, Technology & Innovation, https://doi.org/10.1007/978-3-030-86913-7_14

natural environment, squeaky issues on human health, and massive harm to natural ecosystems (Change, 2007).

The issue here that only a single sector consumes an enormous amount of energy to achieve thermal comfort, generally the building sector (residential, commercial, and industrial constructions) would contribute to lowering its power consumption with the proper insulation techniques. An efficient insulation preserves energy and accordingly necessitates less energy in the space cooling process in the summer side by side with less heating energy to keep the house snug in winter (Xu et al., 2005). The consequences of this energy efficiency technique implementation decrease consumption of fossil fuels like gas and petroleum reserves, which are utilized for energy generation. As such, it decreases the generation of greenhouse gases (Dixon et al., 2010). Insulation in constructions is considered a simple but highly effective energy-saving technique that could be applied to the residential sector. The thermal insulators are composed of a single substance or composite substances that could be obtained through the characteristic of high thermal resistance, which shows the ability to reduce the heat flow in and out of the spaces (Al-Homoud, 2005). Consequently, building insulation has the ability to keep the heat/cool within the spaces and reduce heat flux with the environment (Al-Sallal, 2003). Many substances like fiberglass, foam, mineral wool, and many other materials are usually implemented as an insulator. Another key benefit of building insulator is cost-saving, this is businesslike since the insulation participates in balancing net energy through a massive amount of saved energy through the insulation implementations than the energy used to produce the insulation itself (Ostrom, 2010). Furthermore, implementing thermal insulation offers other benefits like fire defense, buildings' users' comfort, and sound control. However, many studies were focused on the relationship between thermal insulation and the quality of indoor thermal environment characteristics in KSA like Al-Sanea and other researches in 2016 that shed light on the optimum R-values for insulators that should be used in Saudi Arabia, but their study never focused on the feasibility of the implementation of any insulations (Al-Sanea et al., 2016), another study that was also made in Saudi Arabia was Abdul Mujeebu study, which was focused on the implementation of nano-technology insulation in openings and windows as the case study there was about the Aerogel; however, the study also never shed the light on the feasibility of implementing those materials (Mujeebu, 2018).

1.1 Residential Buildings Sector in KSA

Over the last few decades, the Gulf Cooperation Council (GCC) region has experienced huge and largely steady growth in the construction sector in terms of expansion in size and modernity. As KSA aims to become one of the leading Middle Eastern financial centers and major trading centers in the world, the construction boom in there over the 10–15 years has been a trendsetter for other countries in the region. In terms of construction projects, KSA is a major player in the region as it accounts for almost 72% of the total construction projects in the GCC region as shown in Fig. 1 (Deloitte, 2010). Of the total construction projects underway in the whole Middle East region, worth $2.5 Trillion, 60% of the work is reported to be carried out by Saudi Arabia and the United Arab Emirates (McCullough, 2014).

As stated by the Middle East Economic Digest (MEED), KSA is at the top of the GCC market in terms of investment with a share of 43% which accounts for 1$ Trillion worth of projects that are either planned or underway as of 2013. According to an analysis of the construction sector, majority of the construction works are residential buildings that are meant to meet the demand for new homes (MOMRA, statistical data on construction sector, 2011).

As per the MEED, the major beneficiary of the aforementioned investment is the residential sector with a 29% share as suggested in Fig. 2. With the increase in population, at the rate of 2.5% annually, a significant growth in the residential sector in KSA is expected to take place. In spite of this, only 24% of the Saudi nationals own their homes (Deloitte, 2010). While around two-thirds of the population is under the age of 30 years (CDSI, n.d.), estimates suggest that to meet the needs of the growing population, the country should build 2.32 million new homes by 2020 (Arabia, 2009).

The last decade witnessed a significant growth in the number of licenses being issued for buildings where residential and commercial buildings account for most of these licenses according to an analysis of the Saudi construction sector carried out by the Saudi Ministry of Municipal and Rural Affairs (MOMRA, Construction licenses Statistics, 2013).

2 Case Study Characteristics

The case study was selected from the Hetyen neighborhood in Al Riyadh City as one of the high-income housing areas in Al Riyadh City. The selected case study in this research is a corner villa on Marakish and Mahboba streets, with a total area of 1000 m^2, the case study location and layout are demonstrated in Fig. 3 (Google maps, 2019).

The selected case study villa is a three-floor villa with a single entrance and a private garden, the villa is 30.2 * 36.5 m in layout dimensions with a total high 15 m. The ground floor is containing the reception areas, kitchen, and office. The second floor contains the four master bedroom areas while the roof floor contains the service room. The floor plans are shown in Fig. 4.

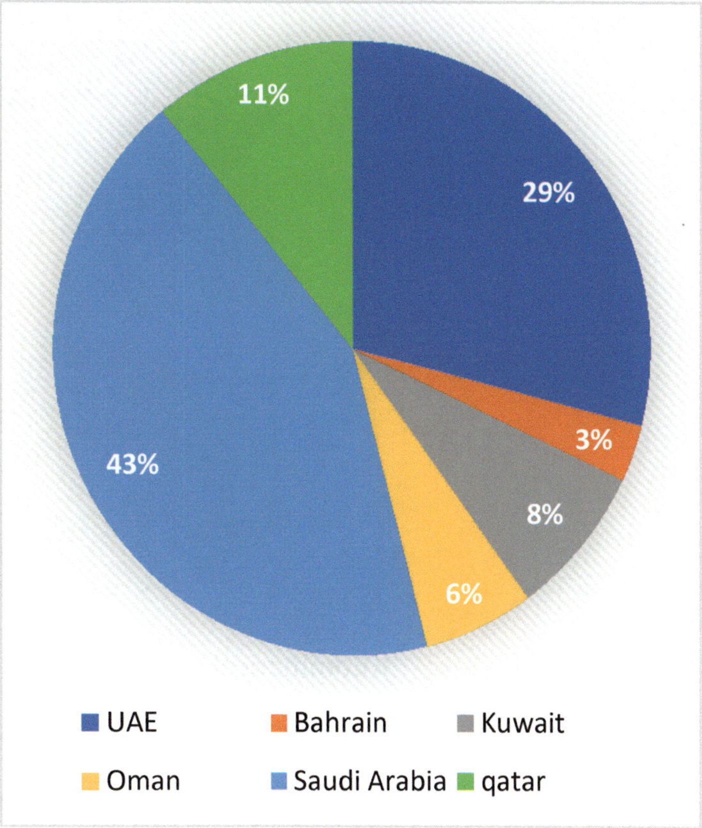

Fig. 1 Construction investment in the GCC region

The current building style in the case study doesn't contain any types of insulations, as the materials in the outer shelter only consists of the main core and plasters, the outer roof materials are—from outside to inside—ceramic tiles, cement plaster, sand, reinforced concrete, and cement plaster. While the materials of the outer walls are—from outside to inside—outer paintings, cement plaster, bricks, cement plaster, and painting.

3 Methodology

To evaluate the effects of changing the construction's parameters for a residential building in KSA. *Firstly*, a defined prototype (it's a detached private house). *Secondly*, Design-Builder was selected to evaluate the cooling loads for the whole buildings and the indoor mean radiant temperature (MRT) based on the current buildings' parameters identified through its simulation analysis tool. *Thirdly*, construction parameters were changed with adding nano-technology-based insulation materials into the outer shell of the building (roof, outer walls, glazing) and retrieve the changes into the cooling loads and indoor mean radiant temperatures. *Fourthly*, the current cooling loads and simulated cooling loads (after adding insulation) were compared and transformed to annual energy bills from the rates that were retrieved from Saudian Electricity Authority (SEA). The whole methodology process is demonstrated in Fig. 5.

3.1 Why Design-Builder (DB) Simulation Software

EnergyPlus is one of the most authoritative software that has been extensively used and validated by the research community. It includes several user interface add-ons such as Design-Builder. It is a progressive software tool used to check building's energy, lighting, carbon, and thermal comfort performance. The software enables a quick comparison of building designs and delivers timely and low-budget results as it is developed to facilitate the building simulation. Moreover, the mean radiant temperature is one of the simulated characteristics of the building in design-builder, which could be used as an indicator of the indoor thermal characteristics as the decrease in it should be reflected by a decrease in the cooling loads and an increase in the users' thermal comfort inside the space (Tran et al., 2017).

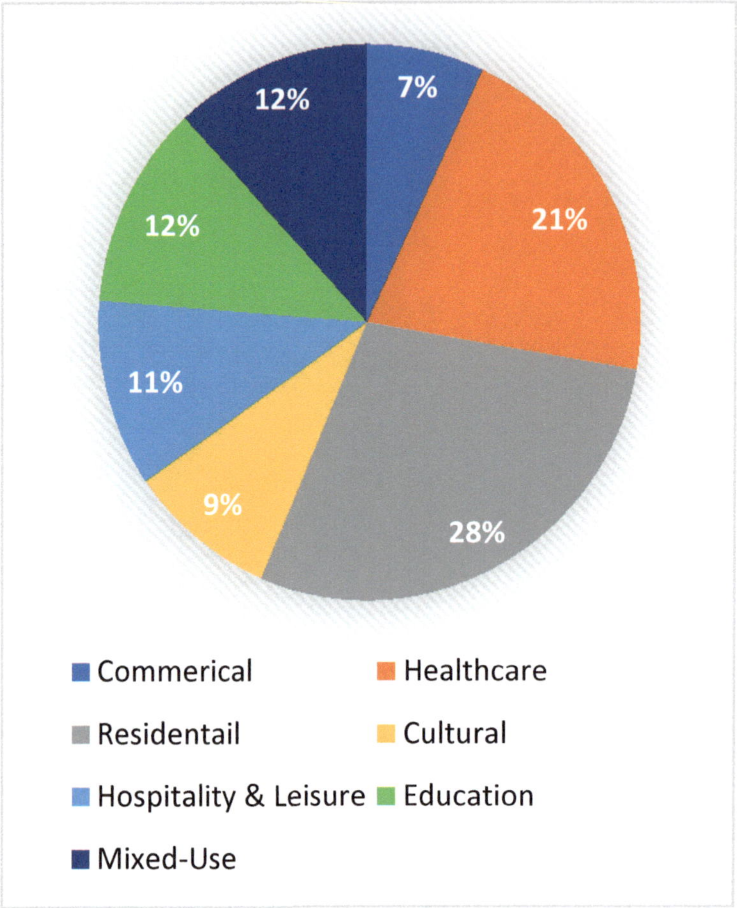

Fig. 2 Construction spend in terms of segments in KSA

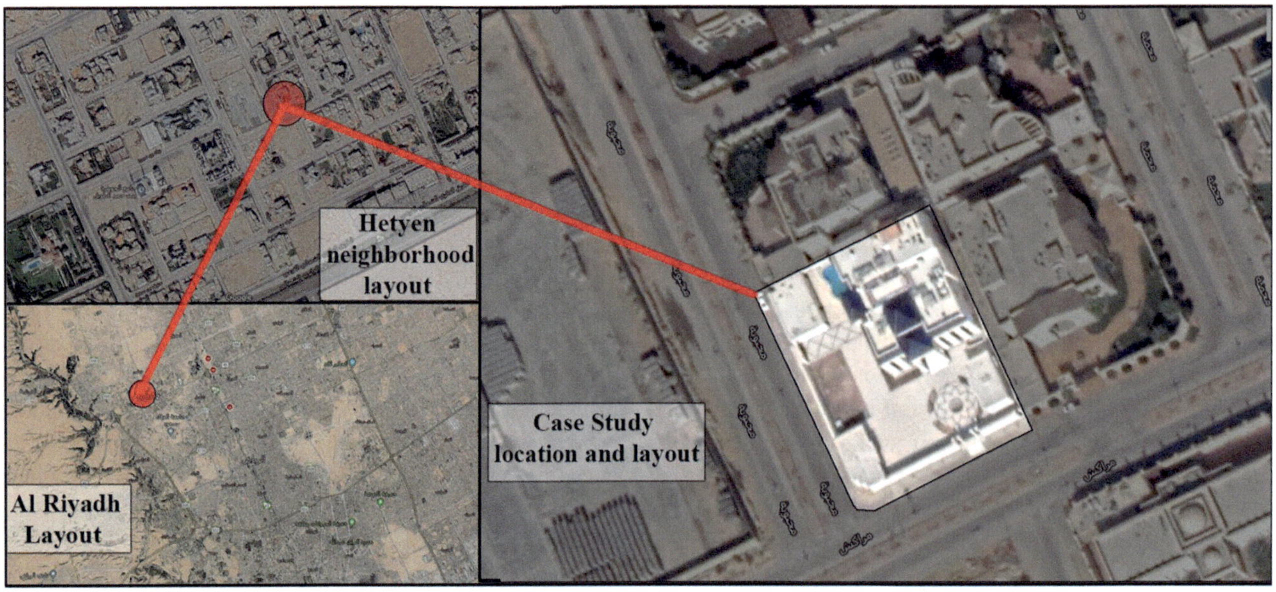

Fig. 3 Case study layout (Google maps, 2019)

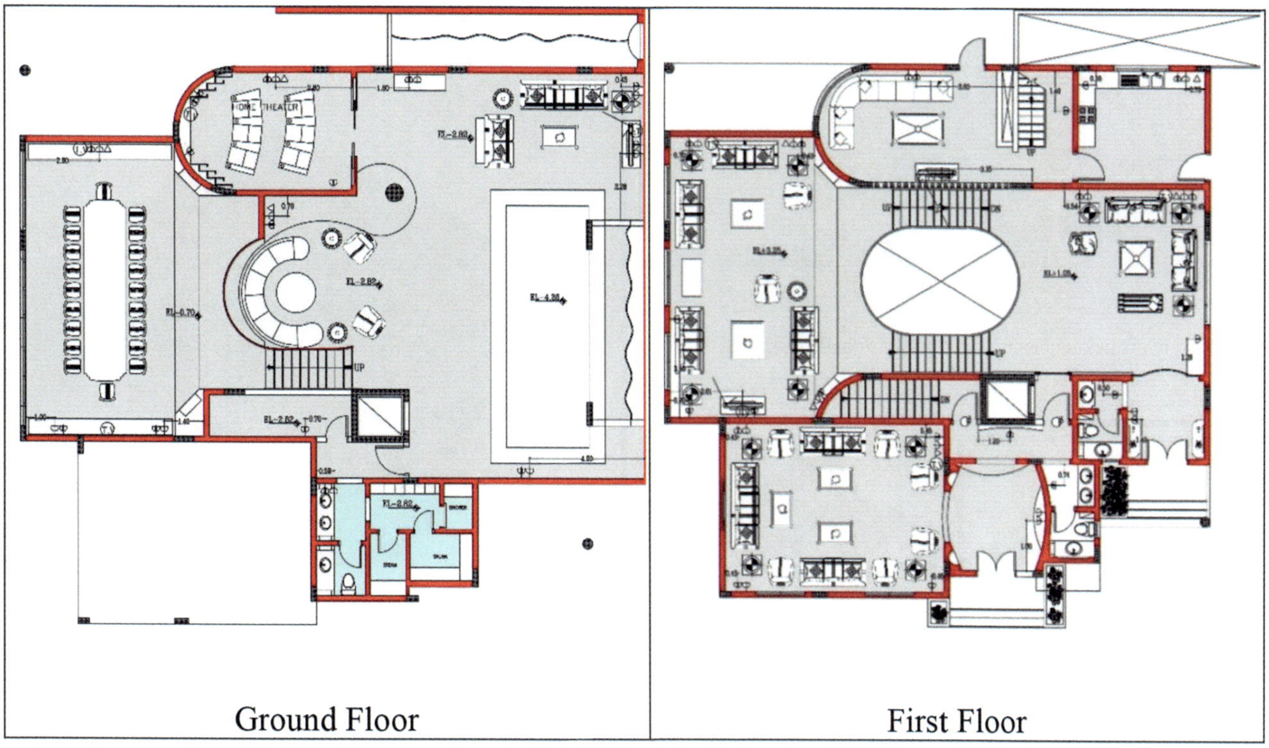

Fig. 4 Villa N perspective and plans (Author's own)

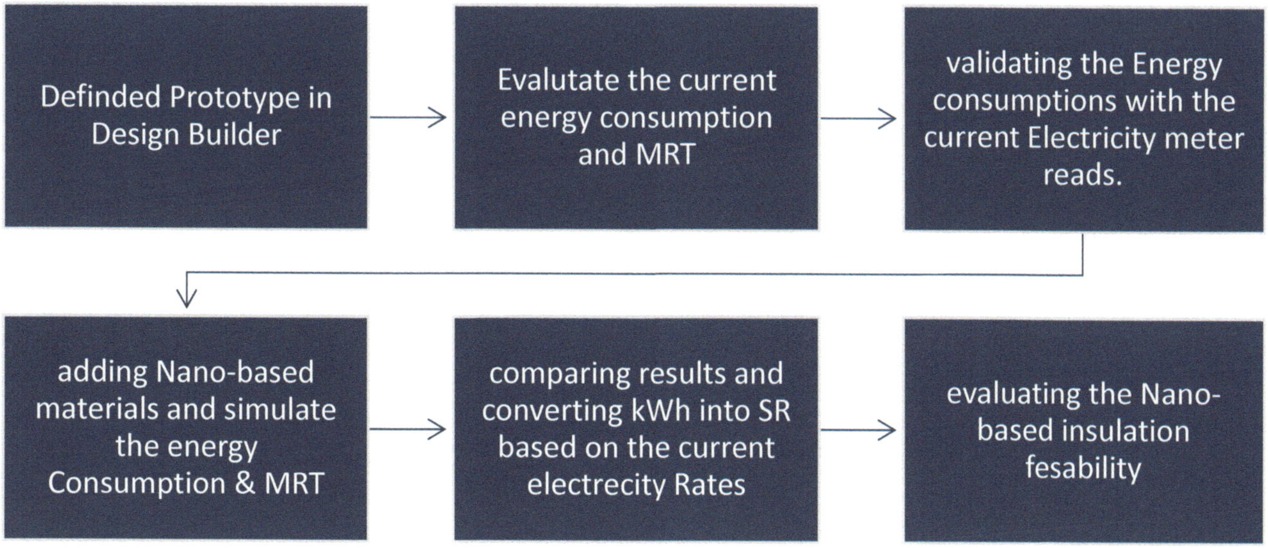

Fig. 5 Methodology steps

3.2 Building a Prototype in Design-Builder (DB)

A prototype with the current building construction buildings' parameters are building geometry, orientation, construction materials, users' activities, and working hours profiles. The climate data was set to the Al Riyadh climate data as the case study is located within the same climate zone. The glazing type was set to be single clear glass glazing in all the windows, the whole spaces were considered as thermally conditioned spaces as the space cooling system implemented in the case study is a central cooling system with air blowers in all the space ceilings. All the data are demonstrated in Figs. 6, 7, and 8. All the simulations that were done for a specific day which was the summer design

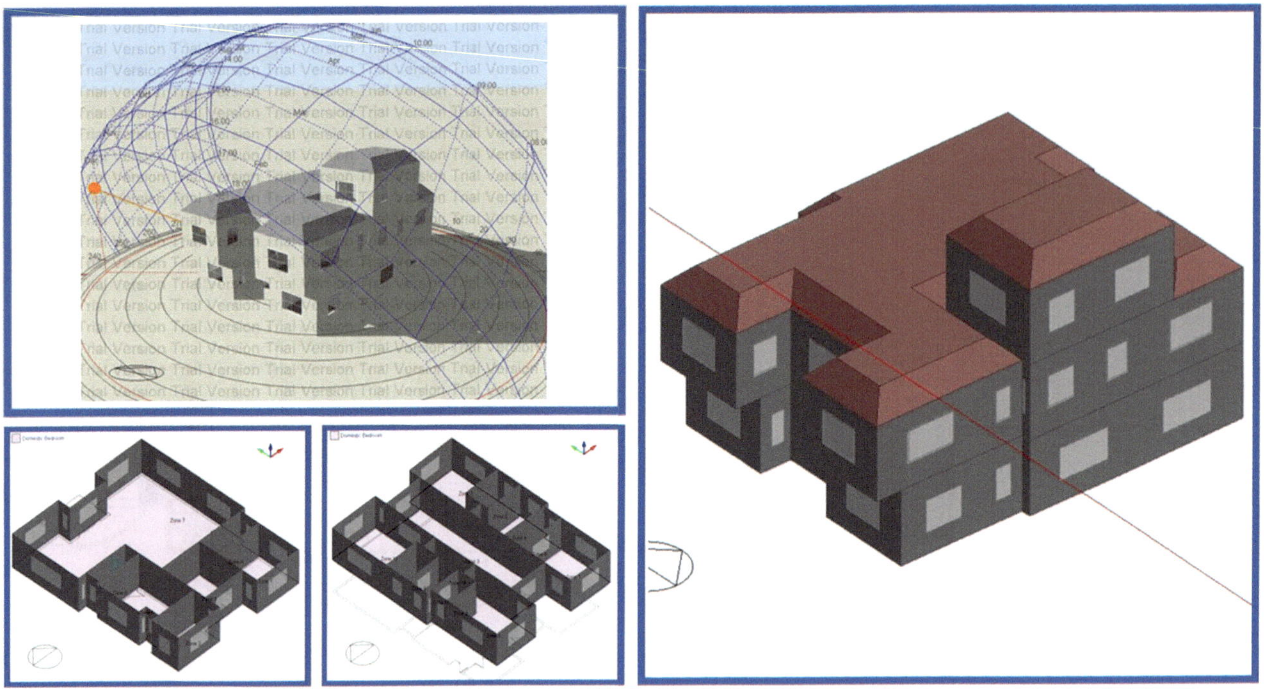

Fig. 6 Building geometry and orientation

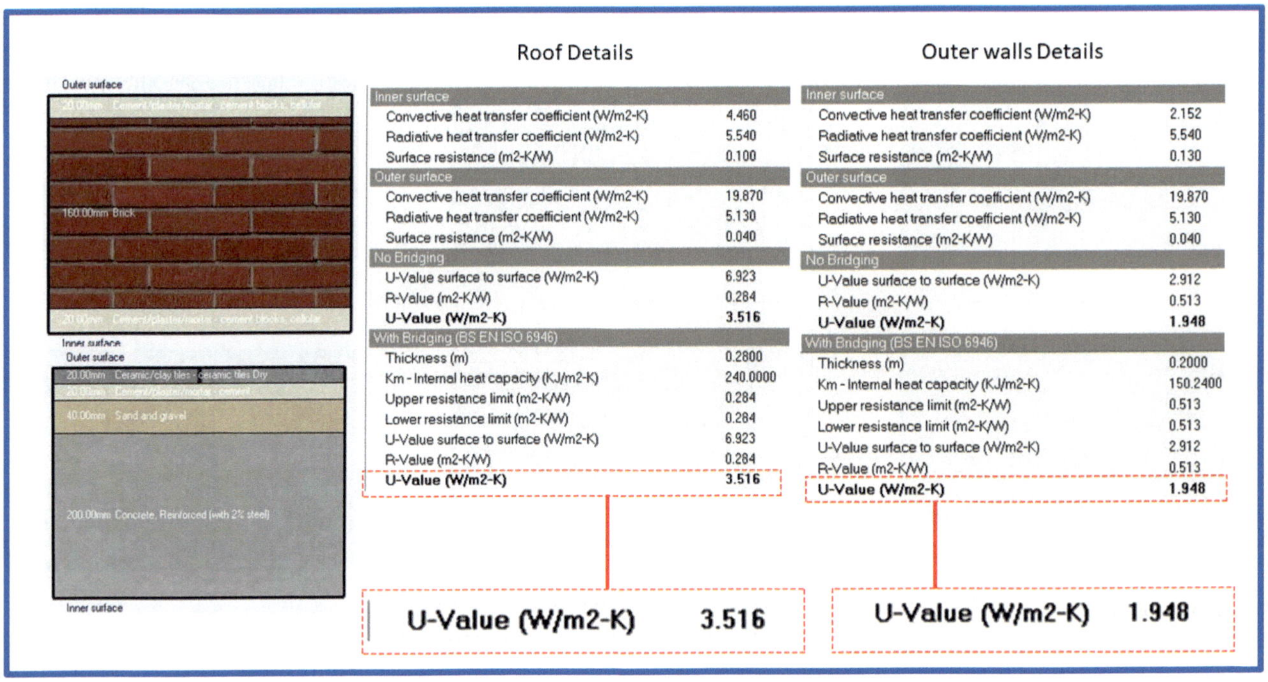

Fig. 7 Building material details

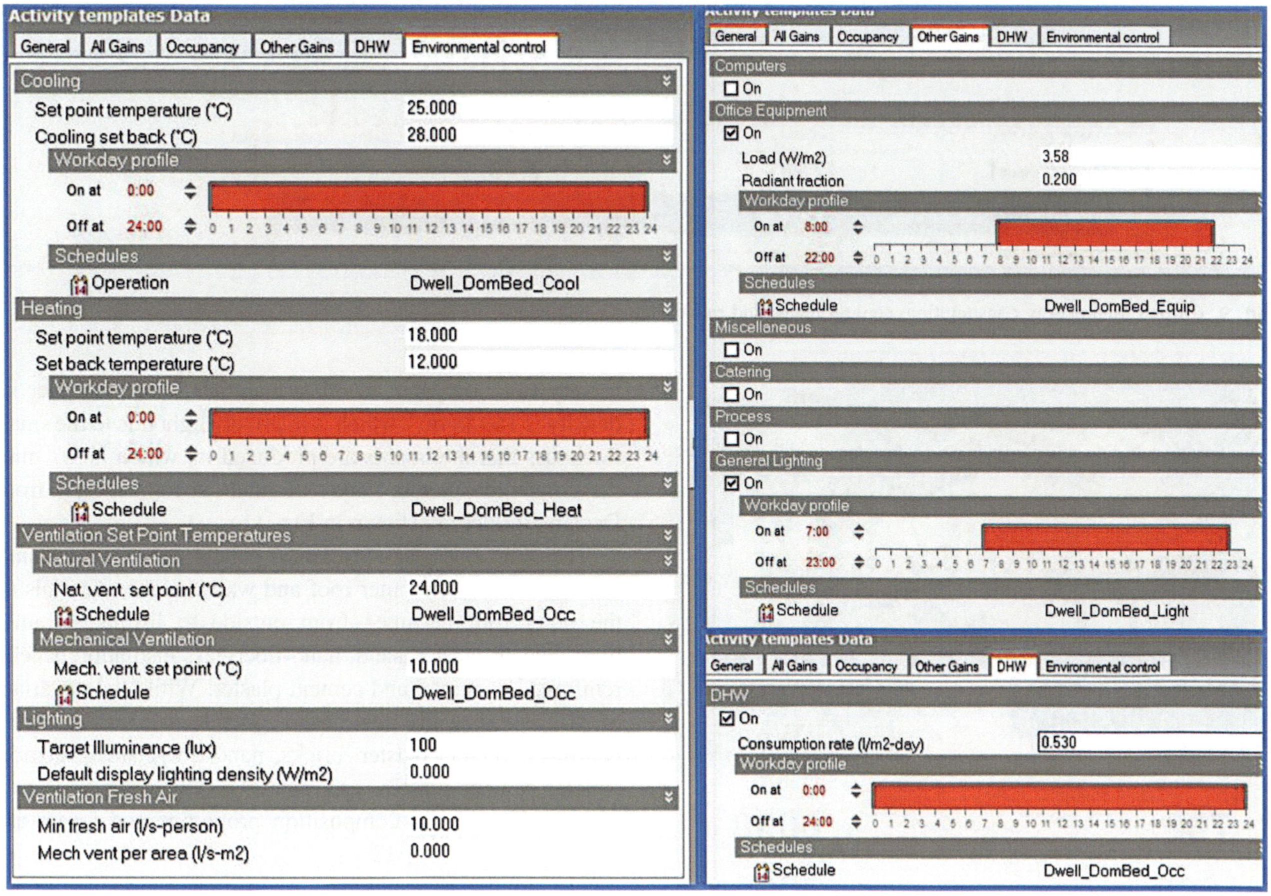

Fig. 8 Space usage profile, lighting, and HVAC profiles

day 15 July, as the summer months in Al-Riyadh were June to August with a middle and peak day for outdoor temperature in the summertime is 15 July (El-Gebeily et al., 2010).

3.3 Current Energy Consumption and Mean Radiant Temperature

The current total energy consumption, cooling loads in the summer design day—the summer design day was set to 15 July in the climate data—and the mean radiant temperature were simulated in Design-Builder and the results were referring to a total cooling load of 77.74 kW at the summer design day (15 July), the cooling system design capacity was 89 kW, the total energy cooling load over the year was 54,993 kWh, and the highest recorded radiant temperature was 38.26 °C. The radiant temperature over the day is demonstrated in Table 1, while the total cooling loads, total energy consumption, and cooling capacity are demonstrated in Fig. 9.

3.4 Validating Current Energy Usages

The total energy usage that was simulated from the Design-Builder was referring to 11,463 kWh during the whole year, which means that the average usage for each month is 955 kWh. Based on the kWh prices that were published by Saudian Electricity Authority that the 900:1000 kWh has an average price for 900 SR—Saudi Riyal (SEA, 2018). The validation process was done through one of the

Table 1 Radiant temperature in the case study—without insulation at 15 July (summer day)

	1:00	2:00	3:00	4:00	5:00	6:00	7:00	8:00	9:00	10:00	11:00	12:00
Radiant Temperature	35.77	35.3	34.85	34.41	33.99	34.08	34.55	35.05	35.52	35.78	35.95	36.06

13:00	14:00	15:00	16:00	17:00	18:00	19:00	20:00	21:00	22:00	23:00
36.43	36.94	37.44	37.89	38.23	38.26	37.84	37.71	37.59	37.3	36.82

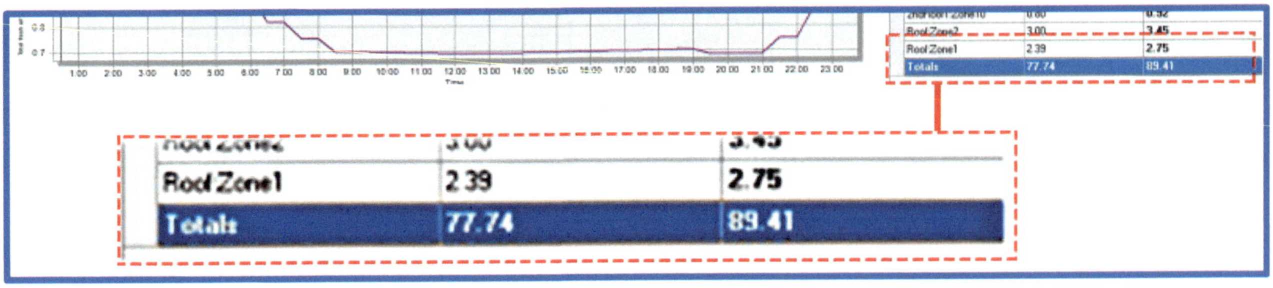

Fig. 9 Current total energy consumption, cooling loads, and radiant temp. simulations

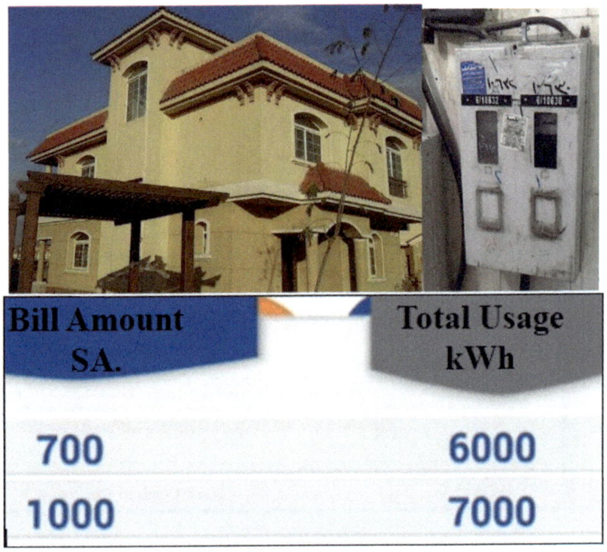

Fig. 10 The actual validated villa in Al Riyadh and the installed electricity meter

actual villa users who stated that they pay around 800:1000 SR each month for the electricity bills, with a total 10,600 SR as total bills for the period July 2017:June 2018, which refers to an average bill each month of 833 SR, and average usage of 912 kWh each month, which is in the same identified range from the simulation done by the Design-Builder. However, the actual bills of the electricity couldn't be accessed by the researchers. The actual villa-installed electricity meter readings were validated as shown in Fig. 10.

3.5 Adding Nano-based Insulation Material Simulation

The same building geometry was kept, only layers of the outer shell materials were redefined. As the selected layer to be inserted as nano-insulation material was "Nano-Fiberglass Insulation Panel", this material has magnificent thermal insulation properties, its conductivity is 0.036 w/m k, which is very low compared to other materials, and moreover its density is 160 kg/m^3, which is relatively light due to the small thickness that the boards are produced in, which is 5:7 mm, the material picture and thermal specifications from Design-Builder are shown in Fig. 11.

The nano-fiberglass insulation panels were inserted into the materials of the outer roof and walls, as the materials of the outer roof became—from outside to inside—ceramic tiles, cement plaster, sand, nano-fiberglass insulation panels, reinforced concrete, and cement plaster. While the materials of the outer walls are—from outside to inside—outer paintings, cement plaster, bricks, nano-fiberglass insulation panels, cement plaster, and painting. The outer roof and outer walls material composition properties and layers are demonstrated in Fig. 12.

The simulation results were referring to a reduction in the cooling power, total energy and electricity consumption, and radiant temperature. As the total cooling loads—for the summer day 15 July—was 73 kW, while the cooling system design capacity was reduced to 84 kW and the total cooling energy for the whole year was 52,012 kWh and the highest recorded radiant temperature during the summer design day was 34.4 ° C. The radiant temperature over the day is demonstrated in Table 2, while the total cooling loads, total energy consumption, and cooling capacity are demonstrated in Fig. 13.

3.6 Comparing Results

The reduction in results was referring to better energy efficiency and improved thermal characteristics for all the case study indoor spaces. As demonstrated in Fig. 14, the whole building cooling energy consumption was reduced from 54,993 to 52,012 kWh, so the reduction in the cooling loads is 2981 kWh for each year which is around 5.5% of the whole cooling loads. Moreover, the design capacity for the cooling system was reduced by 5.4% from 89.41 to 84.5 kW. And the same with the cooling load for the summer design day which was reduced by 8% from77.74 to 71.4 kW. However, the highest recorded radiant temperature in the indoor spaces was reduced from

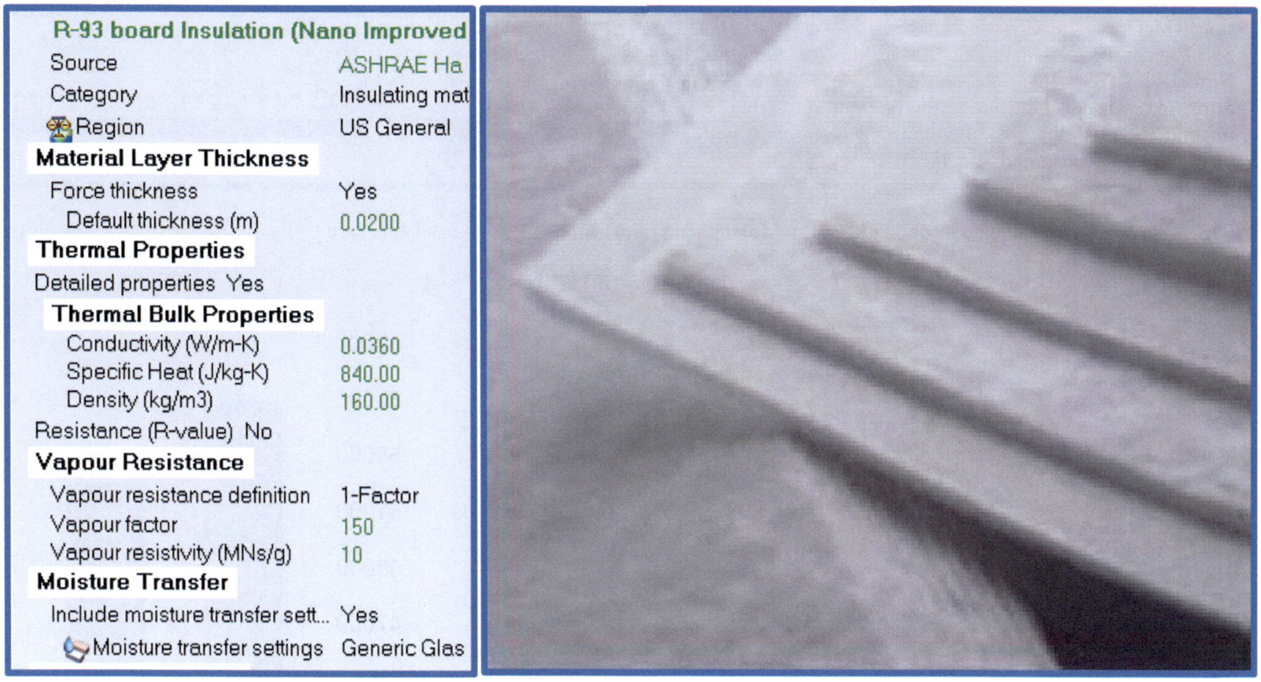

Fig. 11 Nano-fiberglass insulation panel picture and specification

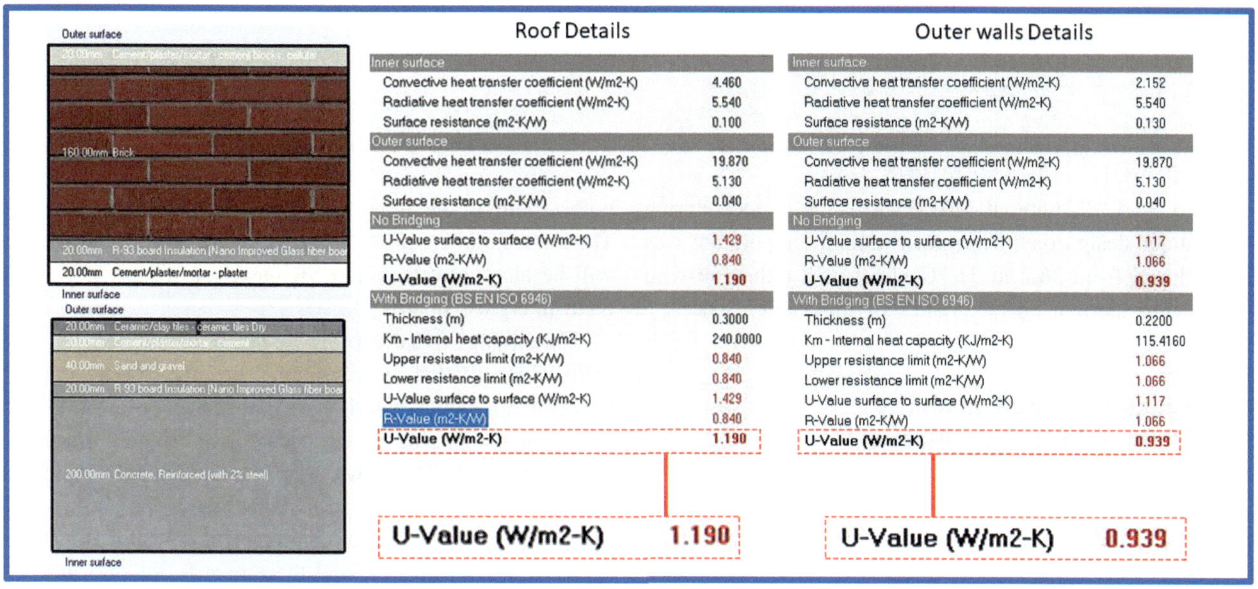

Fig. 12 Materials layers and specifications after adding nano-insulation

Table 2 Radiant temperature in nano-insulation materials case on 15 July (summer day)

	1:00	2:00	3:00	4:00	5:00	6:00	7:00	8:00	9:00	10:00	11:00	12:00
Radiant temperature	29.38	29.08	28.82	28.58	28.36	28.65	29.03	29.26	30.03	30.77	31.23	31.54

13:00	14:00	15:00	16:00	17:00	18:00	19:00	20:00	21:00	22:00	23:00
32.08	32.74	33.38	33.95	34.38	34.49	34.11	32.55	31.34	30.64	30.13

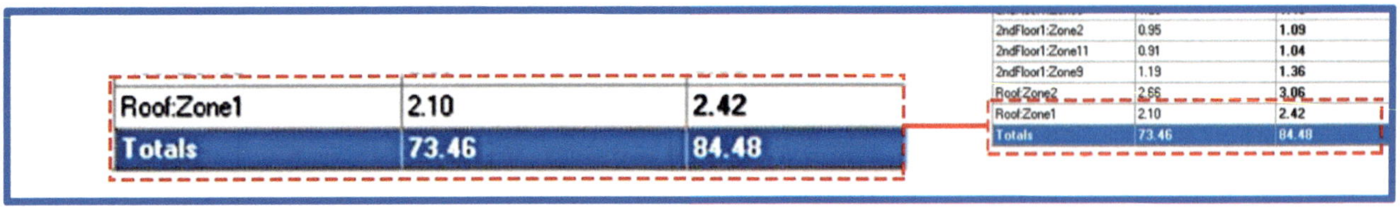

Fig. 13 Total energy consumption, cooling loads, and radiant temp. simulations for nano-insulation case

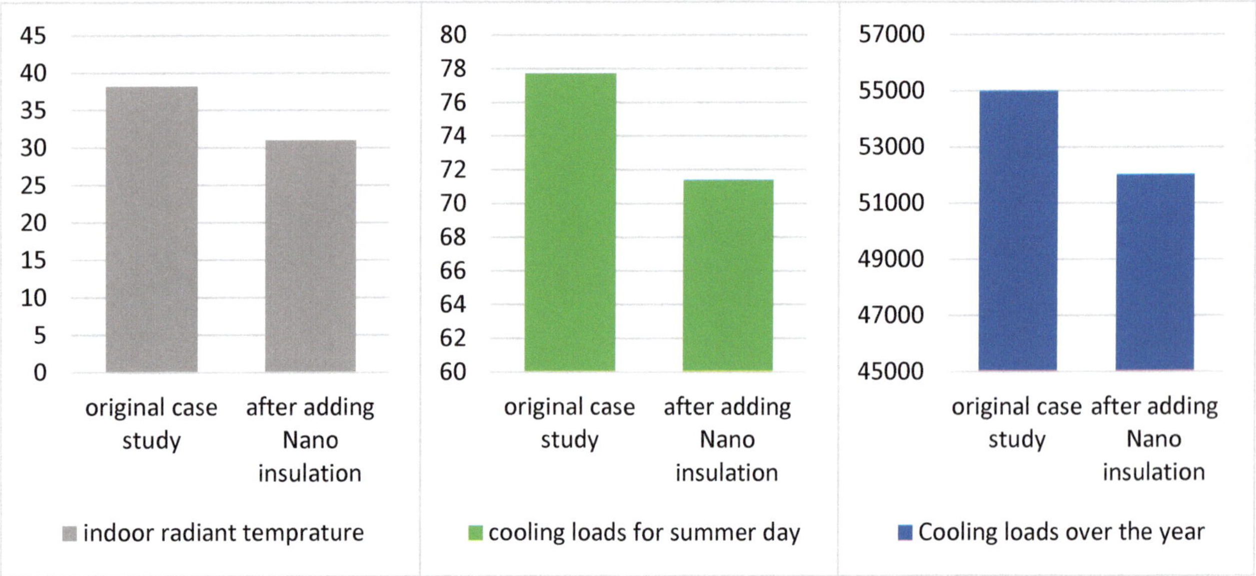

Fig. 14 Total energy consumption, cooling loads, and radiant temp. difference between original case study and nano-insulated case

38.26 to 34.4 °C, without any HVAC interference during the summer design day. The mean radiant temperature was also reduced from 36.2 to 31 °C which is a reduction with 5 °C in the thermal characteristics of the indoor spaces.

3.7 Evaluating the Feasibility of Nano-insulation Material

Evaluating the feasibility will be based on measuring the reduction in the cooling loads with kWh during the whole year, then convert it into SR with the kWh announced prices from Saudi Electricity Authority in July 2018. This will shed light on the saved money from installing the insulation in the outer shell of the building during the building operation. Then the number of installments of nano-insulation in m^2 will be also converted into SR based on the pricing of the nano-fiberglass boards.

The saving in the cooling power over the year is 2981 kWh when divided by 12 months, the monthly saving will be 285 kWh, which could be bought each month for 335 SR based on the pricing of electricity in the Saudi Electricity Authority (SEA, 2018), which means the annual saving will be 4024 SR. The amount of installed nano-fiberglass boards is 523 m^2 based on the simulations from Design-Builder as demonstrated in Fig. 15, and these boards are sold in many online places for a fixed price that is around 0.5$ for the board that is 1 * 1 m^2 (inchuangthermal, 2019), which

Materials Embodied Carbon and Inventory	Area (m2)	Embodied Carbon (kgCO2)	Equivalent CO2 (kgCO2)	Mass (kg)
R-93 board Insulation (Nano Improved Glass fiber board)	522.9	0.0	0.0	1673.3

Fig. 15 Nano-glass fiber boards installation in the building—simulated by DB

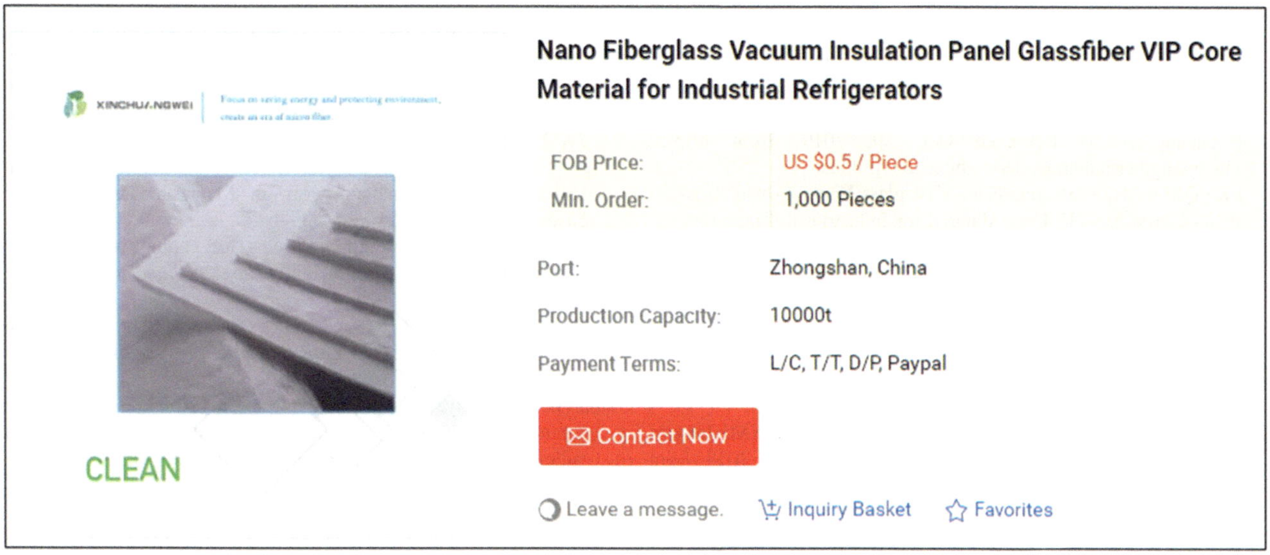

Fig. 16 Nano-fiberglass insulation panels rates in China

means that the 523 m^2 will be installed for 262$ or 4702 SR based on currency pricing (misr, 2018). Figure 16 shows the nano-fiberglass board's rates and specifications.

All this leads to the fact that after the end of the construction phase and starting to operate the building within 14 months, the amount of paid money for extra insulation will be saved from the electricity bills, and after that the saving will be feasible for the building user. The feasibility could be changed when shipping and construction manpower rates of installing applied; however, in the long term, the implementation of nano-insulation materials will always be feasible as it saves operating wasted energy for the cooling.

4 Conclusion

As an overall look, the implementation of thermal insulations wasn't obligatory by the construction laws in Saudi Arabia for a long time, but the recent updates in the construction codes and laws make it mandatory to every new construction to become energy efficient through many environmental techniques and measurements like the thermal insulators. Moreover, the implementation of nano-technology-based materials isn't yet habitual by the constructors in KSA since many recent researches are inviting constructors to implement nano-technology materials in the construction process. On the other hand, nano-glass fiberboards mightn't be the optimum insulators for the thermal characteristics; however, compared to the pricing, the feasibility of its implementation was profitable after 14 months only, which means that other materials with better thermal characteristics might also be profitable more even with longer time compared to the initial cost. Moreover, the Saudi government is willing to continue the subsidy removal (the governmental financial support), which will lead to major increase in the energy prices for the consumers, and more profitability for passive techniques in the construction technologies in Saudi.

References

Al-Homoud, M. S. (2005). Performance characteristics and practical applications of common building thermal insulation materials. *Building and Environment, 40*, 353–366.

Al-Sallal, K. A. (2003). Comparison between polystyrene and fiberglass roof insulation in warm and cold climates. *Renewable Energy, 28*, 603–611.

Al-Sanea, S. A., Zedan, M., Al-Mujahid, A., & Al-Suhaibani, Z. (2016). Optimum R-values of building walls under different climatic conditions in the Kingdom of Saudi Arabia. *Applied Thermal Engineering, 96*, 92–106.

Arabia, H. t. (2009). Sidawi, Bhzad. *Emirates Journal for Engineering Research, 14*(1), 73–82.

CDSI. (n.d.). *Population statistics database*. (Central Department of Statistics and Information) Retrieved March 25, 2019, from https://www.stats.gov.sa/en.

Change, I. P. (2007). Climate change 2007: The physical science basis. *Agenda, 6*, 333.

Deloitte, G. (2010). *powers of construction 2010: Construction sector overview*. Deloitte & Touche.

Dixon, G., Abdel-Salam, T., & Kauffmann, P. (2010). Evaluation of the effectiveness of an energy efficiency program for new home construction in eastern North Carolina. *Energy, 35*, 1491–1496.

El-Gebeily, M. A., Rehman, S., Al-Hadhrami, L. M., & AlMutawa, J. (2010). Continuous and discrete wavelet transforms based analysis of weather data of North Western Region of Saudi Arabia. *World Journal of Science, Technology and Sustainable Development, 7*(4), 369–389.

Google maps. (2019). *Al Riadh Map*. Retrieved 3 25, 2019, from https://www.google.com/maps/@24.7476395,46.6056076,235m/data=!3m1!1e3.

Hui, S. C., & others. (2000). Building energy efficiency standards in Hong Kong and mainland China. In *Proceedings of the 2000 ACEEE summer study on energy efficiency in buildings*.
inchuangthermal. (2019). *Nano fiberglass vacuum insulation panel glassfiber VIP core material for industrial refrigerators*. (inchuangthermal) Retrieved Oct. 20, 2019, from https://xinchuangthermal.en.made-in-china.com/product/hKYQSFmEHZWn/China-Nano-Fiberglass-Vacuum-Insulation-Panel-Glassfiber-VIP-Core-Material-for-Industrial-Refrigerators.html.
Lau, L. C., Tan, K. T., Lee, K. T., & Mohamed, A. R. (2009). lau2009comparative. *Energy Policy, 37*, 4771–4778.
McCullough, D. (2014). Saudi Arabia green decree brings hopes of sustainability. *The Guardian, 12*.
misr, b. (2018, October 7). *Currency rate*. Retrieved October 7, 2018, from http://www.banquemisr.com/ar/exchangerates.
MOMRA. (2011). *Statistical data on construction sector*. (Ministry of Municipaland and Rural Affairs) Retrieved March 25, 2019, from http://www.momra.gov.sa/GeneralServ/statistics.aspx.
MOMRA. (2013). *Construction licenses statistics*. Ministry of Municipal and Rural Affairs.
Mujeebu, M. A. (2018). Nanotechnology in eco-efficient construction: Materials, processes and applications. In *Department of building engineering, college of architecture and planning, Imam Abdulrahman Bin Faisal University, Dammam, Saudi Arabia*(Woodhead Publishing), 417.
Oh, T. H., & Chua, S. C. (2010). Energy efficiency and carbon trading potential in Malaysia. *Renewable and Sustainable Energy Reviews, 14*, 2095–2103.
Ong, H., Mahlia, T., & Masjuki, H. (2011). A review on energy scenario and sustainable energy in Malaysia. *Renewable and Sustainable Energy Reviews*, 639–647.
Ostrom, E. (2010). Polycentric systems for coping with collective action and global environmental change. *Global Environmental Change, 20*, 550–557.
SEA. (2018, September). (Saudian Electricity Authority) Retrieved March 25, 2019, from https://www.se.com.sa/24Consumption/BillData.aspx.
Tran, D. Q., Nguyen, L. D., & Faught, A. (2017). Examination of communication processes in design-build project delivery in building construction. *Engineering, Construction and Architectural Management*.
Xu, X., Zhang, Y., Lin, K., Di, H., & Yang, R. (2005). Modeling and simulation on the thermal performance of shape-stabilized phase change material floor used in passive solar buildings. *Energy and Buildings, 37*, 1084–1091.

Achieving Thermal Comfort by Simulating the Ecosystems of the Desert Environment

Esraa Ahmed Hamdy

Abstract

Egypt has more than 95% of desert land, with a fluctuating climate and varying characteristics. The desert has its unique features that forced mankind to create architectural and urban patterns to adapt to the surrounding environment by building cities and developing indoor spaces, in order to protect themselves from extreme climatic conditions. Advanced technology in construction aided with many difficulties that face designers connecting internal spaces with the external environment, through providing a harmonious system that achieves thermal comfort and prevents the consumption of energy resulting from the use of conventional methods of cooling and heating, achieving permanent control over the amount of energy flowing under climate change throughout the year. The impact of our current lifestyles in the desert environment leads us to a tragic state of depletion of energy, and desert environment has become more vulnerable to pollution and damage, where the need to seek solutions should be inspired by nature itself. "Ecological simulation" is the major key to find solutions for design problems through the observation of nature, using biological research and information available on the ecosystems and organisms in the desert environment. Nature simulation is a process that begins with a design problem and ends with solutions for that problem through the designer.

Keywords

Thermal comfort • Ecological simulation • Desert ecosystems

1 Introduction

Since the beginning of creation, the main concern of man was trying to adapt to his environment; he lived in caves and huts that were built in primitive ways during prehistoric times to protect himself against the danger of animals and to overcome the various climatic changes. Through his understanding of the phenomena of the environment and its geographical and climatic characteristics, the early architect could develop his habitat and use different types of housing from the common types, relying on previous experiences in developing the construction methods and masonry materials, to create a suitable residential environment, considering the atmosphere as the main player (Hassan, 2016).

Ancient Egyptians were the first to inspire the world with the idea of environmental adaptation. Egypt has had the natural ecological components that influenced architecture to be compatible with the nature of climatic factors that prevailed in the country. There were differences in architectural designs between palaces of kings and houses of the common and poor, but in general, they were made up of rooms that clustered around a courtyard (Okasha, 2008). Nevertheless, the Assyrian civilization created a new technique that was established in the style of domes, and they also used flat roofs and courtyards in houses that was very useful in summer nights to enjoy the breezy atmosphere, escaping the heat stored inside the building in the day (Almalky, 2011).

In Islamic architecture, domes and vaults appeared in ceilings, wind catchers, and thick walls built with mud and stones as well as timber in hybrid architectural openings (windows) (Mashrabya), and others to achieve thermal comfort inside buildings such as mosques, gates, public baths, and schools (Konya, 2011). In the modern era, *Hassan Fathy* developed the use of mud and mud bricks in construction, he used the concept of domes to achieve thermal comfort inside buildings, the most famous examples "*Architecture of the poor*" and the village of "*the new Garna*" (Fahmy, 2012).

E. A. Hamdy (✉)
Faculty of Arts & Design, October University for Modern Science and Arts "MSA", October City, Egypt

F. Rosso et al. (eds.), *Advances in Architecture, Engineering and Technology*, Advances in Science, Technology & Innovation, https://doi.org/10.1007/978-3-030-86913-7_15

Desert is a fascinating place, but it is difficult to follow standard rules for classification. Desert regions differ from one region to another in their characteristics of heat, humidity, brightness, and rain, but these properties and rates fluctuate in the same desert region during short periods between night and day and change completely from one season to another ("CONSERVE ENERGY FUTURE", 2016a). Regardless of the common idea that living organisms cannot survive arid regions, various creatures managed to survive through the scarcity of vegetation and unfavorable conditions. While desert occupies one-fifth of the surface of the globe and in the mind of everyone a desert stands as a hot, arid area, it also tends to be cold in some parts of the world. Despite the location, all deserts are usually cold at night and gets almost no rain, yet, vegetation and animals in these regions have adjusted to such physical conditions ("Conserve Energy Future", 2016a). Desert climate is generally unstable due to the open environment of the desert, besides the rapid change in climatic factors from one spot to another, and the nature of the land or the soil which is characterized by its ability to gain and lose heat rapidly (Sheba, 2012).

Thermal comfort for living organisms is linked to the need to keep up consistent conditions, and it is one of the highlights that link the facilities with nature (Fajal, 2002). Keeping up a suitable temperature is basic for the parts of biochemical and physiological capacity, living organisms consume a lot of energy regulating body temperature, and the ideal thermal comfort is the physical conditions transmitted by the tangible and mental pathology of man (Azmi, 2016).

Regardless of the stunning specialized advancement accomplished by mankind, we are still chasing the technology of nature. The observer of human technology finds that nature preceded it many millions of years ago. Plane, submarine, radar, and sonar are models of nature. While human innovation depends on energy and materials to solve the issues confronting us, nature seeks to solve its problems through processing information and surroundings adaptation, this is evident in the comparison between the philosophy of nature in solving its problems and the philosophy of human beings in solving their problems. Where the philosophy of nature is the delicate balance between energy, materials, methods of construction, information, space, and time, human philosophy is based on energy, especially in the problems that occur in the nanometer, while the role of construction is reduced and the role of information is inconclusive, which confirms the ability of nature to solve its problems (Sheba, 2012).

2 Purpose of the Work

The severe change of desert climate from heat, humidity, rain, and wind throughout the day or between seasons is a major feature and one of the most important problems related to the desert climate. With the increasing demand for desert activities for many purposes such as working in oil extraction, conducting various biological and geological researches as exploratory missions, recreational and relaxation purposes such as camps and meditation, expeditions and safari, finding effective design solutions and applications compatible with user needs and to achieve thermal comfort without damaging the environment became essential. Traditional methods of lodging like tents, caravans,.. etc. do not achieve the appropriate thermal comfort in desert facilities against extreme weather conditions. The architectural treatments of high temperatures during the day may be useless for temperature sharp drop in the night, which may reach freezing as well as the appropriate treatments for lack of rains may become catastrophic when the floods occur. The way to address this situation is to simulate inherently sustainable nature.

The evolutionary processes in the desert environment have created multiple solutions for the architectural design, such as embodying the shape and adapting to external stimuli and thus enables us to transfer the functional principles of desert beings to architectural applications. Different functional and structural aspects must be combined in a coherent system by applying ecosystems simulation and structural formation on buildings through simulating the mechanisms inspired by desert organisms to achieve needed thermal comfort, especially for workers in these arid areas who are forced to travel long distances back and forth daily through the arid desert, which leads to high economic costs, dangerous roads, and a waste of time taking into account the difficulty of attracting labor to go to these arid areas daily. By studying desert organisms and cases of buildings that managed to adapt to the desert climate, designers can create new housing styles with technologies inspired by desert creatures, which will make a revolution in the desert facilities sector that will positively affect the economic, social, and environmental aspects.

3 Methodology

An analytical descriptive approach includes the study of the environment, its ecosystems, and their reflection on architectural designs, as well as case studies analysis of environmental buildings adaptation based on ecosystems simulation. Desert ecology is the investigation of how living organisms live and what they do, it is the interaction between both biotic and abiotic processes in arid regions, it is characterized by interactions between living organisms, the climate in which they live and, any other non-living effects on the natural surroundings. Deserts comprise several abiotic factors–sand soil, drought, and extreme temperatures. Several biotic aspects that affect deserts, including living organisms, flora, and fauna. Despite the simplicity of the desert ecosystem and the limited number of organisms, the ecology of desert remained through the harshness of the desert and the lack of biodiversity and limited resources, desert organisms follow special strategies to confront the features of desert ecology from environmental fluctuation and scarcity of resources to achieve balance (Sheba, 2012). Ecosystems differ according to desert type, several factors create a desert ecosystem, and some of them are.

- Structure
- Features and properties
- Living organisms.

4 The Background Analytical Study of the Desert Being Adaptation Strategies

From the study of the desert, the main problem of the climate beside it is very harsh, was that the thermal range varies between night and day, summer and winter, making any desert facility in this climate is a mistake that forces us to use more mechanical means to fix the variation in climate factors. Desert beings have been able to achieve balance through their strategies that have ensured their survival and reproduction for millions of years, and some of these strategies are.

4.1 Physical Form Strategies in the Desert Environment

All desert beings realized that their vital organisms were perfectly designed to be closed out and opened internally; to create transitional circles between the inside and outside, limiting the direct encounter with the harsh environment that is constantly changing and fluctuating. Using their bodies (Physical form) to maintain constant thermal comfort is one of their adaptation and survival methods.

4.1.1 Water Vapor Condensation Capacity and Conversion

The Namibian Desert beetles follow a unique approach of drinking water (Fig. 1a–c) exposing their bodies in the early morning to steam-laden winds and condensing them, because of the nature of their corrugated bodies in a way that collects small droplets and turns them into larger droplets that beetles can pick up at their mouths (Maglic, 2012).

4.1.2 Capacity of External Surface Shaping to Reduce the Effect of Sunlight

The outer surface of some desert plants covered with various forms of (scales, hairs, or thorns), where external hairs are an extension of skin cells, they reduce the effects of the sun and break the intensity of the surface of the leaf (Sheba, 2012), it is also covered with a layer of cellulose and wax (Fig. 2). Most desert arthropods have a waxy layer covering the surface of the skin, which helps them to retain water and acts

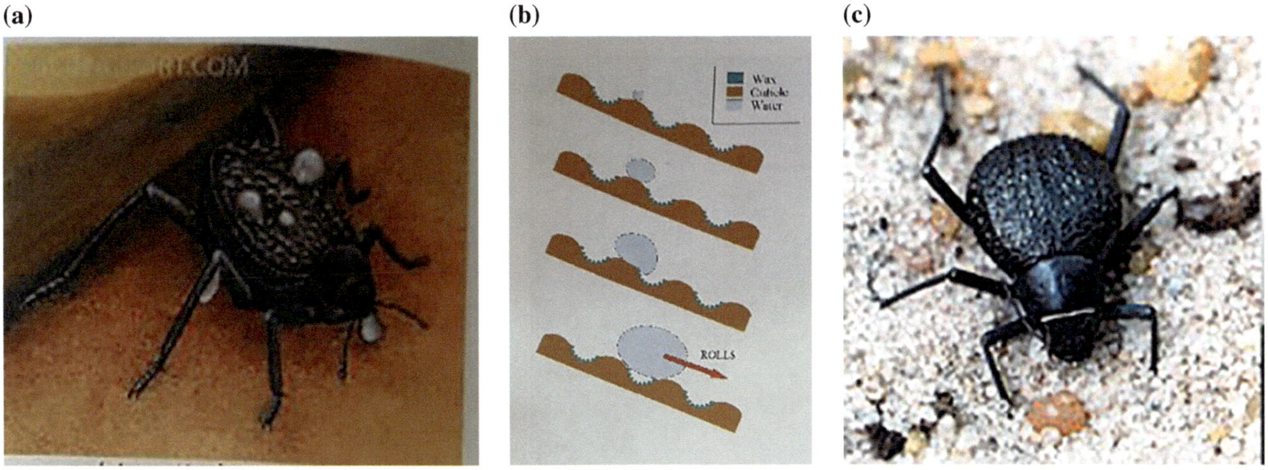

Fig. 1 a, b, c The Namibian desert beetles' technique of drinking water. *Sources* Sheba (2012, p. 138) & (greeninterns.blogspot.com)

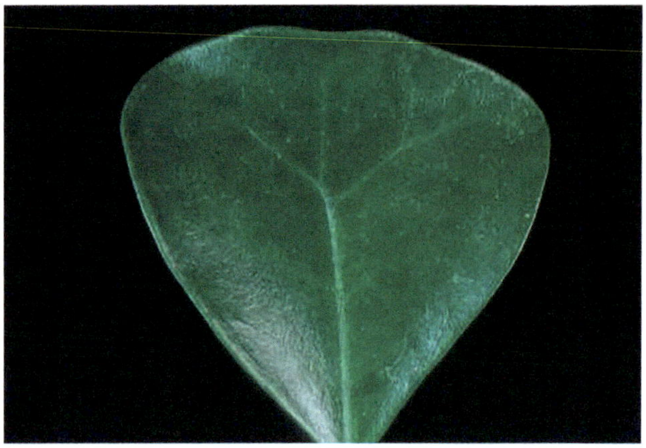

Fig. 2 Leaf covered by a layer of waxy cuticle. *Source* https://homeguides.sfgate.com

as a heat insulator. Wax or Cuticle wax helps to reduce sun penetration of the skin, which reduces water loss and maintain it as possible (Elshazly, 1999).

4.2 Behavioral Reactions Strategies in the Desert Environment

The living organism is created with a programmed behavior that enables it to survive without training or loss of time. The patterns of behavior are divided into two types; the first pattern is defined as the fixed patterns inherited by the living organisms as they inherit their organs, these patterns do not change fundamentally, but they can be modified. The second type is the patterns acquired by learning, mastermind thinking, and experimentation, the higher the intelligence of the living organism, the more expanded his tank of intelligent behavior patterns.

4.2.1 Ability of Transformation

The "Armadillo Lizard" is an example of transformation ability, this fascinating creature can wriggle easily because of its body shape, the body length is between 15 and 17 inches, the tail is between 14 and 16 inches, and it has a flattened head as same as its body (Fig. 3a, b). It can take a defensive posture in predation pressure and risk, and severe climatic conditions such as high temperature and precipitation, by rolling itself up with its tail held in its jaws (Broeckhoven et al., 2018). The strategy of transformation is one of the most prominent strategies that may seem unreal when applied to architecture; where the stability of mass, function, and the environment are essential features of the architecture. This is what has become the theory of bio-architecture to change and revolutionize those fixed concepts in a fast-paced world where human aspirations are accelerated. The inability of the architecture to transform, is the beginning of the end of those urban patterns of unusable functional facilities, as opposed to living organisms that have been able to mutate to meet the changes in their environment through re-shaping to adapt the variables, which is an effective strategy, the activation of this strategy could take more time and technology to be applied to a more adaptive architecture. A vision of a changeable architecture can cope with environmental changes to achieve its functions, adapt to its surroundings, and transcends critical periods for a future that is difficult to be predicted in the face of looming crises.

4.2.2 The Ability to Control the Outer Layer: Changing the Outer Surface Colors

Desert animals are characterized by different colors that match the environment and vary between sand, earthy, yellow, and white. All colors of desert animals are like the environment, reflecting a large amount of solar radiation

(a) (b)

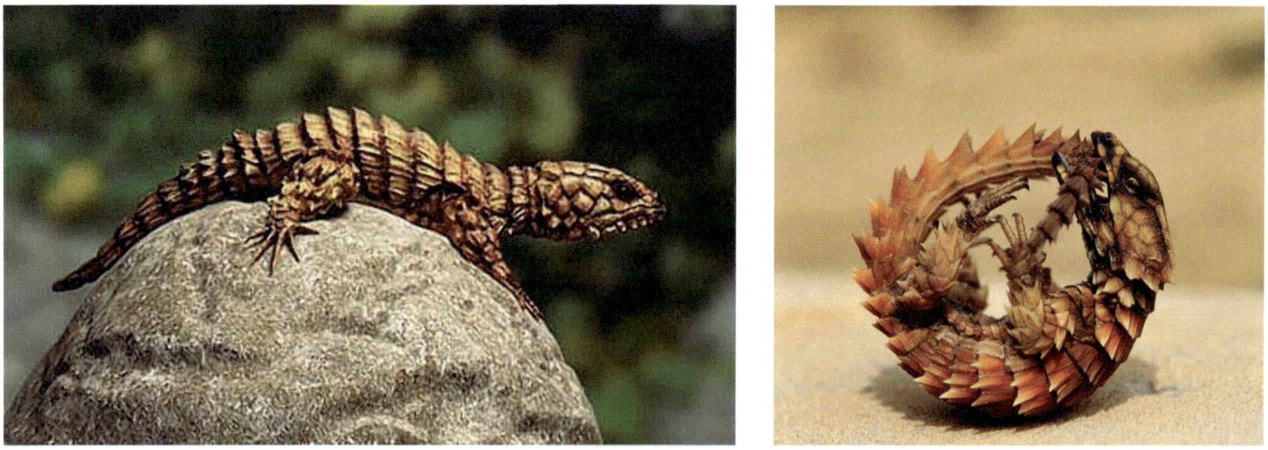

Fig. 3 **a**, **b** Armadillo lizard body postures. *Source* http://animalia-life.club/other/armadillo-girdled-lizard-skull.html

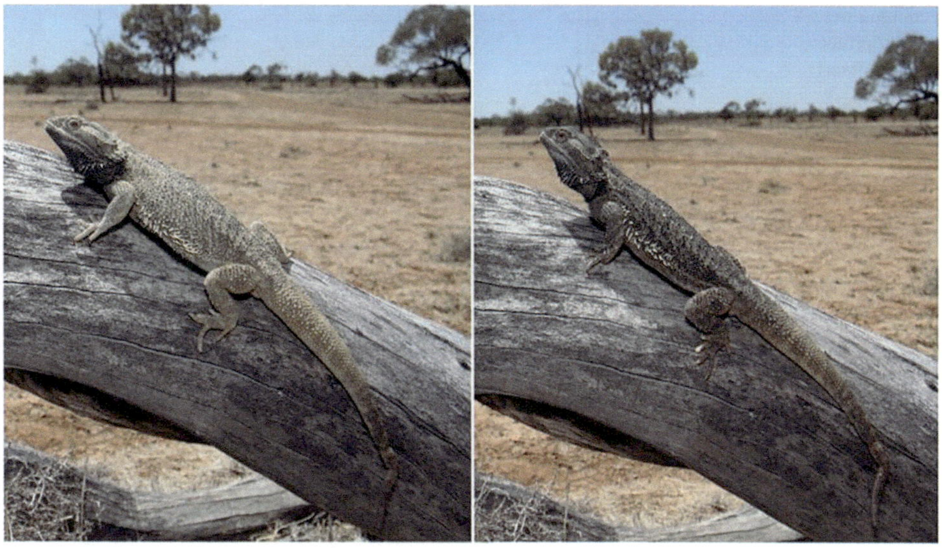

Fig. 4 **a** Color change in the desert lizard skin. **b** Lizard adaptation to the desert environment. *Source* https://phys.org/news/2017-03-lizards-local-colour.html

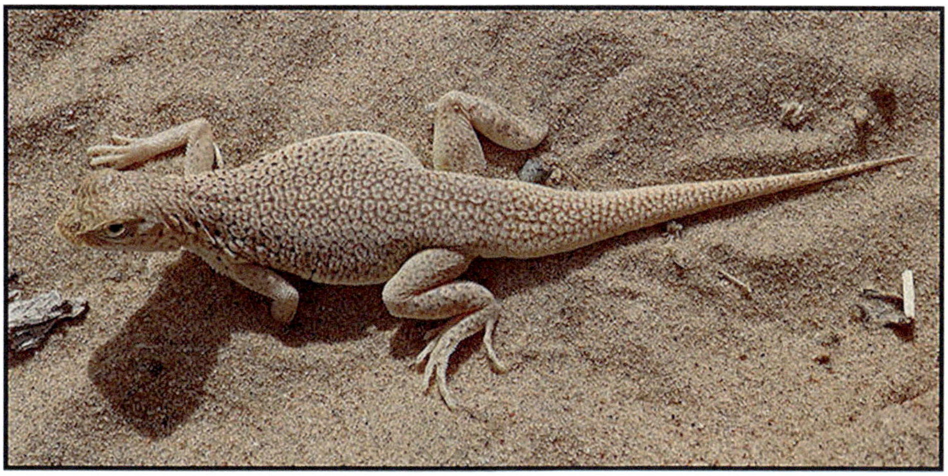

Fig. 4 (continued)

falling on their bodies and reducing heat gain in hot seasons. While animals change colors during winter, the outer surface becomes dark, helping them to gain the amount of heat to overcome the temperature drop during that season (Sheba, 2012). Lizards are the most fascinating animals as desert reptiles (Figs. 4a, b).

4.2.3 Ability to Avoid External Environment to Reduce Transpiration by Wrapping or Rotating

Many desert plants avoid direct exposure to the harsh conditions of the environment by wrapping themselves around, thus isolates the surface of the leaf which contains pores, away from direct environmental effects, while the non-porous surface is exposed to the environmental conditions. Due to the plants' ability of convergence to contain the cells of the lower surface, the motor cells are shrinking during droughts. The leaf edges bend to protect the upper surface from direct sun radiation, where the stomata are abundant in contrast to other kinetic cells, thus decreases transpiration and loss of water to 40% compared to the amount lost in the case of leaf expansion (Fig. 5) (Sheba, 2012).

Another example is the holly plant, the leaves are held at 70-degree angles which means that the sun just hits the sides. The leaves also have a soft salt coverage that helps in reflecting the sun off the plant (Fig. 6). During the sunset or on winter days, the whole leaf is exposed ("Conserve Energy Future", 2016a). Many animals also avoid exposure to climatic conditions when they become unsuitable for their climate comfort zone by shading.

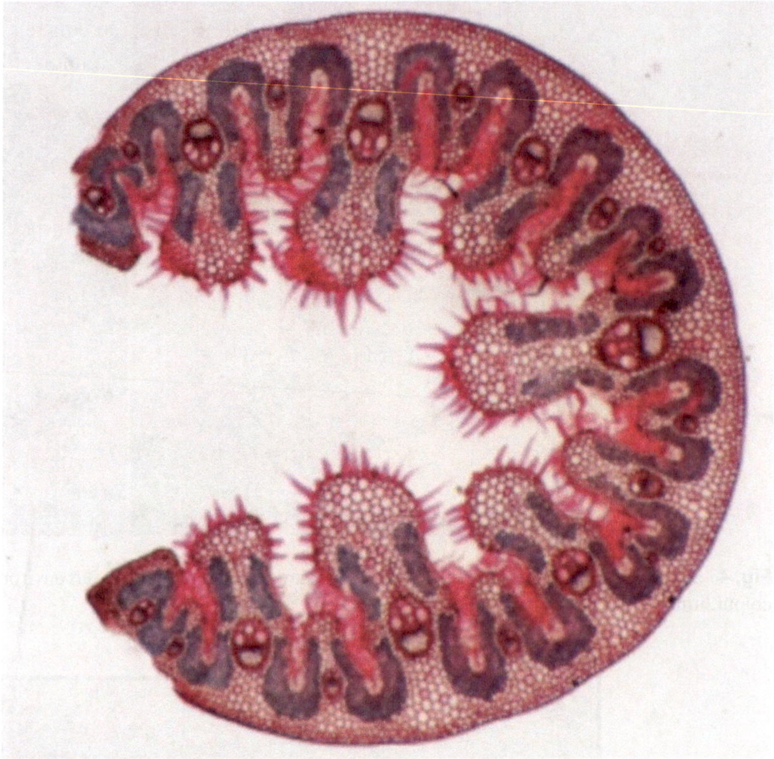

Fig. 5 Wrapping technique to isolate the top surface and protect the pores (Sugarcane sand plant). *Source* Sheba (2012)

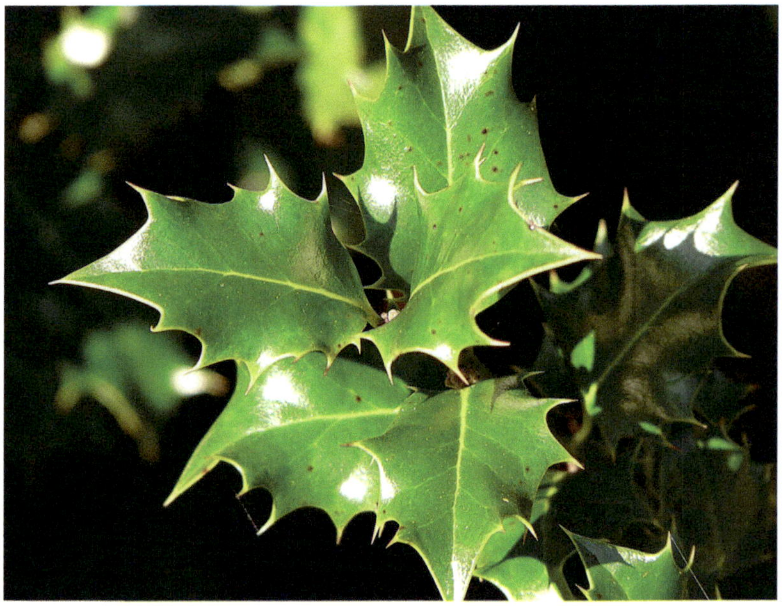

Fig. 6 The holly plant's leaves. *Source* https://pixabay.com/en/holly-leaf-plant-green-sting-3121/

4.3 Ecosystem Strategies: Gaining or Losing Heat Strategies

The mechanisms of desert beings, whether physical or behavioral, have proved their ability to achieve strategies and vital principles, which should be used and adopted as mechanisms for a vital architecture that mimics nature and achieves a balance with the environment. Desert beings control their temperatures using two strategies. *Ectotherms*, include invertebrates and reptiles, acquire the vast majority of their body heat straightforwardly from external sources. *Endotherms*, for example, mammals and birds, obtain warmth through metabolic processes and can keep up constant body temperature (Azmi, 2016).

4.3.1 Keeping Warm

Minimizing heat loss is obtained by some physiological methods. Hollow hair fibers for added insulation that improves insulation performance like that of the reindeer's fur; a thick under-layer of fur traps air against the skin to minimize heat loss, while hair reduces coldness by holding off the water (Azmi, 2016).

4.3.2 Keeping Cool

Ectotherms reptiles utilize their surroundings to control their body temperature through interaction between the skin properties and behavior, as an example, "*the side-blotched lizard*" which has a color-changing pattern in its skin, dark on the back to absorb daylight and light on the stomach to reflect heat gained from the sand. It can change the position of its body dependent on the daylight to absorb or to avoid heat gain by limiting the body range from contacting the ground (Mazzoleni, 2010). Some smaller desert creatures, tunnel beneath the outside of the ground to avoid the high temperatures at the surface. These techniques include many warm-blooded animals, reptiles, insects, and all the desert amphibians. Rodents also block the passageways to their tunnels to avoid hot, drying up air ("DesertUSA", 2019b).

- **Cooling by liquids evaporation**: Liquid evaporating from a surface has a cooling effect. In animals, when a liquid touches the hot skin of the animal, the blood in the vessels near the skin surface is cooled by it, then is circulated throughout the body to regulate the body temperature. Desert coyotes panting and fluttering gular in owls are two examples of this technique.
- **Heat Conduction and Radiation**: Animals can conduct heat from their bodies to the environment by reducing the insulating value of the outer skin, and Coyotes lose their thick winter coats in late spring; their early summer coats are relatively thin ("The Encyclopedia of Earth", 2019a). Many desert animals are pale in the outer envelope; feathers, fur, scales or skin, to take in less warmth from the environment and to be less obvious to predators in the bright, pale environment ("DesertUSA", 2019b).

5 Case Study of Environmental Building Adaptations Based on Ecosystems Simulation

5.1 Eastgate Building in Zimbabwe (Office and Shopping Complex)

- Architect: Michael Pearce and partners
- Year: 1995
- Location: Harare, Zimbabwe
- Area: 55,000 sq m.

The project concept was the sustainability of the design (Fig. 7), keeping up an environmental solution that would be suitable for Zimbabwe's climate; the structure is based partially on the technique of passive ventilation and temperature regulation that have been noticed in termite mounds (Fig. 8), to create a thermally comfort interior environment.

Eastgate consists of two parts side by side attached by a roof made of glass. The building is made of precast concrete, brushed to expose the granite aggregate that matches up the lichen-covered rocks in Zimbabwe's landscape, these elements increase the internal surface area of the building to minimize heat gain during the day and improve heat loss at night (Maglic, 2012). The termites achieve the stable temperature by drawing in the air through a breeze-catchers at the base of the structure, cool it by pulling it through cells at the base of the wet mud, while hot air ascends through vents at the top of the mound. The same strategy has been achieved in this building (Fig. 9) by designing a system under the office floors, a mezzanine plant room contains 32 banks of low and high-volume fans pull air from the atrium through channels. This air is pushed up through the supply area of vertical pipes in the focal spine center of each office wing, from the channel the air is bolstered through the hollow floors to low-level grilles under the windows. The hot air results from human behavior in the space ascend to the vaulted roof where it is pulled out through the exhaust ports at the end of each vault to the exhaust areas of the focal vertical stacks ("Eastgate Building Harare", 2016b).

5.2 Hydrological Center for University of Namibia, Namibia

- Architect: Matthew Parkes of KSS Architects
- Location: University of Namibia, Namibia

This building is inspired from the Namibian Desert beetle's organism that was stated previously in "Sect. 4.1.1.", (Fig. 10a, b). The climate in Namibia is so harsh, and it is one of the most arid lands on Earth. *Matthew Parkes* while talking to native people in Namibia discovered the strategy of the Namibian beetle in catching fog, he was impressed by the beetle shell design, so he used this technology to build a fog catcher to turn water droplets collected from fog into usable water. The building consists of a series of pods located behind a tall, curved nylon mesh screen used to collect water by being positioned toward the ocean to capture as much moisture as possible from the fog. Because of the mesh wall form and direction, the water runs down the mesh into a gutter system located at the bottom of the screens then into large tanks that keep the water at a suitable cooler temperature so that the water does not evaporate (Maglic, 2012).

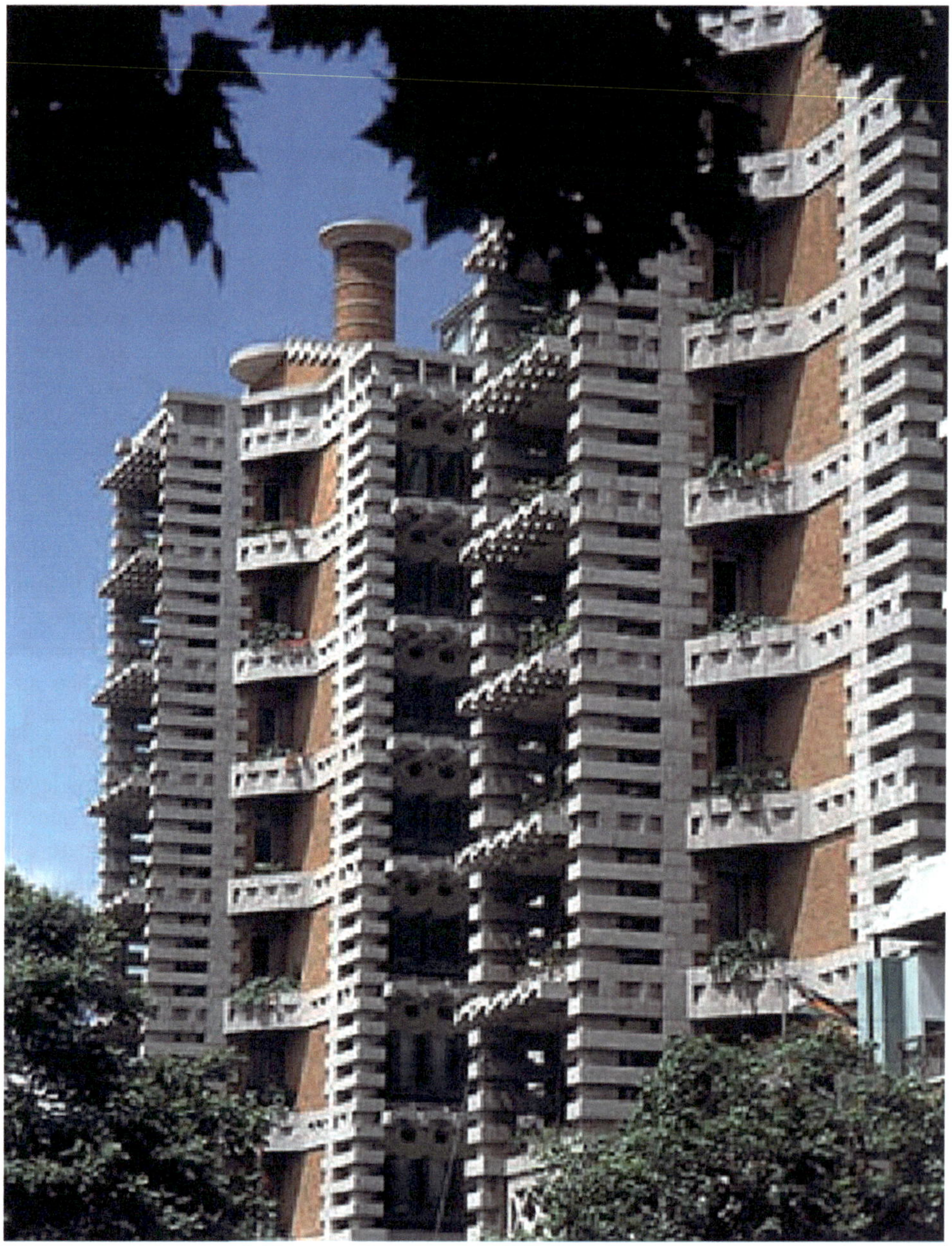

Fig. 7 Eastgate is zimbabwe. *Source* http://www.mickpearce.com/Eastgate.html

6 Case Study of Environmental Adaptation Techniques and Its Similarities with the Desert Ecosystems

6.1 Primary School in Gando/Kéré Architecture

- Project by: Kéré Architecture
- Location: Gando village, Burkina Faso
- Architect: Diébédo Francis Kéré
- Area: 318 m^2
- Year: 2001, Extension: 2008.

The school design developed from many variable factors including cost, climate, materials accessibility, and feasibility of building. Clay/mud hybrid construction was essentially utilized (Fig. 11). Clay is available in that area and is traditionally used in the construction of buildings where there is integration between passive solar design and cross-ventilation. The materials are cheap, durable, and easy to make and provide thermal protection (Cardenas, 2016). The ceiling is made of compressed earth blocks supported by 12 mm steel bars. Walls are made of sun-dried mud blocks; a large overhanging light metal roof protects the walls from the damaging rain that last for four or five months in the

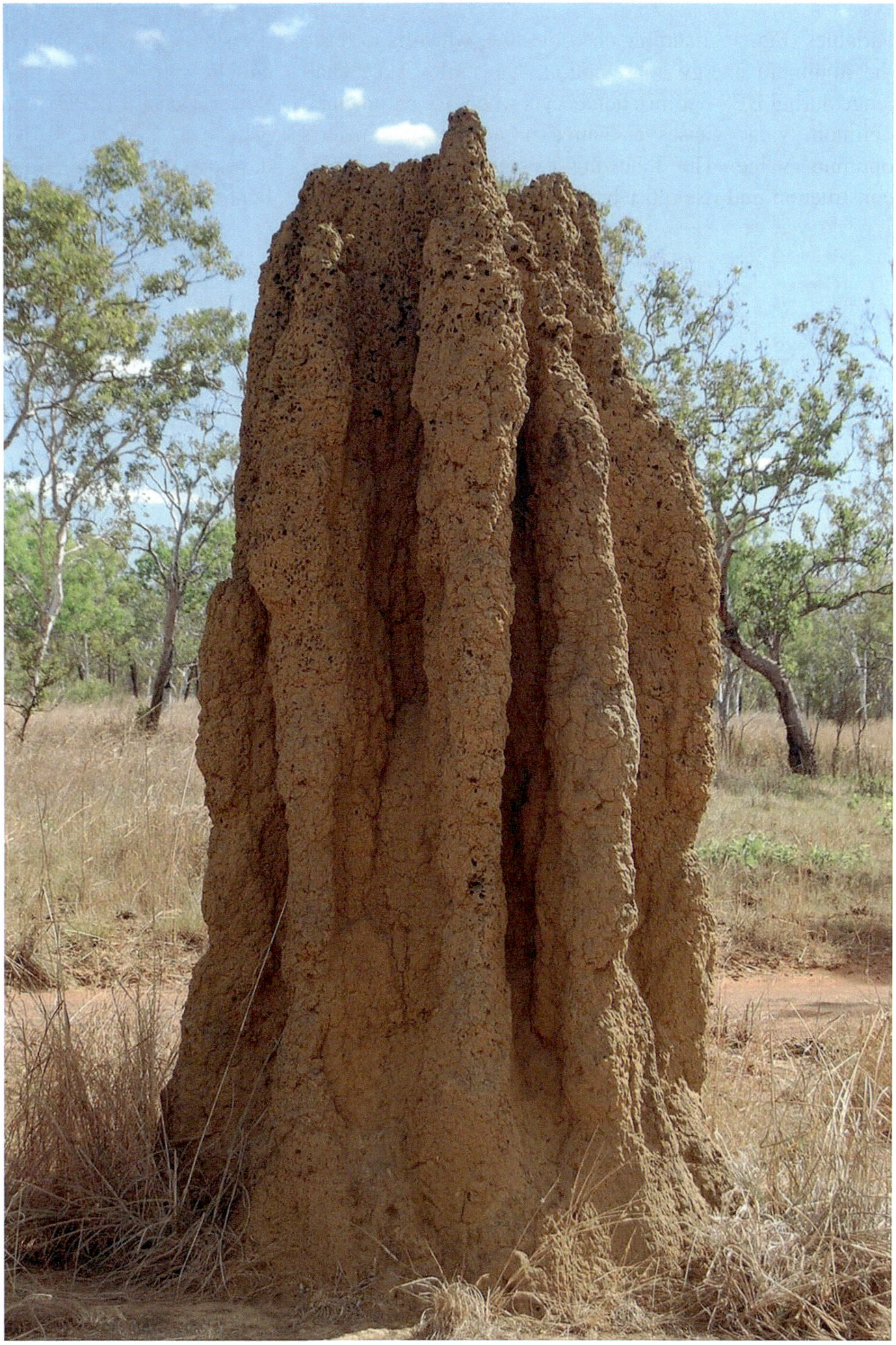

Fig. 8 Termite mounds. *Source* https://www.pinterest.com

rainy season (Fig. 14). The roof design was intended to direct rainwater into a channel then used to irrigate the vegetable garden (Varanda, 2004). The roof, wall, and ceiling construction were intended to maintain the thermal comfort of the inner space.

By pulling the roof away from the clay ceiling a wind channel has been formed, which expels hot air. At the end of each classroom, there is a 10 cm gap between the ceiling blocks which allow light and air to enter (Fig. 13). Shutters provide natural lighting and ventilation control. The thick clay walls act as a storage unit for the cooled air. The simple plan consists of three classrooms linearly (Fig. 12), separated by covered outdoor zones, that link the building to the surrounding landscape (Varanda, 2004). Learning from the desert snail that curved surfaces in a dry hot environment keep lower temperatures and reflect heat more than flat

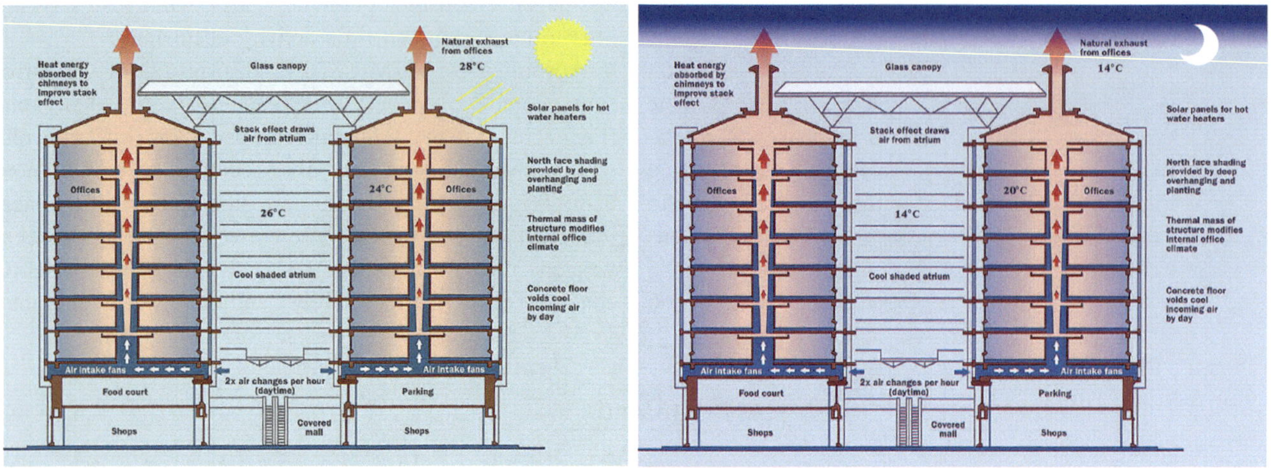

Fig. 9 The system of Eastgate building. *Source* http://www.mickpearce.com

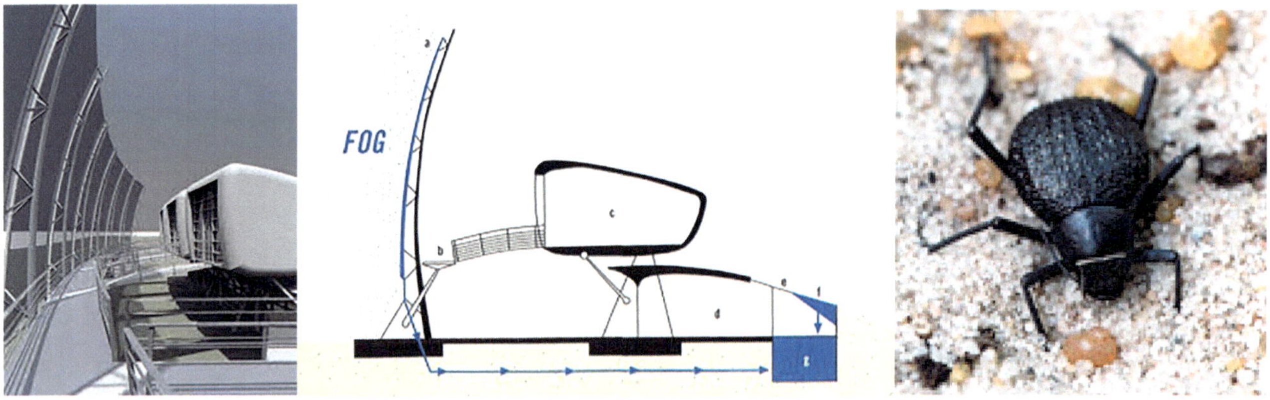

Fig. 10 a, b Hydrological center, University of Namibia. *Sources* Adebisi et al. (2015; greeninterns.blogspot.com)

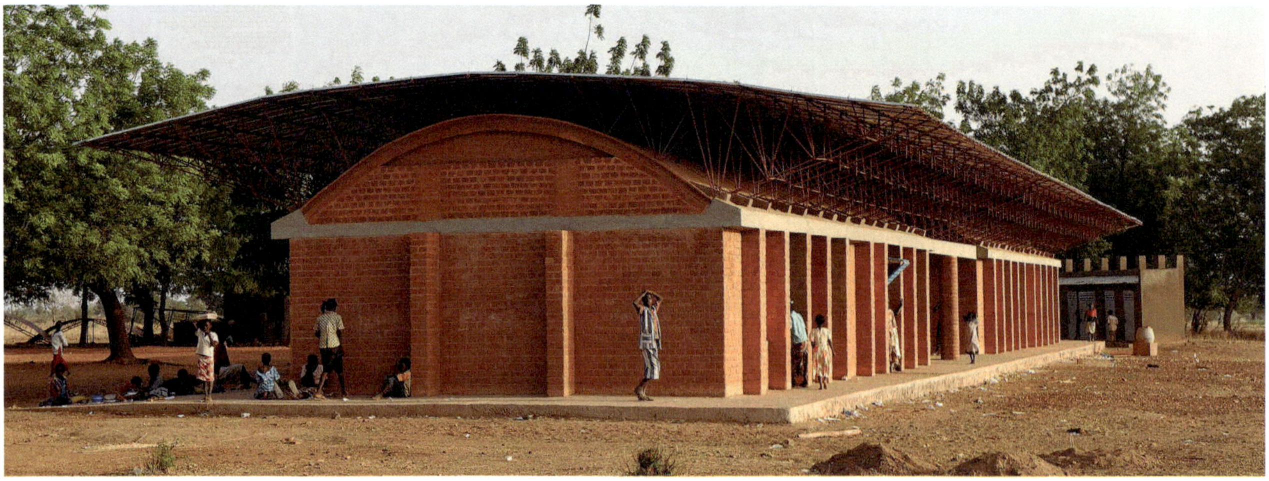

Fig. 11 Primary school in gando. *Source* www.afritecture.org/architecture/gando-primary-school

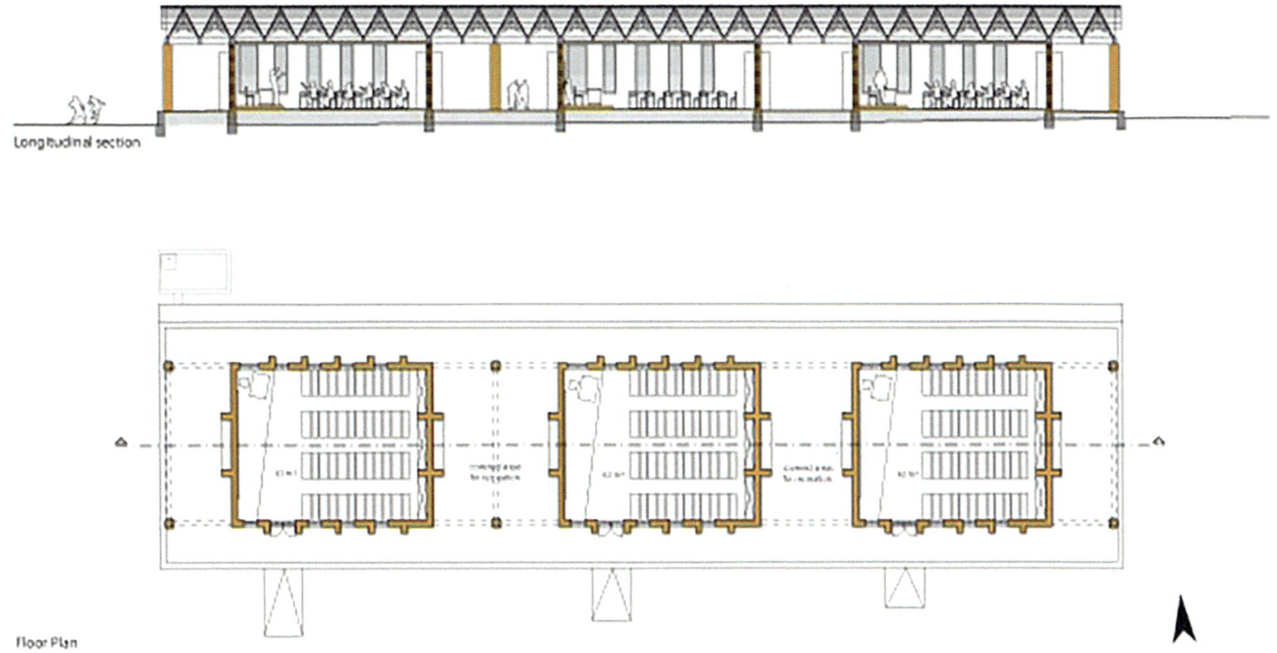

Fig. 12 Plan of the school. *Source* www.afritecture.org/architecture/gando-primary-school

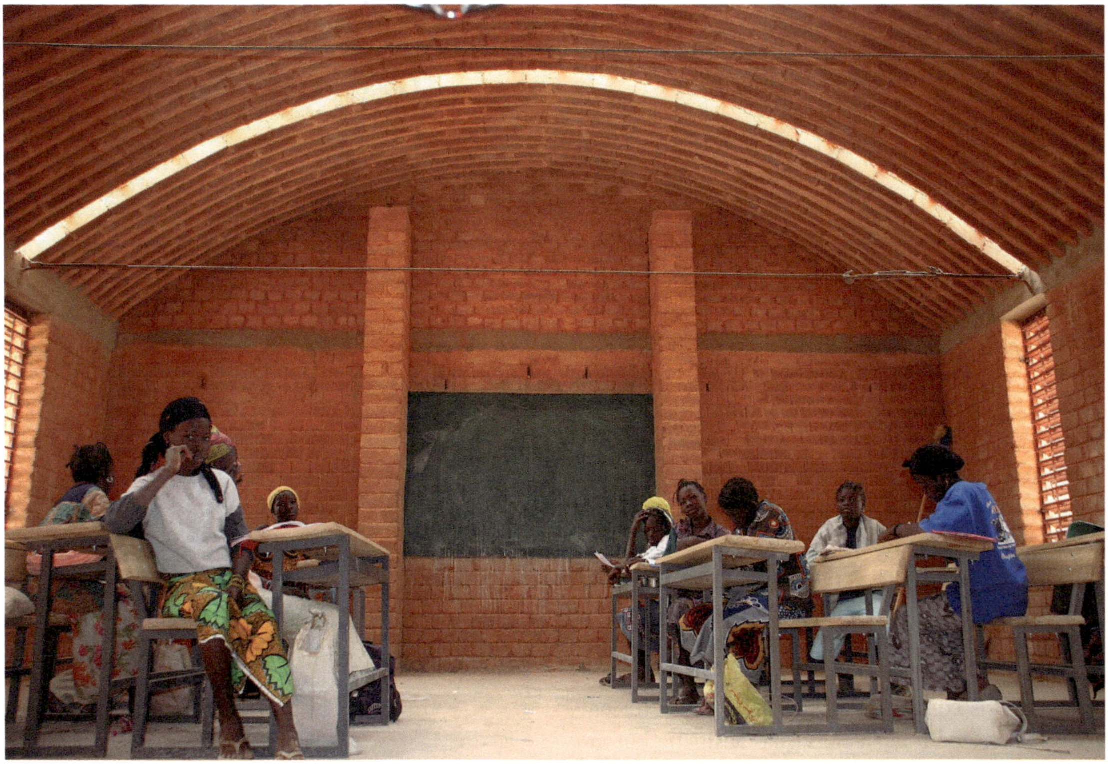

Fig. 13 Classroom interior ceiling. *Source* www.afritecture.org/architecture/gando-primary-school

surfaces. The intensity of solar radiation is distributed over a larger surface, meaning that heat transmission to the interior is minimized. Because of the snail shell curvature, it allows air to flow in and be captured, creating an insulating air cushion, that allows the snail to keep constant body temperature and not overheat (Hadavand& Yaghoubi, 2008).

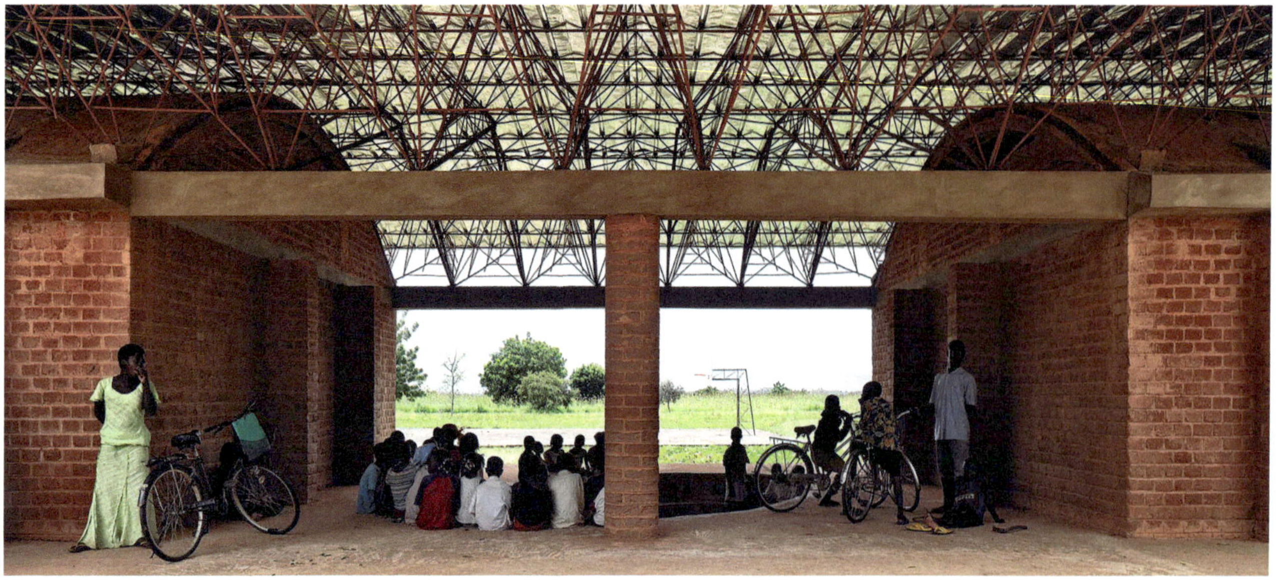

Fig. 14 The roof of the school. *Source* www.afritecture.org/architecture/gando-primary-school

Based on the findings that have been observed in the snail, we can apply the same principles, but with a different interpretation forming a new architectural expression.

6.2 The Arab World Institute in Paris

- Architect: Jean Nouvel, Gilbert Lézénès & Pierre Soria
- Location: Paris, France
- Year: 1987.

The facade detailing was the main concept in this project. The main feature and creative component is the progressed responsive metallic Sun blocker on the south facade (Fig. 15a, b). It was inspired by the traditional Arabian lattice that was made to protect the inhabitants from the sun and maintain confidentiality. The concept of the system was well-received for its originality in Arabic architecture—the hybrid architectural openings (windows) "Mashrabiya" (Winstanley, 2011). The system consists of hundreds of light-sensitive diaphragms that regulate the amount of daylight entering the building. The lens has several phases, a changing geometric pattern is shaped and displayed; squares, circles, and octagonal shapes are created in a smooth movement as light is controlled at the same time. Not only have the devices created a fantastic visual aesthetic but also functional from an environmental control point of view. Solar gain is moderated by shutting or minimizing the lens opening proportion (Winstanley, 2011). Mechanically

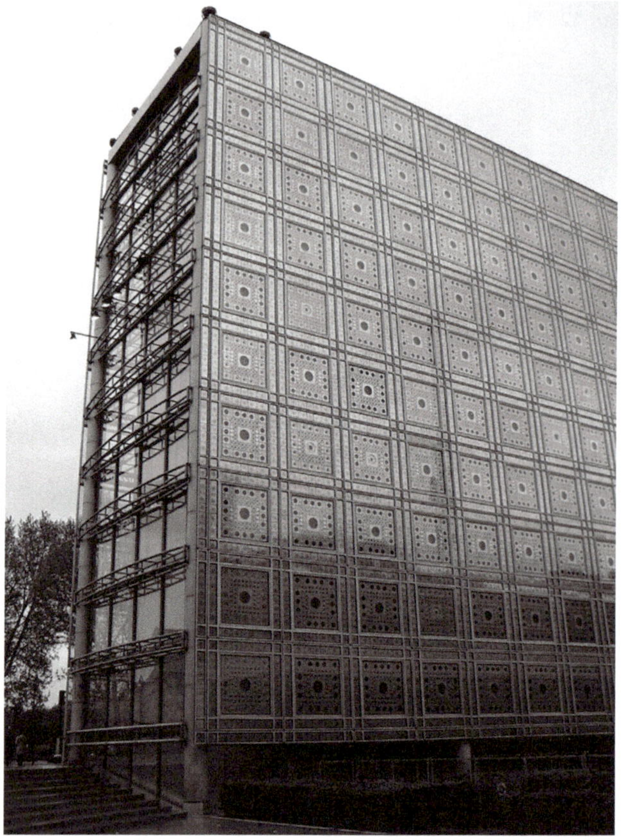

Fig. 15 **a** The Arab World Institute in Paris. **b** The Arab World Institute in Paris. *Source* www.archdaily.com

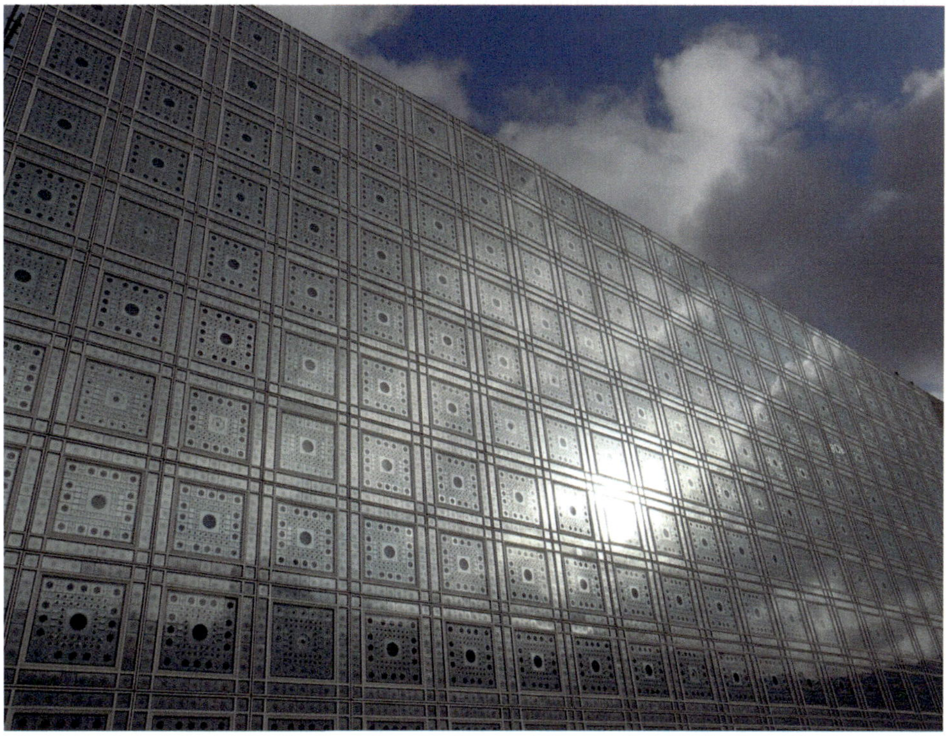

Fig. 15 (continued)

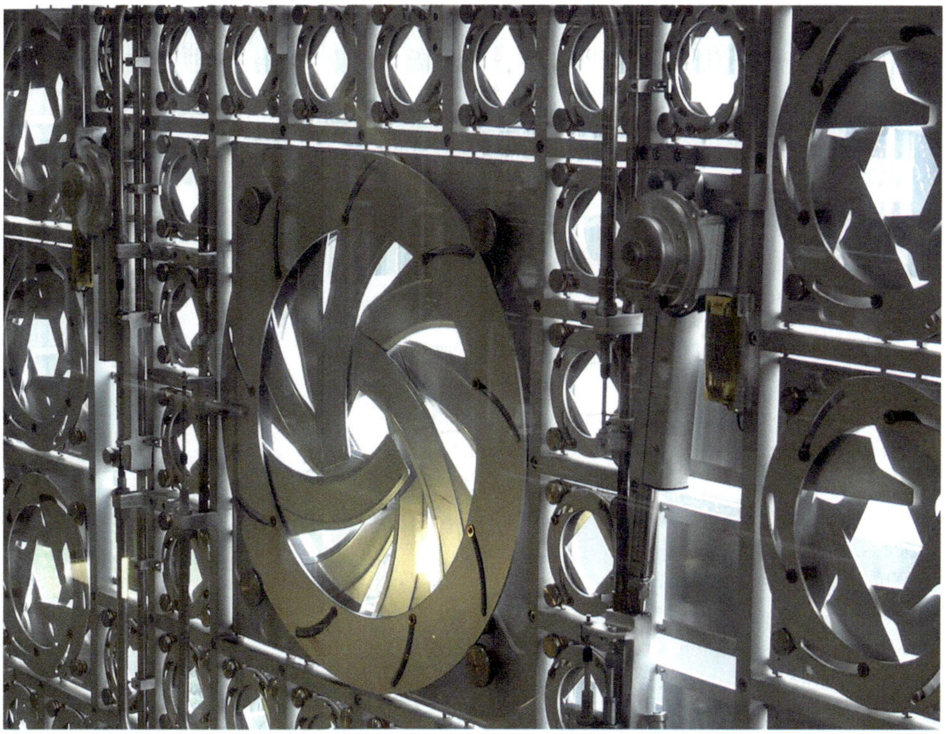

Fig. 16 **a** The mechanical adaptation of the system. **b** The mechanical adaptation of the system. *Source* archdaily.com

adaptations in facades change the appearance and characteristics of the outer surface by mechanically altering its components. The hybrid architectural openings (windows) "Mashrabiya", this symbolic characteristic of the building was designed to open and close upon the amount of daylight (Fig. 16a, b). This mechanism is like some desert plants' ability to control the outer layers to reduce the impact of the sunlight (Fig. 17).

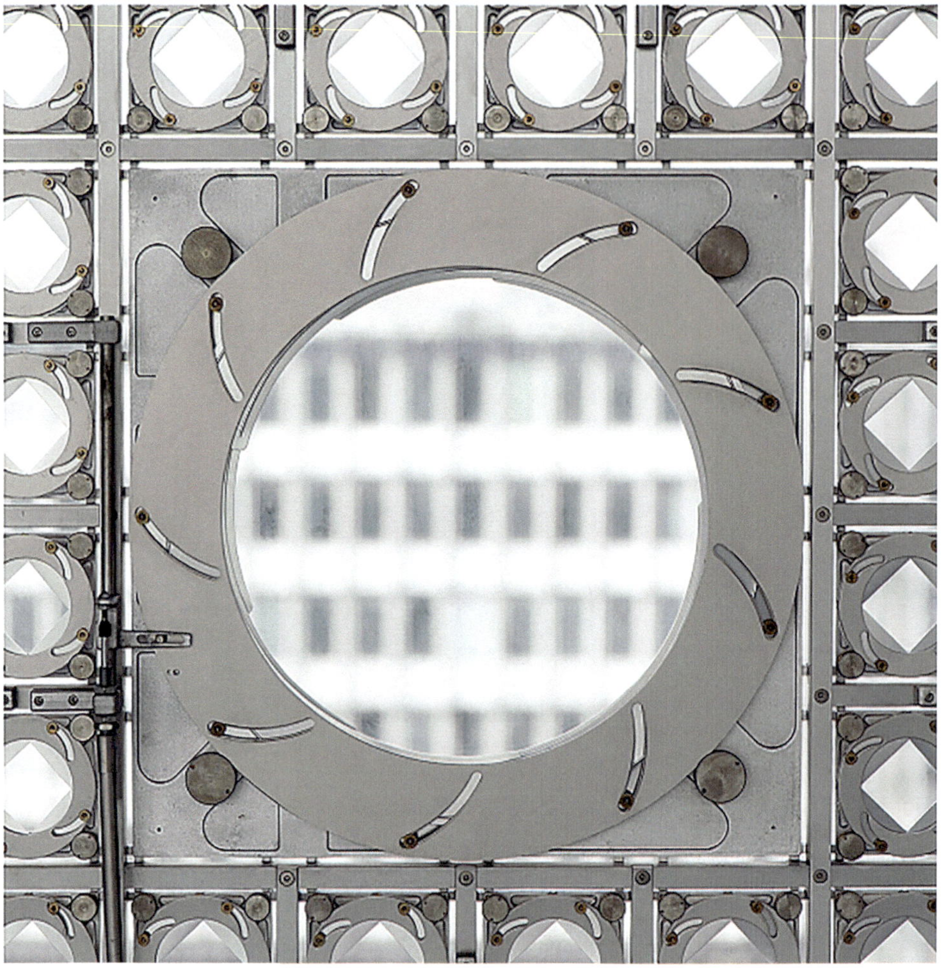

Fig. 16 (continued)

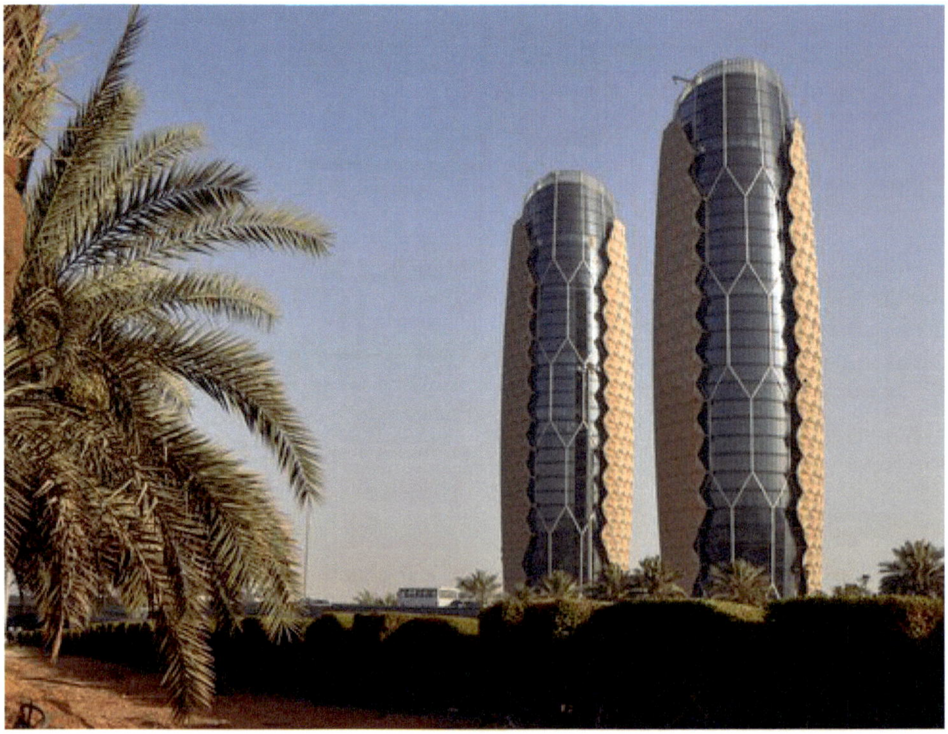

Fig. 17 Al Bahr towers. *Source* https://www.archdaily.com/270592/al-bahar-towers-responsive-facade-aedas

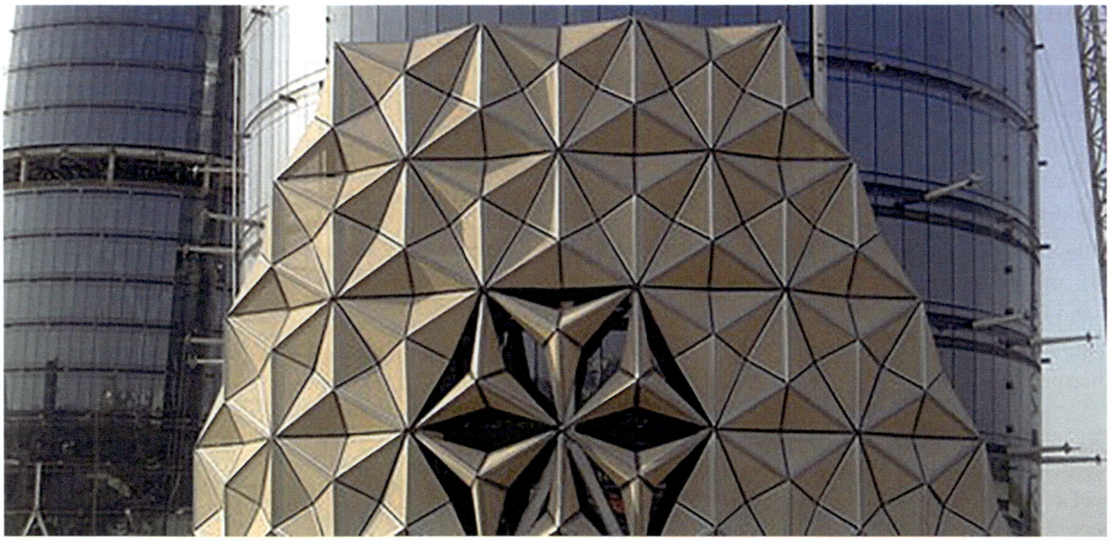

Fig. 18 The "Mashrabiya" shading system. *Source* https://www.archdaily.com/270592/al-bahar-towers-responsive-facade-aedas

6.3 Al Bahr Towers

- Architect: Abdul Majid Karanouh, AEDAS Architects.
- Year: 2012.
- Location: Abu Dhabi, UAE.

In the extreme weather conditions of *Abu Dhabi*, a responsive facade inspired by the "Mashrabiya" was designed by *AEDAS* architects, the facade integrates the building into its cultural context, the traditional shading device inspired from Islamic architecture. Architects were upon a tough challenge because of the climatic requirements of the region. Architects used a highly developed technology; parametric and algorithmic to control the geometry of the facade panels. The "Mashrabiya" form of the screens was able to simulate the response to sun trajectory at different points during the year. Each screen works as a high-performance, reflective curtain-walls, two meters from the exterior facade. The giant lattice surrounds the two towers, except for the part of the facades that face North (Cilento, 2012).

Each screen consists of triangles coated with micro-perforated glass fiber and programmed to respond to the sun reducing heat gain (Fig. 18). At night, all panels will fold. In the morning, the "Mashrabiya" on the east side of the building will begin to close and as the sun moves around the building, the whole vertical strip of the hybrid architectural openings (windows) "Mashrabiya" will expand and contract depending on the sun position (Mahmoud, 2017). Noticing the mechanism will find that the system is like the strategies of some desert plants and animals in the process of gaining or losing heat mentioned previously in Sect. 4.3.

7 Discussion

Ceiling thermal insulation is the most important action to take care of in a hot climate, and it can reduce or stop the amount of heat from the outside getting through the buildings skin. Using light materials with low thermal capacity in walls with the use of insulating layers in the outer walls and to avoid low temperatures at night, thermal curtains are placed on the openings. These techniques can be applied on the external casing of buildings, several moving layers can be applied as well, where one or all of them can be kept according to the temperature of the external atmosphere, which is also inspired by some desert creatures and also through human behavior, wearing light summer clothes in warm weather, medium clothes in autumn, and heavy clothes of several layers in winter.

Applications:

- Adaptive facades: building facades can adapt to climatic changes, and with internal different conditions of multiple uses and user activities by folding, wrapping, shutting, or minimizing openings proportions.
- Respondent ceiling: where the ceiling responds to the change occurring in sunlight, it opens and closes according to the degree of brightness and radiation of the sunlight.
- Respondent space: where the structural system fully responds and changes according to the activity occurring in the inner space.

8 Conclusion

The simulation of the desert ecosystems can be applied on buildings by achieving the ecosystems techniques of desert organisms, by using the physical strategies such as the structural formation of buildings to achieve the optimum thermal comfort of space, or using the behavioral strategies such as rotating some parts of the building, controlling the outer skin, or applying thermal insulation using suitable materials and fixation that reduces the energy amount needed for keeping the Inner space of the hot climate in the status of needed thermal comfort.

Acknowledgements This paper is an approach for a master thesis with the title:" Ecosystems Simulation in the desert environment to achieve thermal comfort in lightweight structure facilities". The data and the views expressed are solely under the responsibility of the author. The author would like to give appreciation in acknowledging the support of my thesis supervisor, Dr. *Ahmad Ibrahim*.

References

Adebisi, G. O., Onuwe, J. O., Sani, A. M. (2015, April). Assessment of biomimicry design concept adoption in architecture: Towards a sustainable built environment in Nigeria. *IOSR Journal of Environmental Science, Toxicology and Food Technology (IOSR-JESTFT), 9*(4), Ver. II, 41–46. e-ISSN: 2319-2402, p-ISSN: 2319-2399. https://doi.org/10.9790/2402-09424146. Retrieved from: http://iosrjournals.org/iosr-jestft/papers/vol9-issue4/Version-2/G09424146.pdf.

Almalky, Q. (2011). *History of architecture*. Oman, Jordan: Dar Al-Manahej for Publishers and Distributers. Retrieved from https://archive.org/details/432556667/page/n1/mode/2up.

Azmi, M. (2016). *Biomimetic architecture as a new approach for energy efficient buildings*. Ph.D thesis. Cairo, Egypt: Faculty of Engineering, Cairo University.

Broeckhoven, C., Mouton, P. L. F. N., & Hui, C. (2018, April 30). Proximate causes of variation in dermal Armour: Insights from armadillo lizards. *OIKOS, 127*(10), 1449–1458. Retrieved from https://onlinelibrary.wiley.com/doi/epdf/https://doi.org/10.1111/oik.05401.

Cardenas, D. (2016, April 22). Primary school in Gando/Kéré architecture. *Archdaily*. Retrieved from https://www.archdaily.com/785955/primary-school-in-gando-kere-architecture.

Cilento, K. (2012, September 5). Al Bahr towers responsive facade/Aedas. *Archdaily*. Retrieved from: https://www.archdaily.com/270592/al-bahar-towers-responsive-facade-aedas.

Elshazly, S. A. (1999). *Technology of feeding, fertilizing and irrigating fruit trees in desert lands*. Cairo, Egypt: The academic library.

Fajal, K. (2002). *Environmental architecture in the hot desert regions*. Cultural House for Publishing.

Fahmy, M. (2012). *Dialectic of the formation of the outer envelop of the building: From the perspective of thermal comfort of the (Master thesis)*. Faculty of Engineering, Cairo University.

Hadavand, M., & Yaghoubi, M. (2008, August). Thermal behavior of curved roof buildings exposed to solar radiation and wind flow for various orientations. *Applied Energy, 85*(8), 663–679, P. 4). https://doi.org/10.1016/j.apenergy.2008.01.002. Retrieved from https://www.researchgate.net/publication/223063885_Thermal_behavior_of_curved_roof_buildings_exposed_to_solar_radiation_and_wind_flow_for_various_orientations.

Hassan, S. (2016, December). Urban development factors and its definition (urban geography). *Word press*. Retrieved from https://sarahhassanblog.wordpress.com/2016/12/.

Konya, A. (Translation and Commentary by Dr. Ahmed Al Khatib). (2011). Original book (1984). *Design primer in hot climates*. Cairo, Egypt: Anglo Egyptian Library.

Maglic, M. J. (2012). *Biomimicry: Using nature as a model for design (Masters theses 1911–February 2014. 871)*. Retrieved from: https://scholarworks.umass.edu/theses/871.

Mahmoud, S. E. (2017). *Towards the theory of biodynamic architecture (PHD thesis)*. Cairo, Egypt: Faculty of Engineering, Cairo University.

Mazzoleni, I. (2010, June). Biomimetic envelopes. *Disegnarecon, 3*(5). https://doi.org/10.6092/issn.1828-5961/1944. Retrieved from https://www.researchgate.net/publication/307648464_Biomimetic_Envelopes.

Okasha, A. (2008). *Islamic architecture in Egypt*. Bardi for publication.

Sheba, A. S. (2012). *Bio-architecture towards an environment-friendly architecture*. Ph.D. thesis). Cairo, Egypt: Faculty of Engineering, Cairo University.

Unknown. (2019a, January). Adaptations of desert birds and mammals. *The Encyclopedia of Earth*. Retrieved from https://editors.eol.org/eoearth/wiki/Adaptations_of_desert_birds_and_mammals.

Unknown. (2019b, January). Desert animal survival. *DesertUSA*. Retrieved from https://www.desertusa.com/survive.html#ixzz5Z1xcxuPp.

Unknown. (2016a, December 25). Desert ecosystem. *Conserve Energy Future*. Retrieved from https://www.conserve-energy-future.com/desert-ecosystem.php.

Unknown. (2016b). Eastgate building harare. Mick pearce architect. Retrieved from http://www.mickpearce.com/Eastgate.html.

Unknown. (2009, May 22). Gando primary school. *Afritecture*. Retrieved from http://www.afritecture.org/architecture/gando-primary-school.

Varanda, F. (2004). Primary school Gando, Burkina Faso, On-site Review Report, edited by *Aga Khan Award for Architecture, (2920. BOU)*. Retrieved from: https://s3.amazonaws.com/media.archnet.org/system/publications/contents/1221/original/FLS1252.pdf?1384750131.

Winstanley, T. (2011, October 2). AD classics: Institut du Monde Arabe / Enrique Jan + Jean Nouvel + Architecture-Studio. *Archdaily*. Retrieved from https://www.archdaily.com/162101/ad-classics-institut-du-monde-arabe-jean-nouvel.

Effects and Requirements of Energy

Friction Stir Spot Welding

N. Gadallah and A. S. Wanees

Abstract

The present paper studies the friction stir spot welding (FSSW) of different materials. This work is considered as a study aimed to analyze the impact of the rotational speeds (n) and dwell times (T) on the mechanical behavior and the microstructural characteristics of the produced FSS welded joints. Al 2024 specimens have been welded by FSSW versus A 2024 specimens and versus polycarbonate specimens. The influences of the rotational speed of the tool and dwell times on the mechanical and microstructural characteristics of the welded joints have been studied. For tensile shear studies undertaken on lap joints universal test machine utilized for the two following conditions: at room temperature and also at 200 °C for AL/AL specimens. Vickers's micro hardness tests executed on the joints of welded of AL/AL and shore D hardness test tests have been executed on the joints of welded AL/PC, the micro- and macro-structure of the welded joints have been studied. Polycarbonate specimens versus polycarbonate specimens have been welded, and the same parameters have been studied with aim of applying tensile shear test on universal tensile machine, shore D hardness test and observing the macrostructure of weld joints to conclude the weld characteristics.

1 Introduction

Friction stir welding (FSW) is capable of welding various materials and identical and dissimilar metals. Due to some benefits over another welding procedures, this method has attracted extensive interest (Gadallah et al., 2020; Ibrahim Sabry et al., 2020). As an extension of FSW for joining aluminum alloys, a novel solid-state welding mechanization called FSSW has recently been advanced. Compared to Resistance Spot Welding (RSW) for aluminum, FSSW reported a significant decrease in energy exhaustion and equipment expenditure. Since FSSW is a solid-state soldering operation, there is no need for compressed air or coolant. In comparison, as in RSW, less energy is needed. Spot Welds have better strength, improved fatigue longevity, lower distortion, lower residual stress, and better resistance to corrosion (Yang et al., 2014).

Unlike FSW, since plunging a spinning non-consumable instrument into the workpieces, there is no traverse movement. The tool utilized for FSSW has two elements, a pin and an arm. The pin disrupts the faying surface of the workpieces, shears, and transports the material around it and in the thick workpieces induces deformation and frictional heat. The tool shoulder creates a lot of the workpieces' surface and subsurface regions with frictional heat. Furthermore, the shoulder limits plasticized material movement and creates the downward forging operation (Ibrahim Sabry & El-Kassas, 2019; Tozaki et al., 2010). In existing industries, plunge style FSSW is most widely used. During FSSW plunge style, a tool of rotation by a protruded pin is plunged inside the workpieces from the upper surface to a predetermined depth, and after a assured dwelling time it is withdrawn and a keyhole is left (Yang et al., 2014). At the tool-workpiece interface, the frictional heat produced softens the embracing material, and the spinning and moving pin persuade the flow of material in together the peripheral and axial directions. The forging pressure exerted by the tool shoulder and the mixing of the plasticized material allows the stable bond area to shape (Fig. 1).

FSSW has multiple advantages over typical ones. Without porosity, fractures, and contamination, high joint strength. During FSSW, no substance melting takes place. Higher intake of energy (Nguyen et al., 2011). Higher investments in facilities. No large-scale supply of electricity

N. Gadallah (✉) · A. S. Wanees
Manufacturing Engineering and Production Technology, Modern Academy for Engineering and Technology in Maadi, P.O. Box Cairo, 11571, Egypt

F. Rosso et al. (eds.), *Advances in Architecture, Engineering and Technology*, Advances in Science, Technology & Innovation, https://doi.org/10.1007/978-3-030-86913-7_16

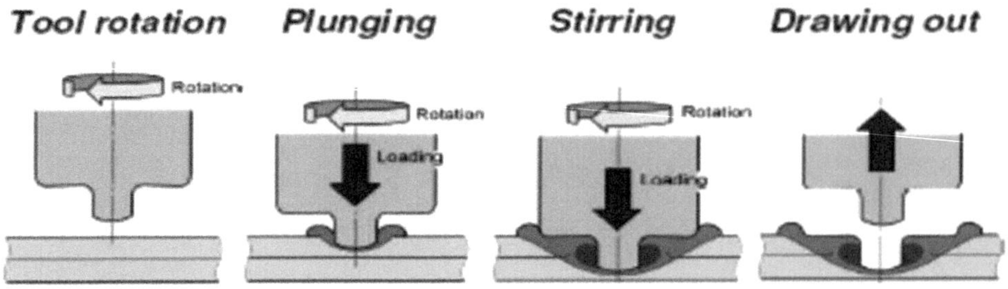

Fig. 1 Plunge FSSW (Gadallah et al., 2020)

is required and FSSW instruments are non-consumable. No harmful and eco-friendly pollution little deformation welding. Because of its simple joining mechanism with few method parameters, high repeatability, and accuracy. Lower servicing (Tozaki et al., 2010; Yang et al., 2014).

2 FSSW of 2024 Aluminum Alloys

The present work studies the FSSW of Al 2024 wrought aluminum alloy. The effect of the tool rotational speed on the mechanical and microstructural characteristics of the welded joints has been studied. Tensile tests have been performed on lap joints using universal testing machine at room temperature and test at 200 °C. Vickers's hardness tests have been carried out on the welded joints. The micro- and macro-structure of the welded joints have been studied. The results showed the possibility of welding a 2024 aluminum alloy with 3 mm thickness using FSSW. Joints friction stir spot welded at 1300 RPM, 1500 RPM, and 2000 RPM with dwell times 6 s, 8 s, and 10 s. The material used in the current investigation was wrought 2024 (Al-Mg-Si) Al alloys. The chemical composition of the alloy is listed in Table 1. The material has been received in the form of Rolled sheets having different dimensions. The material has been cut to dimensions of 50 mm (length) × 140 mm (width) × 1.5 mm (thickness). The material where divided into two categories: one was (O) heat treatment applied to it, and the other was (T6) heat treatment applied to it.

2.1 Process Parameters

Two rolled sheets of 2024 Al alloy, each has dimensions of 200 mm (length) × 140 mm (width) × 1.5 mm (thickness), were joined using FSSW. Figure 2 illustrates the FSSW process. The tool consists mainly of a pin and shoulder. The pin has a cylindrical shape with 5 mm diameter and 2 mm height. The shoulder has 25 mm diameter, and the height of the tool is 110 mm. The tool was from K100 tool steel with a chemical composition listed in Table 2. An illustration of schematics of the used tool is shown in Fig. 3. The FSSW was carried out using an universal milling machine shown in Fig. 2 at three different tool rotational speeds (rpm) typically 1300, 1500, and 2000 rpm and dwell time (sec) is 6, 8, and 10 s (Fig. 4).

Table 1 Al-Alloy chemical structure of 2024 (wt.%), Egypt aluminum company

Element	Mg	Si	Fe	Mn	Cr	Cu	Pb	Al
Composition	0.02	0.10	0.35	0.70	0.01	3.8	0.04	Bal

2.2 Microstructural and Macrostructural Examinations

The macro- and micro-structure characteristics of FSSW sheets have been investigated using an Olympus Optical Microscope of Metallurgical. Each sample was ground underwater on a (Metasery Grinder 2000) Using silicone carbide abrasive disc, spinning disc with increasing fineness (120–2000 grit), and polished with 0.3 μm alumina suspension. A chemical solution was used to micro-etch and macro-etch for 2 min at ambient temperature. Micro-etching and macro-etching were done at ambient temperature using a chemical solution for 2 min.

Illustration (Fig. 5).

In Figs. 6, 7, and 8, grains appear to be finer equiaxed close to the welding nugget zone and start to become bigger and elongated due to exposing to plastic deformation and thermal energy the effect its clearly appears in the micro-hardness test that moves apart from the stirring zone, the micro hardness values start to decrease, as shown in Figs. 9, 10, 11, and 12.

2.3 Tensile Shear Test

The tensile-shear force difference versus rotational speed of tools having different dull time and heat treatment is shown in Figs. 13, 14, 15, and 16. The FSSW welded joints' maximum overall tensile shear strength were 5.3 KN at 1500 rpm, dwell time 8 s, and O Heat Treatment. The minimum tensile shear strength of the FSSW joint is 1.1 KN

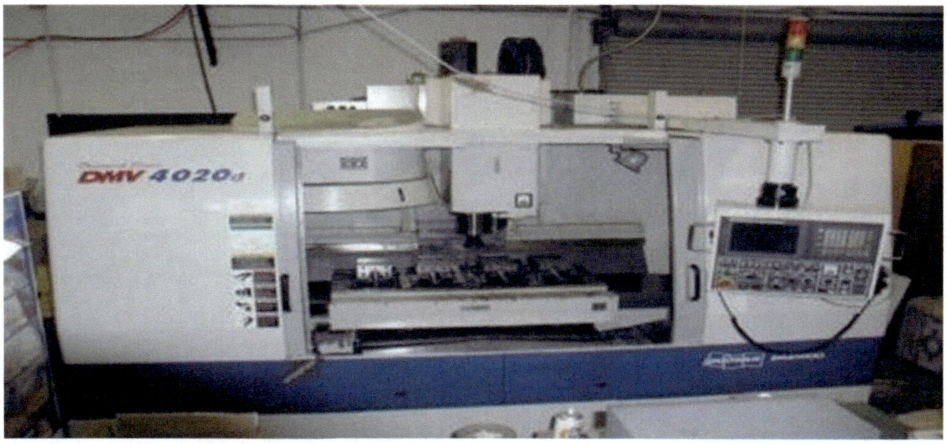

Fig. 2 CNC milling machine

Table 2 K100 tool steel of chemical composition, ASTM A681

Alloy	Chemical composition (wt.%)				
	C	Si	Mn	Cr	Fe
K100	2.01	0.24	0.34	10.89	Bal

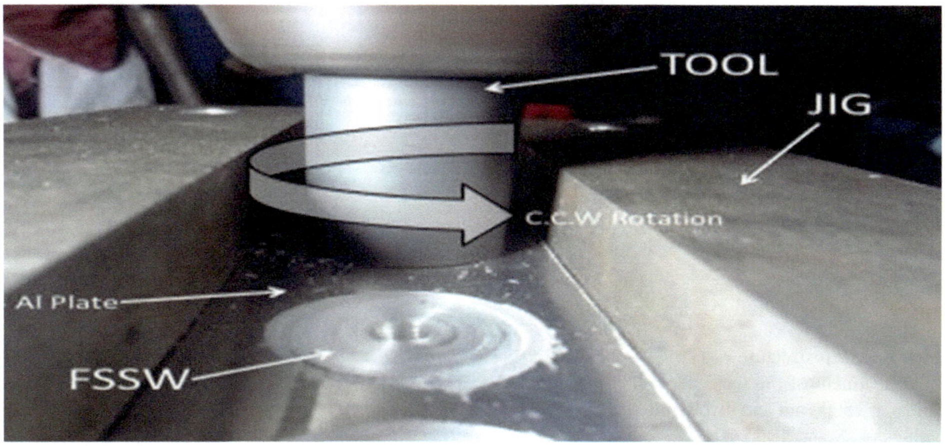

Fig. 3 FSSW process illustration

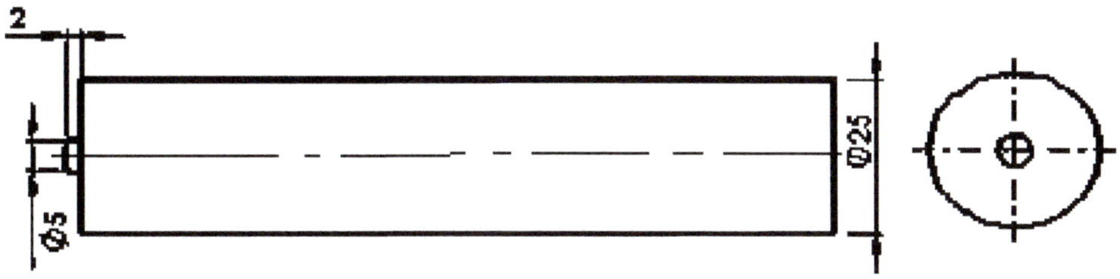

Fig. 4 A schematic drawing for FSSW tool

at 1300 rpm, dwell time 8 s, and T6 Heat Treatment. When experiments T6 Heat Treatment and O Heat Treatment at dwell time 8 s were compared, it is clear that the tensile shear strength increased from 3.1 KN at 4 KN, as seen in Figs. 13 and 14. When experiments T6 Heat Treatment at 200 °C and O Heat Treatment at 200 °C at dwell time 8 s were compared, it is clear that the tensile shear strength increased from 4.3 KN at 4.6 KN, as seen in Figs. 15 and 16. While the experiments T6 Heat Treatment at 200 °C and O Heat Treatment at 200 °C at dwell time 10 s and 1500 rpm were compared, it is clear that the tensile shear strength increased from 5.1 KN at 5.3 KN, as seen in Figs. 15 and 16. The FSSW joints that were obtained with best resulted in a higher tensile shear strength at Treatment at 200 °C at dwell time 10 s and 1500 rpm. The greater volume of bonded materials will clarify these outcomes as the pin height

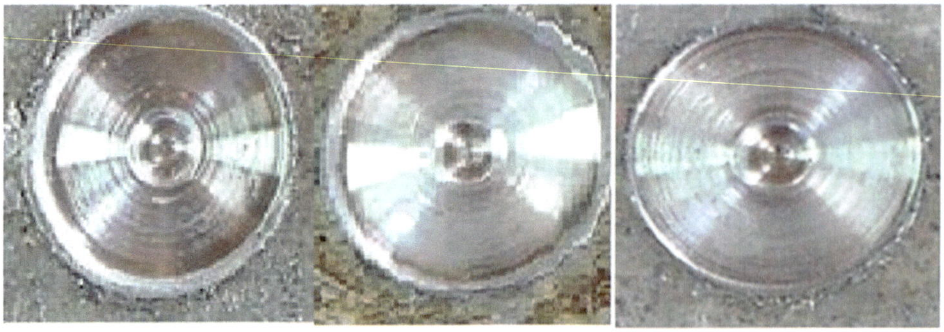

Fig. 5 Plates FSS welded at: **a** 1300 rpm, **b** 1500 rpm and **c** at 2000 rpm

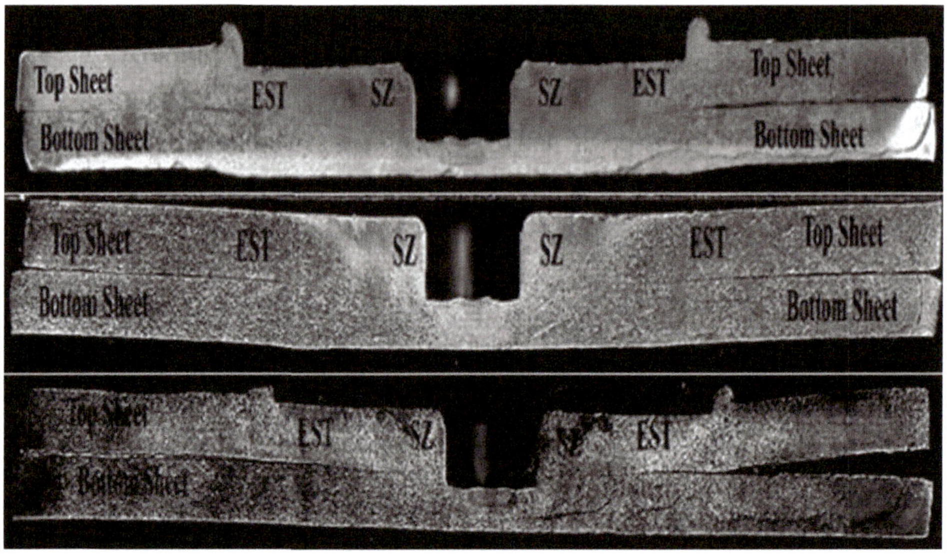

Fig. 6 Macrostructure of (O) heat treatment of the welded joints at rotational speed 1500 RPM and dwell time values (6, 8 and 10 s)

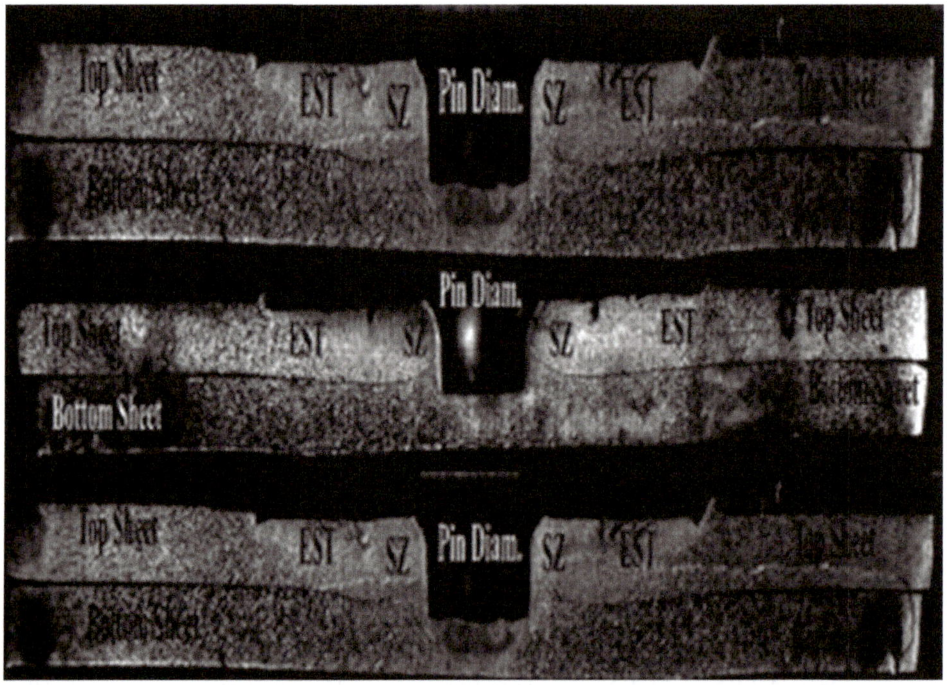

Fig. 7 Macrostructure of (T6) heat treatment of the welded joints at 1500 rpm and dwell time values (6, 8 and 10 s)

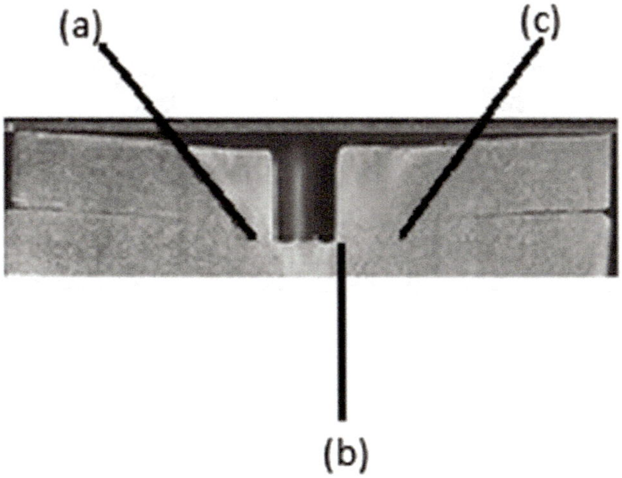

Fig. 8 Illustration of microstructure regions on cross- section of weld joint

increases. The tensile shear strength has an appositive impact on increasing the dwall time at 10 s at 1500 rpm height. These findings illustrate that the mechanical properties of the FSSW joints are enhanced by the stirring effect and the refined structure. As the welding time increased from 6 to 10 s, the tensile shear strength of the FSSW joints decreased. The tensile strength of the FSSW joint is greater. The structure obtained with the FSSW approach may explain this strength change. The studies in the literature claim that the FSSW microstructure is a refined structure, while the cast structure of the RSW welds is 1, three mechanical properties of the FSSW joint are enhanced by the stirring effect and the refined structure (Figs. 17, 18, 19, 20, 21 and 22).

- For 0 heat treatment results obtained shows that as Rotational Speed increased from 1300 to 1500 rpm, maximum Hardness values achieved when dwell time is increased from 8 to 10 s. However, when increasing

Fig. 9 Microstructure of (O) heat treatment at 1500 rpm, dwell time 8 s

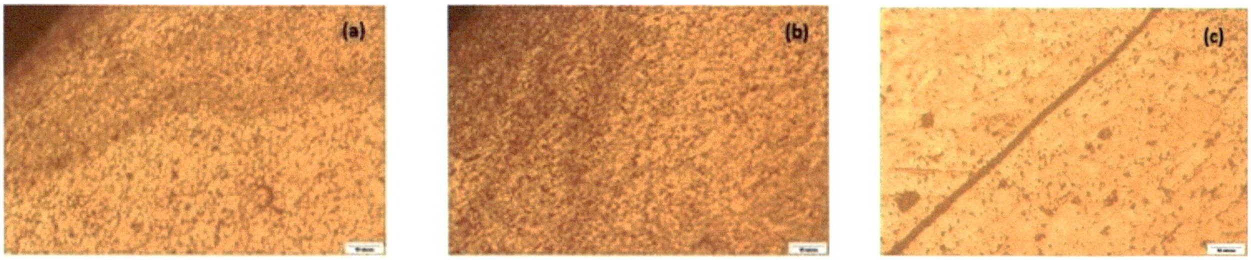

Fig. 10 Microstructure of (T6) heat treatment at 1500 rpm, dwell time 8 s

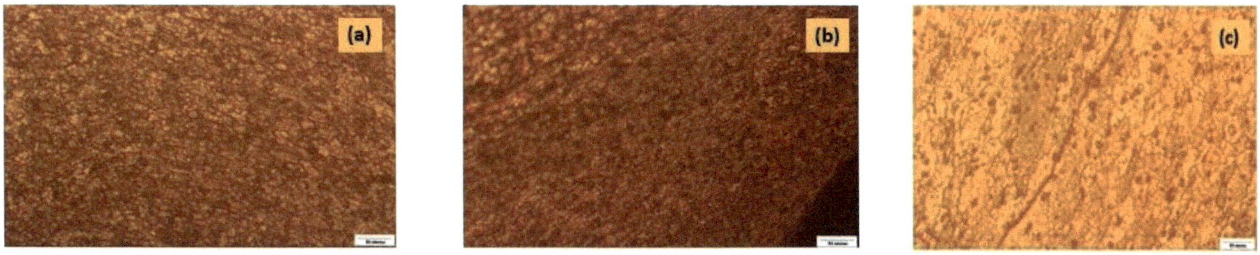

Fig. 11 Microstructure of (T6) heat treatment at 2000 rpm, dwell time 6 s

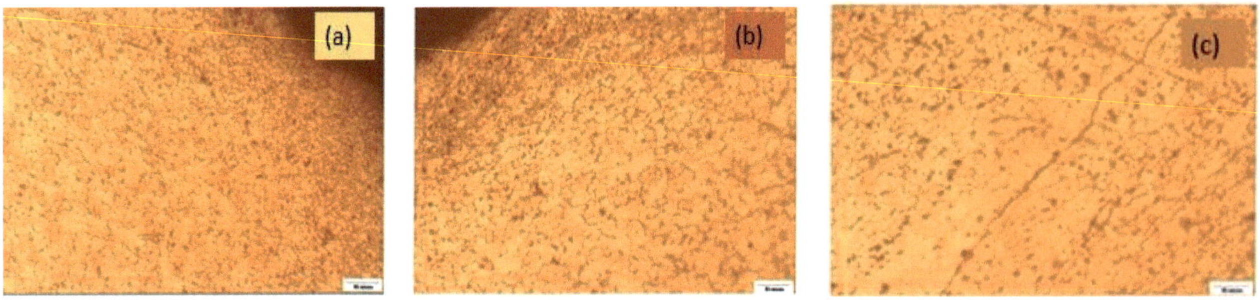

Fig. 12 Microstructure of (O) heat treatment at 2000 rpm, dwell time 6 s

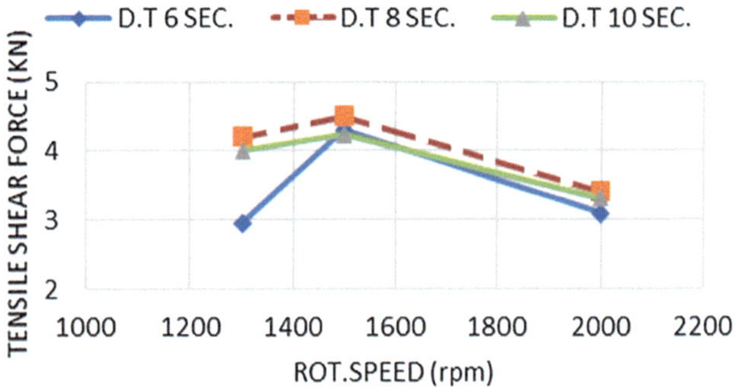

Fig. 13 Tensile shear force versus rotational speed for O heat treatment

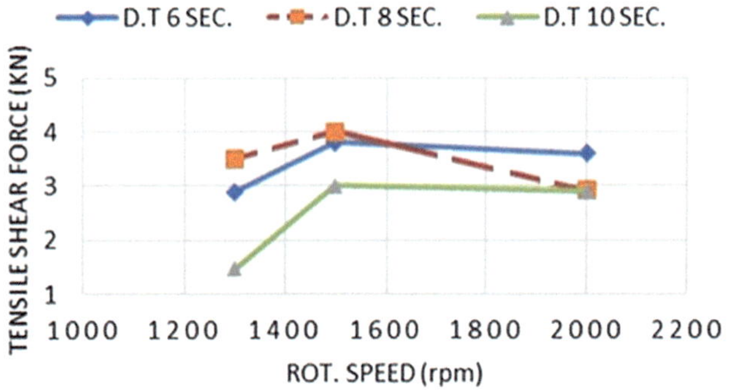

Fig. 14 Tensile shear force versus rot. Speed for T6 heat treatment

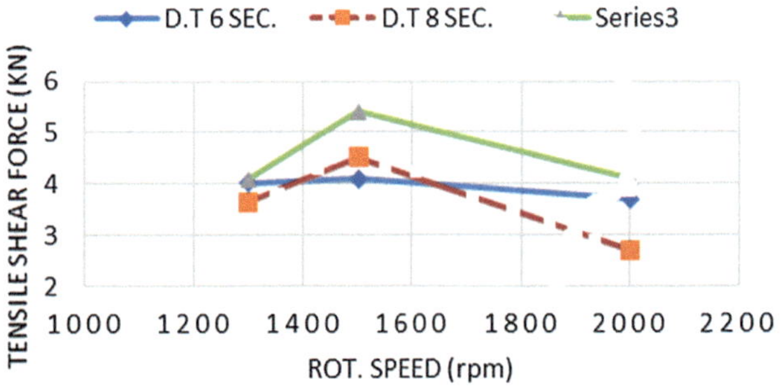

Fig. 15 Tensile shear force versus rot. Speed for T6 heat treatment at 200 °C

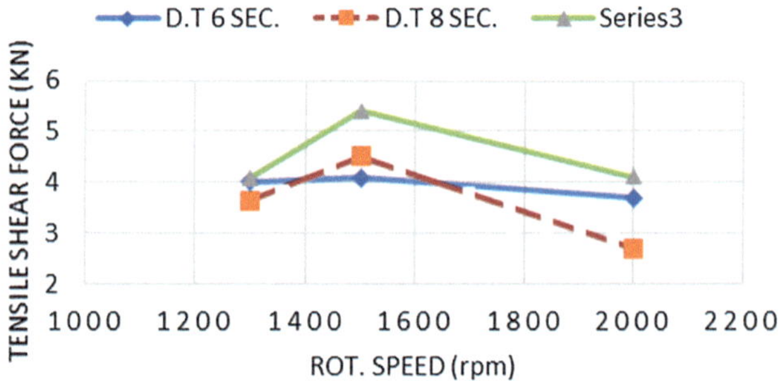

Fig. 16 Tensile shear force versus Rot. Speed for O heat treatment at 200 °C

Rotational Speed from 1500 to 2000 rpm, maximum Hardness achieved at minimum dwell time 6 s.

- For T6 heat treatment results obtained shows approximate results as in case of (O) heat treatment.

3 FSSW of PC/PC

The present work studies the friction stir spot welding (FSSW) of polycarbonate (PC)/polycarbonate (PC). The influences of the rotational speed of tool on the mechanical, microstructure characteristics, and Tensile tests were performed on lab joints. Joints FSSW at 365 rpm, 840 rpm, and 1425 rpm with dwell time 12 s, 16 s, 24 s, and 32 s.

The Chemical composition polycarbonate (PC) work piece is $C_{15}H_{16}O_2$. The material was cut to dimensions of 120 mm (length) × 25 mm (width) × 2.6 mm (thickness) (Tozaki et al., 2010).

3.1 Process Parameters

The tool content is primarily of a shoulder and a pin. The pin has a cylindrical shape with 4.7 mm diameter and 3.5 height. The shoulder has 18.75 mm diameter, and the length of tool

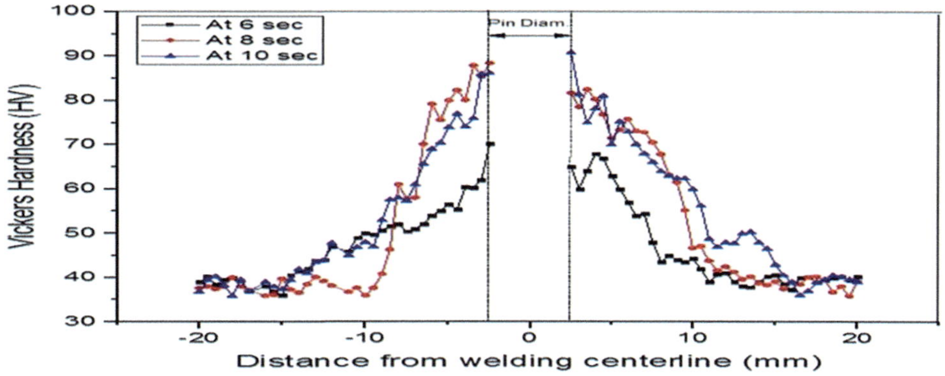

Fig. 17 Hardness values at 1300 rpm and dwell times (6, 8 and 10 s).

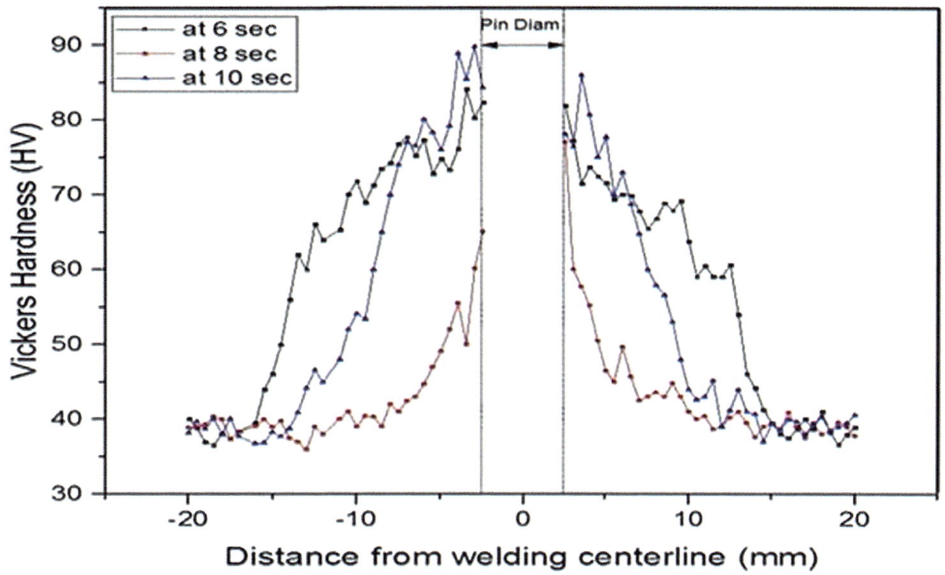

Fig. 18 Hardness values at 1500 rpm and dwell times (6, 8 and 10 s).

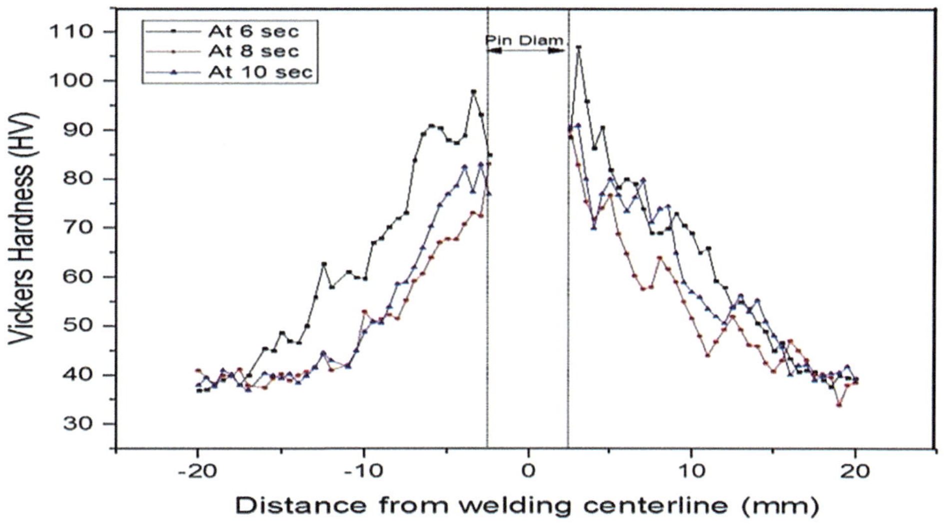

Fig. 19 Hardness values at 2000 rpm and dwell times 6, 8 and 10 s.

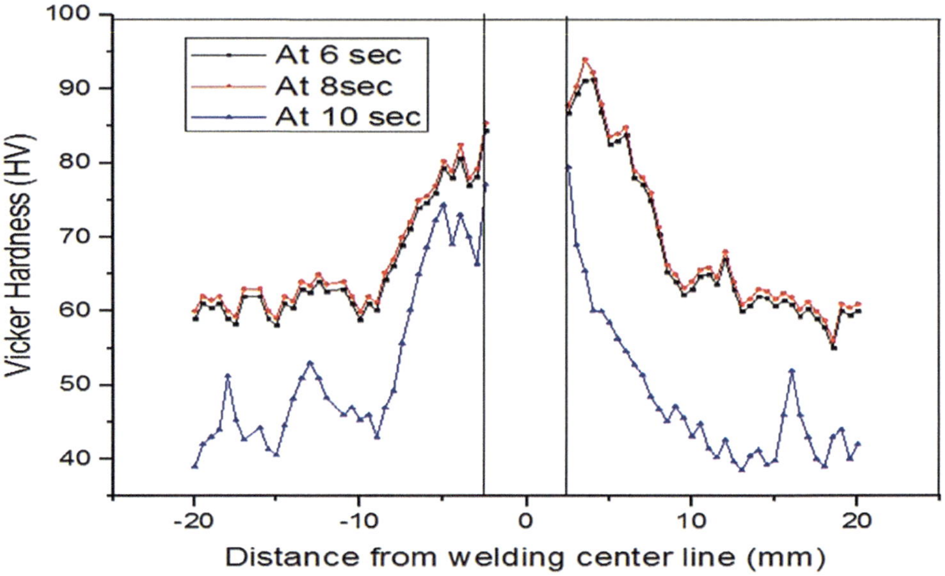

Fig. 20 Hardness values at 1300 rpm and dwell times 6, 8 and 10 s.

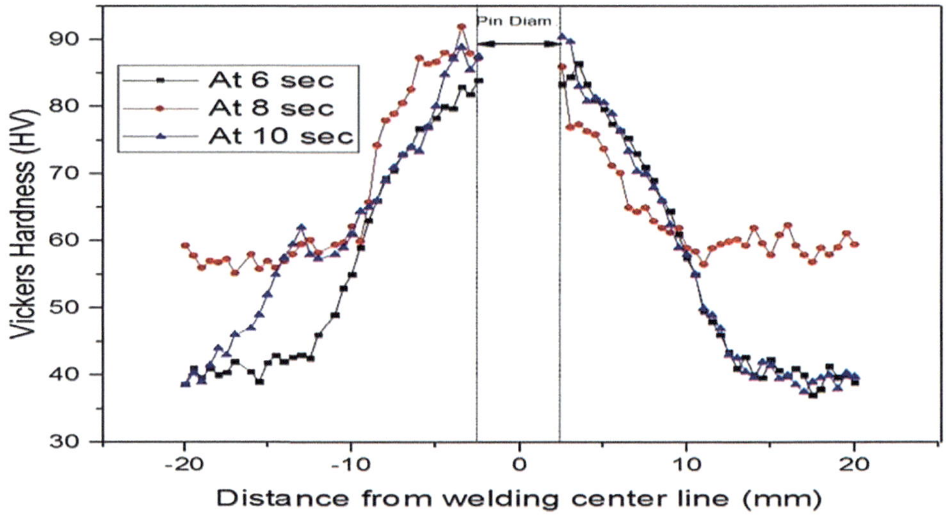

Fig. 21 Hardness values at 150 rpm and dwell times (6, 8 and 10 s).

is 103.5 mm. The tool was from k110 chemically composed tool alloy listed in Table 3. The FSSW was performed using a universal milling machine at three rotational speeds namely: 365, 840, and 1425 rpm and dwell time is 12, 16, 24, and 32 s (Fig. 23).

3.2 Macrostructure Test

3.3 Tensile Shear Test

Figure 24 shows that the tensile shear increases with increasing rotational speed, especially when 840 rpm is applied. It may be explained as the high rotational speeds supply more friction and consequently more heating is produced. When molten thermoplastic materials is heated to a high temperature, a decrease in the molecular weight of the material occurs and decreases the mechanical properties of the thermoplastic. The strength decreases with increasing the dwelling time.

3.4 Shore-D Hardness Test

Figure 25 shows that the hardness measurements increase with increasing rotational speed at 840 rpm. However, above 840 rpm at constant dwell time the hardness value decreases. It may be explained as high rotational speeds supply more friction and more heating is produced. When molten thermoplastic materials is heated to a high temperature, a decrease in the molecular weight of the material shows a smaller chains of material and mechanical

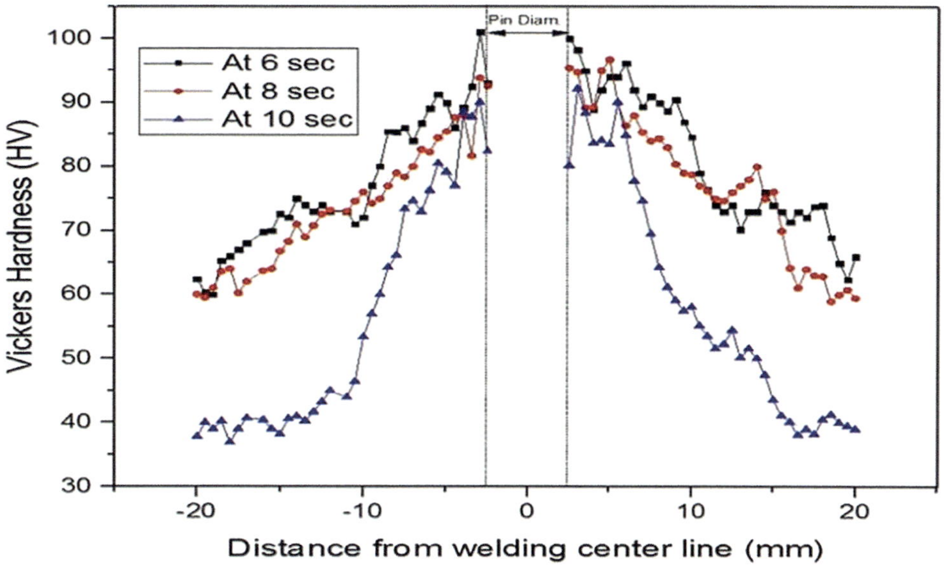

Fig. 22 Hardness values at 2000 rpm and dwell times (6, 8 and 10 s).

Table 3 Chemical composition of tool K110 tool steel (Yang et al., 2014)

Element	C	Cr	Mo	V	Ni	Mn	Si	Cu	P	S
Composition	1.50	12.50	0.99	0.85	0.4	0.5	0.4	0.2	0.02	0.02

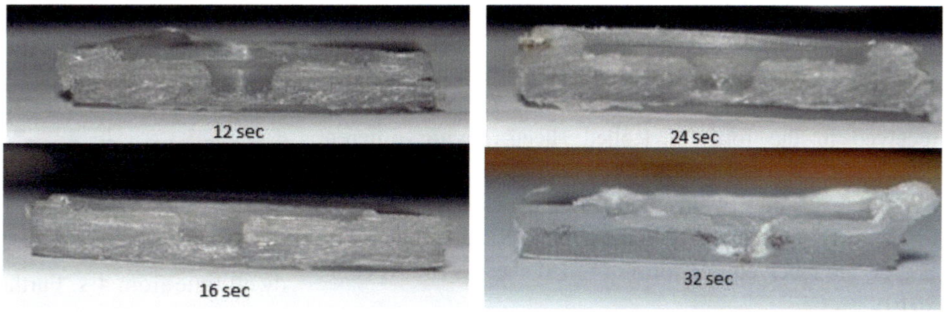

Fig. 23 Macrostructure of the welded joints at rotational speed 840 RPM and dwell time values (12, 16, 24, and 32 s).

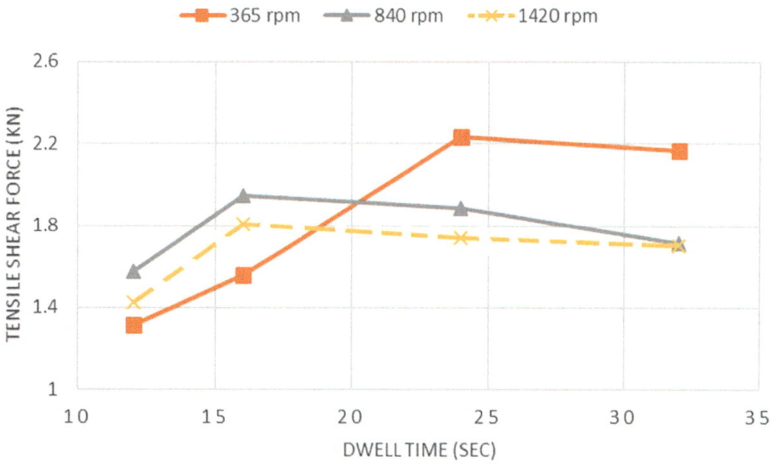

Fig. 24 Tensile shear force versus dwell time

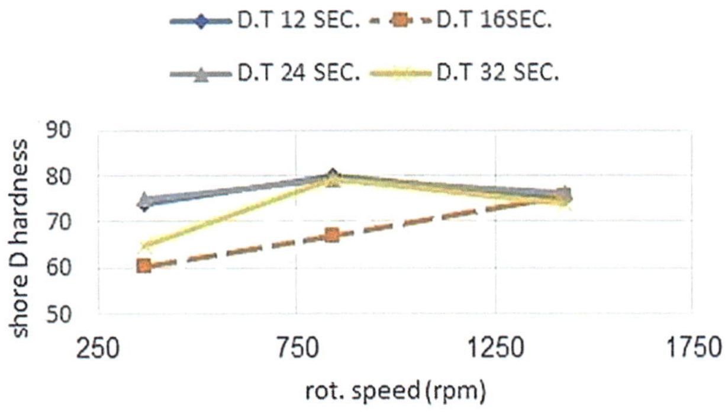

Fig. 25 Shore D hardness versus rotational speed

properties. Hardness value at constant rpm and changed dwell time shows best results at 24 s due to good overlap of welding between the two plates.

4 FSSW of 2024 Al Alloy/Polycarbonate

The analysis involved the effect of rotational speed and dwell time variation on the mechanical characteristics of joints (carried out by means of shear tensile test), hardness test, and an observation of the microstructure.

The material used in the current investigation was wrought 2024 (Al-Mg-Si) Al alloys. The chemical composition of the alloy is listed in Table 4. The material was received in the form of Rolled sheets having different dimensions. The material was cut to dimensions of 120 mm (length) × 25 mm (width) × 1.5 mm (thickness).

Polycarbonate sheets with a chemical composition of $C_{15} H_{16} O_2$ are received in a form of rolled sheets (Tozaki et al., 2010).

4.1 Process Parameters

Specimens were joined using FSSW. The tool consists mainly of a pin and shoulder. The pin has a tapered shape, and the shoulder is cylindrical and flat. The tool was from K110 tool steel with a chemical composition listed in Table 5.

The FSSW was carried out using a universal milling machine at three different tool rotational speeds typically 1300, 1500, and 2000 rpm and dwell time (sec) is 6, 8, and 10 s.

Table 4 Chemical composition of 2024 Al-Alloy (wt.%) (Ibrahim Sabry et al., 2020)

Element	Mg	Si	Fe	Mn	Cr	Cu	Pb	Al
Composition	0.02	0.10	0.35	0.70	0.01	3.8	0.04	Bal

4.2 Macro- and Micro-structural Examinations

Using an Olympus Optical Metallurgical Microscope, the macro- and micro-structure features of FSSW sheets were analyzed. Specimens were ground under water on a revolving disc (Metasery Grinder 2000) using an abrasive disc of silicon carbide with an improved fineness of 120–2000 grit. They were then polished using an alumina suspension of 0.3 μm. Using a chemical solution, Kellers etched 190 ml of purified water, 5 ml of nitric acid, 3 ml of hydrochloric acid, and 2 ml of hydrofluoric acid for 2 min at room temperature. Micro-etching and macro-etching were carried out.

The grains appear to be finer equiaxed close to the welding nugget zone and start to become bigger and elongated due to transforming to plastic deformation as shown in Figs. 26, 27, and 28. However, the effect of thermal energy its clearly appeared in the micro-hardness test that as we move apart from the stirring zone the micro-hardness values start to decrease.

4.3 Tensile Shear Test

From the above results it is noticed that the maximum tensile shear value occurs at speed 1425 rpm by increasing the dwell time from 4 s. Further increase in dwell time leads to decrease in tensile shear value. It may be due to the increase of generated heat leading to course grains. This type of grains is soften leading to low tensile shear stress.

The maximum tensile shear value occurs at speed 1500 rpm. Increasing the speed to 2000 rpm, the tensile shear force decreases, while further increase in dwell time leads to a decrease in tensile shear value as shown in Fig. 29.

4.4 Shore-D Hardness Test

Figure 30 shows the highest value of the hardness test occurred at 2000 rpm and dwell time 4 s, while lowest value occurred at 1425 rpm and 8 s dwell time.

Table 5 Chemical composition of K110 tool steel (Yang et al., 2014)

Element	C	Cr	Mo	V	Ni	Mn	Si	Cu	P	S
Composition	1.50	12.50	0.99	0.85	0.4	0.5	0.4	0.2	0.02	0.02

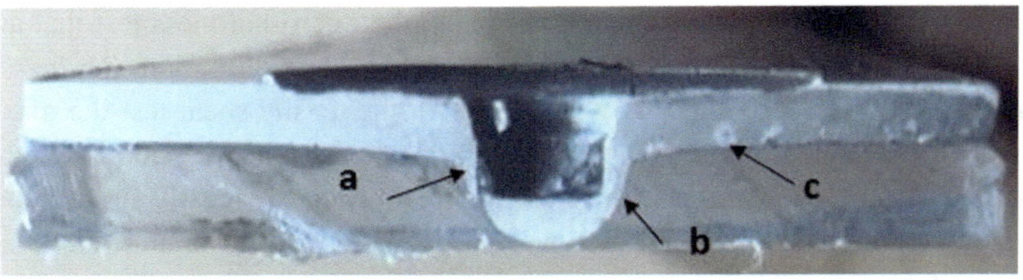

Fig. 26 Illustrations of microstructure regions on cross section of welded joints (macrostructure examination)

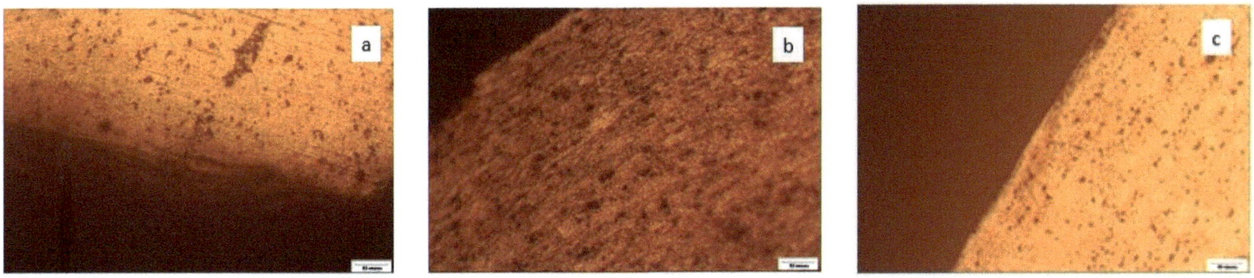

Fig. 27 Microstructure at 1425 rpm rotational speed and 6 s dwell time

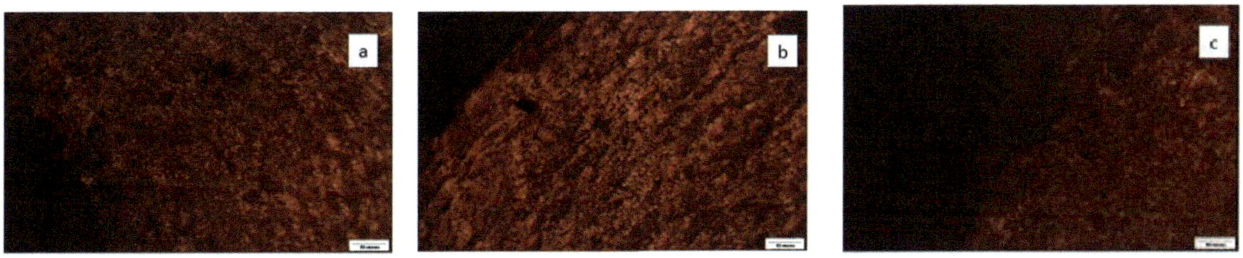

Fig. 28 Microstructure at 2000 rpm rotational speed and 6 s dwell time

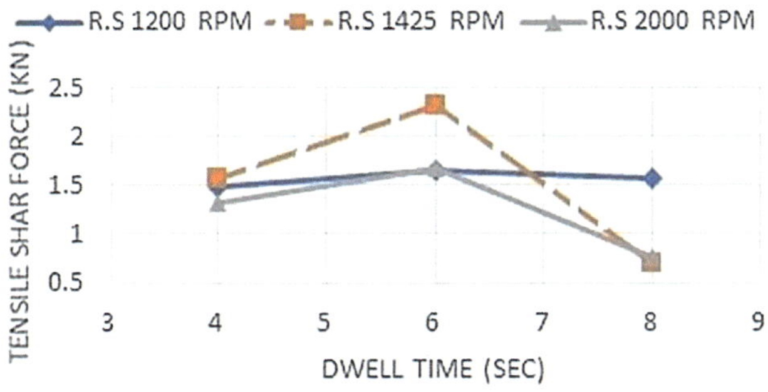

Fig. 29 Tensile shear force versus dwell times

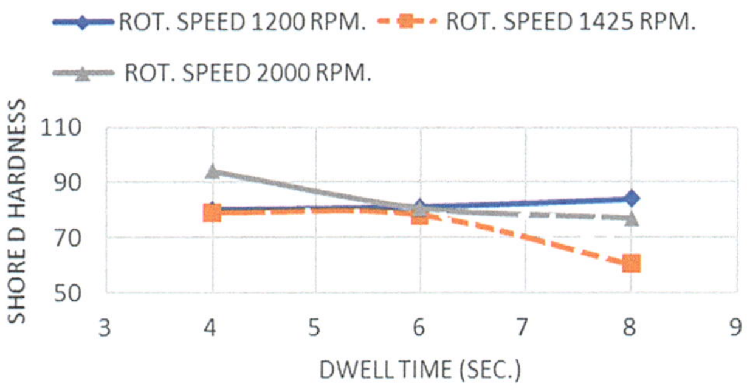

Fig. 30 Shore d hardness test versus dwell time

5 Discussion and Conclusion

- From the previous results it has been proved that 2024 aluminum alloy is successfully joined using Friction Stir Spot Welding. FSSW shows significant increase in mechanical properties proved by Micro-structure examination where the grains appear to be finer equiaxed close to the welding nugget zone and start to become bigger and elongated due to exposing to plastic deformation and thermal energy the effect its clearly appears in the micro-hardness test that as we move apart from the stirring zone the micro-hardness values start to decrease. The tensile shear test applied on two temperatures at room temperature and 200 °C for (O) and (T6) heat treatment the maximum tensile shear value occurs at speed 1500 rpm by increasing the dwell time from 6 to 8 s further increase in dwell time lead to decrease in tensile shear value it may be because the increase of generated heat. For (O) and (T6) heat treatment at 200 ° C, the maximum tensile shear value occurs at speed 1500 rpm by increasing the dwell time from 6 to 10 s further increase in rotational speed lead to decrease in tensile shear value it may be because the increase of generated heat and friction, micro Vickers's hardness test applied and for (0) heat treatment results obtained shows that as Rotational Speed increased from 1300 to 1500 rpm, Max. Hardness values achieved when Dwell time increased from 8 to 10 s. But when increasing Rotational Speed from 1500 to 2000 rpm. Maximum hardness achieved at minimum dwell time 6 s. For (T6) heat treatment obtained an approximate as the same results for (O) heat treatment. And these show that type of heat treatment does not has a major effect on weld's properties while using FSSW.

- Also, it has been proved that polycarbonate plates are successfully joined using Friction Stir Spot Welding with significant increase in mechanical properties specially in joint region than base material proved by results of tensile shear test tensile strength increases with increasing rotational speed, especially when 840 rpm is applied and shore D hardness test shows hardness value at constant rpm and changed dwell time shows best results at 24 s due to good overlap of welding between the two plates
- For aluminum and polycarbonate plates are successfully joined using Friction Stir Spot Welding with sufficient mechanical properties proved by micro-structure examination where the grains appear to be finer equiaxed close to the welding nugget zone and start to become bigger and elongated due to transforming to plastic deformation. However, the effect of thermal energy its clearly appeared in the micro-hardness test that as we move apart from the stirring zone the micro-hardness values start to decrease, for the tensile shear test the maximum tensile shear value occurs at speed 1500 rpm. Increasing the speed to 2000 rpm, the tensile shear force decreases, while further increase in dwell time leads to a decrease in tensile shear value. Hardness test shows that the highest value of the hardness test occurred at 2000 rpm and dwell time 4 s, while lowest value occurred at 1425 rpm and 8 s dwell time.

Acknowledgements The authors would like to express thanks to Dr. Ibrahim Sabry from the Department of Manufacturing Engineering, Modern Academy for Engineering and Technology, for this valuable help during the experimental work and final review for this research.

References

Gadallah, S., Sabry, I., & Ghafaar, M. A. (2020). A summarized review on friction stir welding for aluminum alloys. *International Journal on:The Academic Research Community Publication*. https://doi.org/10.21625/archive.v4i1.695.

Nguyen, N.-T., Kim, D. Y., & Kim, H. Y. (2011). Assessment of the failure load for an AA6061-T6 friction stir spot welding joint. *Proceedings of the Institution of Mechanical Engineers B: Journal of Engineering Manufacture*, 1746–1756.

Sabry, I., & El-Kassas, A. (2019). An appraisal of characteristic mechanical properties and microstructure of friction stir welding for aluminium 6061 alloy – silicon carbide (SiCp) metal matrix composite. *Journal of Mechanical Engineering and Sciences.*

Sabry, I., Gadallah, N., & Abu-Okail, M. (2020). Optimization of friction stir welding parameters using response surface methodology. *IOP Conference Series: Materials Science and Engineering*. IOP Publishing. 710.1088/1757-899X/973/1/01201.

Tozaki, Y., Uematsu, Y., & Tokaji, K. (2010). A newly developed tool without probe for friction stir spot welding and its performance. *Journal of Materials Processing Technology*, 844–851.

Yang, X. W., Fu, T., & Li, W. Y. (2014). Friction stir spot welding: A review on joint macro-and microstructure, property, and process modelling. *Advances in Materials Science and Engineering*, 1–11.

Corrosion Assessment Methodologies of Steel Reinforcement in Concrete Treated with a Chemical Inhibitor

S. Abdulsalam and A. Al-Arbeed

Abstract

The high ambient temperature, humidity, and severe weather condition in Kuwait provide an extremely aggressive environment for reinforced concrete structures. Corrosion is the most common source for premature failure in reinforced concrete buildings. Using corrosion inhibitors in concrete is considered a simple and cost-effective method in reducing the chloride-induced corrosion. However, the efficacy of some of these chemical inhibitors under the prevailing environmental and marine conditions in Kuwait has not been well studied. This paper will present the methodology applied in a study conducted to assess the effectiveness of adding a commercial chemical to concrete mixtures in order to provide protection against the corrosion of embedded steel rebars in concrete and to study the effect of using sulfate resistant cement (SRC) in contrast with ordinary Portland cement (OPC) as cementing in chloride-rich environment. The study was conducted on two concrete mixtures of different water-to-cement ratios, w/c of 0.4 and 0.6. It was performed at three levels of exposure zones: tidal zone, atmospheric zone, and submerged zone. Concrete specimens were subjected to long-term regular corrosion potential testing, ASTM C1202 Test, ASTM G109, chloride diffusivity test, and the corrosion health monitoring test.

Keywords

Chloride-induced corrosion • Steel reinforcement • Electrochemical corrosion testing • Chloride diffusivity • Migrating corrosion inhibitor

S. Abdulsalam (✉) · A. Al-Arbeed
Energy and Building Research Centre, Kuwait Institute for Scientific Research (KISR), P.O. Box 24885 13109 Safat, Kuwait City, Kuwait
e-mail: ssalam@kisr.edu.kw

Nomenclature and Acronyms

OPC	Ordinary Portland cement
SRC	Sulphate resistant cement
w/c	Water-to-cement ratio
EIS	Electrochemical impedance spectroscopy
MCI	Migrating corrosion inhibitor
Cl	Chloride ions
Dc	Diffusion coefficient
IM4	Concrete with OPC, MCI, and w/c of 0.4
IM6	Concrete with OPC, MCI, and w/c of 0.6
IP4	Concrete with OPC, no inhibitor, and w/c of 0.4
IP6	Concrete with OPC, no inhibitor, and w/c of 0.6
VP4	Concrete with SRC, no inhibitor, and w/c of 0.4
VP6	Concrete with SRC, no inhibitor, and w/c of 0.6
KISR	Kuwait Institute for Scientific Research

1 Introduction

Studies revealed that the deterioration of reinforced concrete structures in Kuwait usually occurs within 20–25 years after construction when they are typically designed for a service life time of 50 years (1994b; Attiogbe et al., 1994a). Corrosion of the steel reinforcement rebars in concrete due to chloride ingress and concrete carbonation, induced by the reaction of atmospheric carbon dioxide with cement hydration product Ca(OH), is the major causes of failures in reinforced concrete structures in a marine environment. Steel reinforcement is usually protected by the concrete cover which provides high alkalinity due to the formation of stable passivation film Fe_2O_3, while also slowing down the ingress of chlorides, sulphates, and carbon dioxides. In the presence of moisture and oxygen, reinforcement corrosion results from the failure of the passivation film, usually as a result of electrochemical action involving chloride ions or the

F. Rosso et al. (eds.), *Advances in Architecture, Engineering and Technology*, Advances in Science, Technology & Innovation, https://doi.org/10.1007/978-3-030-86913-7_17

breakdown of the concrete alkalinity surrounding the steel reinforcement. The time before the passivation film breaks down and corrosion begins is very important to predict the service life time of buildings. This time is determined by the ability of concrete to resist the diffusion of chloride ions and loss of alkalinity, in addition to the impact of the environment (Abdulsalam et al., 2015). The corrosion process could be efficiently reduced by limiting salts in concrete ingredients, lowering the w/c in concrete mixes, having adequate concrete cover, and using corrosion protection techniques. Incorporating corrosion protection systems in concrete, such as corrosion-inhibiting chemical admixtures and supplementary cementitious materials, increases the cost of construction. However, this cost is marginal compared to the cost of repair of the structure as a result of premature deterioration. Using corrosion inhibitors in concrete is considered a simple and cost-effective method to delay the onset of the chloride-induced corrosion in concrete.

The objective of this project was to assess the effectiveness of mixing a commercial corrosion inhibitor, entitled migrating corrosion inhibitor (MCI), with ordinary Portland cement (OPC) concrete for marine structures under Kuwait's local environmental conditions. Additionally, this study evaluated the effectiveness of using sulfate-resistant cement (SRC) in concrete in a chloride rich environment, as SRC is occasionally used as an alternative to OPC, to reduce corrosion in some marine projects in Kuwait. Furthermore, there is no clear information available on the real efficacy of some of the chemical inhibitors under the harsh environmental and marine conditions in Kuwait (Albahar & Hussain, 2015), and there are no studies highlighting the advantages and disadvantages of these materials under hot weathering to be addressed in the national construction guidelines and code of practice; therefore, this assessment was conducted.

The MCI is a water-based organic corrosion inhibiting admixture that can be applied either directly on the concrete surface or as an admixture while mixing concrete (Holloway et al., 2004). It diffuses through concrete to form a monolayer film at the interface between concrete and steel reinforcement, which supports both anodic and cathodic corrosion reactions by preventing the passage of aggressive Chloride ions (Cl) and other corrosion-induced species through this film. Morris and Vazquez (2002) stated that the effect of using the MCI inhibitor in concrete is highly dependant on the initial chloride ion content in concrete; it can reduce the corrosion of the steel reinforcement when the initial concentration of chloride is less than 0.2% by weight of cement. The most common commercial type of MCI is mainly composed of mino carboxylates or mino alcohol (Holloway et al., 2004). Unlike other corrosion inhibitors (Franciŝković et al., 2006), MCI protects the reinforcement steel in both oxidation ranges: Cathode and anode. It penetrates through the concrete and prevents the diffusion of O_2 and H_2O to protect the steel from the oxidation process.

In this study, concrete specimens were cast with different concrete mixes. Subsequently, the mechanical properties of the fresh concrete mixes were studied. After the concrete specimens hardened, they were assessed in the laboratory and at the field by long-term corrosion monitoring with corrosion potential measurements (ASTM G109-7, 2013) test, which determines the effect of admixtures on steel corrosion in concrete exposed to chloride, chloride diffusivity test and (ASTM C1202) test, responsible for providing an indication of the ability of concrete to resist the chloride ions ingress. The concrete specimens in the field were exposed to two exposure zones: tidal zone and atmospheric zone, while the chloride diffusivity specimens were submerged into the sea. Chloride ions penetrate the concrete pores either by capillary action, diffusion, or both (Charis A. Apostolopoulos et al., 2013). Most of the literature use Fick's second law to model the chloride transport through concrete, neglecting the interaction of chloride with the solid phase. In this study, Fick's second law was used to model the results of chloride diffusivity. After two years of field exposure for the reinforced concrete specimens, the concrete and steel reinforcement were comprehensively and morphologically inspected. Electrochemical corrosion testing functioned by utilizing corrosion rate sensors and corrosion instrumentation through special corrosion test probes to allow direct measurement of inhibitor performance in concrete.

2 Methodology

2.1 Concrete Preparation

As presented in Table 1, six different concrete mixtures were established, comprising of two cement types: ordinary Portland cement (OPC) and Sulfate resistant cement (SRC) and two levels of w/c ratios to resemble normal to stringent concrete quality. The MCI was used with each set of concrete mixed with OPC. Enough specimens were cast for laboratory and field investigations.

Concrete specimens were prepared and cast according to standards and specifications. The testing of mechanical and physical properties was performed on fresh concrete for the determination of slump, slump loss, air content, setting time, and unit weight.

Table 1 Concrete mix design per meter cube for six different mixes

Mix number	1 (IP4)	2 (IP6)	3 (VP4)	4 (VP6)	5 (IM4)	6 (IM6)
Design strength	K400	K250	K400	K250	K400	K250
MCI (lt)	–	–	–	–	0.6	0.6
w/c ratio	0.4	0.6	0.4	0.6	0.4	0.6
Cement type	OPC	OPC	SRC	SRC	OPC	OPC
Cement (kg) =	450	450	450	450	450	450
Dry Sand (kg) =	640	590	640	590	640	590
Aggregate 20 mm kg	770	660	770	660	770	660
Aggregate 10 mm kg	380	300	380	300	380	300
Free water (l) =	180	270	180	270	180	270
Total water (l) =	191	281	191	281	191	281
Plasticizer minimum (l/m^3)	5.0	5.0	5.0	5.0	5.0	5.0
Plasticizer maximum (l/m^3)	7.5	7.5	7.5	7.5	7.5	7.5

2.2 Experimental Outline

2.2.1 Electrical Indication of Concrete's Ability to Resist Chloride Ion Penetration Test (ASTM C1202, 2010)

In this test, a concrete cylinder of 100 mm in diameter and 200 mm in height was prepared and cured for 28 days. A concrete disc of 5 cm height was cut in the middle of the cylinder and was epoxy coated on the side. This disc was specially prepared to ensure that all concrete pores were saturated with water. A cell of concrete disc with sodium hydroxide on one face and sodium chloride on the other was constructed and connected to 60 V power supply. The total charge passed in six hours across the concrete disc was measured to provide the basis for assessing the ability of different concrete mixes in resisting chloride ion ingress, by comparing the total coulombs passed through the concrete per ASTM C1202, as described in Table 2 (2010).

Table 2 Penetration of chloride ion based on coulombs passed (ASTM C1202-10, 2010)

Charge passed (coulombs)	Chloride ion penetrability
>4,000	High
2,000–4,000	Moderate
1,000–2,000	Low
100–1000	Very low
<100	Negligible

2.2.2 Standard Test Method for Determining Effects of Chemical Admixtures on Corrosion of Embedded Steel Reinforcement in Concrete Exposed to Chloride Environments (ASTM G109, 2013)

This test method is used to asses the effect of admixtures added to concrete, on the corrosion of steel rebars inside the concrete. Reinforced concrete specimens (control and inhibited concrete) were prepared according to the relevant ASTM standards of mixing and preparing concrete specimens. Each specimen has a size of 279 mm × 152 mm × 114 mm, with one steel bar at the top and two at the bottom. A corrosion macrocell was established by connecting the top and bottom steel with a 100 Ω resistor. The specimens were exposed to cycles of wetting (ponding) with 3% NaCl solution for two weeks and drying for another two weeks. The voltage between the two layers of the reinforcement was measured at the end of the first week of ponding. The derived current (I = V/R) indicated the actual corrosion occurring. The test is terminated when the average current is greater than 10 μA for the control specimens (plain concrete) and no less than half the specimens showing macrocell current, greater or equal to 10 μA.

2.2.3 Corrosion Rate Testing

Triplicate concrete samples of size 380 × 200 × 76 mm were cast from each mix. These concrete specimens were designed in order to get the best impedance results, which simulated the actual harsh condition of steel rebars inside concrete. Then the concrete specimens were exposed to two

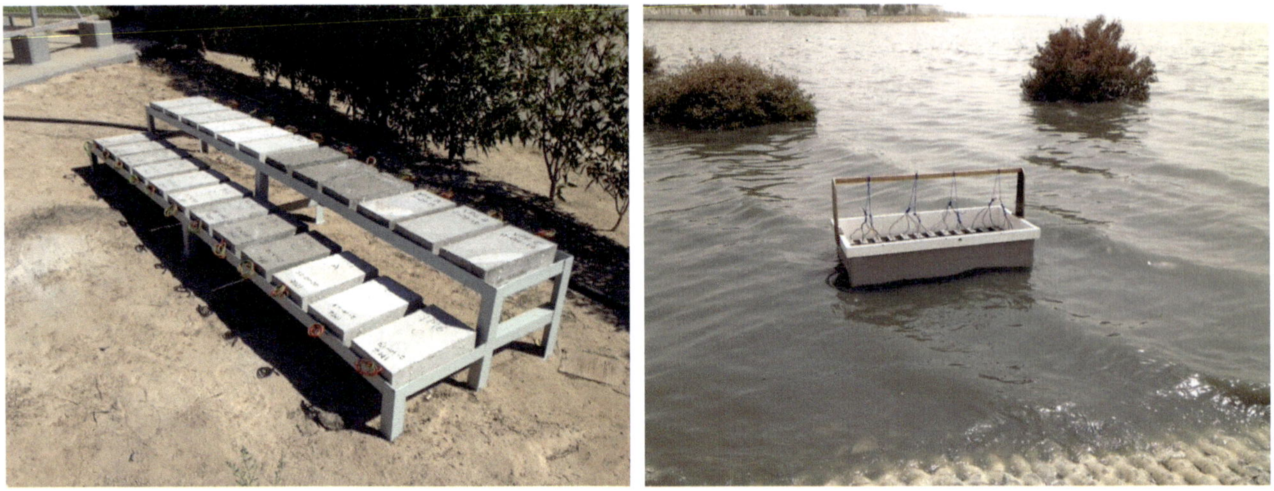

(a) (b)

Fig. 1 Concrete specimens at the atmospheric zone (**a**). Concrete specimens at the tidal zone (**b**)

different zones in the field under the local harsh environmental condition of Kuwait. The two exposure zones are the following (Fig. 1):

- Atmospheric zone: 18 concrete specimens were placed on racks exposed to humid coastal conditions and salt-laden winds (Fig. 1a).
- Tidal zone: 18 concrete specimens were exposed to wetness and dryness, corresponding to high and low tides (Fig. 1b).

For both the tidal and atmospheric zone concrete specimens, corrosion rate readings were measured according to the ASTM C876-09 (2009).

2.2.4 Chloride Diffusivity Test

In this test, concrete cylinders with a diameter of 100 mm and a height of 200 mm were cast, cured, and then epoxy-painted on all sides, with the exception of the bottom side, to allow one-dimensional ingress of chloride ions inside the concrete. The aim of this test was to asses the integrity of concrete cover in resisting the diffusion of chloride ions. The cylinders were submerged in seawater at KISR premises and monitored. In the case of a saturation state of concrete, the chloride ions penetrate through the concrete due to the concentration gradients at the surface of the concrete and the solution of the pore inside the concrete. This process is usually expressed by Fick's second law of diffusion (Martin-Perez et al., 2000):

$$\frac{dC(x,t)}{dt} = D_c \frac{d^2[C(x,t)]}{dx^2}, \quad (1)$$

where C(x, t) is the concentration of chloride at x (depth) and t (time) and D_c is the diffusion coefficient. The solution of this differential equation of chloride diffusion, using the error function, is from the surface (Thoft-Christensen, 2002):

$$C(x,t) = C_s\left\{1 - erf\left[\frac{x}{\sqrt{4D_c t}}\right]\right\}, \quad (2)$$

where C_s = The concentration of chloride on the concrete surface given by percentage weight of cement and erf = The error function.

At different submersion ages, cylinders were recalled to extract powder samples at different depths (0–10 mm, 10–25 mm, 25–35 mm, and 35–50 mm) after drying. This resembles the chloride diffusivity rate through the concrete cover protecting the steel reinforcement. Using Fick's second law, the chloride diffusivity coefficient and the expected variation of chloride content in concrete over time were calculated and correlated for each mix. The chloride diffusivity rate model was developed using MATLAB.

2.2.5 Corrosion Health Monitoring

Two concrete blocks of size 600*600*750 mm were cast from each of the six mixes in Table 1. The sensors were installed on the rebar prior to the casting of concrete, as shown in Fig. 2a. Following concrete hardening, the sensor was ready to monitor the corrosion inside the concrete and transmit the data reflecting the actual corrosion situation and the proper functioning of corrosion protection measures.

Facing wind direction, one specimen from each mix was placed in tidal zone (Fig. 2b) and the other in atmospheric zone at KISR site, and the corrosion monitoring was carried out continuously using corrosion sensors installed inside the

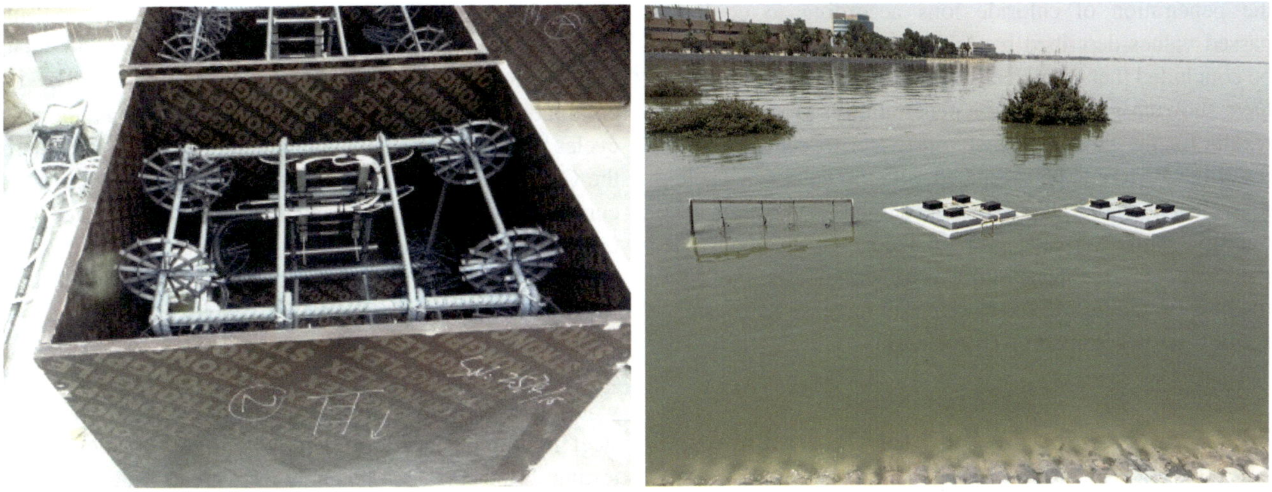

(a) (b)

Fig. 2 Installation and setting up of corrosion sensors (**a**). Concrete specimens with embedded sensors under testing in the tidal zone (**b**)

block. The purpose of installing corrosion sensors in the structural reinforced concrete elements was to monitor parameters important to long-term corrosion monitoring including corrosion potential, corrosion current, concrete resistivity, linear polarization resistance (LPR), concrete relative humidity, and concrete temperature. The complete setup for the installation of corrosion sensors in this study included anode ladder type corrosion sensor, corrosion monitoring data logger (CMDL) for corrosion signals, wireless bridges, and electrical protective enclosures. The data was collected by a wireless connection through a global system for mobile communication (GSM) line and were continuously stored by a data acquisition system available in the laboratory. These measurements provided information to monitor the ingress rate of substances, such as chlorides and carbon dioxide, into the concrete. These measurements over time can be compared to the predictions of the service life models and provide basis for the prediction of corrosion onset at the reinforcement level.

3 Results and Discussions

3.1 Electrical Indication of Concrete's Ability to Resist Chloride Ion Penetration Test (ASTM C1202, 2010) Results

Figure 3 shows the total coulombs passed in six hours across the concrete discs. From Fig. 3, at the low w/c ratio, the MCI concrete demonstrated the highest resistance to the penetration of chloride ion while the OPC concrete showed the lowest resistance. When comparing these results (Fig. 3) to

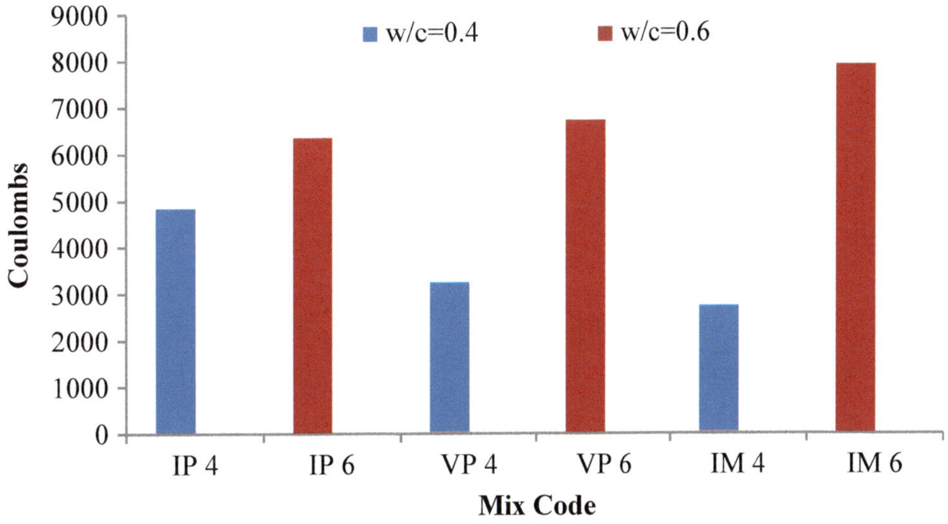

Fig. 3 Number of coulombs passed through each concrete disc after 6 h

the penetration of chloride ions according to coulombs passed values described by ASTM C1202 (Table 2), it can be concluded that all mixes with water to cement ratio of 0.6 have shown high chloride ion penetrability. Moreover, if the results for the three mixtures with water to cement ratio of 0.4 (Fig. 3) were to be compared with the values in Table 2, it is clear that the OPC concrete demonstrated high penetrability, while the SRC and MCI concretes demonstrated moderate chloride penetrability. This indicates the effectiveness of adding MCI to concrete to increase its resistance against the penetration of chloride ions.

3.2 Standard Test Method for Determining Effects of Chemical Admixtures on Corrosion of Embedded Steel Reinforcement in Concrete Exposed to Chloride Environments (ASTM G109, 2013) Results

Figure 4 shows the macro-cell current readings during the 24 cycles since the beginning of the test, where each cycle is equal to four weeks. It demonstrated high macro-cell current readings for the control concrete, SRC concrete, and MCI concrete, all with water to cement ratio of 0.6, reflecting the corrosion activities that were taking place in those three mixes. However, the macro-cell current readings of MCI concrete of the same w/c of 0.6 was demonstrated to be lower than the other two mixes with the same w/c. All mixes with w/c of 0.4 showed a very low macro-cell current with a maximum current value not exceeding 2 μA, which was recorded for the OPC concrete mix of 0.4 w/c. This could highlights the significance of lowering the w/c ratio to obtain stronger and more durable concrete. Figure 5 shows the combined results of half-cell potential readings after 24 cycles using the Great Dane corrosion meter. The results of the Great Dane coincided with the results of the macro-cell current. The corrosion potential of a steel-reinforced concrete specimen, according to ASTM-C876-09, demonstrated the occurrence of corrosion activity in the embedded reinforcing steel bar (ASTM-C876, 2009). ASTM-C876-09 established −200 mV and −350 mV as thresholds. If the corrosion potential was less than −200 mV, then there would be a more than 90% probability that no corrosion of steel reinforcement is present in that part at the time of testing. If the corrosion potential were greater than −200 mV and less than −350 mV, then corrosion of the reinforcing steel at that part would be uncertain. If the corrosion potential were greater than −350 mV, then there would be a more than 90% probability that corrosion is present in that area of steel reinforcement at that time of testing.

From Fig. 5 and according to ASTM C876 criteria, it can be concluded that the OPC concrete and SRC concrete, both with w/c of 0.6, passed the region of probability of 90% that the corrosion of reinforcement is taking place at the 8th cycle, whereas MCI concrete of 0.6 w/c passed this threshold at cycle 19. However, all the concrete mixes with 0.4 w/c fell within the range of 90% probability of no corrosion of steel reinforcement exist, except the plain concrete that passed the threshold to the area where corrosion was uncertain at cycle 9.

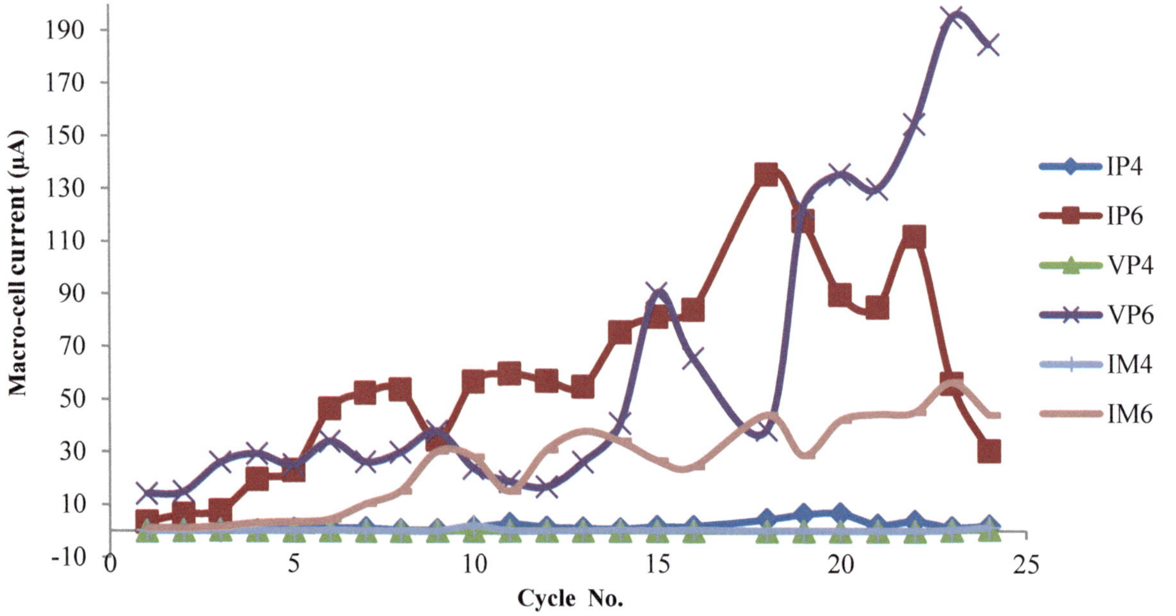

Fig. 4 Macro-cell current readings after ASTM G109 test for six different concrete mixes

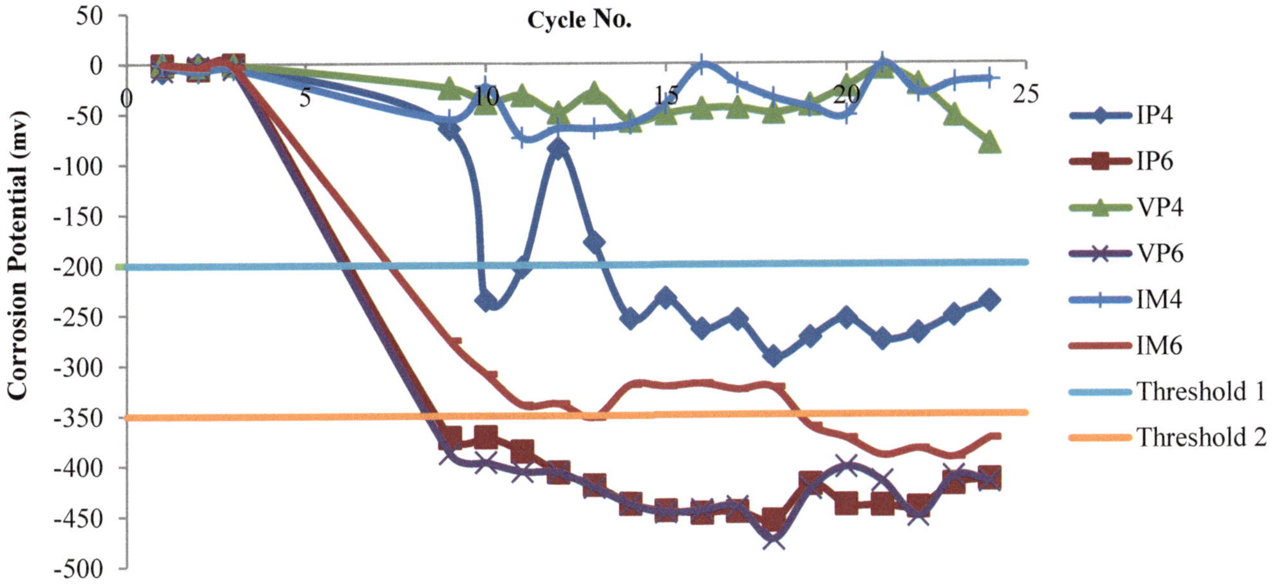

Fig. 5 Half-cell potential readings after ASTM G109 test for six different concrete mixes

3.3 Corrosion Potential Results

Figures 6 and 7 present the corrosion potential results versus time for atmospheric zone specimens and tidal zone specimens respectively. The corrosion potential measurements were recorded using the Great Dane device.

According to ASTM-C876 (2009), results of corrosion potential for concrete specimens in the atmospheric zone (Fig. 6) demonstrate that there was more than 90% probability that no corrosion of steel reinforcement was happening in all atmospheric specimens. Due to the lack of exposure to water, the atmospheric specimens showed no evident signs of corrosion activity occurring. The specimens with a 0.6 w/c ratio had shown to have more negative corrosion potential values thus making it evident that an increase in the w/c would increase the probability of occurrence of corrosion activity. The concrete specimens with the combination of chemical inhibitor and a 0.4 w/c ratio had, comparatively, less negative corrosion potential values, while also showing little to no sign of corrosion activity occurring.

According to ASTM-C876 (2009), results of the corrosion potential for concrete specimens in the tidal zone (Fig. 7) indicate that IM4 concrete specimens (MCI concrete with w/c of 0.4) were the only specimens to have a more than 90% probability that no corrosion activity of the steel reinforcement was present. The corrosion of the steel rebar was uncertain in specimens IP4 and VP4 (OPC and SRC concrete of 0.4 w/c respectively), and there was a more than

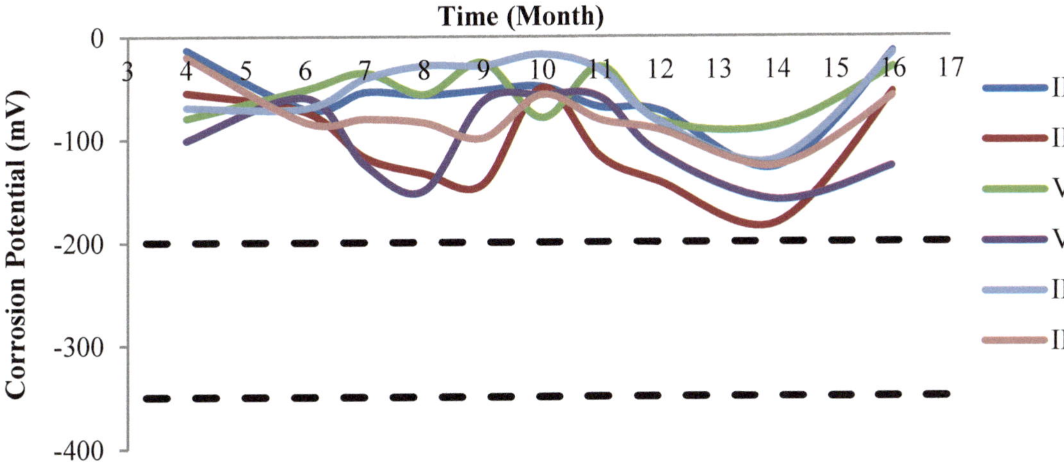

Fig. 6 Corrosion potential results of atmospheric zone concrete specimens

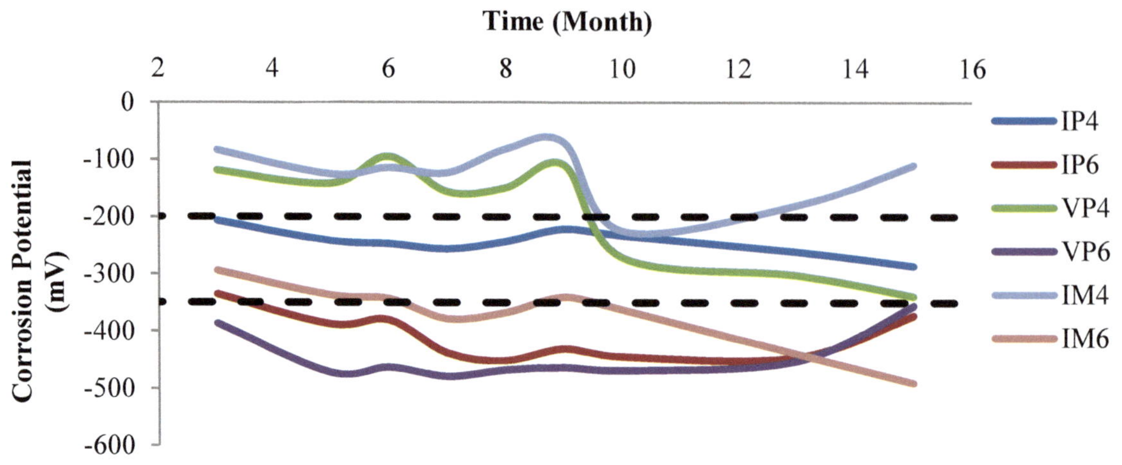

Fig. 7 Corrosion potential results of tidal zone concrete specimens

90% probability that corrosion is taking place in the reinforcing steel of specimens IM6, IP6, and VP6 (concrete specimens with w/c of 0.6), according to Fig. 7. However, as shown in Fig. 7 and according to ASTM-C876 (2009), the curve of MCI concrete of w/c of 0.6 has passed the threshold of 90% probability of corrosion activities presently occurring at the tenth month of testing, while the other two mixes of w/c of 0.6 have passed this threshold at the beginning of the testing. This indicates that the corrosion inhibitor postpones the corrosion onset and hence can extend the lifetime of the structure.

3.4 Chloride Diffusivity Results

Figures 8 and 9 show the chloride diffusivity results for the concrete specimens, of water to cement ratios of 0.6 and 0.4, respectively, submerged in seawater for 1.5 ys. The chloride diffusivity test results demonstrated that concrete specimens with 0.6 w/c (Fig. 8) have a higher chloride content than those with 0.4 w/c (Fig. 9). This would demonstrate that the high values of w/c in concrete mix could increase the amount of chloride diffused into concrete specimens. From Fig. 8, MCI concrete of high w/c presented the lowest chloride content, followed by SRC concrete, and OPC concrete showed the highest chloride content. Also, from Fig. 9 for low w/c ratio, the result of chloride content in MCI concrete was the lowest especially at higher depth, followed by SRC concrete, then OPC concrete, demonstrating the highest chloride content at all depths. However, from Fig. 9, the chloride content at a shallow depth of concrete (10 mm) in MCI concrete of w/c of 0.4 was high.

After modeling the chloride content using Fick's second law by a MATLAB program, the results in Table 3 and Fig. 10 were obtained. Table 3 illustrates the results from modeling for the chloride diffusivity coefficient for each mix after immersing the concrete specimens for 1.5 ys in the sea water. Figure 10a, b, c present the expected chloride profile for different mixes at various depths. The MCI concrete with w/c of 0.4, in Table 3, shows the least chloride diffusivity coefficient among all concrete mixes. Also, from the graphs of chloride concentration, time versus depth for different mixes of concrete after 1.5 year of immersing in seawater in Fig. 10 a, b and c, MCI concrete demonstrates lower chloride ion diffusion than OPC and SRC concrete.

4 Conclusions

Conclusive points of this study are encapsulated in the following:

- The field test results of concrete specimens have shown the effectiveness of adding corrosion inhibitor to concrete to enhance its resistance to the diffusion of chloride ions and corrosion of reinforcing steel.
- The specimens with 0.6 w/c have shown more negative corrosion potential values, thus, making it evident that using higher w/c ratio in concrete mix would increase the susceptibility of corrosion activity occurrence.
- Results of the ASTM C1202 test demonstrated the effectiveness of using the corrosion inhibitor in concrete.
- Chloride ion diffusion was lower in inhibited concrete compared to OPC concrete.
- Corrosion inhibiting admixtures are a second line defense of protection for steel rebars against corrosion and chloride attack and nothing will compensate for proper high performance concrete mix design. At 0.40 w/c ratio, concretes with MCI corrosion inhibiting admixtures have

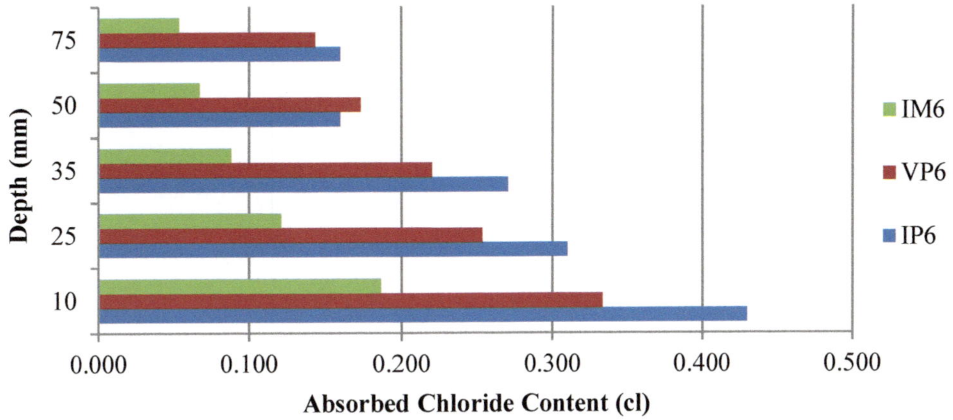

Fig. 8 Chloride diffusivity results of concrete mixes with w/c of 0.6

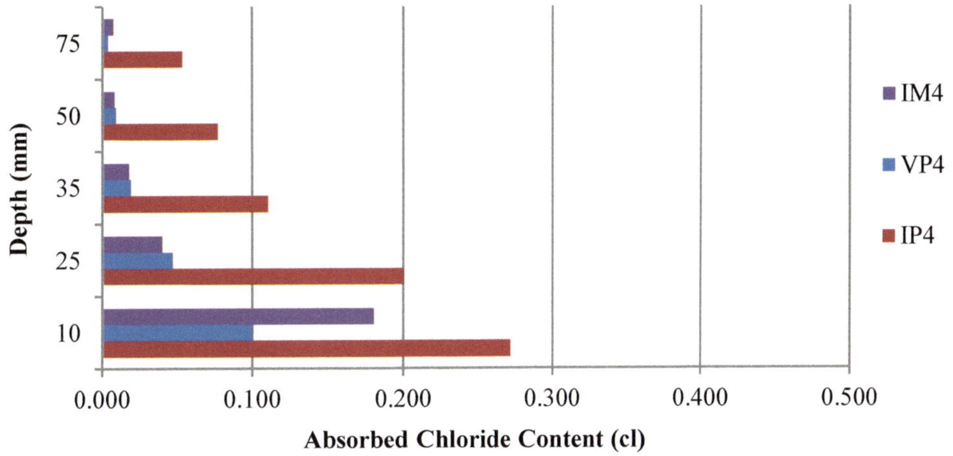

Fig. 9 Chloride diffusivity results of concrete mixes with w/c of 0.4

Table 3 Diffusion coefficient for different concrete mixes after 1.5 ys of exposure

Sample ID	Diffusion coefficient (m^2/s)	Surface chloride content % by wt. of cement	Diffusion coefficient at t = 1 year (m^2/s)
VP4	4.93E−12	0.659	5.49E−12
VP6	4.49E−11	2.25	5.00E−11
IP4	1.09E−11	2.18	1.22E−11
IP6	1.97E−11	3.25	2.19E−11
IM4	2.37E−12	1.47	2.64E−12
IM6	2.04E−11	1.17	2.27E−11

performed better with regard to resisting chloride penetration. Nevertheless, concretes at 0.40 w/c and lower were seen to perform very well in resisting chloride penetration.

- Using SRC had shown slight, and sometimes negligible, improvement to the concrete durability exposed to a chloride-rich environment.
- The establishment of a field research station at KISR's exposure sites provided information that is highly representative of a real life setting on the performance of the chemical inhibitor, highlighting its importance.
- The corrosion health monitoring system has proved to be an effective method for monitoring the corrosion in concrete, especially in marine structures.

Fig. 10 Chloride concentration versus time for different depths in three types of concrete mix with 0.6 w/c ratio after 1.5 year of immersing in Seawater: OPC concrete (**a**), SRC concrete (**b**), and MCI concrete (**c**)

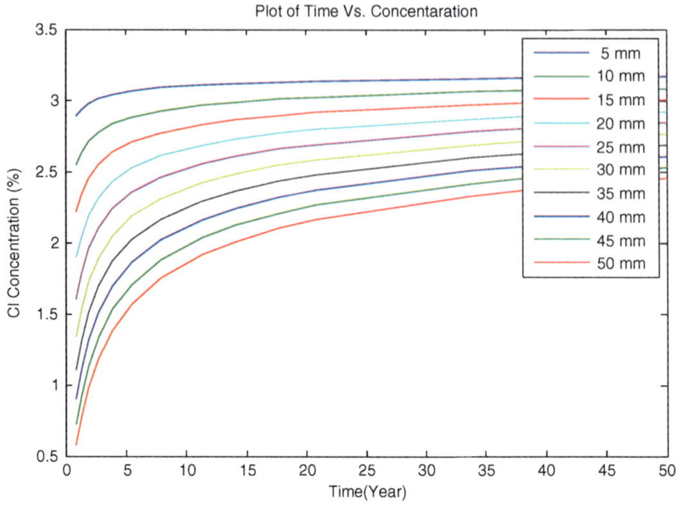

(a)

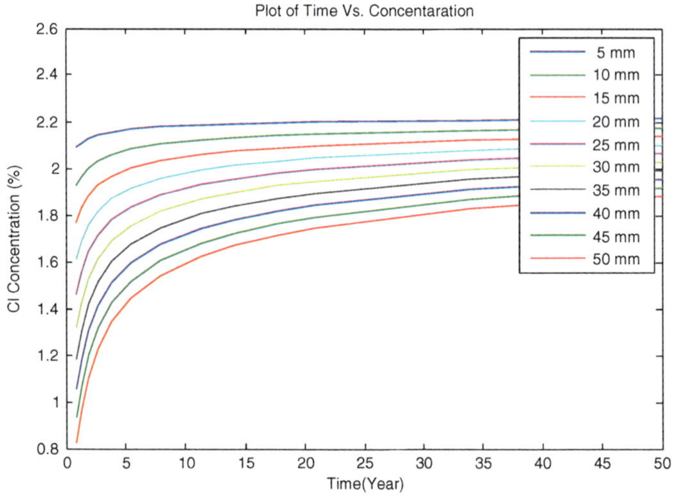

(b)

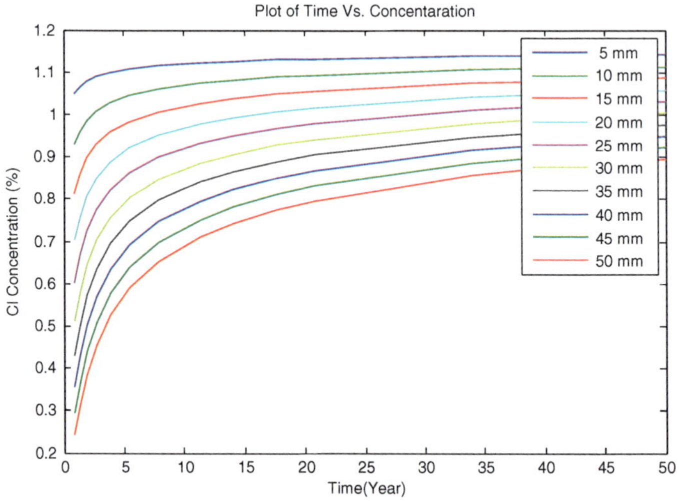

(c)

Acknowledgements The authors are greatly thankful to the Kuwait Oil Company (KOC) and the Kuwait Institute for Scientific Research (KISR) for the financial support to conduct this research project. Also, the authors' gratitude is offered to High Slant Precast Factory for using its casting facilities in this study.

References

Abdul Salam, S., Husain, A., Al-Bahar, S., Chakkamalayath, J., Eid, O., Al-Aibani, A., Al-Fahad, F., Abdulsalam, M., Al-Baghli, H., Hamada, O., Al-Arbeed, A., & Makherjee, T. (2015). *Final report on: Assessment of the durability and corrosion behavior of concrete treated with chemical inhibitors for Kuwait Oil Company (KOC)*. Kuwait Institute for Scientific Research. KISR Report No. 13147. Kuwait.

Al-Bahar, S., & Husain, A. (2015). Impact of aggressive exposure conditions on sustainable durability, strength development and chloride diffusivity of high performance concrete. *Structural Monitoring and Maintenance*, *2*(1), 35–48.

ASTM C876-09. (2009). *Standard test method for corrosion potential of uncoated reinforcing steel in concrete*. American Society for Testing and Materials, United States.

ASTM C1202-10. (2010). *Standard test method of test for electrical indication of concrete's ability to resist chloride ion penetration*. American Society for Testing and Materials, United States.

ASTM G109-7. (2013). *Standard test method for determining effects of chemical admixtures on corrosion of embedded steel reinforcement in concrete exposed to chloride environments*. American Society for Testing and Materials, United States.

Attiogbe, E. K., Al-Bahar, S., & Kamal, H. (1994a). *Evaluation of a concrete floor system, Safir International Hotel*. Kuwait Institute for Scientific Research, Report No. KISR 4535, Kuwait.

Attiogbe, E. K., Al-Bahar, S., Kamal, H., & Abduljaleel, A. (1994b). *Evaluation of a seawater exchanger structure at Shuaiba Refinery*. Kuwait Institute for Scientific Research, Report No. KISR 4575, Kuwait.

Apostolopoulos, C. A., Demis, S., & Papadakis, V. G. (2013). Chloride-induced corrosion of steel reinforcement – mechanical performance and pit depth analysis. *Construction and Building Materials, 38*, 139–146.

Franciŝković, J., Mikŝić, B., Rogan, I., & Tomičić, M. (2006). Protection and repair of reinforced concrete structures by means of MCI-inhibitors and corrosion protective materials. In *Structural Engineering Conferences: International Conference on Bridges*, Dubrovalik, Croatia, May 21–24, 2006 (pp. 553–560).

Holloway, L., Nairn, K., & Forsyth, M. (2004). Concentration monitoring and performance of migratory corrosion inhibitor in steel-reinforced concrete. *Cement and Concrete Research, 34*, 1435–1440.

Martin-Perez, B., Zibara, H., Hooten, R. D., & Thomas, M. D. A. (2000). A study of the effect of chloride binding on service life predictions. *Cement and Concrete Research, 30*, 1215–1223.

Morris, W., & Vazque, M. (2002). A migrating corrosion inhibitor evaluated in concrete containing various contents of admixed chlorides. *Cement and Concrete Research, 32*, 259–267.

Thoft-Christensen, P. (2002). Deterioration of concrete structures. In *The 1st International Conference on Bridge Maintenance, Safety and Management*, Barcelona, 14–17 July 2002.

Smartly Designed Playgrounds to Renew Energy

Dalia Wagih A. ElHalim

Abstract

Lack of energy is a major problem in many countries. Researchers found that fast movement that resulted from moving objects can be used in producing energy that in turn can be converted to electricity. Kids' movement whether playing, jumping, kicking, or running is considered a valuable source in producing energy. The present paper aims to prove that kids' movement can utilized as a source to produce electricity. To achieve this objective, the paper is divided into two parts: theoretical and analytical. The theoretical part explores the different renewable sources that can produce energy. Hence, it focuses on playgrounds because of its importance to kids, the different ways by which they can be used in renewable energy, and how to establish this concept in both the private and public fields. This part ends by selecting the most efficient devices that produce energy from kids' movements in playgrounds. Moving to the analytical part, the paper analyzes some selected playgrounds for kids and suggests a proposal in developing them to produce electricity using the renewable energy. Discussion of the findings aims to point out the difference between the usage of energy before and after the suggested proposals.

Keywords

Kids' playgrounds • Renewable energy • Smart open spaces • Kinetic energy

D. W. A. ElHalim (✉)
Mataria Faculty of Engineering, Architectural Department, Helwan University, Cairo, Egypt
e-mail: Dalia_Said@m-eng.helwan.edu.eg

1 Introduction

Electricity generation supplies carbon dioxide that has negative effects on the environment. Therefore, clean energy is the solution as it produces energy without any harmful effects (David, 2017). There are three keys for the future energy systems that can be summarized as follows: low energy usage by applying the transformation technologies, low carbon emissions by producing energy, and low transport distance by the usage of the local applicable energy. Landscape can be a source to produce energy by the usage of planting areas to be employed, such as biomass energy that can be used to generate electricity. Every region has to find out the local opportunities which can be deployed to produce different types of clean energy: solar, wind, biomass, etc. (Blaschke et al., 2013). Moreover, kinetic energy is a valuable source to generate energy, by converting the mechanical energy to an electrical one which can be applied by different devices. Open spaces have a great role in landscape design; they are considered the main theater where users can enjoy their time. Kids are the most active actors on the stage of landscape projects; they always move by running and jumping. These continuous rapid movements produce a great amount of energy that is unfortunately wasted. This dilemma is the main spark that inspires the concept of the research to find the way to utilize this energy especially in public places. These open places, such as clubs, usually receive large numbers of kids whose kinetic energy can be converted into electricity that can aid in providing different places with the needed electric power.

2 Methodology

The research hypothesis is concerned with the idea of producing electricity through the usage of the kinetic energy of kids, as clean renewable energy, that will contribute in solving the dilemma of pollution and lack of energy even in

F. Rosso et al. (eds.), *Advances in Architecture, Engineering and Technology*, Advances in Science, Technology & Innovation, https://doi.org/10.1007/978-3-030-86913-7_18

poor countries. Thus, the research introduces many sources of renewable energy derived from kinetic energy and pinpoints the importance of establishing the idea of renewable energy by the cooperation of governments, experts, and citizens under restricted regulations. As playgrounds are considered the most attractive areas for kids, the research studies different ways to use their kinetic energy to produce electricity. The research analyzes different international playgrounds as case studies that produce electricity using the kinetic energy of kids in different ways. Then the research chooses one of these ways to apply as a national case study to prove the benefit of using the kinetic energy of kids in playgrounds to produce a different form of energy.

2.1 Renewable Energy

Energy sources can be classified into nonrenewable and renewable; the nonrenewable are those produced from resources that can be depleted such as fossil fuels, and the renewable are those produced from resources that are continuously available as they are produced from nature such as biofuel produced from food crops. Renewable types are preferable for their limited negative impact on the environment; they are divided into five main types: solar energy, wind power, hydroelectricity, geothermal energy, and biomass. Solar energy comes from the sun and is considered the most practical renewable energy source for homeowners, its cost is dropping that can save money, and it does not pollute or release fossil fuels which means that it has a less environmental impact than other sources. Wind power can be attained by turbines to generate electricity; it is suitable for the sites that are characterized by high-speed wind. Hydroelectricity is the source that can produce energy from moving water; it can be made by the usage of turbines that spin to produce electricity, but it needs adequate water speed like that in waterfalls and dams. However, this source affects the environment negatively as it can change water levels, currents, and migration paths for fish and freshwater life. Geothermal energy can be obtained from the massive energy source of the earth to produce electricity instead of being wasted; this source can be realized by using the steam of the heated water to spin turbines to generate electricity. Moreover, it can directly provide heating or cooling to buildings by pumping a fluid below the ground surface to be heated or cooled. Biomass is the source that can be produced from any living organic matter like plants or animals; it is a renewable energy source as plants can regrow quickly by the effect of the sun (Marsh, 2018). All of these sources except biomass are kinetic sources as they are based on motion.

The idea of renewable energy has to be established by both the awareness of local governments and the cooperation of people, whether residents or users. The awareness of governments is of less importance than people, because if residents are included in taking care of a project in their neighborhood, the project can fulfill its goals, and its success will have a direct impact on them. Moreover, they will become more aware of their actions toward energy. Projects of renewable energy have to be directed to achieve certain goals. Some researches argue that renewable energy should be used to support some services such as education. Others have focused on the interaction between energy production and new technologies in certain regions (Benedeka et al., 2018).

Engaging communities in encouraging renewable energy is regarded as highly important because of its various benefits. Meetings and discussions with the public, for example, save time and effort of decision-makers to achieve their target by collecting the experiences of different people especially who live in the region of development. Many ideas can be gained from the discussion with people—especially elders—by knowing the advantages and disadvantages of different types of energy in the past and what can be reached in the future, in addition to finding out both their positive or negative visions and threats toward different projects, and ways to produce energy. Non-profit organizations can propagate the idea of renewable energy among people through different advertisements, meetings, questionnaires, conferences, etc., and this can be achieved with the aid of volunteers and experts in different fields. Moreover, governments have a great role in encouraging people to use renewable energy by many means such as providing people with experiences, decreasing taxes on the projects that use renewable energy, and buying stored energy from people by suitable prices. If residents find that applying the concept of renewable energy will have a positive impact on their life, they will willingly share and become keen enough to establish and preserve these types of projects (David, 2017).

Regulators have to establish regulations and policies for renewable energy (RE) by measuring the efficiency of any project using it and the efficiency of the produced energy (EE). The cost of renewable energy projects faces difficulty in measuring its efficiency and how to deal with these projects in the programs of tax. This problem forces regulators and governments to analyze any renewable energy project in three phases: the cost of using renewable energy, the efficiency of using it, the taxes that can be reduced from such projects that make it difficult to give the license for these types of projects, and results in reducing its spread (Bjork et al., 2011).

Each country has its factors and goals to achieve the efficiency of energy. The agencies established four keys for the efficiency policies that have to be recognized by governments and regulators which are energy security to secure the projects, the energy efficiency to perform its goals, economic development that encourages investing in these types of projects, climate change that should be considered carefully for the future, and finding out solutions for these changes to deal with it in different projects, and finally public health by reducing different types of pollution (FAQs, 2018).

There are many barriers that can obstruct the performance of energy efficiency projects the most important of which are market barriers as the cost of developing projects is higher than the investment in energy savings, financial barriers as the primary cost prevents investors from financing such projects, awareness barriers because people and investors lack enough awareness to the benefits of such projects, regulatory barriers for the energy tariffs discourage the investors, and finally technical barriers as there is a lack of suitable technologies for local conditions (FAQs, 2018). These barriers have to be removed to encourage the concept of spreading renewable energy either in public or private projects.

2.2 Playgrounds

The playground is an invaluable field for establishing the concept of renewable energy; therefore, it is important to study its needs in design. To design a playground, the designer has to fulfill the needs of users that can differ according to age and gender. According to researches, playgrounds for kids need to fulfill eight requirements: safety, free movement, running, playing, touching, jumping, pushing and pulling, and hearing capability (Amouzegar, 2010). These requirements can be smartly designed for using renewable energy as follows:

- Safety: the playground has to be safe for kids, first, in its design by creating safe paths in its finishing, illumination, clearness, and suitable portions for the target age and gender. Second, it has to be safe in the materials and surfaces used for different objects such as using soft toys. In the case of producing energy, the designer has to place the batteries that store energy in safe places away from kids.
- Free movement: kids have to feel free in space while playing to enjoy themselves. This can be reached by designing open spaces that include joyful toys like race cars. The movement of the cars on smartly designed paths that include spots can be used to store energy as well.
- Running: kids like to run and play their games. Their running can be traced to find out their movement that can be acquired by means of spots on the floor that collect their energy to be stored in special batteries.
- Playing: swings, slides, teeter-totter, and other means of joyful games for kids can be used to store energy through smartly designed games to benefit from them in producing electrical energy.
- Touching: kids especially young ones learn about the world from the sense of touch. This can be a source of energy by designing interactive smart walls and toys that convert their rapidly touching and playing activities to energy. This can be achieved by designing special toys and walls that attract kids to generae energy whether through slow or rapid touches, and at the same time kids can enjoy these playthings by different means such as using appealing sounds or attractive lights when the child uses them.
- Jumping: kids always jump, especially the hyperactive ones, their continuous jumping up and down can be stored through smartly designed trampoline, floors with spots that illuminate with attractive colors or lovely cartoon faces. Their mechanical movement can be converted into electrical energy to be stored for later use.
- Pushing and pulling: this can be found in different toys and games like the see-saw that can be connected to special batteries which capture their action to be converted to energy.
- Hearing capability: this can be fulfilled by using different instruments like musical ones. The vibration produced from these instruments can be captured and stored in special batteries.

The designer has to encourage kids to use smartly designed toys to gain the maximum benefit from kids' energy that can be stored in batteries and converted to electrical energy for later use. This can be achieved by designing attractive shapes, colors, and sounds that encourage kids to use the different parts of the project. For example, the child can be encouraged for rapid and continuous jumping on smart tiles by the appearance of different lovable characters, colors, and sounds in every jump. Meanwhile, these toys must be safe for kids to use them. Therefore, the designer has to select suitable materials and designs that provide safety for kids without any threats that could hurt kids while playing.

2.3 Examples of Smart Equipment in Playgrounds

Kids' energy can be gathered during their playing activities to mix fun with a great benefit by producing electricity which can be used for different purposes. There are different equipment that converts the mechanical energy from the motion of playing into electrical energy, below are some examples:

- The Sustainable Light Swing: designed by Guilherme Pena Costa and Ingrid Gabor. It is a lighted swing using LEDs that attracts people at night (Fig. 1). Its idea is to generate energy from the swinging action by which the energy is captured to provide electricity for public places (Costa, 2012).

- The Piezoelectric Playground: designed by Margot Krasojevic for a park in Belgrade, Serbia. It is a temporary structure that produces energy from the movement and vibration on it whether from kids or adults or even from the vibration that is caused by trucks rattling or airplanes overhead (Krasojevic, 2012).
- Kinetic Playground: it is designed by three industrial designers, Andrew Simoeni, Joel Lim, and Funfere Koroye. They designed some attractive objects that provide not only fun to play but also produce energy. They designed a tire-shaped dynamo (Fig. 2) that can be connected to the equipment to produce about 31.5 W/h of play. They designed five pieces of equipment: the see-saw, the skipping rope, the swing, the teeter-totter, and the roundabout. The see-saw consists of the rounded tire with a rope that is held from two sides to provide swinging of two kids in front of each other (Fig. 3a). The skipping rope is modulated to pass the rope by the tire connected to a battery that stores the energy from the movement of the rope (Fig. 3b). The swing is connected to two tires by which passes the movement of the robe of the swing that allows the tire to capture energy (Fig. 3c). The teeter-tooter in which the axis of movement passes by the tire as if the movement is faster, the produced energy is increased (Fig. 3d). The roundabout was designed as a simple tire on which the kids can move directly whether slow movement or fast one like jumping (Fig. 3e) (Simeoni et al., 2012).

Giraffe Street Lamp: it is a mix of a swing and streetlight. Its idea is to capture the kinetic energy produced from the swing movement to power a LED light (Fig. 4). Moreover, it uses an elevated solar panel to soak up the solar variety storing in a battery for later use (Seth, 2012).

- Trains: scientific research found that trains can be good sources for producing clean energy that can be captured through wind, solar, and wastes. This theory can be applied to kids' area using rapidly moving toys such as toy trains. As the train moves rapidly forward, it produces wind flow in the opposite direction, and this wind can be directed to wind turbines that can generate energy. Moreover, the roof surface of the train can be designed to add solar cells to generate energy (Prasanth & Sudheshnan, 2011). But in the case of kids' toys, the theory of reusing wastes of users cannot be applied because there is no wastes resulted from the toy train for it does not include bathrooms as the public train.

Fig. 1 Sustainable light swing

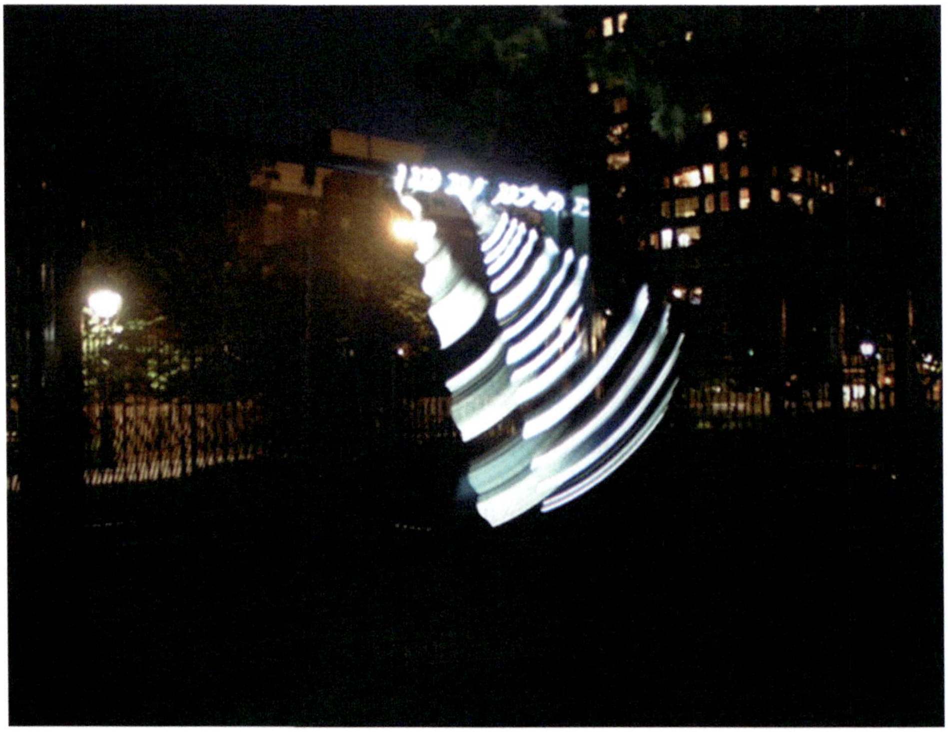

Fig. 2 The dynamo tire

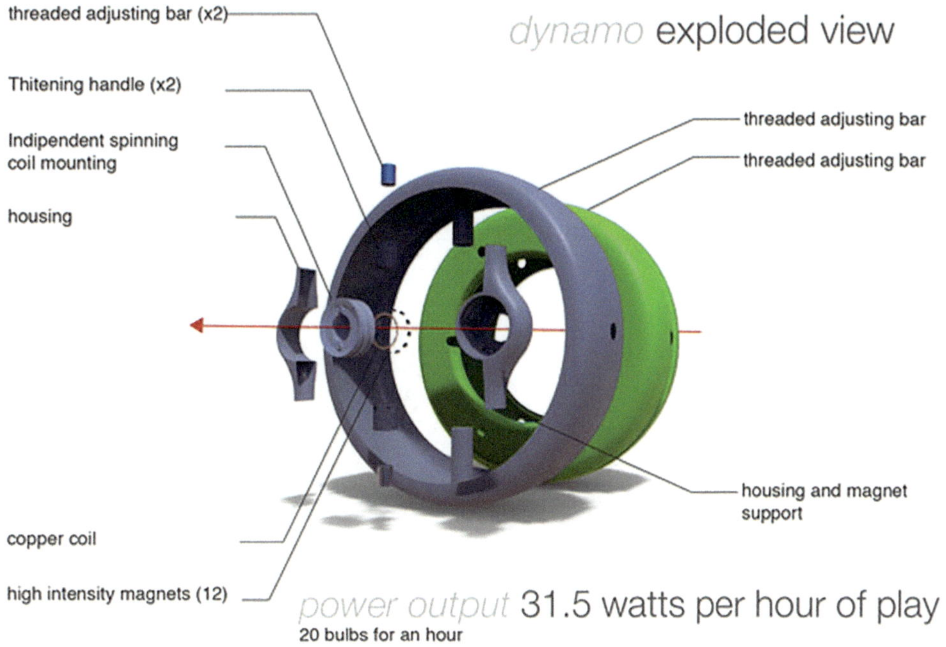

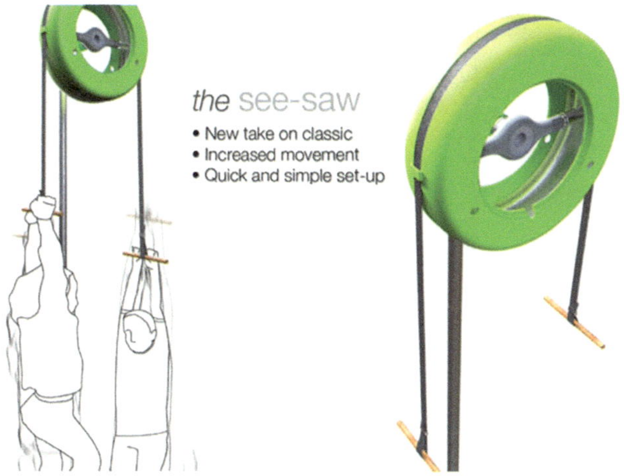

Fig. 3a The see-saw

- Merry-go-round equipment: in Ghana, children spend a long time outdoor from school to jobs as they work after school either in farms or houses to finish certain chores. Since 2008, a non-profit organization that is called Empower Playgrounds Inc. (EPI) has established many electricity-generators that can capture the energy produced from children to be stored and generate electricity. These generators are established in a piece of equipment called merry-go-rounds (Fig. 5) that was located in many playgrounds (Empower, 2009). This merry-go-round equipment was used to provide children not only amusement and fun but also energy to charge batteries that can be used for illuminating houses to enable children to do their home works after returning home at night every day (Danielle, 2017; Empower, 2009).

2.4 Encourage Kids to Energetic Play

To capture the kinetic energy of kids, their continuous play must be encouraged especially in areas that use the idea of producing energy from their movement. Kids always move, and their physical activity is very important for their health as it improves their bones, skeletal properties, psychological, and cognitive health. In Canada and Australia, a recommendation was established which mentioned that children from 2–4 years old should spend minimum 180 min daily in playing, while those who are more than 3 years can spend about 120 min daily. Researchers found that children spend about 14% of their outdoor activities in energetic play that is very important to improve both their physical and mental health. Finally, the 24 h Movement Guidelines in Canada and the United States of America suggested that preschool children must have minimum of 40 min/day as outdoor energetic play. While in South Carolina, there is a policy for childcare centers which states that each child must spend 90–120 min through 2–3 intervals daily in outdoor energetic play (Driediger et al., 2018). From the previous instructions, we find that the minimum time for each child is 40 min/day in energetic play that has to be encouraged by smartly designed outdoor spaces to be used in producing energy. If we use the smart toys that can capture energy such as the

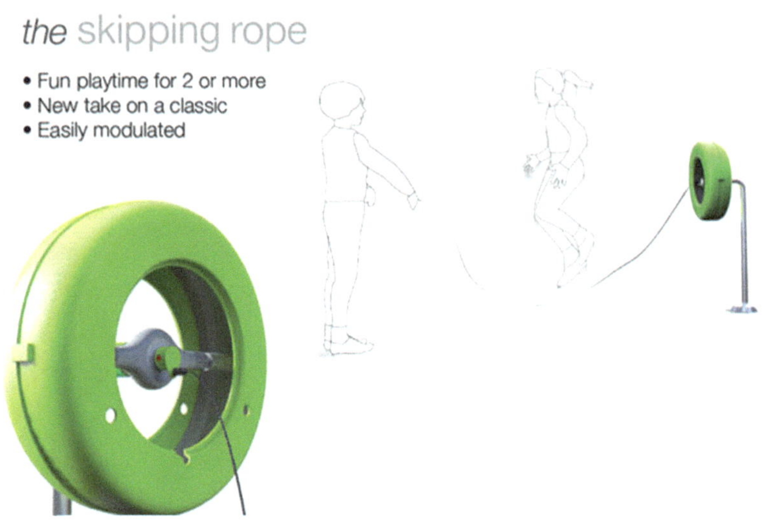

Fig. 3b The skipping robe

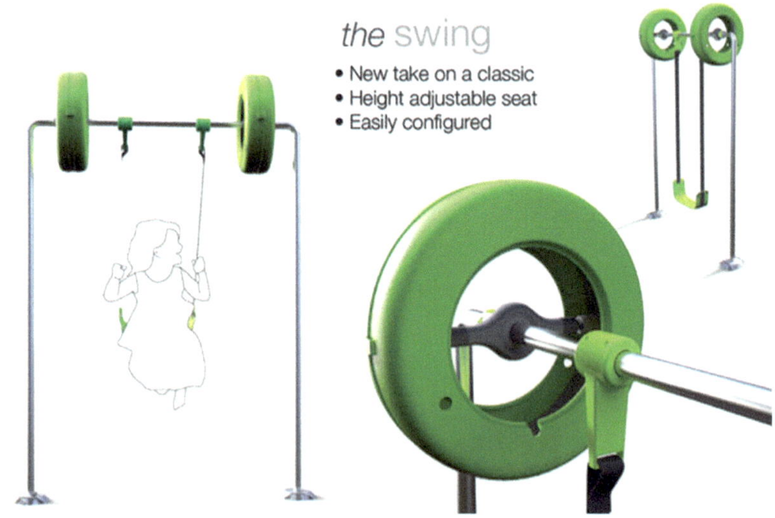

Fig. 3c The swing

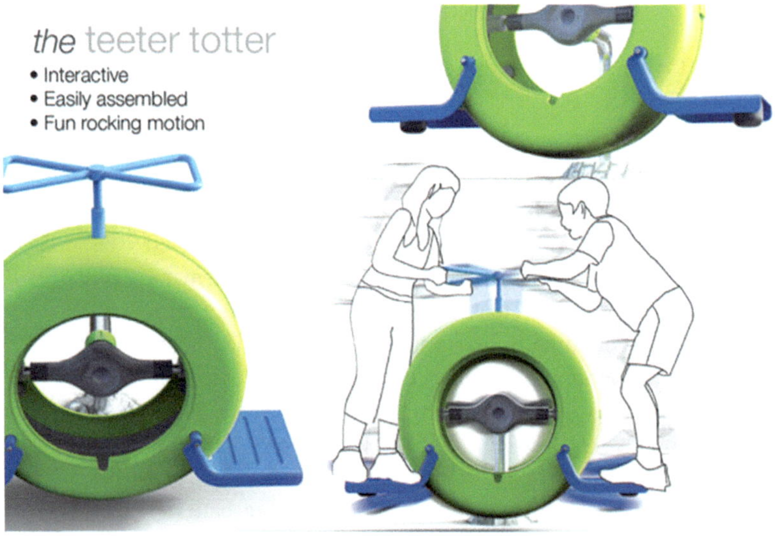

Fig. 3d The teeter-totter

Fig. 3e The roundabout

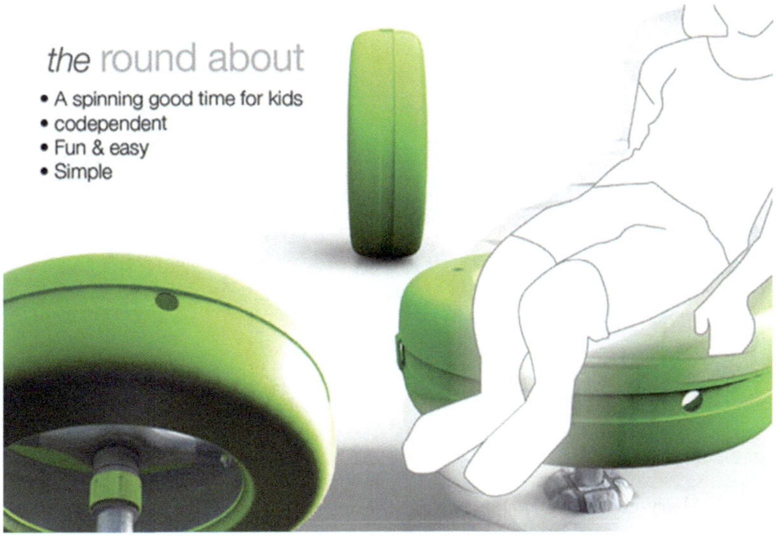

Fig. 4 Giraffe street lamp

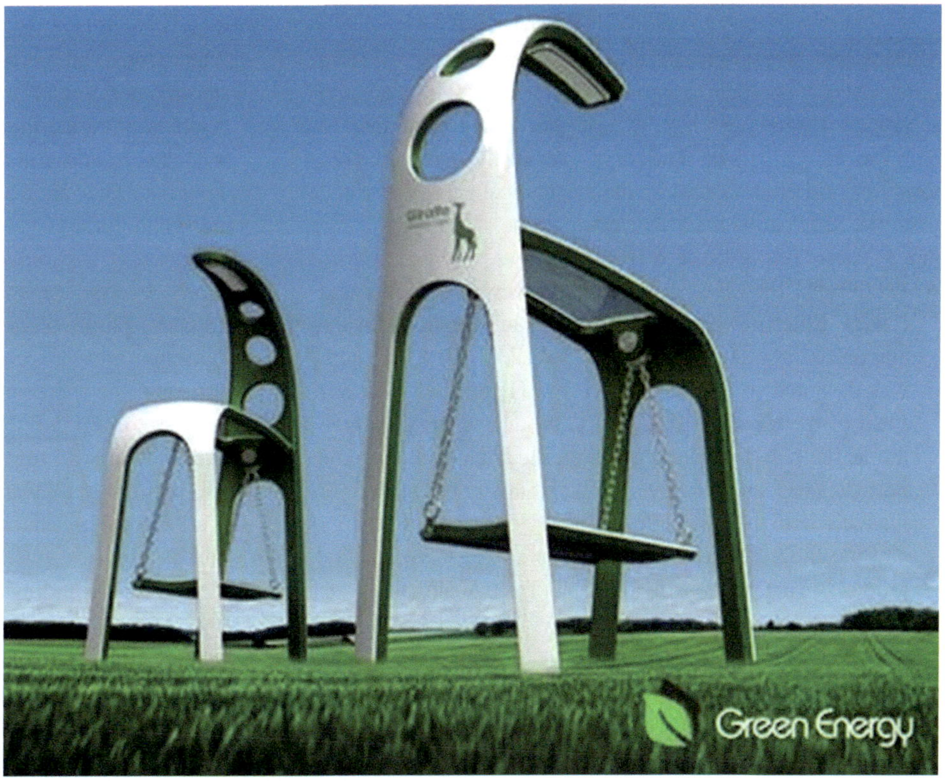

dynamo tire (mentioned in the project of kinetic playground), we can benefit from about 31.5 W of energy produced per hour. If one child plays 40 min/day, in other words 0.67 h/day, then the minimum energy produced from one child daily is as follows:

$$\textit{The minimum energy from one child} = 31.5 \times 0.67 = 21\ \text{W/day}$$

3 Analytical Study

Many projects established the idea of producing energy from movement. This study analyzes some of these projects to find out different ways in producing energy using kinetic energy.

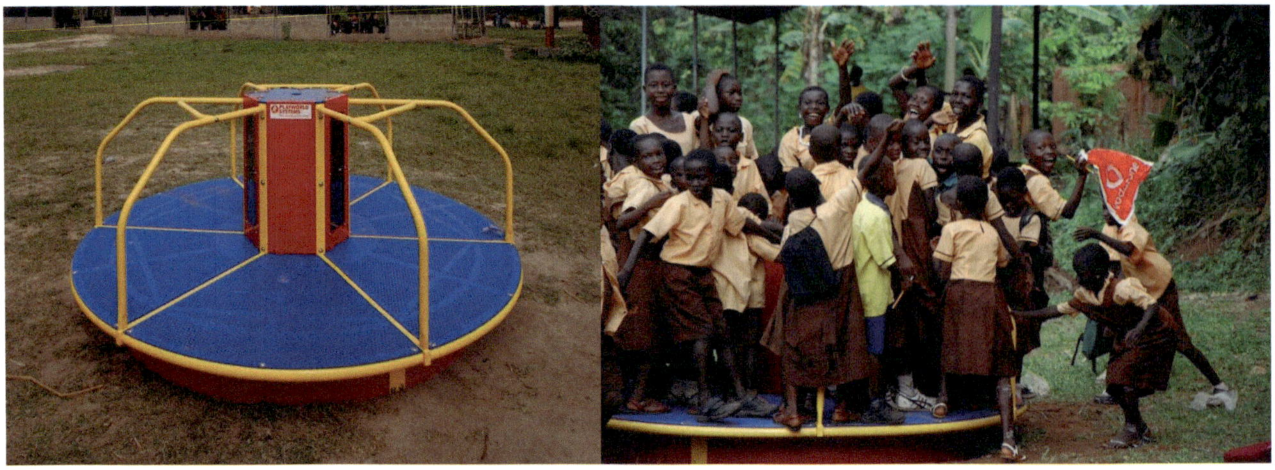

Fig. 5 Merry-go-rounds equipment

3.1 International Case Studies

Many projects established the renewable energy to store kids' energy inorder to provide electricity. The research selects two parks, through which the designers aimed to emphasize the idea of producing energy by using playing toys for children, and convert the mechanical energy resulted from the kids' movement to electrical one.

- Park (1): In Brazil, a project called Renewable Energy and Energy Efficiency Park was designed to take place in the city of Tucurui, the state of Para`, Brazil, in 2013. The park was designed to be located near the Hydroelectric Power Plant of the city. It is designed to produce energy through the active play of children. Children can enjoy their play, besides, producing energy. The park aims to be opened for kids in school to teach them the benefit of renewable energy and the different technical ways that can be used for this purpose. The park is divided into three main areas: an external area with a car race, another for playing toys, and a third as a small lake (Fig. 6) (Martins et al., 2014).
  - The external area is designed to be a place for learning about energy and training, aiming to represent the sources of renewable energy such as solar and wind systems, in addition to emphasize this concept and its benefits. This area also includes a speedway that is designed for RC car racers with batteries which are charged by a photovoltaic system to charge cars and store energy (Fig. 7). The photovoltaic system includes photovoltaic panels that are used to recharge batteries and are connected to the LCD screen to indicate the percentage of the charged battery to recharge it through these panels. At the end of the first area, there is a ramp to enable handicaps to move to the second area (Martins et al., 2014).
  - The second area is a building called Efficient House that is used for dwelling. The building is designed to depend on providing clean energy through bioclimatic

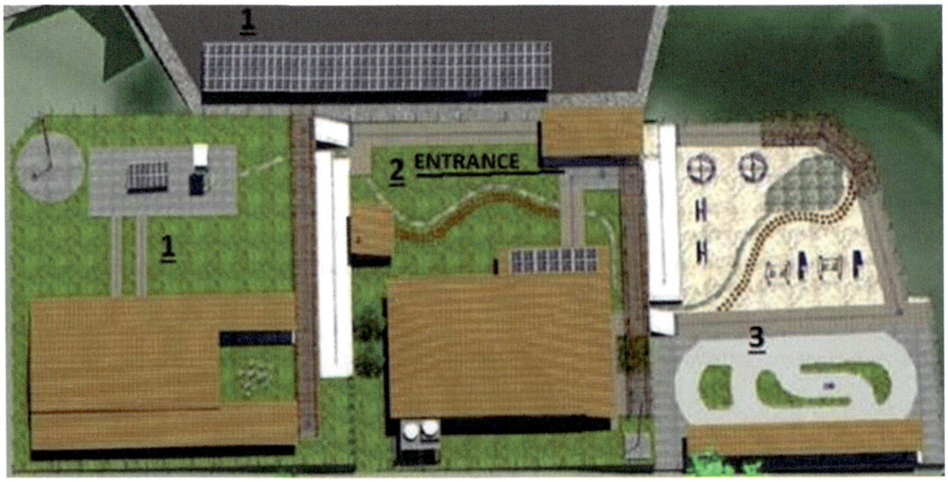

Fig. 6 Layout for the park

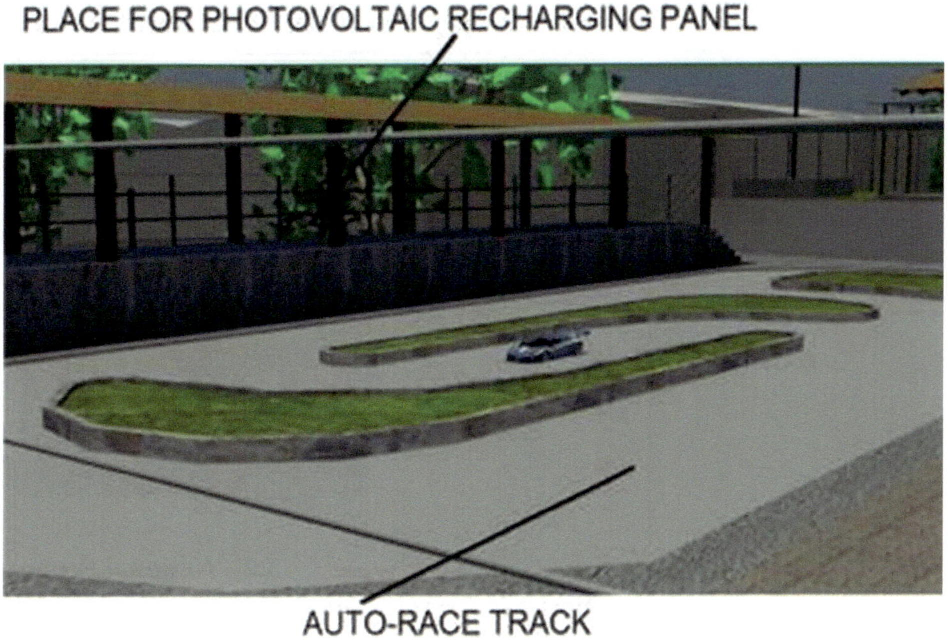

Fig. 7 The auto-race track

architecture, amazons typology, and energy efficiency. The building uses a photovoltaic generator, and solar water energy that is installed on the ceiling of the house (Martins et al., 2014).

- The third area is the playground that includes toys such as slide, swing, teeter-totter, and carousel which are connected to batteries to store the energy that is produced from the movement of the children (Figs. 8a, 8b, 8c, 8d and 9). The toys have electromechanical energy conversion system that can multiply the speed to generate more power, DC generators, and electronic system. Children while playing find that the more the speed of the toy, the more power that can be stored. This action reinforces the concept of competition between children to win more produced energy than his/her friends. Moreover, there is a small lake that includes toys which are driven by photovoltaic energy to allow the movement of toys through the lake. The lake has a photovoltaic pumping system that allows the circulation of toys through the lake using gravitational means. These parts of toys are connected to lead-acid batteries, converters, and DC/AC inverters (Martins et al., 2014). This system allows producing energy that can be used in many services for the users of the park such as recharging electronic portable device like tablets and cell phones.

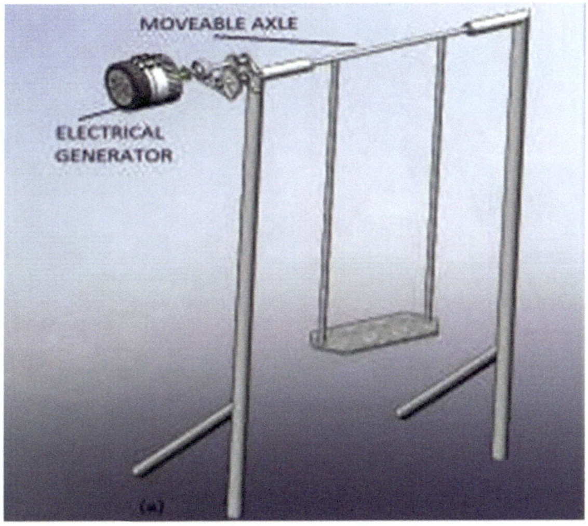

Fig. 8a The swing

- Park (2): A project called Nature Energy Park was introduced to teach children the concept of renewable energy using toys in playgrounds. It is considered as an experimental playground, through which the child can recognize the different types of renewable energy while playing. The designers are You Song Young, Jin-Soo Yeon, Ahn Ho-Sang, and Lee Sung Jae. They divided the toy into five parts, three of which are designed to generate energy which are a pinwheel cycle that generates electricity to illuminate a traffic sign, a see-saw that generates power used to turn a waterwheel, and a pedal that used to move a flying airplane. Moreover, there are three another parts to teach children some scientific principles which are videophone periscope, a kite of Benjamin Franklin, and a pinhole camera (Fig. 10) (Young et al., 2018).

Fig. 8b The teeter-tooter

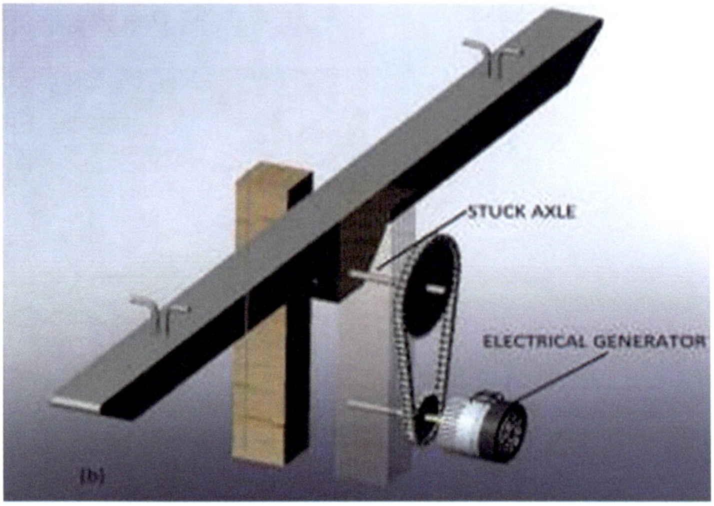

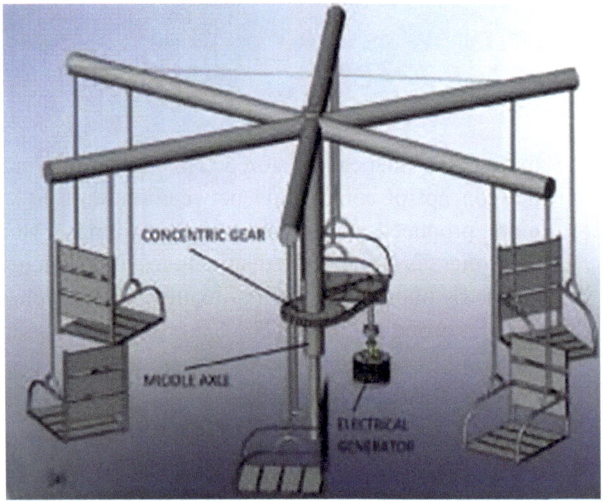

Fig. 8c The carousel

3.2 National Proposed Case Study

The study proposes a future development for the Egyptian clubs as they are considered as beneficial source for producing energy through redesigning the playgrounds of kids. The study selected one of the famous social clubs in Egypt to apply the concept of renewable energy on its kids' gardens aiming to measure the benefits of this development in providing some of the needed energy to the club.

The Egyptian Shooting Club in Douqi, Giza, was built since 1939. The club passed by many stages of development, the most famous one started in 1992 by the architect Hussein Sabour. Nowadays, it is passing with new development for the kids' area by the United Consultants office, architects Esam Abaza and Sahar Attia.

The previous kids' area consisted of many toys that could be developed to save the kinetic energy of kids to produce energy. But the club decided to redesign this area

Fig. 8d The slide

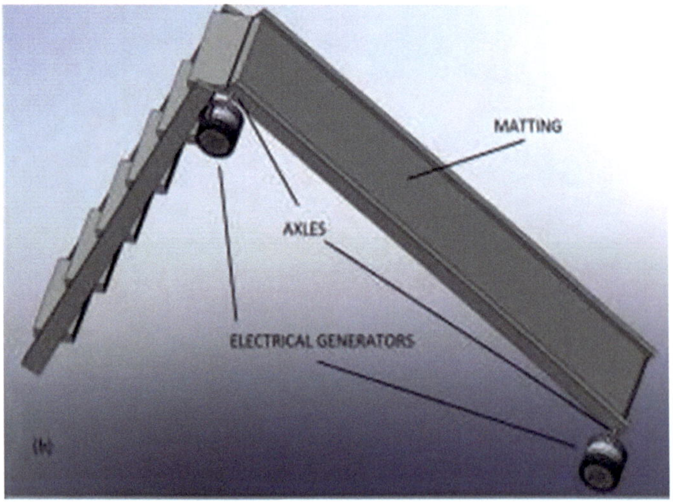

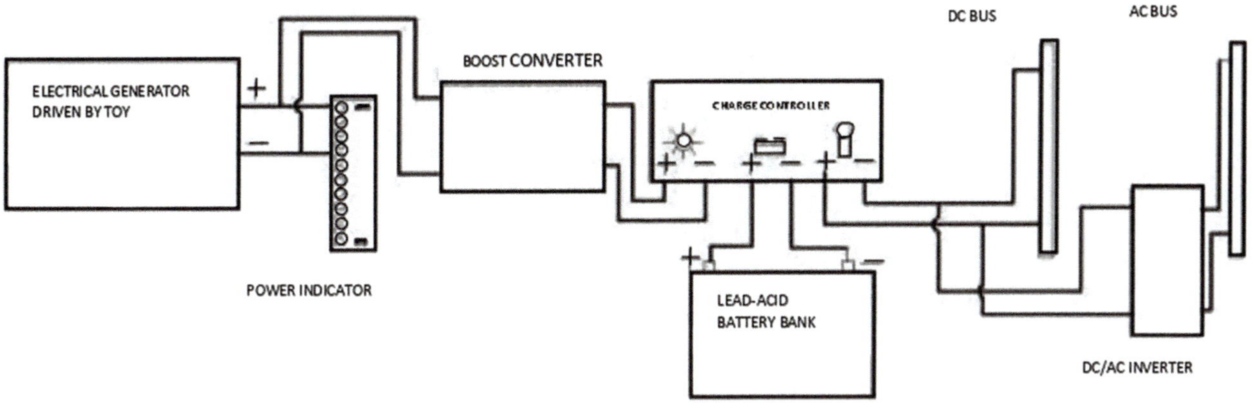

Fig. 9 Electrical system for toys

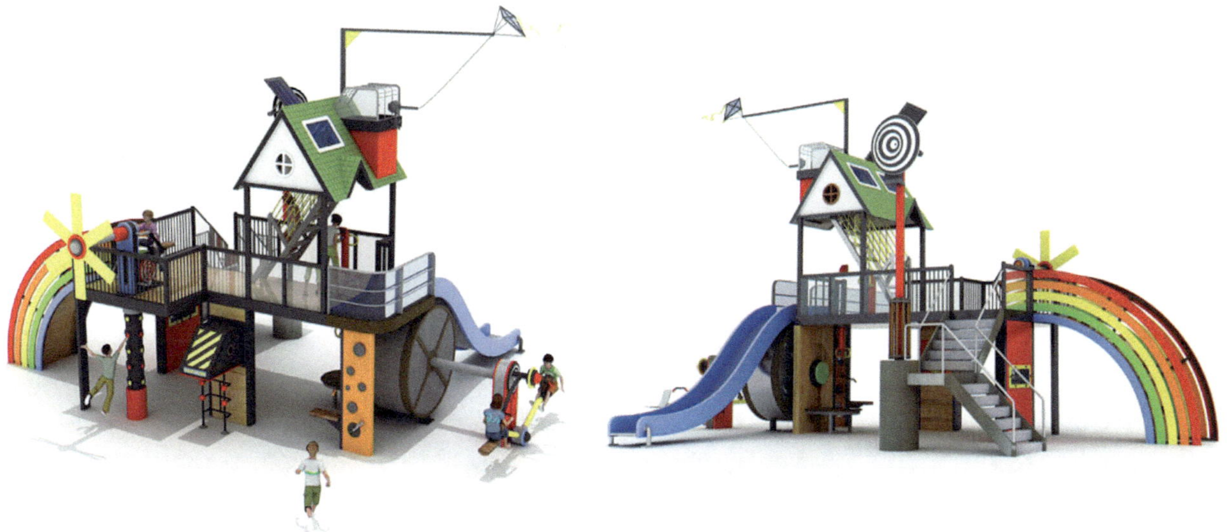

Fig. 10 Nature Energy Park

and replace the toys with other ones. In the following, the research demonstrates the two cases of the kids' area: the previous and the new design. Through representing the two cases, the research calculates the quantity of energy that can be captured in each design, if the club benefits from the kids' movement, and how it can be a great aid in providing electricity and save money.

Previously, the kids' area included many toys: swings, see-saws, sliding, rolling game, and small wooden houses. Moreover, there was a large area of about 15 × 15 m for cycles and surfing activities. From the actual visit to the club, I have calculated the numbers of each type of toys that can be summarized in the following:

23 swings (Fig. 11), each for one child; this means *23* children.

6 see-saws (Fig. 12); one of which was a big one that took a shape of a boat its capacity was about *12* children, one see-saw for *3* children, four see-saws for 2 children each; that means 4 × 2 = *8* child, two see-saws for 4 children each; that means 4 × 2 = *8* child.

12 sliding (Fig. 13); each passed from a house or a bridge; that means *12* children.

One rolling game for about *4* children.

5 small houses (Fig. 14) for 4 children each; that means 5 × 4 = *20* child.

One large area (Fig. 15) for bikes, running and surfing of area about 225 m2. It occupies minimum *10* children.

Most of the time especially in vacations, the toys become completely occupied by children. By a simple calculation if the kids' toys are full then:

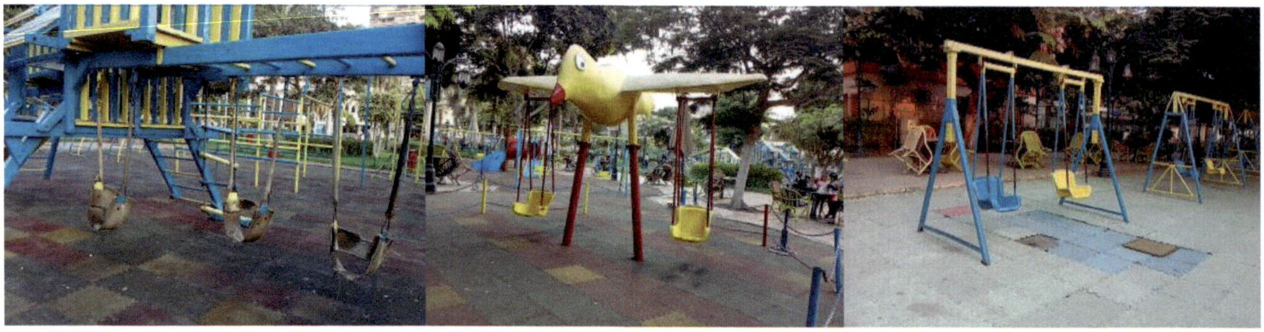

Fig. 11 The swings of different types

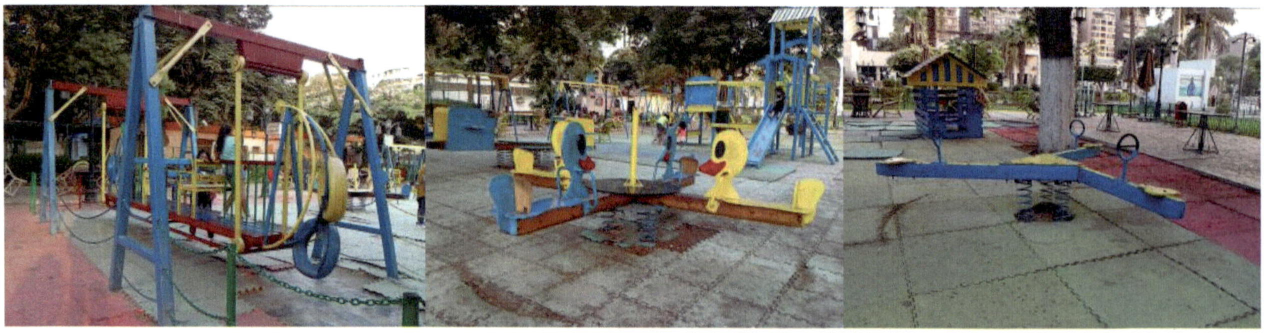

Fig. 12 The see-saws of different types

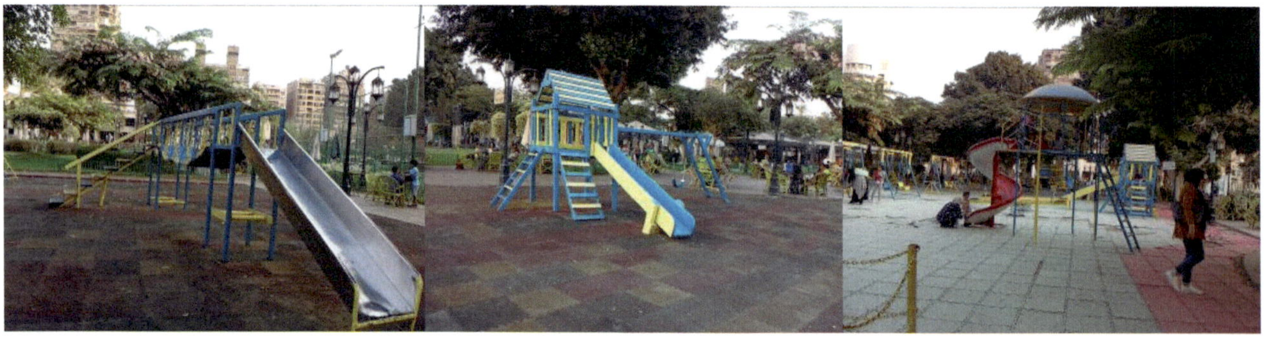

Fig. 13 The slidings of different types

The total number of children at the same time = 23 + 12 + 3 + 8 + 8 + 12 + 4 + 20 + 10 = 100 child.

Due to the guidelines of many countries (as mentioned in this research), each child can produce minimum 21 W/day, if he/she plays on the toys for 40 min only, then the minimum quantity of energy produced from 100 child at the club is as follows:

$$\textit{The minimum quantity of energy} = 100 \times 21 = 2100\,\text{W/day}$$

The club opens its gates from 7:00 am till midnight. From an actual tracking to the playing activity for children in the club, I found that the toys are completely occupied for a minimum two hours/day that is mean 120 min/day. If each child can play for minimum of 40 min/day, then how many times can the club receive children?

$$\textit{The number of receiving children} = 120/40 = 3\ \textit{times}$$

Then the club can receive from the playing of kids how much electricity?

$$\textit{The minimum quantity of energy daily} = 2100 \times 3 = 6300\,\text{W/day.}$$

$$\textit{The minimum quantity of energy monthly} = 6300 \times 30 = 189000\,\text{W/month.}$$

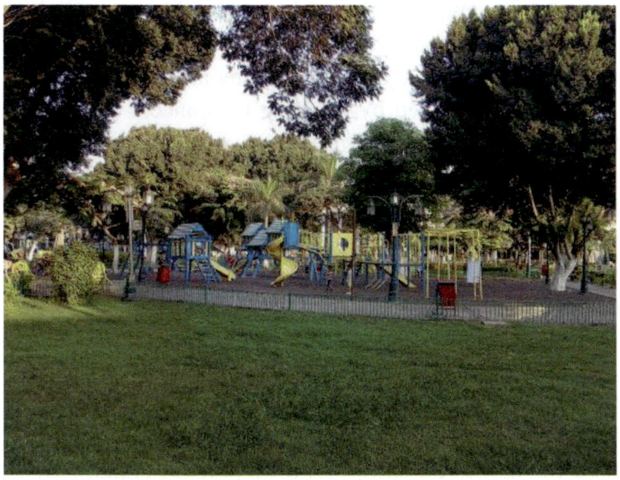

Fig. 14 The small houses

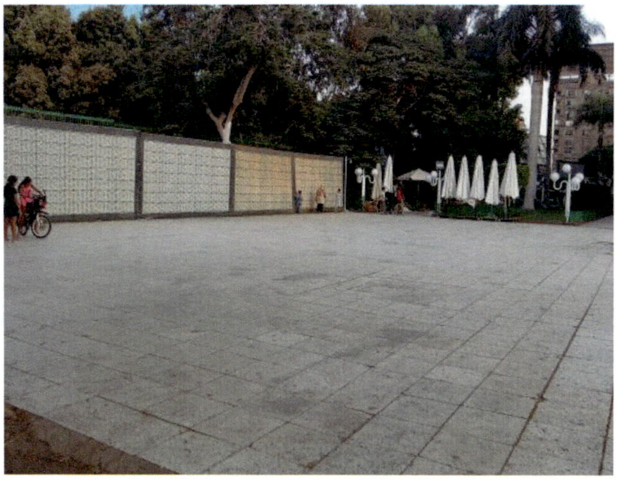

Fig. 15 The large area for bikes

If the amount of the bill of electricity is one pound/watt, then the club will save 189,000 L.E/month. This is the minimum amount of money if the playing areas receive 100 children only 3 times/day, in other words 300 children, as a total, occupy the toys for 120 min/day.

At present, it is not clear from the new design (Fig. 16) the types of toys and their capacities. But we can find approximately the number of toys are about 40 toys in different activities. If we predict that each toy can be used at the same time by two children, that is mean about 40 × 2 = 80 child. And if we suppose that the track of bikes located around the garden occupies minimum 10 children, then the total number of children will be 90 children which is near to the number of children in the old design.

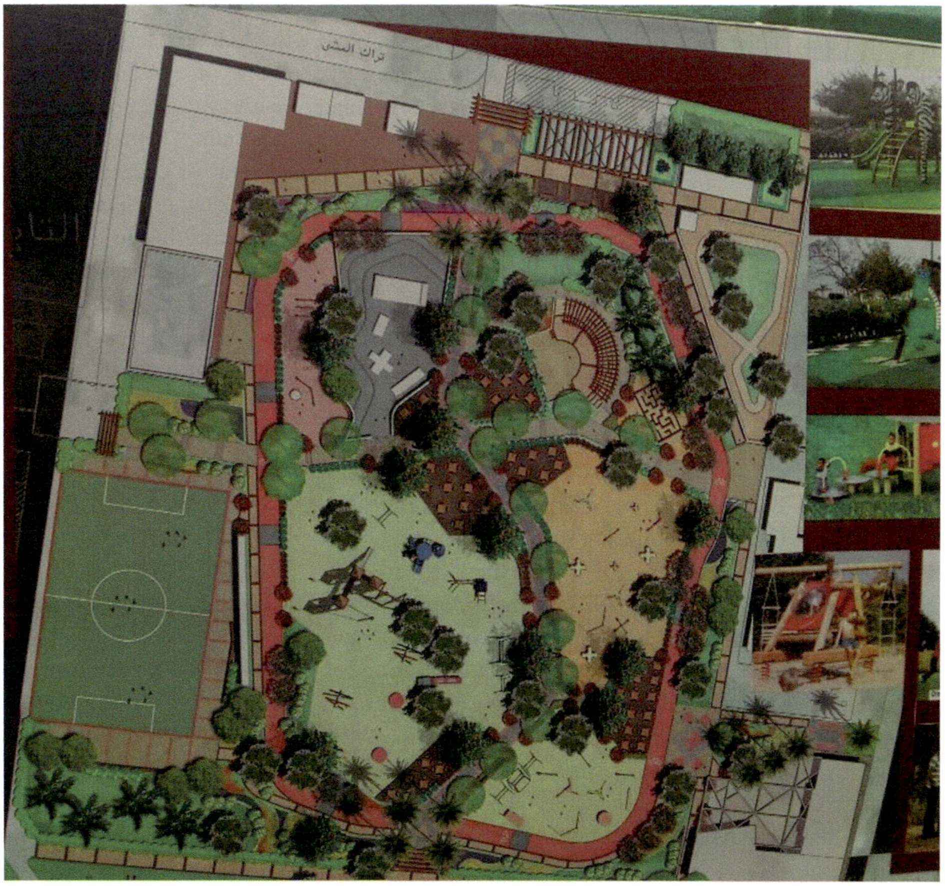

Fig. 16 The new design

Then the minimum quantity of energy produced from 90 children at the club is as follows:

$$\textit{The minimum quantity of energy} = 90 \times 21 = 1890\,\text{W/day}$$

If each child can play for minimum of 40 min/day, then how many times can the club receive children?

$$\textit{The number of receiving children} = 90/40 = 2.25\,\textit{times}$$

Then the club can receive from the playing of kids how much electricity?

$$\textit{The minimum quantity of energy daily} = 1890 \times 2.25 = 4252.5\,\text{W/day}.$$

$$\textit{The minimum quantity of energy monthly} = 4252.5 \times 30 = 127575\,\text{W/month}.$$

If the amount of the bill of electricity is one pound/watt, then the club will save 127,575 L.E/month. This is the minimum amount of money if the playing areas receive 90 children only 2.25 times/day. And this is due to the proposed available design that will exceed these calculations.

To fulfill the aim in capturing the kids' valuable energy, the research suggests to use smart toys as those were used in the project of Energy Efficiency Park (using the tires that were designed to capture kinetic energy), moreover, to use the smart tiles in the track of bikes to collect energy through the fast movement of bikes, and skating wheels on the track. This collected energy can illuminate the kids garden and maybe other parts of the club.

4 Discussion

The research is based on the hypothesis that kinetic energy from kids can be the lifejacket for providing clean energy. The results of the research are concerned with capturing the energy and saving money by using the kinetic energy of kids to produce an electrical one. But, the quantity of the produced energy and saving money are not expected. I think it would be less than the explored quantities that are resulted from applying the concept of using smart tires, as those were used in the Energy Efficiency Park, on the toys of kids. Moreover, the data from the 24 h Movement Guidelines in Canada and the United States of America, which mentioned the minimum needs for each child to play about 40 min/day, were helpful to calculate the quantity of energy that can be captured from each child per day by applying the smart tires on the toys of kids, which resulted in the expectation of minimum 21 W/day for each child. These calculations helped in finding out the total quantity of energy that can be captured per day if the toys of kids in the Shooting Club are connected to smart tires, which reveals the amount of saved money from the budget of the club/month. The dilemma of energy can be solved by applying the concept of capturing kinetic energy especially in kids areas, as they are considered a valuable and durable source for renewable energy. Governments, experts, residents, and all citizens in general have to establish the trend of using kinetic energy especially from kids in producing electrical one to gain clean energy, benefit from kids' energy, and save money.

5 Conclusion

Lack of energy is a serious problem in many countries; each country tries to find out solutions for producing energy especially the clean one. Renewable energy is the most effective type of energy for their limited negative effect on the environment. The research represented many types of renewable energy such as solar energy, wind power, hydroelectricity, geothermal energy, and biomass. Renewable energy has to be supported by establishing suitable policies and regulations that encourage the use of clean energy by giving great benefits to any organization that uses it. The paper discussed the kinetic energy as the safest type to produce clean energy by converting mechanical energy, which can be produced from movement, to an electrical one. As kinetic energy based on capturing the movement of any living or non-living thing, the research stressed the idea of using kids in producing energy as they always move, run and jump. This continuous movement produces a great amount of energy that should be captured and saved to be used later in electricity. Kids always move in open spaces especially in the playgrounds; therefore, in designing playgrounds, the designer has to fulfill both the needs of children, and the suitable ways in producing energy from kids' toys. The paper shows many ideas for converting the mechanical energy to electrical one; these ideas are used in inventing some equipment that are used in many projects. Previous studies found that the child must play a minimum of 40 min per day that can produce about 21 W per day. To have great benefit from the kinetic energy of kids, there should be public places that receive large numbers of children, such as the clubs. Depending on these findings, I have calculated the minimum amount of watts that can be produced per day from the play of kids in a club. These calculations depend on proposing the usage of invented tires that can be connected to the toys of children to capture the mechanical energy of kids and store it in special batteries to be converted later on to electrical energy. The study calculates the minimum energy that can be captured per day in the Shooting Club in Egypt, and this amount of energy saves not only energy but also money. In conclusion, the research reinforces the idea of using the kinetic energy of kids in public places as their continuous movement is considered an essential source to produce energy. Moreover, capturing kinetic energy from the playgrounds in clubs such as football, volleyball, and basketball playgrounds could produce huge quantities of electricity and save more money. This could be studied through separate research that has to focus on using spot sensors on the playground to capture the kinetic energy from the players' rapid movements.

References

Amouzegar, Z. (2010). Design principle of playgrounds' equipments and spaces for children. *Procedia - Social and Behavioral Sciences, 2*(2), 1968–1971. Elsevier Ltd. https://www.sciencedirect.com/science/article/pii/S1877042810003058.

Benedeka, J. et al. (2018). Evaluation of renewable energy sources in peripheral areas and renewable energy-based rural development. *Renewable and Sustainable Energy Reviews, 90*, 516–535. Elsevier Ltd. https://www.sciencedirect.com/science/article/pii/S1364032118301205.

Bjork, I. et al. (2011). *Encouraging renewable energy development: A handbook for international energy regulators* (pp. vii–138). Prepared by Pierce Atwood LLP for NARUC. USAID and National Association of Regulatory Utility Commissioners.

Blaschke, T. et al. (2013). Energy landscapes: Meeting energy demands and human aspirations. *Biomass and Bioenergy, 55*, 3–16. Elsevier Ltd.

Costa, G. (2012). The LED sustainable light swing harnesses kinetic energy to light up playgrounds. Project. https://www.penacosta.com/sustainable-light-swings.

Danielle, T. (2017). Power to the kids: Energy-generating playgrounds harness the energy produced by children's play and convert it to wider community benefit. *Parks & Recreation*.

Driediger, M. et al. (2018). Encouraging kids to hop, skip, and jump: Emphasizing the need for higher-intensity physical activity in childcare. *Journal of Sport and Health Science, 7*(3), 333–336. Elsevier Ltd. https://doi.org/10.1016/j.jshs.2018.03.003.

David, H. (2017). Evolving energy landscapes in the South Wales Valleys: Exploring community perception and participation. *Energy Policy, 108*, 818–828. Elsevier Ltd. https://doi.org/10.1016/j.enpol.2017.04.028

Empower Playgrounds, Inc. (2009). Empower Playgrounds, Inc. enhances educational opportunities for children in deprived villages in Ghana. West Africa through providing light and learning in innovative ways. Project. http://www.empowerplaygrounds.org/mgr
.

FAQs on Renewable Energy and Energy Efficiency. (2018). Body knowledge on infrastructure regulations. http://regulationbodyofknowledge.org/faq/renewable-energy-and-energy-efficiency/.

Krasojevic, M. (2012). Piezoelectric playground: The interactive Lumia Canopy in Pioneers Park Belgrade. Project. http://www.evolo.us/piezoelectric-playground-the-interactive-lumia-canopy-in-pioneers-park-belgrade/.

Marsh, J. (2018). Five types of renewable energy sources: Best alternatives to fossil fuels. https://news.energysage.com/five-types-of-renewable-energy-sources/.

Martins, F. et al. (2014). Use of alternative energy generation in playgrounds as a means of knowledge spreading. *Energy Procedia, 57*, 1024–1033. Elsevier Ltd. https://doi.org/10.1016/j.egypro.2014.10.068.

Prasanth, G., & Sudheshnan, T. (2011). A renewable energy approach by fast moving vehicles. In *Proceedings of the National Seminar & Exhibition on Non-Destructive Evaluation NDE* (pp. 641–665). Tirupur: Angel College of Engineering & Technology. https://www.ndt.net/article/nde-india2011/pdf/2-18D-7.pdf.

Seth, R. (2012). Swing along green power. Project. https://www.yankodesign.com/2012/02/17/swing-along-green-power/.

Simeoni, A. et al. (2012). Kidetic project teaches children the relationship between energy, interaction and fun. Tuvie. Project. www.tuvie.com/kidetic-project-teaches-children-the-relationship-between-energy-interaction-and-fun/.

Young, Y. et al. (2018). Nature energy park: Playground that teaches children about renewable energy. Tuvie. Project. http://www.tuvie.com/nature-energy-park-playground-that-teaches-children-about-renewable-energy/.

Study of Shading Device Parameters of the Mixed-Mode Ventilation on Energy Performance of an Office Building: Simulation Analysis for Evaluating Energy Performance in Egypt

Ayah Mohammed

Abstract

Investigations have supplied that double skin facades (DSFs) are capable of developing structure strength performing the usage of the stack effect (buoyancy effect) to dispel heat in addition to shading device to lower the solar warmth transmission in this simulation analysis. The goal of the study is not simplest to research the parameters of shading devices in combined mode ventilated buildings but additionally to lower the annual realistic cooling load of the building energy saving. In times of horizontal shading gadgets, the air temperature of an outdoor dealing with wall's air void area was more than that of vertical shading strategies. The daylighting operation was estimated by the usage of the daylight factors and effective daylight illuminance (UDI). The conclusions supplied a 38% saving in cooling loads due to the shading device. Observing the daylight component, more points were controlled in a suitable daylight thing of 2–5%, which was uncovered to upward the UDI to 500–2,000 lx. Therefore, the reduction of the cooling loads of constructions at some stage in a shading device is estimated to steer the reduction of the strength intake of buildings. With regard to annual energy consumption, it was found that horizontal shading devices were more effective to reduce energy more than 85% than the base case with a single skin façade.

Keywords

Double skin facades • Thermal comfort • Energy performance • Annual sensible cooling loads

1 Introduction

Building skin is the fence between the users of the construction and the external outdoors of the construction and indoor atmosphere from the outdoor environment, where the outdoor has its outcome on the indoor assuming the indoor behaviors and the human comfort (Chowdhury, 2014).

Skin and structure technology permits building with a huge glazing surface which submits transparency, daylight in the visual outdoors as well as increases the amount of solar heat gain. These vertical areas of glass furthermore perform as a heat trap. The thermal effect of the glazing envelope relies on the shading devices and thermal performance of glass (Murray, 2009).

In-office buildings, almost 50% of cooling loads demand to appear over the building skin, and almost 50% is created by solar rays. Consequently, the regulation of solar radiation consuming external shading devices to decrease cooling loads throughout the building. Incidentally, the primary goal of the paper is to estimate the decrease of sensible cooling demand in an office structure that has extreme pressure for cooling demand by consuming outer screening strategies. The office building classification is often characterized by wide-ranging use of full glazing façades (Danny & Tsang, 2008).

The incidence of huge transparent surface and the function of shading devices have typically been the aim of examination because solar gain rates can mainly affect the structure in both the thermal and visual comfort, thermal energy consumption in summer and winter terms, and the lighting demand. Some authors have evaluated the performance of several forms of glazing and shades on the cooling and heating energy demands in office buildings with a parameters aspects in demand to estimate various opaque facades arrangement, floor shape ratio, windows area (Aldawoud, 2013).

Façade is an idea that can assist to resolve lighting difficulties and extremely glare of daylight. The reason for

A. Mohammed (✉)
Faculty of Engineering, Environmental Architecture, Modern Academy, Cairo University, Cairo, Egypt

F. Rosso et al. (eds.), *Advances in Architecture, Engineering and Technology*, Advances in Science, Technology & Innovation, https://doi.org/10.1007/978-3-030-86913-7_19

research to inspire the thermal overall performance of airflow by using the parameters of a double facade in hot climates. This research aim on the performance of double skin facades in workplace structures in hot weather. Adaptation of this arrangement is looked-for to decorate its overall functioning in operating areas in relatives of daylight (Hamza et al., 2007).

2 Problem Definition

Cooling loads has ever been a problem facing architects who try to provide natural ventilation in the space without depending on artificial ventilation which is energy-consuming. So the most difficult issue in the design is the determination of openings design aspects that affect the amount and distribution of the daylight in the zones. The main problem in the design process is the lack of understanding of the natural ventilation strategies and lighting techniques inside the space in the initial design stages.

3 Aim and Methodology

The study objectives downgrading heat benefit in the double façade air depth via growing a second layer as a pattern that affords self-shading thus lowering direct sun radiation. The paper recommences the exam anywhere the diversifications are built for increased daylight levels inside the spaces. To simulate the thermal performances and daylight simulation of the model, they are evaluated to those of the modern creation envelope single pores and skin facade (SSF) building and to a Double skin facade building (DSF) to understand whether or not or no longer they have enhanced. These correlations were carried out in each South (S) and West (W) directions.

The approaches consumed in this investigation are as follows.

- To begin with, the exciting seasons, which contain shading devices, have been simulated and external shading techniques for both direction (South and West) orientation all through this season were calculated. The overheated benefit of the carried out shading device turned into testing. The next aim of this investigation is to grow indoor daylighting functioning in constructing with a massive window-location ratio through making use of shading devices. For example, this paper purposes to decorate the visual consolation in construction with the aid of allowing daylighting indoors even as delaying extreme sunrise.
- Second, the zone realistic cooling demands for the shading devices have been simulated to estimate the zone sensible cooling produced with the aid of delaying solar radiation.
- Third, operation of the daylighting for the shading strategies was examined. The daylighting working hours for each space was estimated by the daylight factor and valuable daylight illuminance (UDI).
- In order to the exam, the good-sized difference among the influences of double skin façade as opposed to single-skin façade on the daytime overall functioning of running zones, for example, observes of working zones are investigated.

A methodology with the flow chart short the investigation is proven in Fig. 1.

4 Literature Review

Hamza et al. (2007) associated the thermal and daylight implementations of a DSF type with a corridor-type wherein there had been going via paths (40% opened air cavity) each floor. They have air vents at the top and lowest of the cavity but no shading devices were consumed. Just the corridor-type had air openings inside the outside envelope layer at individually ground layer. Simulations have been completed on a summer daytime in the weather of Cairo. The assessment among the thermal and daylight rendering of the 2 case studies in South and West directions. The consequences confirmed that direct solar radiation is the principal heat gain supply as via to outdoor radiation.

The corridor type air depth temperature becomes 1.5 °C lower than the base case. These guides display the position of various vent for ventilation via the height of the envelope and not just at the higher and lowest (Boake, 2003). United Arab Emirates have additional an examination in the extremity warm weather. Radhi et al. (2013) *shaped* a status structure with a double façade, and in the assessment with a single façade, the double façade had air vents at both flooring layer, and at the top and lowest of the air depth.

They executed the critiques in important directions (S, W, E, and N). This assessment adapts various glazing materials and hollow space depths to verify their effects on depth temperature, cooling masses, and airflow. The double façade hooked up in lessen cooling demand in all directions not included in the North in which a 3% growth changed into calculated. Air depth temperatures were typically little distinct than the external air temperature because of the window wall ratio via the hollow space height. They planned hollow space depths among 0.78 m and 1.5 m and that the U-cost

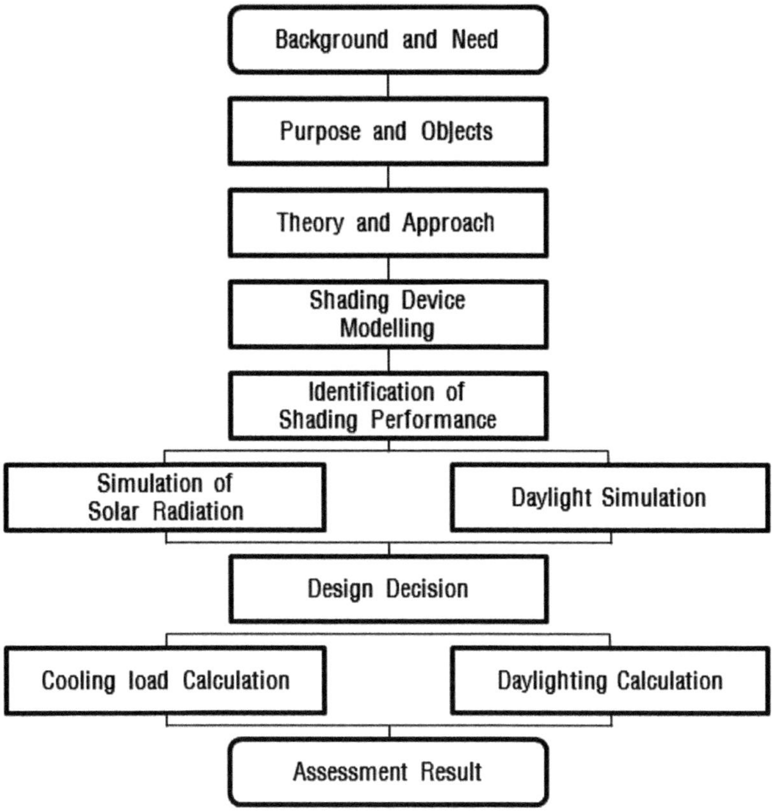

Fig. 1 Examination methodology

and solar heat benefit coefficient (SHGC) of the outside glazing layer may be actual claimed. The predictation consumed 19% lower in zone sensible cooling demand via making use of a double-skin façade system.

Laura Belia et al. described that annual power consumption can be decreased with the aid of as lots as 20% via changing the duration and form of the shading device. D. Selena W. et al. tested cooling-power masses consuming the approach and show a discount as massive as 20% in numerous models. Abdelsalam Aldawoud researched the mixture of shading gadgets and electrochromic glazing, clarifying that this aggregate can decrease the sun heat benefit by using 30–50%. Gon Kim et al. evaluated overhangs, blinds, light-shelves, and investigational shading system. Overhangs develop the overall performance of energy with the aid of way of 18% (Bellia, 2015).

The 20% of the cooling load may be devoured by the light-shelves. Investigational shading systems have lowered the cooling load by up to 70%. Nihal et al. measured the variations in the temperature of the indoor air by contrasting overhangs, louvres, and egg crates, describing that egg crates produced the largest decrease in the inner temperature of the inner air (Nihal & Fadzil, 2011). The thermal performance in step with the mind-set of the louver and demanded that the greatest shade approach be inspired by manner of the region and meteorological conditions.

5 Shading Systems

Affecting difficulties includes overheating and glare complications; however, keeping an appropriate daylight operation has to prevent straight sunlight from diffusing into areas (Vartiainen, 2001). Accordingly, utilizing a comfortably designed shading system to avoid thermal and visual discomfort and consequently decreasing energy consumption measured being a widely consumed useful plan in office buildings (Fig. 2). The shading devices typically adapt to the external envelope to limit extra natural light. Also, combining the shading systems with the modern concept, highly jointed envelope needs allowing for the aesthetics and environmental features through the primary design steps.

6 Horizontal Shading

Horizontal shading strategies are one of the traditional systems utilized as shading designs to reduce direct solar gain for extreme sun radiation (Fig. 3). In addition, horizontal shading strategies are able to provide complete screening depend on the sun's location. In spite of this, wrapped visual communication with the exterior for the occupants has measured a weakness of these forms of shading systems.

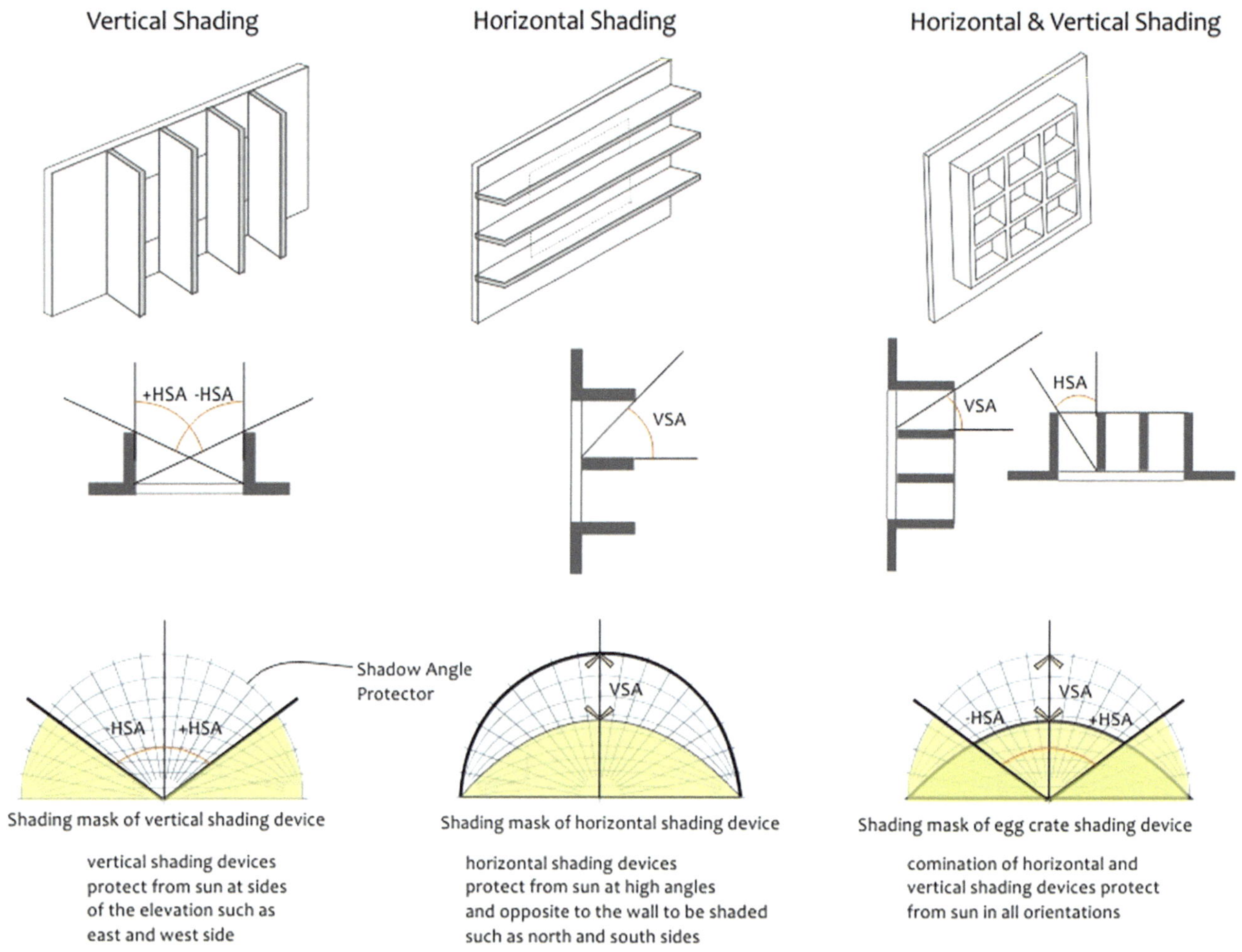

Fig. 2 Shows horizontal shadow angle (HAS) and vertical shadow angle (VSA) device respectively

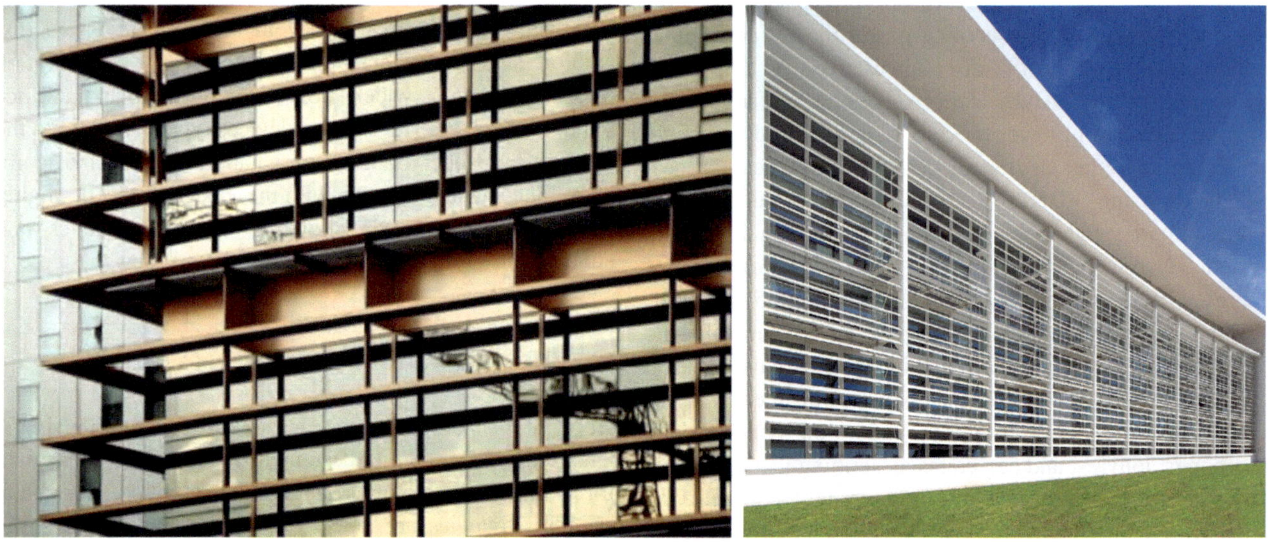

Fig. 3 The horizontal shading devices within the Head Office constructing of the telecommunications market commission (CMT), Barcelona, Spain

7 Vertical Shading

Vertical Shadings are effective in preventing straight sunlight at a short angle sun location, for example at the early sunrise, and the sunset time, furthermore to the east and west envelope as they block short solar radiation arriving from the sideways (Fig. 4). The weakness of Vertical Shadings types is decreasing the outside view (Ruck, 2000).

8 Horizontal and Vertical Rotating Shading Units

Louvers are the most well-known component to be powered and regulated either manually or mechanically through computerized techniques to operate daylight inside interior zones. A function of the application of net-zero houses can be got by mixing the circulated responsive system of skins, as accessible by the north house responsive envelope which is design to match the near northern climates (42-55 latitude) where heating is demanded and the daylight time is short where the is a great need to increase the daylight inside the interior places as shown in Fig. 5 (Trubiano, 2013).

As measured at the period of the structure, the south façade for the Arab world institute is a modern-day version to the traditional Arabic sun shades but in a dynamic method with changing small units in a space, style is keeping different patterns of daylight and shades to the interior of the building as shown in Fig. 6 (Kronenburg, 2007).

Digitally created and designed moved from the vernacular architecture in UAE (Fig. 7), the dynamic sun shades are presenting a unique atmosphere to the Al-Bahr Towers, and regarding the external environment by following the sun to control the opening and closing of a complicated dynamic origamic 1048 units (Karanouh, 2016).

9 External Shading Devices-Typology and Materials

Shading devices can be partitioned into two obvious types: interior and exterior types. Exterior shading devices are 30–35% more helpful than interior shading strategies. Shading devices are generally categorized into three forms based on its combination with the window. They are coverable or removal, adaptable or movable and fixed shading devices. Fixed shading devices are of three types depending on their physical methods. They are vertical shading devices, horizontal shading device, and integration of the two vertical and horizontal shading devices (egg-crate) as shown in Fig. 8.

10 Daylight and Double Skin Façade

Element, exterior reflected element, and internally reflected element (Abdulsalam, 2014). The direct sunlight causes access to solar heat gain and rises ultraviolet of sunlight that produces extra glare, office area in command to act the clearest demand to reject the direct sunlight and be contingent on inner and outer reflected features.

Double skin facade is a double Skin of facade divided with the aid of an air hole that adjustments in its smallest depth selections between 0.2 m and 2 m. It generates a skin that has stored a buffer sector for regulating and operating the relation between indoor and outside thermal comfort conditions (Rahmani et al., 2012).

Double Skin Façade can dismiss extreme direct sunlight and inform reflected light, to limit inflowing daylight and heat gain. Admission of incessant huge daylight is crucial for office building for two aims, to avoid excessive night-time to prevent extreme glare produced by the extreme pass of daylight at specific times of the day to admit their harmful

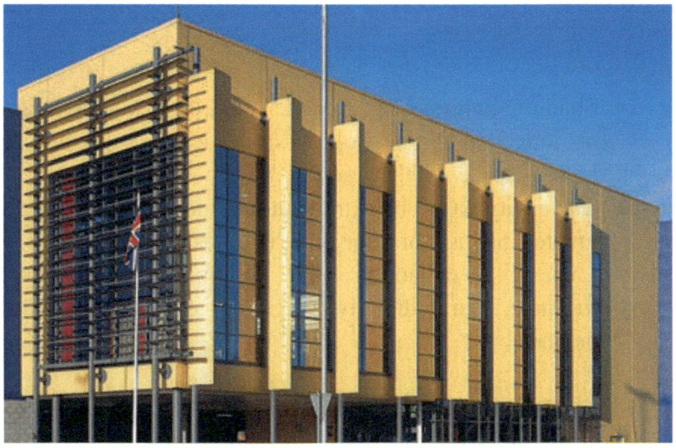

Fig. 4 Vertical fins involved in the curved building facade, England

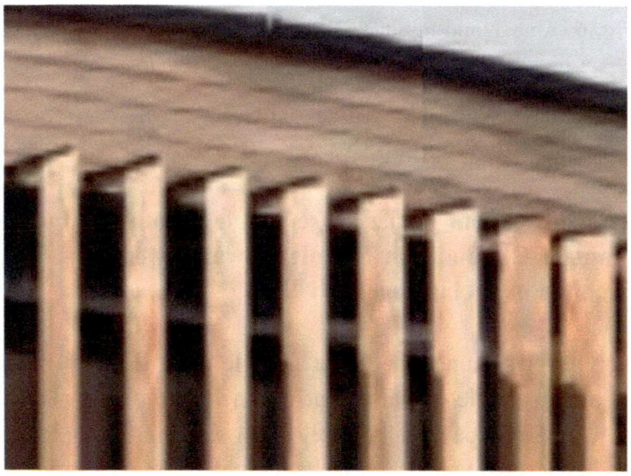

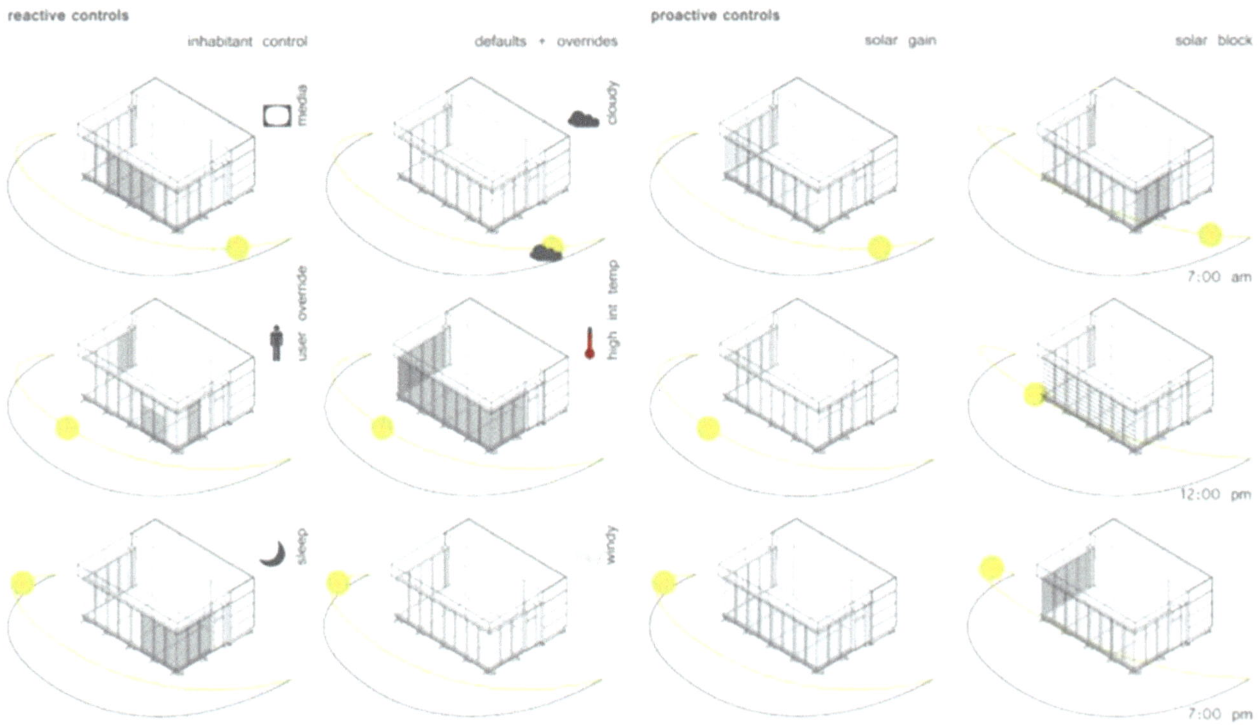

Fig. 5 Exterior shade configuration scenarios based on relative exterior environmental conditions and related responsive envelope reactions. *Source* by Perkins and Will research journal

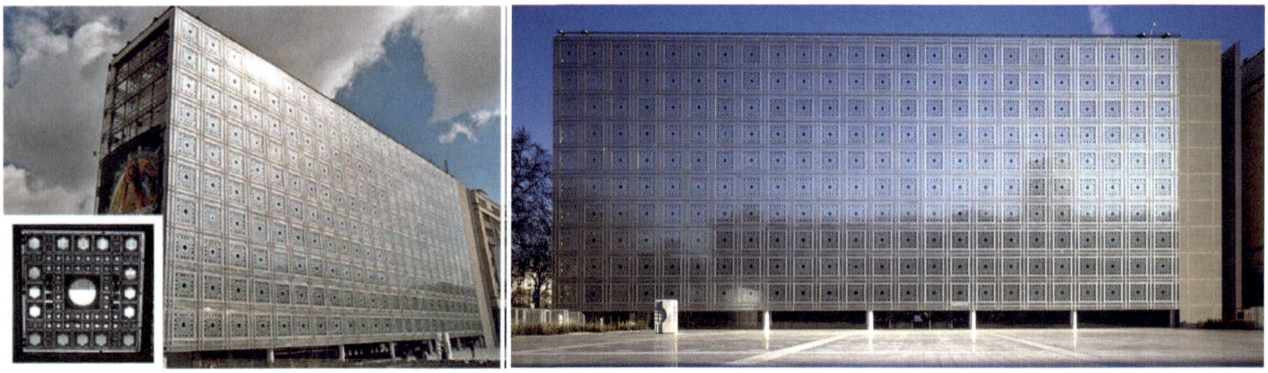

Fig. 6 Arab World Institute, Paris, France, 1989, by Jean Nouvel. *Source* flexible, architecture that responds

influences, with decreasing expended energy to achieve demanded lighting aim (Napier, 2015).

The subsequent respects are visible because of the most giant of Double Skin Façade in warm weathers for the architectural challenge:

- The glass function of the Double Skin Façade (balance among little Solar Heat Gain Coefficient and low U-Value) specifically if rejection shading is employed.
- Numerous vents along the height of the external skin layer to avoid overheating problems.
- Shading components are rather placed outside layer and no longer in the air hollow space depth. Suggested air hollow space width variety: 0.8 m to 1.5 m.
- The cavity has to be extra than the floor ground plan and greater than a roof floor plan. (Barbosa et al., 2015)
- Later solar radiation is the principal heat advantage resource, extraordinarily suggested to practice shading devices.

The study confirmed by Fallahi et al. (2010) provided that the solution of thermal mass (concrete) on the outside layer (b) and at the shading device (c) available comparable

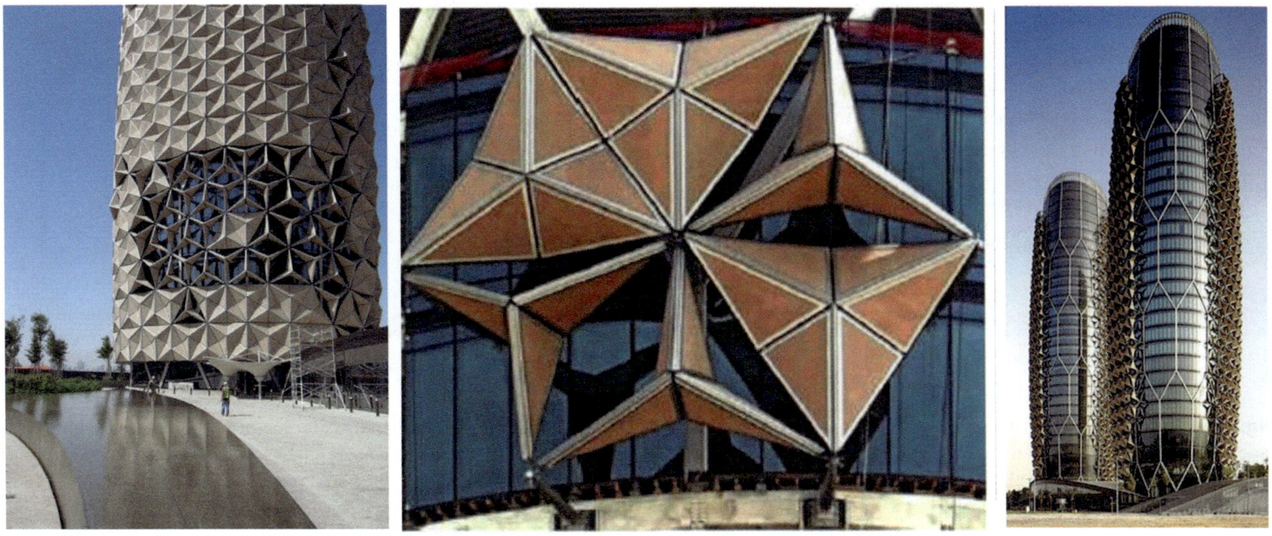

Fig. 7 The Al-Bahr Towers-Abu Dhabi by Aedas-UK (now AHR) in collaboration with Arup. *Source* Karanouh, Abdulmajid

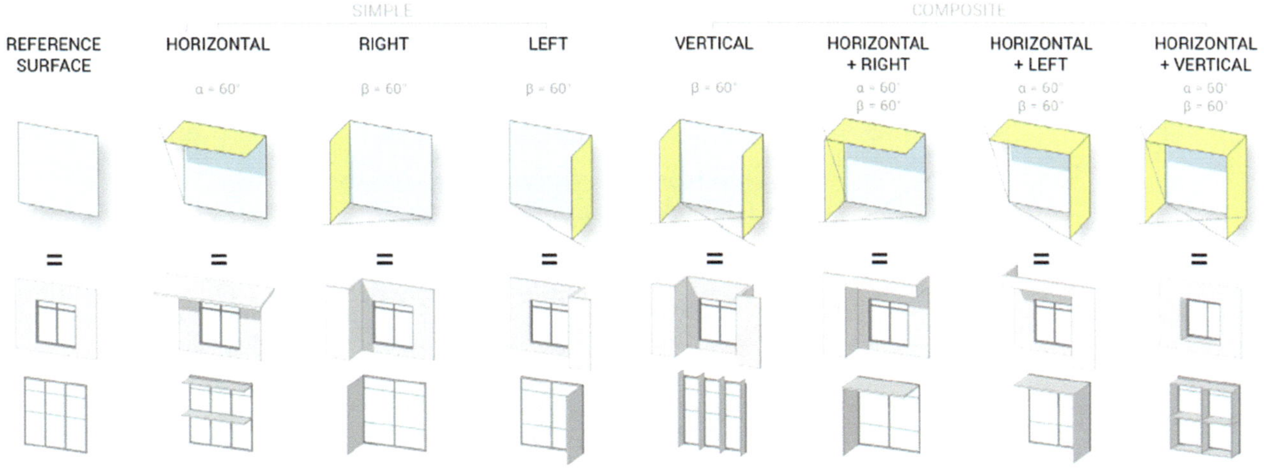

Fig. 8 Types of external shading device-**a**. Vertical, **b**-Horizontal, **c**-Eggcrate

cooling masses demand in the summer and much less than the amount consumed inside the base case without DSF (a). But the case in which the thermal mass was realistic at the internal layer (d) had the most cooling load demand as shown in Fig. 9.

11 Technique

In demand for observing the import difference between the impact of double skin façade (DSF) versus single-skin façade (SSF) on sunlight hours houses of the office building, a case look at of two workspaces is simulated, an assessment between their result is showed.

The investigation of the base case was simulated by DesignBuilder was an appropriate technique that offers acceptable decision devices to examine the shading system. Figure 4 shows the movement of the suggested workflow method as shown in Fig. 10. The numerical template that DesignBuilder displays the operator is comparable to the standard case studies. DesignBuilder is employed with typical weather data and occupant activity (climate consultant).

12 Simulation Setup

This takes a look at indicates that simulations and analyses were made with DesignBuilder. With the item to reach accurate and measurable records and results, the submission of simulations with ordinary calculation methods is essential.

Design Builder software (V.5) was chosen for the reason of this have a look at primarily based on its established

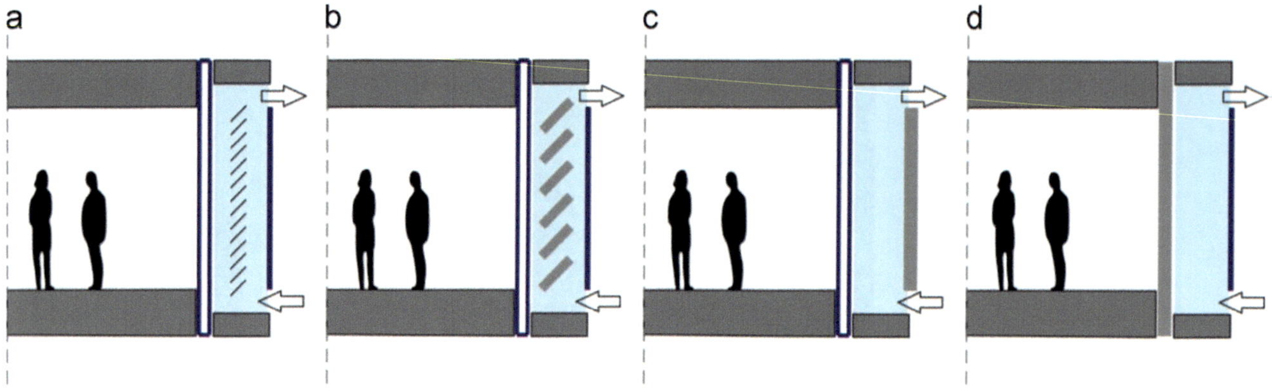

Fig. 9 Models the usage of thermal mass (concrete) proposed by the usage of Fallahi et al.: **a** traditional DSF with internal layer manufactured from double glazing, outer layer set as single glazing and aluminum Venetian blind; **b** aluminum blind replaced with the aid of a concrete blind; **c** outer layer changed by way of concrete; **d** inner layer changed by the usage of concrete pane

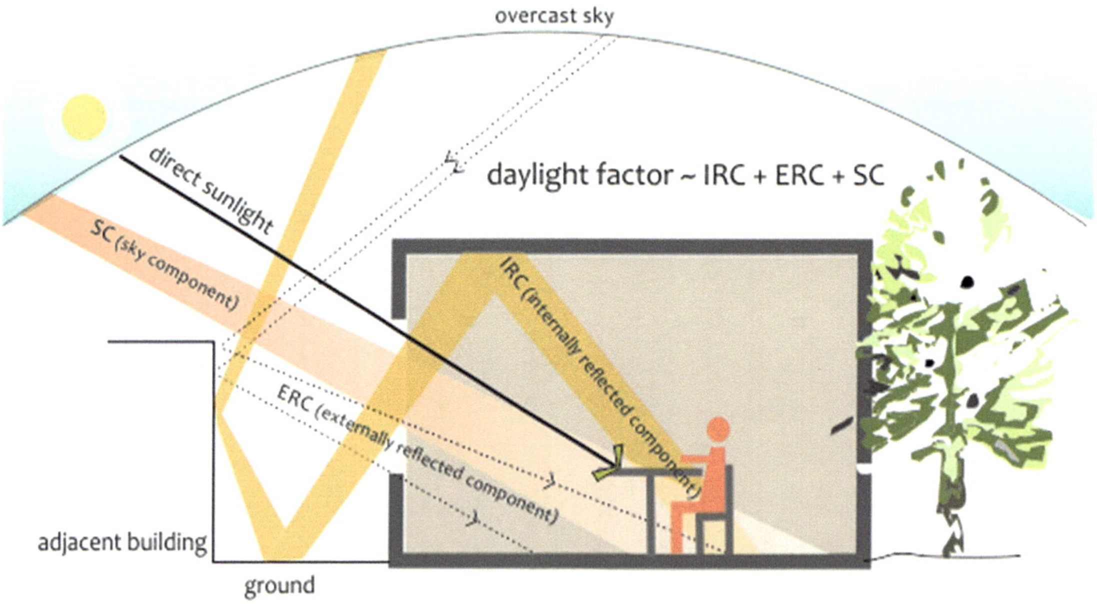

Fig. 10 Daylight factor is used for figuring out daylight. Its equivalent to the sum of the diffused skylight (SC), internally pondered light (IRC) and externally reflected light (ERC)

power performance assessment functionality as compared with similar strength simulation tools. A comparison between predicted basecase and retrofit of the basecase by adapt shading device (see Figs. 11 and 12).

Egypt is categorized as a hot arid area, which is taken into consideration via hot summers and cold winter seasons. The most up to date months are from May to September with the high-quality temperature going up to nearly 45 °C (100 °F).

An entire constructing simulation the usage of Design-Builder changed into a suitable technique that keeps a suitable results device that makes use of a shading evaluation in a traditional office building.

The main expertise that employs the power model has greater to do with the building's construction, operations, and settings of HVAC systems. Furthermore, DesignBuilder is operating with common weather records of Egypt. Daylighting performance was analyzed using a 500 lx (46 FC) to observe what percentage of the office space would be adequately daylighting for each simulation model (EnergyPlus, 2014). The methodology taken in the research is shown in Fig. 13. See Tables 1 and 2.

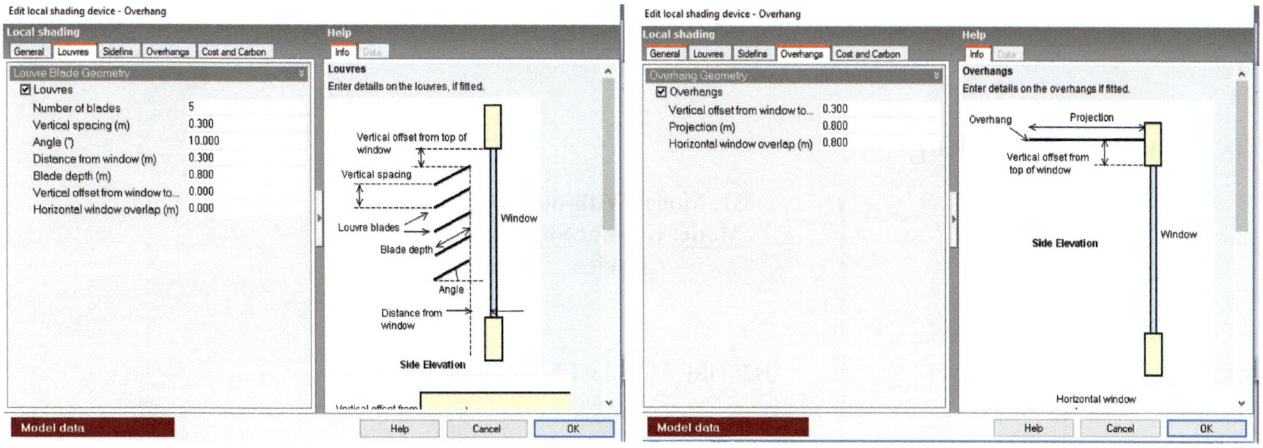

Fig. 11 Set-points adaptation (18 °C:24 °C) & Façade glazing system adaptation, design builder screen shoot

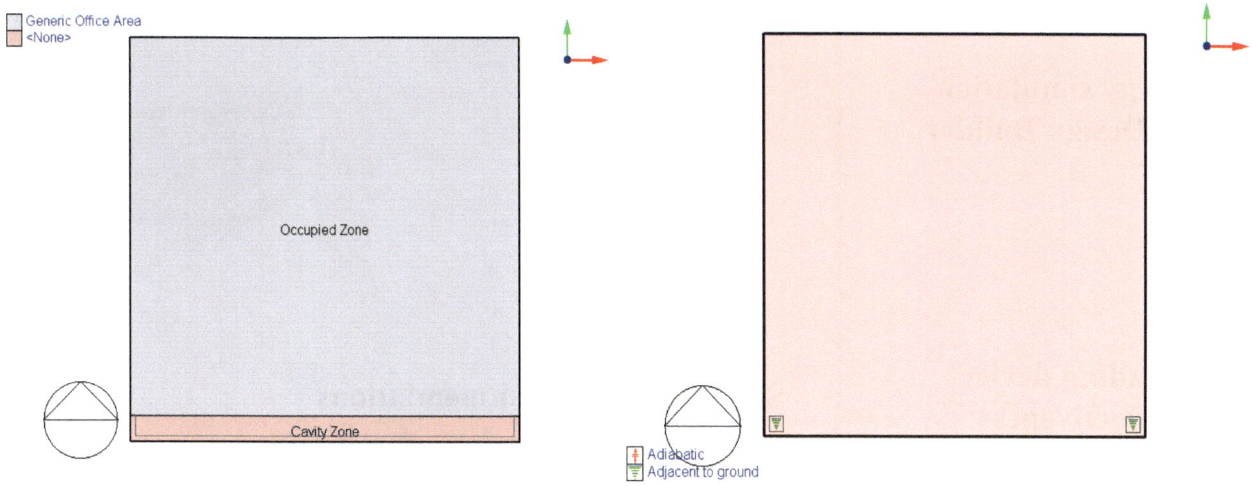

Fig. 12 Architecture plans of single skin Façade (SSF) & double skin façade (V.DSF), design builder screen shoot

13 General Office Building Design Data

14 Daylighting Results

The daylight penetration in Fig. 14 produced by Design-Builder simulation modeling shows that 70% of the floor receive adequate daylighting to perform office-related tasks almost between 768–4374 lx for the base case while 0.5 m DSF achieve 137–2711 lx, 1.0 m DSF achieve 219–2615 lx, 1.5 m DSF achieve 96–2672 lx, 2.0 m DSF achieve 105–2715 lx, and 2.5 m DSF achieve 113–2699 lx. However, the base case arises to the final 75% of the ground as it receives over 4374 lx; this means that these regions will suffer from glare problems or at least receive excessive daylighting that do not add to any gain to the desired LUX ranges to perform office-associated functions as proven in Fig. 14.

The effects supplied a 38% lower in cooling demands by means of applying shading devices. Concerning the sunlight hours element, more mind was concerned with a proper daytime issue of 2–5%, which was provided to increase the UDI to 500–2,000 lx.

15 Results and Discussion

Figure 15, it is quite clear the most valuable configuration was the corridor DSF with an intermediary horizontal sun breaker followed by the in saving energy% while the double-skin facade without shading device the minimum value in saving energy%. This was clear when compared to the base case of a single skin façade.

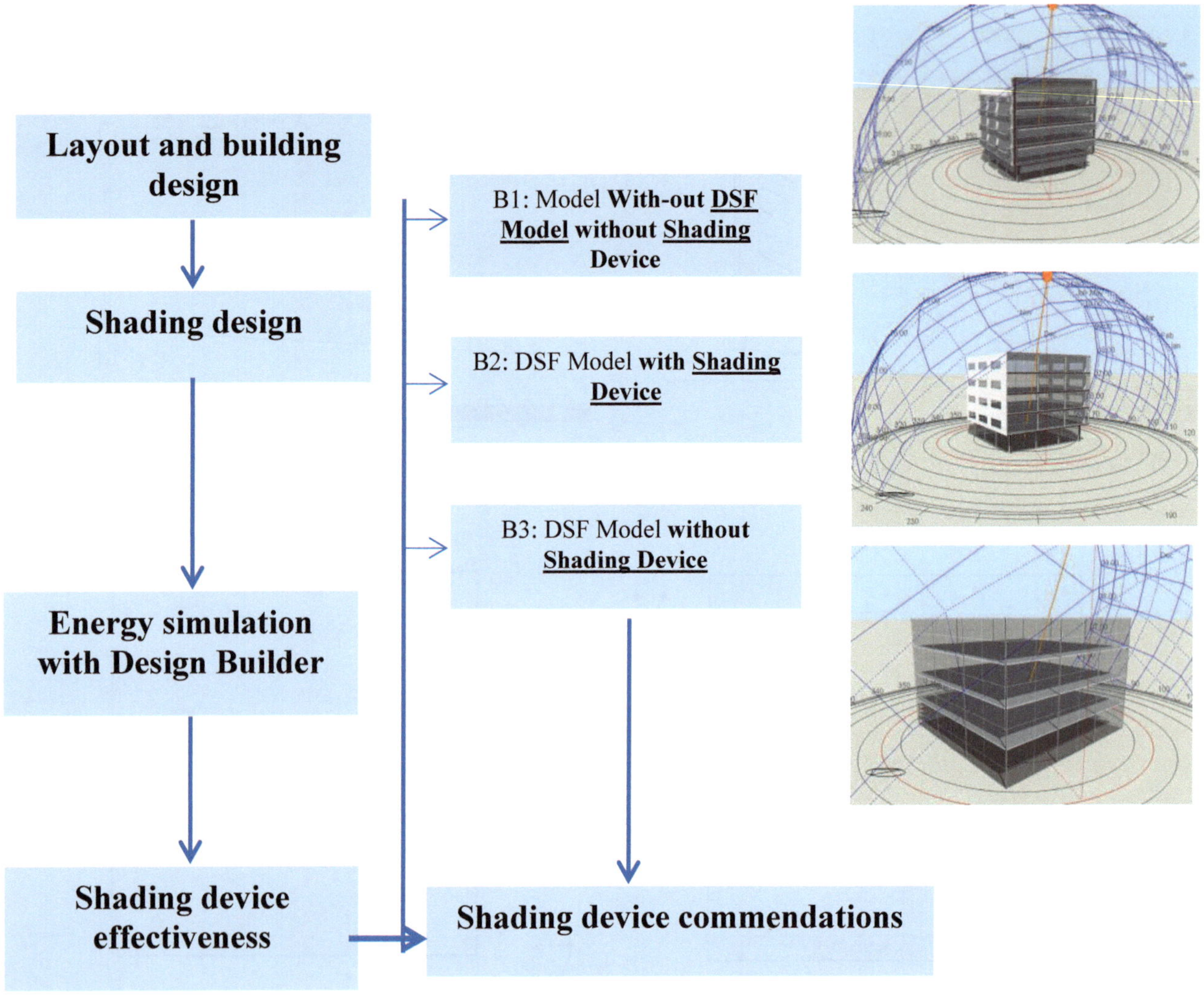

Fig. 13 Methdology for a predictable residential structure with numerous shading devices

Table 1 DesignBuilder simulation participation data

Variables	Values	
Air temperature	Cooling set-point	27 °C
Luminance objective	–	150 lx
Least fresh air	Working spaces	8.5 (l/s-person)
Weather data	Egypt	–

The graph shows that the zone sensible cooling with the optimum air cavity width of 2.0 m of ventilated corridor DSF with a horizontal south shading device achieved the best enhancement of sensible cooling of 9000 (kWh) while corridor Ventilated DSF without shading device, the zone sensible cooling increased to 18000 (kWh) when compared to the base case model which have the highest value of zone sensible cooling of 25000 (kWh) at July as shown in Fig. 15

Table 2 Office building architectural data

Building type	Office building
Zone 1/occupancy	12 person/100 m^2
Structure level numbers	4 floors
Typical office floor height	4.0 m (slab to slab), 3.70 m ~ 3.0 m (clear height)
Ground floor area	15.0 m 15.0 m = 225 m^2
Ventilation system	– For "Base-Case" building: artificial mechanical ventilation (full HVAC) gadget
	– For ventilated DSF building: 'Mixed-Mode' (hybrid) machine totally all the year

In employing the Fanger PMV on the various configurations, it is obvious that the modify cases set among positioning of (−1 to 1) SET (Fig. 16). Although the base case

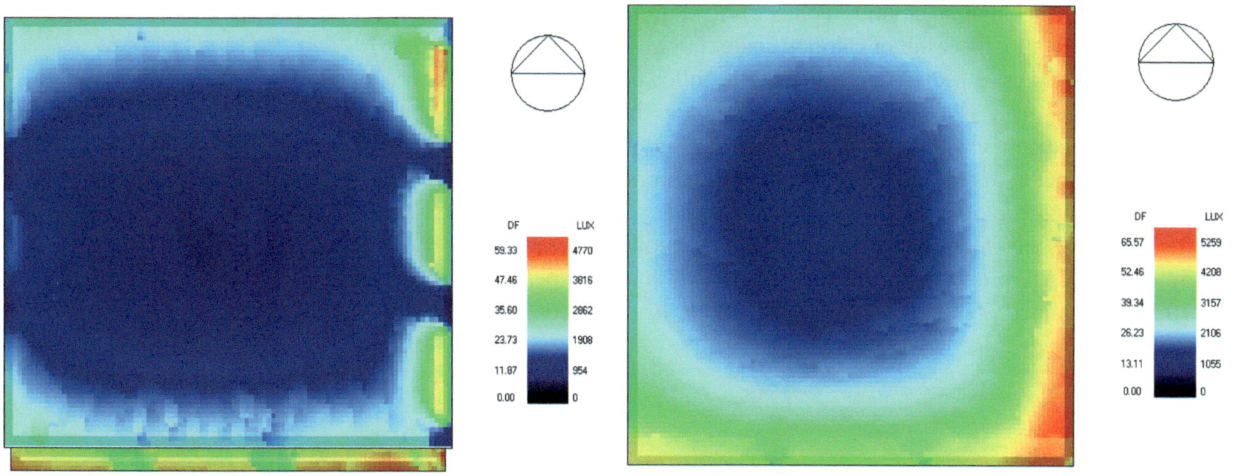

Fig. 14 Brief of daylight factor distribution and illumination element for base case and DSF for south orientation, processed by DesignBuilder with radiance engine

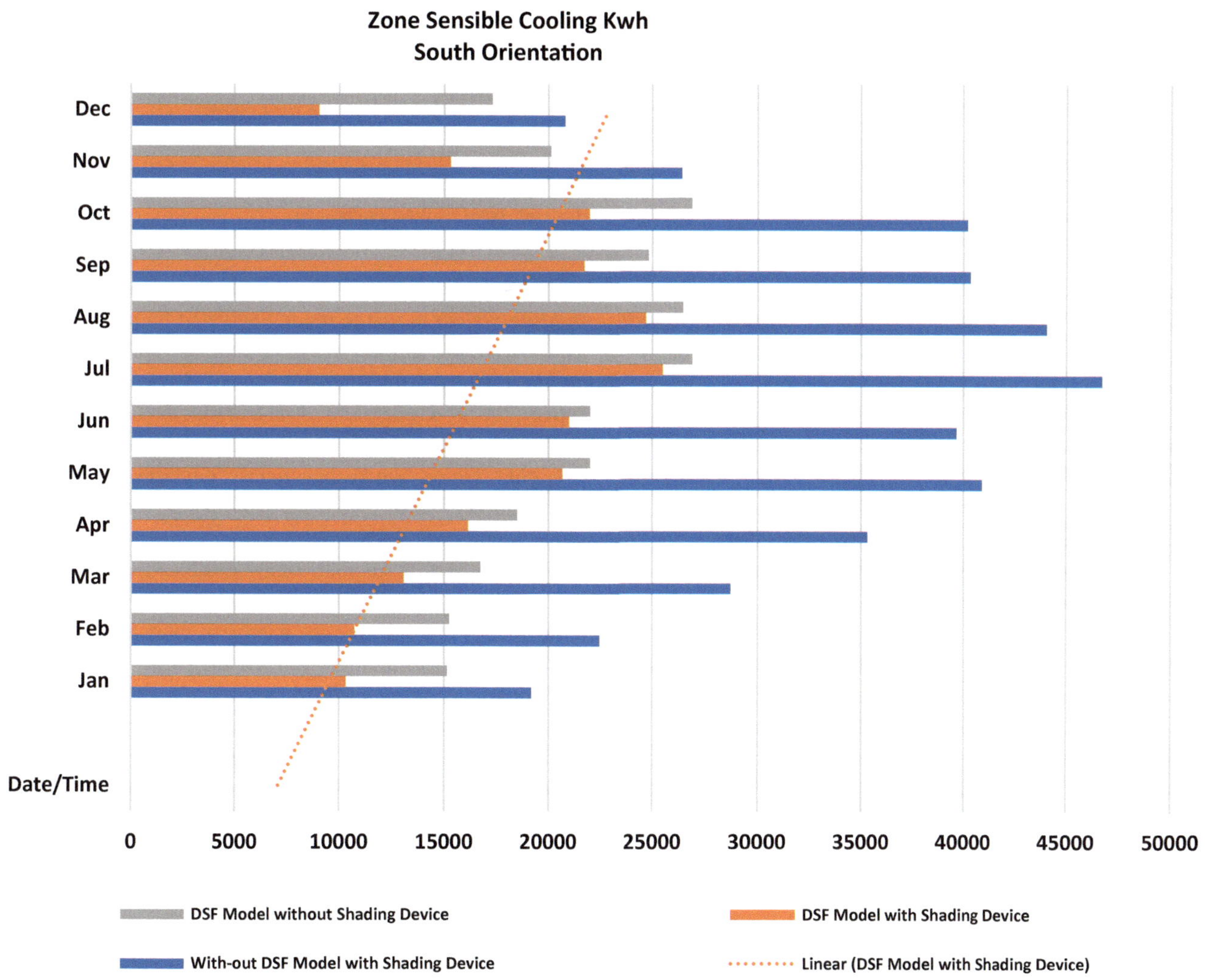

Fig. 15 Periodic monthly outcome of zone sensible cooling (kWh) at south facade

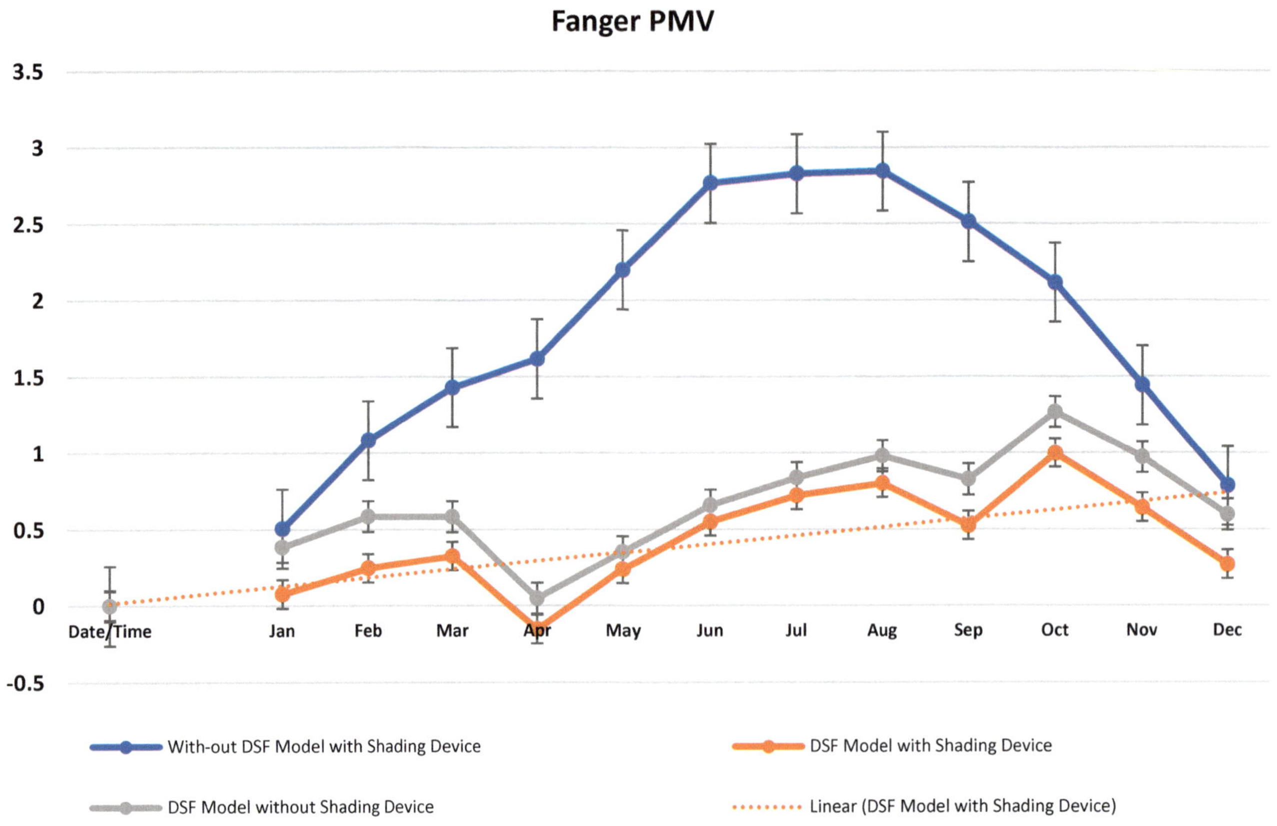

Fig. 16 Periodic monthly outcome of Fanger PMV for a shading device for DSF types at West orientation

achieved above 4 SET, the most active type was the corridor Double Skin Facade with a horizontal south shading device.

16 Summary

16.1 The Result from the Simulation Analysis

16.1.1 Effect of the Intermediate Shading Devices of the DSF on Peak Summer Cooling Loads

- It was clear that there was a reduction of almost 67% in the cooling energy consumption with the shading devices at the west façade and 76% at the south facade with the most effective type for DSF, keeping in mind that both skins were 100% transparent of maximizing the view.
- Protective Effects of Shading and Lighting Elements: Damages that may occur due to the external climatic changes are prevented as a result of the installing of shading devices into the facade cavity.

The aims of this paper to ***examine the parameters of the shading system*** in mixed mode ventilated buildings but also to ***decrease the annual sensible cooling and increase the building energy saving***. In cases of ***horizontal shading system***, the air temperature of an external envelope side's air cavity was ***larger than of the vertical shading system***. The daylight factor and beneficial daylight illuminance (UDI) turned into evaluated via using the daylighting implementation.

With respect to annual zone sensible cooling (energy saving), it was found that horizontal shading devices were more effective to decrease the energy of more than 85% than the base case with a single-skin façade (SSF).

Simulations of the daylight factor display that the exterior immovable shading system combined with the internal shading system provides acceptable DFs and a level distribution of the daylight levels in a south envelope office building. Permitting to the customers, external fixed shading is a suitable system for south-facing offices when limited glare problems can be explained by interior roller shades.

References

Abdulsalam, M. (2014). Simulation of day-lighting characteristics in the cavity of double glazed Façade. *International Journal of Innovative Research in Science, Engineering and Technology, 3*(4).

Aldawoud, A. (2013). Conventional fixed shading devices in comparison to an electrochromic glazing system in a hot, dry climate. *Energy and Buildings, 59*, 104–110.

Barbosa, S., Ip, K., & Southall, R. (2015). Thermal comfort in naturally ventilated buildings with double skin façade under tropical climate conditions: The influence of key design parameters. *Energy Build, 109*, 397–406.

Bellia, L. (2015). Evaluation of daylight availability for energy savings. *Journal of Daylighting*, 12–20.

Boake, T. (2003). *Understanding the general principles of the double skin facade system*. University of Waterloo.

Chowdhury, S. (2014). *Study of thermal performance of building envelope of readymade garment factories in Dhaka*. M. Arch thesis, Dept. of Architecture, Bangladesh University of Engineering & Technology, Dhaka.

Danny, L., & Tsang, E. (2008). An analysis of daylighting performance for office buildings in Hong Kong. *Elsevier, 43*, 1446–1458.

EnergyPlus. (2014). *The encyclopedic reference to EnergyPlus input and output*. Illinois, US: Department of Energy.

Fallahi, A., Haghighat, F., & Elsadi, H. (2010). Energy performance assessment of double-skin façade with thermal mass. *Energy & Buildings, 42*(9), 1499–1509.

Hamza, N., Gomaa, A., & Underwood, C. (2007). *Daylighting and thermal analysis of an obstructed double-skin façade in hot arid areas*. Newcastle Upon Tyne, UK: Northumbria University.

Karanouh, A. (2016). Innovations in dynamic architecture. *Journal of Facade Design and Engineering, 3*(4), 85–221.

Kronenburg, R. (2007). *Flexible: Architecture That Responds to Change*. London: Laurence King.

Murray, S. (2009). *Contemporary curtain wall architecture* (1st ed., Vol. ISBN-13).

Napier, J. (2015). Climate based Façade design for busniss buildings with examples from central London. *Building Journal*.

Nihal, A., & Fadzil, S. (2011). The potential of shading devices for temperature reduction in high-rise residential buildings in the tropics. *Procedia Engineering, 21*, 273–282.

Radhi, H., Sharples, S., & Fikiry, F. (2013). Will multi-facade systems reduce cooling energy in fully glazed buildings, a scoping study of UAE buildings. *Energy and Buildings, 56*, 179–188.

Rahmani, B., Kandar, M., & Rahmani, P. (2012). How double skin Façade's air-gap sizes effect on lowering solar heat gain in tropical climate? *World Applied Sciences Journal, 18*(6), 774–778.

Ruck, N. (2000). *Daylight in buildings-a sourcebook on daylighting systems and components*. Berkeley (USA): Lawrence Berkeley National Laboratory.

Trubiano, F. (2013). *Design and construction of high-performance homes: Building envelopes. Renewable energies and integrated practice*. London: Routledge.

Vartiainen, E. (2001). Electricity benefits of daylighting and photovoltaics for various solar facade layouts in office buildings. *Energy and Buildings, 33*(2), 113–120.

GPSR Compliance

The European Union's (EU) General Product Safety Regulation (GPSR) is a set of rules that requires consumer products to be safe and our obligations to ensure this.

If you have any concerns about our products, you can contact us on ProductSafety@springernature.com

In case Publisher is established outside the EU, the EU authorized representative is:

Springer Nature Customer Service Center GmbH
Europaplatz 3
69115 Heidelberg, Germany

Batch number: 10731528

Printed by Printforce, the Netherlands